MODERN COLLEGE ALGEBRA With Applications

MODERN COLLEGE ALGEBRA
With Applications

Ronald D. Jamison
Brigham Young University

HARCOURT BRACE JOVANOVICH, Inc.
New York / Chicago / San Francisco / Atlanta

To my sweet Ann

Figures drawn by Rino Dussi
Cover adapted from photo by Norman Rothschild

Puzzles No. 1 on p. 43, 1 on p. 58, 2 on p. 59, 1 on p. 176, and 1 on p. 278
reprinted by permission of Charles Scribner's Sons from
536 Puzzles and Curious Problems by Henry Ernest Dudeney,
edited by Martin Gardner. Copyright © 1967 Charles Scribner's Sons.

Puzzles No. 1 on p. 233, 2 on p. 278, 1 on p. 365, and 1 and 2 on p. 391
copyright © 1959 by Martin Gardner. Reprinted by permission of Simon and Schuster.

Puzzles No. 3 on p. 88, 2 on p. 176, 1 on p. 334, and 3 on p. 392
copyright © 1961 by Martin Gardner. Reprinted by permission of Simon and Schuster.

Puzzles No. 1 and 3 on p. 123, and 2 on p. 335 copyright © 1966 by Martin Gardner.
Reprinted by permission of Simon and Schuster.

ISBN: 0-15-560415-5

Library of Congress Catalog Card Number: 75-21841

Printed in the United States of America

This book is a result of nearly twenty years of teaching during which I have made a conscientious effort to understand the processes by which students learn. I have tried to reflect what I have learned in ways that provide a pleasant yet effective format for learning: The writing is informal yet precise and addresses the reader directly; a two-color treatment of text and illustrations focuses attention on key elements and procedures and stresses essential concepts and formulas; marginal notes (placed next to their referents) show the rationale for essential steps in proofs and solutions; and five categories of problem sets offer a wide variety of learning experiences.

Preface

Because many students never see beyond algebraic manipulations to the real role of mathematics, Chapter 0 discusses the meaning, purpose, and role of mathematics in today's world. For most students, Chapter 1, on elementary logic and set theory, will be a review; in addition, it establishes certain notational forms that are used throughout the book. Chapter 2 treats basic properties of the real-number system, so essential to arithmetic and algebraic manipulation. This, too, will be a review and can be covered briefly, except for Sections 2.9 through 2.12, which examine more deeply the structure and properties of the real-number system. The remaining chapters may be studied in a sequence determined by curriculum requirements or the individual needs of the student. I suggest that the core chapters (0-4) be read in sequence. Mathematics or physical science majors may study the remaining chapters in any sequence, with the exception of Chapter 7, which should be preceded by Chapter 6. The same is true for business or social science majors, except that Chapter 12 may be omitted, as may Sections 2.9 through 2.12 and 7.10 through 7.11.

With respect to the problem sets, I recommend that all problems in the Reading Comprehension sets be used, since they cover basic principles, manipulation techniques, and frequently misunderstood concepts. A selection of problems from each of the other categories — Skills Development, Theoretical Developments, Applications, and Just For Fun — will provide a balanced learning experience. Because most readers of this book will be interested mainly in its applicability to non-mathematical training or employment, the problems are taken from the business world and the physical and social sciences — both for motivational reasons and to give first-hand experience in solving problems pertinent to student interests. Solutions, with appropriate comments, are provided for all Reading Comprehension problems. For all other problem sets except Just For Fun, answers to one-third (those numbered in color) are worked out in detail, and answers only are given for another third; Just For Fun problems are not answered in the book.

My sincere thanks go to the following, who reviewed the original manuscript and offered many valuable suggestions that have significantly affected the final form of the text: Charles L. Murray and Kenneth Eberhard, both of Chabot Junior College; Robert A. Nowlan, Southern Connecticut State College; Janet Ray, Seattle Central Community College; and Richard Phillips, Michigan State University. I also thank my colleagues Steven Cottrell and Diana Armstrong for proofreading the final manuscript and galleys and for working out all the problems—a monumental task. And finally, my deepest thanks and love to my wife, Ann, who typed the manuscript from rough draft to completion, and without whose help and steady support this project would never have been realized.

Ronald D. Jamison

Contents

Chapter

4   Relations and Functions   125

Chapter

5   Exponential and Logarithmic Functions   179

MODERN COLLEGE ALGEBRA With Applications

Mathematics

CHAPTER

0

Introduction 0.1

Today mathematics is one of the fastest growing and most radically changing sciences. Although new discoveries often render former theories obsolete in many scientific disciplines, nearly all the major mathematical developments of the past 4 to 5 thousand years are still valid and useful. Today, notwithstanding these countless contributions of the past, mathematics is experiencing a deluge of new, fresh ideas and ingenious discoveries unparalleled in all its history. In fact it has been estimated that if one could measure all the mathematical knowledge accumulated from the dawn of time to the year 1950, this amount would at least be equaled by that produced from 1950 to 1964. Moreover, if the present trend continues (and we have every reason to believe it will) we expect that within the next 10 to 15 years the total amount of mathematical knowledge accumulated through 1964 will be redoubled. Each year 18 to 20 thousand mathematical articles published in over a thousand scientific journals attest to this phenomenal growth.

This advancement has both inspired and been inspired by the increasingly widespread use of mathematical systems and logic in fields outside mathematics. For example, Boolean algebra ● has recently been applied in designing telephone circuits and electronic computers. In biology we now find mathematical models of the nerve impulse and of energy transfer in predation. Social economist Richard Stone proclaims:

> In every one of the social sciences it has become increasingly evident that an exclusively verbal description of complex systems and their interrelations results in generalizations that are difficult to analyze, compare, and apply. These difficulties are greatly reduced when *mathematical expressions are substituted for words*. For one thing, a number of problems that had seemed to be completely unrelated prove to be mathematically identical. For another, even in subjects whose concepts are rather vague and in which precise information is hard to find, mathematics can provide a means of obtaining valuable insights. ●

Richard Bellman, commenting on the widespread application of mathematics, says, "In industry, control theory, implemented by the computer, is now widely used to regulate inventories, to schedule production lines and to improve the performance of power stations, steel mills, oil refineries and chemical plants." ●

With mathematical systems and techniques now being applied more extensively than ever before, one would expect most mathematical research to be taking place in applied mathematics. This, however, is not the case. That is, the vast majority of all mathematical advancements today are found in pure mathematics. The distinction between *applied* and *pure* mathematics is not always clear, however. One reason for this is that the two overlap considerably; another is that pure mathematics today often becomes applied tomorrow as more frequent and varied uses are encountered. In general, however, we may consider applied mathematics as that part that encompasses those systems and methods that are utilized in analyzing and solving physical problems — problems related to people and the physical world — whereas pure mathematics deals with abstract systems that may or may not have any immediate application or relevance to the physical world.

Many mathematicians spend months, years, and even lifetimes expanding the frontiers of pure mathematics, and this has been the cause of some ill-conceived ridicule. Those who taunt pure mathematics either have forgotten or are not aware that history, particularly modern history, is replete with incidents of pure mathematical developments becoming the very cornerstones upon which have been built some of our most prestigious and honored accomplish-

● Boolean algebra was developed by the English mathematician George Boole in the late 1840s.

● Richard Stone, "Mathematics in the Social Sciences," in *Mathematics in the Modern World,* W. H. Freeman, San Francisco, 1968, pp. 284–293.

● Richard Bellman, "Control Theory," *Scientific American,* September 1964, Vol. 211, pp. 186–200.

ments. Take, for example, the theory of functions of a complex variable (a variable containing the "imaginary" number i, sometimes denoted $\sqrt{-1}$). For many years these numbers were considered useless—unable to measure or describe anything quantitatively (hence the name, imaginary). Today the theory of complex functions is basic and virtually indispensable in the study of electricity and magnetism, radio and television, and the motion of satellites and planets—and these are but a few of the many areas where complex numbers are utilized.

Another very important example of a pure mathematical system becoming applied occurred early in the twentieth century when Albert Einstein (1870–1955) developed his famous general theory of relativity, using non-Euclidean geometry as an essential mathematical tool. The particular form of non-Euclidean geometry Einstein chose was originally developed by the famous German mathematician Georg Friedrich Bernhard Riemann (1826–1866). This geometry ● does not possess lines of infinite length, although they are endless, nor lines that are parallel—a geometry quite distinct from that to which most of us are accustomed. You can get a feeling for the nature of this geometry by thinking of lines as being great circles on a sphere. (See Fig. 0.1.1.) Note that a spherical triangle of this space can have interior angles with a sum *greater* than 180°.

These are only two of an almost endless array of occurrences where pure mathematics has become applied. Thus, continued research in both pure and applied mathematics offers stimulating and exciting prospects.

In brief, modern mathematics, both pure and applied, is playing an increasingly vital role in nearly every facet of our eternal quest to understand the marvelous world that surrounds us.

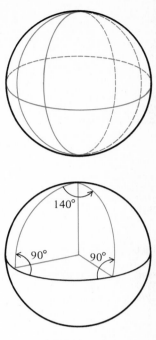

FIGURE 0.1.1

● P. LeCorbeiller, "The Curvature of Space," in *Mathematics in the Modern World*, W. H. Freeman, San Francisco, 1968, pp. 128–133.

The Nature of Mathematics 0.2

The extent of mathematical discovery and application seems, as never before, to be boundless. Yet, surprisingly few people understand the real nature and purpose of mathematics. Many look upon the subject as just a study of various numbers and an attempt to skillfully perform endless calculations. Some envision mathematics as an enormous conglomeration of strange symbols and peculiar looking equations. Others view the mighty computing machines of today as the epitome of mathematical prowess. Mathematics is none of these, although each is used by the mathematician.

Admittedly any attempt to define or describe mathematics would surely fail—fail just as a purely verbal characterization of music or art would fall pitifully short of its intended mark. Mathematics, like music and art, is simply much broader than mere language can adequately describe. To understand these subjects, one must become personally involved with much of the substance of which they are composed.

Although the substance of mathematics fills volumes and can be very difficult, there are some basic and fundamental concepts underlying the very heart of the subject that most everyone can comprehend. Understanding these will add measurably to your vision and appreciation of mathematics. These concepts will also provide a suitable beginning from which we may move more naturally into the main body of this text.

Mathematics, despite its awesome reputation, is the simplest and clearest of studies dealing with the laws of thought and reasoning. We are all mathematicians of a sort. How often, when in a conversation with another, have you found yourself asserting that such and such is true and then following that assertion with a sequence of facts, each designed to help establish or prove your point? It may have gone something like this: "Sure I'm going to do better in algebra this year; I have not only attended every class but also completed every problem assignment." Or perhaps, "I feel confident that Mary Ann will win the election for Studentbody Vice-President — she is a very popular member of the Pep Club, a standout on the debate team, and simply adored by everyone."

Proclaiming something true or valid under certain conditions, then proceeding to gather supporting facts and presenting them in an orderly, logical manner is probably the most fundamental characteristic of mathematics. Throughout history man has always welcomed and even sought out the kinds of problems that would challenge his keenest insights and produce clever feats of ingenuity. Many popular games require reasoning, concentration, and logic in order to consistently win, for example, chess, bridge, three-dimensional tic-tac-toe, and Concentration.

Mathematics in its simplest form, deals with reasoning and the laws of thought, which are usually devoted to the proof of an assertion or to the solution of a problem. This thinking process often takes place without the use of numbers and extensive calculations. In fact, many mathematicians see very little relationship between mathematics and schoolbook arithmetic.

As an illustration, let us take the celebrated problem of the Koenigsberg bridges (Fig. 0.2.1). This problem was first solved by the most eminent of Switzerland's mathematicians, Leonhard Euler (pronounced oiler), in 1735. He wrote,

> In the town of Koenigsberg there is an island called Kneiphof, with two branches of the river Pregel flowing around it. There are seven bridges crossing the two branches. The question is whether a person

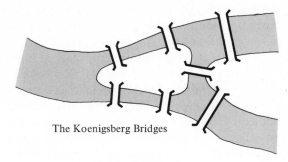

The Koenigsberg Bridges

FIGURE 0.2.1

can plan a walk in such a way that he will cross each of these bridges once, but not more than once. On the basis of the above, I formulated the following very general problem for myself: Given any configuration of the river and the branches into which it may divide, as well as any number of bridges, to determine whether or not it is possible to cross each bridge exactly once. ●

Before reading further, see if you can determine for yourself the answer to Euler's original problem, using, if you wish, Fig. 0.2.1.

If you are like most people, your first impulse is to sketch several paths as you attempt to find one satisfying the required conditions. After each unsuccessful effort, you may begin to suspect that such a path does not exist. To be sure, one way to verify your suspicion would be to test all possible paths, but this manner of resolving the problem would not only be very tedious but would shed little light on how to solve a similar though more general problem, such as that suggested by Euler. One can often save time and effort by first carefully analyzing the problem. This would include identifying its defining characteristics, examining the effect or limitations these characteristics impose upon the situation, and then, by reason and logic, determining a solution, if possible. Let us apply this approach to the above problem (which you may have already done successfully).

Observe that there is an *odd* number of bridges adjoining each portion of land. From this we conclude that a walk could not end where it began. Why? Of course that observation itself poses no difficulty because the other three portions of land where the walk may terminate remain. However, where the path may end does present a problem. Note that if a path does not originate in an area joined by an *odd* number of bridges, then necessarily the path must terminate there. Why? Therefore, *if* a continuous (unbroken) path were to exist crossing each bridge once and only once, the path would have to end in all three remaining but separate portions of land — but that, of course, is *impossible*. Thus we assert that no path exists satisfying the required conditions.

● Leonhard Euler, "The Koenigsberg Bridges," in *Mathematics in the Modern World,* W. H. Freeman, San Francisco, 1968, pp. 141–142.

Euler's argument was more refined and detailed than the one presented here; furthermore, it answered similar questions for considerably more general cases.

An important lesson to learn from this analysis of the Koenigsberg bridges problem is the essential role played by one's own ability to reason carefully in problem solving. In fact, as mentioned earlier, logical and orderly thinking lie at the very heart of mathematics. One may inquire, "How do you develop, improve, and best employ these human virtues?" The answer is—*use them*. Face up to the challenge of a wide variety of problems. For this reason, a plentiful supply of thought provoking and relevant problems is provided in the exercises throughout this book. In addition, many sections and even some chapters exist primarily to provide certain guidelines and ground rules associated with logical thought patterns and carefully organized reasoning sequences.

Before leaving this perspective of mathematics, an additional comment is in order. Despite our best efforts, we all err in our reasoning from time to time. This brings to mind a story of an energetic young man who, after digging a deep hole, decided to fill it back up. Upon finishing the project he was amazed to find dirt left over—dirt that originally occupied space in the hole before any digging took place. Scratching his head and wondering for a moment how this could be, he finally concluded that he simply had not dug the hole deep enough! Well, perhaps his back was a bit stronger than his brain. Yet we are all susceptible to various forms of illogical thinking. A primary objective of this text is to assist you in becoming more adept in the reasoning processes, particularly as they relate to problem solving—and, after all, the primary reason that most people study mathematics is to improve their ability to solve problems.

0.3   The Language of Mathematics

It would be convenient indeed if all problems could be resolved by simply thinking them through carefully as we did with the Koenigsberg bridges problem. Of course, this is not the case. Most of the really significant and challenging problems we face today are difficult even to understand, let alone to solve. We need some kind of device first to help us comprehend the nature of some problems and second to assist us in discovering their solutions. Herein lies the major role of the language of mathematics. As languages go, mathematics would certainly rank among the simplest—there are no conjugations, declensions, tenses, nor genders, and there is only a minimum of grammar. Yet it is a language, a means of communication, with some rather unique characteristics not enjoyed by most other languages.

The essential function of nearly all languages is to communicate in either the spoken or written form. Mathematics, however, is used almost exclusively as a written language. Moreover, rather than providing a means of communication between people, it is generally a vehicle by which problems of various sorts show more clearly and precisely their character and solution to those investigating them.

To illustrate, let us take the following example. Suppose the liquid in the radiator of a car were 18 percent antifreeze and 82 percent water, giving protection, say, down to 18° (Fahrenheit). Further, suppose with winter coming on, it was necessary to have protection down to −10°, which requires a solution that is 36 percent antifreeze. If the capacity of the cooling system of the car were 20 quarts, how much liquid should be drained off and replaced by antifreeze in order to obtain the desired 36 percent concentration? (See Fig. 0.3.1.)

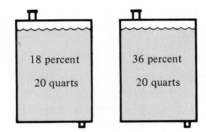

18 percent
20 quarts

36 percent
20 quarts

FIGURE 0.3.1

To solve this problem, we must first be certain the meaning of the given information is clearly understood. That is, a concentration of 18 percent means 18 one-hundredths or eighteen parts in each one hundred. Thus, the amount of antifreeze initially in the radiator would be 18 percent of 20 quarts (0.18 times 20) or 3.6 quarts. Of course, a portion of that amount will be lost in the liquid drained off. Because we do not know, for the moment, how much liquid to release, let us denote that unknown quantity by the letter x. Observe, as you analyze the problem, that the amount of antifreeze remaining in the radiator after draining off x quarts of the original coolant is $0.18(20 - x)$. Further observe, if to this amount we add x quarts of antifreeze, we obtain the desired 36 percent solution. We now *translate* this observation into the language of mathematics. It would read

$$\underbrace{0.18(20 - x)}_{\substack{\text{the amount of antifreeze} \\ \text{left in the tank}}} + \underbrace{x}_{\substack{\text{the amount} \\ \text{added}}} = \underbrace{0.36(20)}_{\substack{\text{the amount} \\ \text{required}}} \tag{1}$$

This is an equation. Such an expression is often called a mathematical model of the problem. One advantage of this form is the elimination of the nonessential verbiage in the English version; it also identifies the more vital elements of the problem. More important, however, is the fact that equations can be written in many different equivalent forms, each, perhaps, bringing to light new and vital information. Let us now examine several equivalent forms of equation (1). The first is obtained by performing the indicated multiplication (0.18 times the quantity $(20 - x)$ and 0.36 times 20), yielding

$$3.6 - 0.18x + x = 7.2$$

Subtracting 3.6 from both sides of the equation gives

$$-0.18x + x = 7.2 - 3.6$$

Upon performing the indicated addition and subtraction, this simplifies to

$$0.82x = 3.6$$

Finally, dividing both sides of this equation by 0.82 and simplifying, we have

$$x = \frac{3.6}{0.82} = 4.39$$

The formerly unknown x is now clearly identified and the problem is solved. If approximately 4.4 quarts (or roughly four and a half quarts) are drained off and replaced by antifreeze, the desired protection to $-10°$ is obtained.

However simple this example may seem, it illustrates two important functions of the language of mathematics. One, this language makes more evident the nature and basic characteristics of a problem—it lays bare the structurally essential elements of the problem. And two, the original mathematical model may be converted into several equivalent forms, each possibly yielding new and valuable information leading eventually to a complete solution of the problem. Of course, one is not always able to reach this ultimate conclusion.

A word of caution: You should always remember that mathematical expressions are abstractions. That is, whole numbers, fractions, and other numbers along with the operations of addition, subtraction, multiplication, and division are abstract creations—they do not exist in nature—they were not discovered like gold or the universal law of gravitation. To be sure, numbers may represent a specific quantity of horses, inches, kilowatts, or what have you, but the numbers themselves do not identify explicitly this information. For this reason, special care is often necessary in properly formulating mathematical expressions that are designed to assist us in problem solving. For example, everyone would agree that mathematically $\frac{1}{2} + \frac{1}{2} = 1$; though if you were to apply this principle in adding a half dress to a half dress, you would not necessarily expect to obtain one whole dress, particularly if the two halves were lower halves.

Consider another example: A woman enters a candy store and purchases four pounds of chocolates at two dollars a pound. The clerk reasons aloud, "Four *pounds* times two *dollars* equals eight *dollars*," and asks for eight dollars in return for the candy. At this point, the woman, perhaps a bit more clever than the clerk, replies, reasoning similarly, "Four *pounds* times two *dollars* is eight *pounds;* thus, I'll pay two dollars and take eight pounds of chocolates." Of course she would not get away with it, unless the clerk had a sense of humor and did not mind losing her job; indeed the woman's reasoning was sound (or unsound) as the clerk's. The clerk's problem was in mixing units — one generally should not multiply pounds by dollars, or inches by quarts, or any other such product of two distinct measures. The clerk could have said more correctly, "One pound at two dollars is equivalent to four pounds at eight dollars." The number 2 can certainly be multiplied by 4 to obtain 8, but this 8 must be interpreted properly in terms of the physical objects being represented and the nature of the problem. The mathematical symbols themselves, being abstractions, do not offer this information.

Problem Sets ● 0.4

Problem Set I ■ *Reading Comprehension*

Determine whether statements 1 through 4 are true or false and give reasons for your responses.

1 Today in mathematics, as in other sciences, most of the old methods and ideas are being replaced by new, modern, up-to-date ones.
2 The major disadvantage of mathematics is that nearly everything of a mathematical nature has now been discovered.
3 Though in years past mathematical techniques and symbols were utilized primarily in the physical sciences, today research and development in nearly all scientific disciplines make use of mathematical methods and reasoning.
4 By and large, mathematics is a study of numbers and various ways they can be manipulated.
5 Explain the difference between "pure" and "applied" mathematics, if there is any.
6 Give at least two examples where pure mathematics has become applied.
7 How does the language of mathematics differ from most other languages?
8 What advantages does the language of mathematics have over most other languages?
9 Criticize the statement, "Five cars times two men equals ten cars."

● Selected answers to problem-set questions are at the end of the book. All Reading Comprehension questions are answered completely; problems numbered in red are worked out or otherwise answered in detail; those immediately following red-numbered problems have answers only; those immediately preceding red-numbered problems are not answered in the text.

1 Can you find a continuous path in the houses in Fig. 0.4.1. that will take you through each door once and only once? If for one or the other of the houses it is impossible, explain why. (HINT: You may want to think of rooms as points and doors as lines leading to and from the points.)

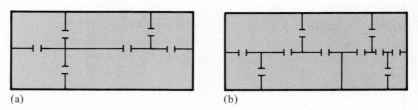

(a) (b)

FIGURE 0.4.1

2 Does there exist a continuous path that would cover each of the line segments in Fig. 0.4.2 once and only once? (Intersection points may, of course, be crossed more than once.)

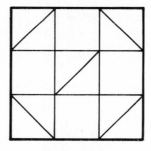

FIGURE 0.4.2

3 Suppose you are given two bottles, one with a 5-cup capacity and the other with a capacity of 7 cups. How can one cup of water be measured exactly using only the two bottles?

4 Suppose 1125 basketball teams are participating in a worldwide single-elimination tournament ("single-elimination" meaning a team is eliminated after its first loss). Because the number of teams participating is odd, one team will have a bye during each round except the final round; this team is determined at each round by drawing lots. How many games will be played in all rounds of the tournament? (HINT: Don't make a hard problem out of one that can be solved quickly.)

5 Three men went to a hotel and requested a single room for the group. The manager had a room for $15, so each man paid $5. Later the manager remembered that the room given the three men cost only $10. Being an honest man, he therefore took five $1 bills from the cash register and told the bellhop to return the money to the three men. The bellhop, who was not quite so honest, gave each man $1 back and kept the remaining $2 for himself. Thus, each man paid $4 for the room. But three times $4 is $12, plus the $2 in the bellhop's pocket, makes a total of $14. What happened to the other dollar of the original $15? (HINT: Don't let the wording mislead you.)

Illustration by John Tenniel
for Lewis Carroll's
Through the Looking Glass

*"I know what you're thinking about," said Tweedledum; "but it isn't so, nohow."
"Contrariwise," continued Tweedledee, "if it was so, it might be; and if it were so,
it would be; but as it isn't, it ain't. That's logic."*

Logic and Sets

Introduction to Logic 1.1

Logical thinking and reasoning, as mentioned in Chapter 0, are fundamental to mathematics; in fact, they lie at the heart of the subject. It is natural, therefore, that at least a brief exposure to modern logic be introduced early in our work. It should be remembered, however, that this presentation represents but a beginning—just enough to assist in some of the logical arguments present in the proofs of certain properties and the solutions of various problems. (Chapters 1 and 2 have primarily two purposes: to establish the notational patterns that will be used in the text and to provide a brief review of certain prerequisite materials. One reading may be sufficient for some classes—or students.)

To illustrate how helpful a few basic rules of logic may be in analyzing certain problems, let us examine the following quote that appeared in a periodical soon after the end of World War II concerning the economic situation in Germany. The article contained the statements:

(1) If there is not a sufficient supply of commercial fertilizer, or if there is not an adequate transportation system in Germany, German farms cannot make available sufficient food to maintain the German population. (2) If there is a shortage of coal in Germany, then there is a shortage of both power and steel. (3) If there is a shortage of power, there is an insufficient supply of commercial fertilizer. (4) If there is a shortage of steel then there is not an adequate transportation system. (5) There *is* a shortage of coal in Germany.

What can we conclude from this information about the economic situation in Germany at that time? At first this problem may seem rather complex. However, after we introduce a few principles of reasoning and logic, we will return to this problem and find it quite simple to analyze and solve.

1.2 Propositions

• A conjecture is a statement based on a supposition or evidence that is insufficient for definite knowledge.

Verbal descriptions of most problems are composed of statements (declarative sentences) that convey either known or conjectured • facts about the problems. These statements usually occur in the form of a simple or compound proposition.

DEFINITION 1.2.1   A proposition is a statement that is either true or false or that has been designated as true or false, but it is not both true and false.

Example 1.2.1   Each week contains 7 days.

This statement *is* a proposition because it is true.

Example 1.2.2   Every month is composed of at least 30 days.

This statement *is* a proposition because it is false – February is a month containing at most 29 days.

Example 1.2.3   This statement is false.

This *is not* a proposition because it is neither true nor false.

Example 1.2.4 $3 - 5 = 2$

This statement *is* a proposition because it is false; $3 - 5 = -2$.

Example 1.2.5   Let's go swimming.

This statement has no attribute of trueness nor falseness. Hence, it *is not* a proposition.

Example 1.2.6   The house is white.

Because there is no information accompanying this statement identifying the house being referred to, that is, information sufficient to determine whether the statement is true or false, this *is not* classified as a proposition. It does, however, represent an extremely important class of statements frequently called open propositions or open sentences. They contain elements called variables, such as the word *house* in the previous statement. This word may stand for any one of many houses. Furthermore, the statement does not become a true or false proposition until this variable is replaced by a specific identifiable house description. For example, take the statement, "The President's house in Washington, D.C., is white." With this replacement of the variable *house* the statement becomes a true proposition. Open sentences and their vital role in mathematics will be studied in detail in Chapter 3.

Example 1.2.7 $3x + 4 = 13$

This is an *open sentence* containing the variable x. The statement is true if the variable x is replaced by 3, whereas for any other replacement it is false.

It is often difficult if not impossible to determine whether or not many statements are propositions. Moreover, the truth or falsity of a statement may depend upon one's point of view. Fortunately, however, mathematical statements, with only rare exceptions, are clearly propositions.

The example propositions above were simple propositions—a proposition composed of a single clause. A compound proposition is made up of two or more simple propositions.

Example 1.2.8   The astronauts of *Apollo 11* returned safely from their historic voyage to the moon *and* brought back several lunar rock samples.

This statement, containing two simple propositions joined by the conjunction *and,* is a *compound* proposition.

Example 1.2.9 $2 < x < 10$ *or* $x^2 - x - 12 = 0$

A *compound open* sentence.

It is often convenient, in analyzing the structure of certain compound propositions to represent their component simple propositions by letters of the alphabet. After a designation has been made, say for the letter A, then whenever A is used within the context of a compound proposition, the simple proposition represented by A is to be understood as if it were there stated in full.

Example 1.2.10 Let A represent the proposition "It is raining outside," and B the proposition "It is wet outside." With A and B thus defined, we form the compound statement, "If A, then B," which means, "If *it is raining outside* then *it is wet outside*."

Propositions of the form "if A, then B" are very common in mathematics and are called conditional propositions or implications. The proposition A is called the antecedent or the hypothesis of the implication, whereas B is called the consequent or conclusion. If a conditional proposition is accepted or proven true, then we understand that the truth of the hypothesis leaves the truth of the conclusion inescapable. That is, whenever the hypothesis is satisfied, the conclusion must inevitably follow. This simple fact is very important and fundamental in mathematical reasoning. An implication is considered *true* if the consequent is true whenever the antecedent is true. If, however, the antecedent is true and the consequent is false, then we say the implication is *false*.

Remark 1.2.1 In *modern* logic, an implication with a false antecedent is considered true no matter what the truth value of the consequent is. Because such a classification is not very intuitive and because for our present purposes we will not have a need for these latter truth value classifications for an implication, we will consider only the two given above.

Example 1.2.11 If $2x = 6$, then $x = 3$.

This implication is *true*. The conclusion $x = 3$ is an inescapable fact given $2x = 6$.

Example 1.2.12 If $2x = 8$, then $x = 3$.

This implication is *false* because if the antecedent $2x = 8$ is true, then the consequent $x = 3$ is false.

Remark 1.2.2 The antecedent in the implication of Example 1.2.11, $2x = 6$, is an open sentence. A true conditional proposition is not obtained until we include the consequent $x = 3$ or any other true statement that would follow from the antecedent.

Determine whether statements 1 through 8 are propositions.

1 In the United States a president is elected every four years.
2 The capital of the state of California is San Francisco.
3 His eyes are blue.
4 Stop the motor.
5 $3 \cdot 4 = 14$
6 $2 \cdot x = 12$
7 $x^2 - 1 = (x - 1)(x + 1)$
8 $x(x + 2)$

Determine whether problems 9 through 11 are true or false and give reasons for your responses.

9 An implication is composed of a hypothesis and an antecedent.
10 An implication is true if the antecedent is true.
11 An implication with a true antecedent and a false consequent is considered false.

● Answers to Sample Problem Set questions are given at the end of the chapter.

Some Fundamental Principles of Logic 1.4

Before presenting these principles, certain notational conveniences will be introduced. Some words and phrases occur so frequently in stating propositions, that it becomes convenient to abbreviate them by shorthand symbols of a sort. Since only a few of these will be used throughout this book, it will not be difficult for you to learn and use them.

Most mathematical statements come in the form of an implication. Hence, to have a *symbol for implication* would be particularly convenient. Let A and B represent two propositions. Then we may represent the following equivalent (having the same meaning) implications:

If A, then B
A implies B
A only if B

by the form

$$A \Longrightarrow B$$

That is, we shall accept the above symbolic expression as the standard mathematical form for a conditional proposition.

We often find in languages several different ways of expressing essentially the same idea. In the language of mathematics every attempt is made to minimize these variations. The following three equivalent verbal expressions have the same meaning as the three above but are expressed in reversed order:

> B is true if A is true
> B is implied by A
> B follows from A

It is natural then to represent these expressions symbolically by

$$B \Longleftarrow A$$

that is, by simply reversing the order and direction of the initial symbolic form. Further, because all these expressions have the same meaning, we shall accept the two symbolic forms

$$[A \Longrightarrow B] \quad \text{and} \quad [B \Longleftarrow A]$$

as being equivalent and hence interchangeable.

Remark 1.4.1 It is important to note in the above equivalent forms that the order of the propositions A and B has been *reversed*. We are *not* saying that the two forms

$$[A \Longrightarrow B] \quad \text{and} \quad [A \Longleftarrow B]$$

are equivalent. In fact, in general they are not.

Now let us examine the verbal expression, "Jim will remain class president if and only if he passes the final exam with a grade B or better." If A represents "Jim will remain class president" and B, "Jim passes the final with a grade B or better," then the above statement may be expressed symbolically as

$$[A \Longrightarrow B] \quad \text{and} \quad [B \Longrightarrow A]$$

that is, "If Jim will remain class president, then he must pass the exam with a grade B or better," and "If Jim passes the exam with a grade B or better, then he will remain class president." Here we have what is called a *biconditional* equivalence — A not only implies B but B also implies A. This idea is very important, and we state it as a definition.

DEFINITION 1.4.1 Let A and B be two propositions. Then A and B are said to be biconditionally equivalent if the implications

$$[A \Longrightarrow B] \quad \text{and} \quad [B \Longrightarrow A]$$

are both true. Such a biconditional equivalence will frequently be expressed more simply as

$$A \Longleftrightarrow B$$

Example 1.4.1 The following is an algebraic biconditional equivalence:

$$3x = 12 \Longleftrightarrow x = 4$$

The truth or falsity of a proposition is critical in any analysis or argument. It is necessary, therefore, to understand the meaning and form of the *negation* of a proposition.

DEFINITION 1.4.2 The negation of a proposition is a proposition that is a direct or immediate contradiction of the given proposition. A proposition is true if and only if its negative is false.

Example 1.4.2 The negation of "They will win" is "They will *not* win."

A slash, denoted /, through a symbol denotes the *negation* of that symbol. For example, $a \neq b$ means "a is not equal to b." Similarly, the mathematical statement

$$A \nRightarrow B$$

means "A does not imply B" or "B does not necessarily follow from A." However, the negation of a proposition A will not be represented by $A\!\!\!/$ but rather by not A.

Although the interrelationships between various implications are numerous, we shall mention here only a few that will suffice for our purposes. We shall call each a fundamental principle of logic. These assertions about propositional forms we shall accept as true — no attempt is made to prove them, although examples will be given to illustrate their validity intuitively. In each, the letters A, B, and C represent propositions.

L-1   FIRST FUNDAMENTAL PRINCIPLE OF LOGIC

The two implications

$$[A \Longrightarrow B] \quad \text{and} \quad [B \Longrightarrow A]$$

(the latter called the converse of the former) are *not* logically equivalent. The truth value of a conditional proposition is, in general, changed when the roles of the hypothesis and the conclusion are reversed. Symbolically, we have

$$[A \Longrightarrow B] \nLeftrightarrow [B \Longrightarrow A]$$

Example 1.4.3 "If a car is out of gas, it will not run." You would not conclude that the *converse* of this implication holds; that is, "If the car does not run, it is out of gas." It could have a dead battery, a broken fuel pump, a short in the ignition system, or any other of several failures causing it to stop running.

We next examine an algebraic representation of this principle.

Example 1.4.4 If a, b, c, represent numbers, then the following implication, called the multiplication law of equality, is true.

$$[a = b] \implies [ac = bc]$$

However, the *converse* of this implication is not true in general, that is,

$$[ac = bc] \;\not\!\!\implies [a = b]$$

Why? (HINT: Let $c = 0$).

Remark 1.4.2 Only when you have a biconditional equivalence

$$A \iff B$$

also written

$$[A \implies B] \quad \text{and} \quad [B \implies A]$$

can you reverse the roles of the propositions A and B without effecting the truth value of the implication.

L-2   SECOND FUNDAMENTAL PRINCIPLE OF LOGIC

The two implications

$$[A \implies B] \quad \text{and} \quad [\text{not } B \implies \text{not } A]$$

(the latter called the contrapositive of the former) are logically equivalent. They both have the same truth classification — they are either both true or both false. We express this equivalence symbolically by

$$[A \implies B] \iff [\text{not } B \implies \text{not } A]$$

Example 1.4.5 "If today is Monday, then tomorrow is Tuesday." The *contrapositive* of this implication is equally valid, that is, "If tomorrow is *not* Tuesday, then today is *not* Monday." If assuming the conclusion to be false leads to a contradiction of that which is known or given true (the hypothesis), then the original implication must be valid.

The following is an algebraic example of this principle.

Example 1.4.6 The true implication

$$[3x = 12] \Longrightarrow [x = 4]$$

is equivalent to

$$[x \neq 4] \Longrightarrow [3x \neq 12]$$

One of the reasons this second fundamental principle is so important is that many problems, when stated as an implication, are more easily solved by using the contrapositive form than the direct form.

In solving problems, we often refer to previously given definitions, principles, or properties to support our conclusions. One convenient way to accomplish this is to indicate the number or name of the property just above or below the principle implication symbol that is introducing the results of that property. For example, if the proposition $A \Longrightarrow B$ is given, and we wish to invoke the Second Fundamental Principle of Logic, denoted L-2, then we may accomplish this as follows:

$$[A \Longrightarrow B] \overset{\text{L-2}}{\Longleftrightarrow} [\text{not } B \Longrightarrow \text{not } A]$$

and we would understand the right-hand implication follows from the left-hand one by L-2.

We introduce the next fundamental principle by the following example.

• You need not memorize the letter or numeral code used to identify a previously given law or property. When a reference is made to a property not initially stated on a given page, the property will be restated in abbreviated form at the side of the page.

Example 1.4.7 "If Karl wins the election tomorrow, he will become class president. Furthermore, the class president has the responsibility of organizing all class parties." From this statement we may conclude, "If Karl wins the election tomorrow, he will have the responsibility of organizing all class parties." To more clearly see the structure of the above statements, let us diagram the remarks. Let A be "Karl wins the election tomorrow," and B, "Karl will become president," and C, "The president has the responsibility of organizing all class parties." The original statement and conclusion, diagrammed, now becomes

$$[A \Longrightarrow B \quad \text{and} \quad B \Longrightarrow C] \Longrightarrow [A \Longrightarrow C]$$

This represents a fundamental principle that we now state formally.

L-3   THE TRANSITIVE LAW OF IMPLICATION

Let A, B, and C denote three propositions. If A *implies* B and B *implies* C, then we may conclude that A *implies* C. Diagrammed, this becomes

$$[A \Longrightarrow B \quad \text{and} \quad B \Longrightarrow C] \Longrightarrow [A \Longrightarrow C]$$

Remark 1.4.3 Notice the principle implication of L-3 goes only in one direction; it is not a biconditional equivalence.

Thus far two of four common connectives have been defined and discussed — implication (denoted $\Longrightarrow$), and both biconditional and logical equivalence (denoted $\Longleftrightarrow$). The other two, well known to everyone, are the conjunction *and* and the disjunction *or*. These have already been used several times in our presentation. There has been no serious danger in this because their meanings in the language of mathematics are almost synonymous with their common everyday usage. However, for completeness we shall now identify more specifically their meanings in mathematics and logic.

DEFINITION 1.4.3   The compound proposition

A and B

formed by the conjunction of two propositions, is true if *both* propositions are true. If either proposition is false or both are false, the conjunction of propositions is false.

Example 1.4.8   Jack's car is blue *and* has white sidewalls.

If it is a fact that Jack's car is both blue and has white sidewalls, then the compound proposition is true. If, however, the car is not blue but has white sidewalls, or is blue but without white sidewalls, or if it is neither blue nor does it have white sidewall tires, the compound proposition is false.

The connective *or* is used in English in both the inclusive and exclusive senses. "The light is either on *or* off" is an example of exclusive use — one or the other but not both. However, in the statement "If Jack *or* Tom plays in the game tonight we will surely win," the word *or* is probably used in the inclusive sense, that is, the proposition would remain true whether one or the other or *both* play in the game. In mathematics, the latter *inclusive*, "and/or" sense of *or*, is the only one used.

DEFINITION 1.4.4   The compound proposition

A or B

formed by the disjunction of two propositions, is true if *one* of the propositions is true. It is also true if *both* propositions are true. If both are false, the disjunction of propositions is false.

A, B	A and B	A or B	$A \Longrightarrow B$	$A \Longleftrightarrow B$	not A
T T	T	T	T	T	F
T F	F	T	F	F	F
F T	F	T		F	T
F F	F	F		T	T

Table 1.4.1
Truth Table

Remark 1.4.4 In some books the symbols $\wedge$ and $\vee$ are used to denote *and* and *or*, respectively. Because the connectives themselves are already sufficiently short, we shall not employ these symbols.

We summarize the meaning of the four connectives *and, or, implies,* and *equivalent,* along with the *negation* of a proposition, in a *truth table*. In Table 1.4.1 A and B represent two propositions, and T and F stand for true and false.

The next fundamental principle provides an important relationship between the four connectives we have discussed.

L-4   DE MORGAN'S LAWS

Let A and B be two propositions. The compound proposition *not* (A *and* B) is equivalent to the compound proposition *not A or not B*. The proposition *not* (A *or* B) is equivalent to *not A and not B* (also written *neither A nor B*). Symbolically, these laws become

[not $(A$ and $B)$] $\Longleftrightarrow$ [not A or   not B]

and

[not $(A$ or $B)$] $\Longleftrightarrow$ [not A and   not B]

Example 1.4.9   Suppose the statement "Tom rises at 6:30 A.M. *and* eats breakfast at 7:00 A.M." is false. By DeMorgan's laws this is equivalent to saying "Either Tom does *not* rise at 6:30 A.M. *or* he does *not* eat breakfast at 7:00 A.M."

Example 1.4.10   If the statement

$[x = 4$ or $x = 6]$

is *false,* then we may conclude by DeMorgan's laws that

$[x \neq 4$ and $x \neq 6]$

is a true statement.

We now state the last of our fundamental principles.

L-5   THE LAW OF DOUBLE NEGATION

If A is a proposition, then the negation of the negation of A is equivalent to A. Diagrammed this becomes

$$\text{not}(\text{not } A) \Longleftrightarrow A$$

Example 1.4.11   To say that a fact is *not ir*relevant is to say that it is relevant.

Example 1.4.12   If a proposition is supposed false but is found to be *not false,* then it is true.

Let us now return to our problem of Germany's economic situation at the end of World War II, which was introduced at the beginning of this chapter. We shall first restate the problem and then proceed with our analysis of it.

> (1) If there is not a sufficient supply of commercial fertilizer, or if there is not an adequate transportation system in Germany, German farms cannot make available sufficient food to maintain the German population. (2) If there is a shortage of coal in Germany, then there is a shortage of both power and steel. (3) If there is a shortage of power, then there is an insufficient supply of commercial fertilizer. (4) If there is a shortage of steel then there is not an adequate transportation system. (5) There *is* a shortage of coal in Germany.

Note the first four statements are *conditional* propositions — the conclusions must follow only when the hypotheses are known. The fifth, on the other hand, is a simple statement of fact. The question is, "What are the implications of that fact — what can be concluded about the economic situation in Germany knowing there *is* a shortage of coal?"

We first diagram the problem. Let A represent, "There is a sufficient supply of commercial fertilizer," B, "There is an adequate transportation system in Germany," C, "Germany's farms can make available sufficient food to maintain the German population," D, "There is a shortage of coal in Germany," E, "There is a shortage of power," and F, "There is a shortage of steel."

Note each of the simple propositions identified above is expressed in the positive sense. This is desirable for clarity in the diagrammed form, though not essential. The entire problem can now be expressed as

(1) [not A or not B] $\Longrightarrow$ not C

(2) $D \Longrightarrow$ [E and F]

(3) $E \Longrightarrow$ not A

(4) $F \Longrightarrow$ not B

(5) D *is* given as a fact

Knowing D, i.e., there is a shortage of coal, what can be concluded? We will reason as follows:

$$D \xrightarrow{\;(2)\;} [E \text{ and } F]$$

$$\xrightarrow{\;(3,4)\;} [\text{not } A \text{ and } \text{not } B]$$

$$\xrightarrow{\;(1)\;} \text{not } C$$

Applying the Transitive Law of Implication, L-3, we come to several conclusions, the final one, not C, being the critical one that the farms of Germany *can not* make available sufficient food to maintain the German population.

This example demonstrates some of the advantages one often obtains from diagramming a problem in the symbolic language of mathematics. Not only are compound propositions expressed in a more compact form, but, when each symbol is clearly understood, the propositions can be read and comprehended with much greater facility — the fundamental nature and basic characteristics of each proposition become more clearly evident and most of the unnecessary verbiage is eliminated, laying bare the essential elements of the proposition. Furthermore, it is significant that in symbolic form the fundamental principles of logic can be applied more easily and with less chance of error, leading to other important properties that may partially or completely resolve a problem. Admittedly, the same results could be obtained from the verbal forms, but such forms are often more cumbersome and awkward to handle.

Before concluding this section, one more concept important to elementary logic will be introduced.

Example 1.4.13 Given the following conjecture: "I believe each student in my algebra class has received at least one passing grade on the four examinations given this semester." Suppose, after a careful check of the records it is discovered that Jim Smith, a member of my algebra class, has failed all four exams. This disclosure of Jim Smith's record becomes a *counterexample* to the above statement. Thus, the conjecture is a false proposition.

In general, if a conjecture is *supposed true* for *all* members of a specific collection and if *one* member of the collection is found for which the conjecture is *false,* then that member becomes a counterexample to the proposition and the proposition is rendered not true.

Example 1.4.14 Suppose one observes that the three odd numbers, 3, 5, 7 are *prime* (a number different from 1 whose only positive integral factors are 1 and itself) and thereby conjectures, "All odd numbers are prime numbers." One certainly would not have to look far to discover a *counterexample* to that conjecture. Take the odd numbers 9 and 15, for example; they factor into $3 \cdot 3$ and $3 \cdot 5$, respectively, and hence are not prime. Thus, these counterexamples render false the statement, "All odd numbers are prime."

1.5 Problem Sets

Problem Set I ■ *Reading Comprehension*

Determine whether statements 1 through 11 are true or false and give reasons for your responses.

1 The statement, "The proposition is both true and false," is necessarily a false proposition.
2 The statement, "The United States is composed of fifty individual states," is a *conditional* proposition.
3 The statement, "We are coming to the party only if invited," is a *conditional* proposition.
4 If $A \Longrightarrow B$, then A and B are said to be *equivalent*.
5 The *converse* of the statement $A \Longrightarrow B$ is $B \Longrightarrow A$.
6 The proposition $A \not\Longrightarrow B$ means, "If A is true then B is not true."
7 The *contrapositive* of $[A \Longrightarrow B]$ is $[\text{not } A \Longrightarrow \text{not } B]$.
8 The meaning of the connective *or* in mathematics is "one or the other but not both."
9 A fundamental principle of logic is

$$\text{not } (A \text{ or } B) \Longleftrightarrow (\text{not } A \quad \text{or} \quad \text{not } B)$$

10 If A is a proposition, it is possible for both A and *not A* to be false statements.
11 A *counterexample* is usually sought if a conjecture suggesting that all elements of some set satisfy certain conditions is suspected false.
12 Determine whether the following statements represent propositions:
   a Leap year occurs every four years.
   b The Western Hemisphere is composed of North and South America.
   c John, where are you going?
   d The one with the red handle.
   e All parallelograms are rectangles.

f The sum of the interior angles of a triangle is 180°.

g $3 \cdot 4 = 16$

h Come on over.

i It is cold and wintery outside.

j The state capital of Utah is Salt Lake City.

13 State the Law of Double Negation and give an example different from those presented in this chapter.

14 Illustrate by example the Transitive Law of Implication.

15 What is meant by diagramming a compound proposition?

16 State some of the advantages of diagramming certain compound propositions in the abbreviated language of mathematics.

17 Find an example illustrating the role of a counterexample different from those presented in this chapter.

Problem Set II ■ *Skills Development*

1 Diagram each of the following propositions and state all the applicable classifications from the following list: conjunction; disjunction; conditional, simple, or compound propositions; and equivalence.

Example If flight 209 does not arrive on schedule, we will miss our connection with flight 226.

Solution Let the simple propositions "Flight 209 arrives on schedule" and "We miss our connection with flight 226" be represented by A and B, respectively. Diagrammed, the example becomes

$$\text{not } A \Longrightarrow B$$

which is a *compound conditional* proposition.

Example Mere possession of a ticket to the game is not sufficient for your admittance. Both a ticket and your activity card are required for admittance.

Solution Let the simple propositions "You have a ticket for the game," "You are admitted to the game," and "You have an activity card" be represented by A, B, and C, respectively. With these designations, the complete statement of the example may be expressed

$$[A \Longrightarrow\!\!\!| \; B] \quad \text{but} \quad [(A \text{ and } C) \Longleftrightarrow B]$$

Thus, within this statement we have

$[A \Longrightarrow\!\!\!| \; B]$ as a *compound conditional* proposition
$(A \text{ and } C)$ as a *conjunction* of propositions
$[(A \text{ and } C) \Longleftrightarrow B]$ as an *equivalence* of two propositions

a Jack and Harry went to the show.

b If Kirt is 12 years old or older, he will have to pay an adult fare.

c Jack is going to the prom only if he gets a tuxedo.

d Susan will pass if she hands in an acceptable term paper.

e Jim's going to the conference follows from the fact that he qualified highest in his class.

f Because the other team is not going to show up, we will win by forfeit if and only if our team arrives on time.

g If Carol has maintained a B average and if she has taken all the required courses, then and only then will she be admitted to graduate school.

h If inflation continues, the prices of almost all commodities will continue to go up. Furthermore, if the prices continue to rise, a recession is inevitable. Thus, the continuing inflation will surely bring on a recession.

i If Jim has the correct address, he will find her house. If he does not find her house, it is simply because he does not have the correct address.

j If Jim has the correct address, he will find her house. However, just because he is able to locate her house, does not mean that he had the correct address.

k To say that Tom is not tall and handsome is to say that he is either not tall or not handsome.

l To not fail is to pass.

2 Identify each of the following as true or false and give reasons for your responses.

Example $3 + 5 = 8$ *or* $2 + 4 = 6$

Solution   True, because the mathematical definition of the connective *or* is the inclusive "and/or" sense.

a $2 \cdot 3 = 6$ or $3 \cdot 4 = 16$ **b** $2 + 3 = 5$ or $3 + 4 = 7$
c $2 + 3 = 5$ and $3 + 4 = 7$ **d** $2 \cdot 3 = 5$ or $3 + 4 = 6$
e $2 \cdot 3 = 6$ and $3 + 4 = 8$ **f** $2 + 3 = 4$ and $4 + 3 = 6$

3 If A represents, "Mr. Politician will win the majority of votes from Mississippi, Kansas, and Oregon," B, "Mr. Politician will become President of the United States," C, "The President of the United States is the Commander in Chief of all our armed forces," and D, "The President of the United States is a member of the Judicial Branch of our government," translate into English each of the following symbolic forms:

a $A \Longrightarrow B$

b $B \not\Longrightarrow A$

c $[A \Longrightarrow B$ and $B \Longrightarrow C] \Longrightarrow [A \Longrightarrow C]$

d $B \Longrightarrow [C$ and not $D]$

e $[C$ and not $D] \Longleftrightarrow$ not(not C or $D)$

f $[A$ or $B] \Longrightarrow [C$ but not $D]$

4 Suppose it is a fact that if all the generators of a city power plant are running simultaneously, the street lights must be on. Furthermore, if the load is over 10,000 kilowatts, all of the generators of the city power plant must be running simultaneously. Can we conclude from this information that if the street lights are on, the load is over 10,000 kilowatts?

Problem Set III   ■   *Challenges in Reason and Logic*

1 Determine whether the following arguments are valid:

a All athletes are well coordinated. All members of my physics class are athletes, therefore, all members of my physics class are well coordinated.

b Some ● triangles are isosceles. No triangle is a parallelogram. Hence, no parallelogram is isosceles.

c All soldiers are well disciplined. Some soldiers are not brave. Hence, some brave men are not soldiers.

● *Some* means
"at least one
and possibly all."

d   Some math courses are not required. Only required courses are worth taking. Some courses worth taking are math courses.

2   Determine whether the following arguments are valid and give reasons justifying your conclusions:

Example   It is not true that Jim is going to join the Army and Tom is not getting married. Jim is joining the army. Therefore, Tom is getting married.

Solution   Let A represent the proposition "Jim is joining the Army" and B represent the proposition "Tom is getting married." The above statements may now be structured as

$$[\text{not}(A \text{ and not } B) \text{ and } A] \Longrightarrow B$$

We wish to determine if B is logically an inescapable consequence of $[\text{not}(A \text{ and not } B) \text{ and } A]$. We reason as follows:

$$[\text{not}(A \text{ and not } B) \text{ and } A] \xleftrightarrow[\text{L-5}]{\text{L-4}} [(\text{not } A \text{ or } B) \text{ and } A]$$
$$\Longleftrightarrow B$$

L-4 $\text{not}(A \text{ and } B) \Longleftrightarrow$ $[\text{not } A \text{ or not } B]$

L-5 $\text{not}(\text{not } B) \Longleftrightarrow B$

The latter equivalence follows from the meaning of *and* and *or*. Hence, the conclusion is indeed valid.

a   It is not true that the company will produce more cars and not hire more men. Either the company will not make trucks, or they will produce more cars. They will produce more cars. Hence, they will hire more men.

b   A commuter waiting for a local train was pleasantly surprised when an express train unexpectedly stopped at the station. As he jumped aboard, the conductor rushed over and said, "This train doesn't stop here, so you can't get on." The commuter replied, "If the train doesn't stop here, then I'm not on it," and walked inside.

Problem Set IV   ■   *Theoretical Developments*

Prove or disprove each of the following propositions justifying each step in your argument by referring to the appropriate definition or property.

D1.4.1 $[A \Longleftrightarrow B] \Longleftrightarrow$ $[A \Longrightarrow B \text{ and } B \Longrightarrow A]$

Example $[A \Longleftrightarrow B] \Longleftrightarrow [\text{not } A \Longleftrightarrow \text{not } B]$

Proof $[A \Longleftrightarrow B] \xleftrightarrow{\text{D1.4.1}} [A \Longrightarrow B \text{ and } B \Longrightarrow A]$
$\xleftrightarrow{\text{L-2}} [(\text{not } B \Longrightarrow \text{not } A) \text{ and } (\text{not } A \Longrightarrow \text{not } B)]$
$\xleftrightarrow{\text{D1.4.1}} [\text{not } A \Longleftrightarrow \text{not } B]$ QED ●

L-2 $[A \Longrightarrow B] \Longleftrightarrow$ $[\text{not } B \Longrightarrow \text{not } A]$

● **QED** (quod erat demonstratum) means "which was to be proved or demonstrated."

Example $[A \text{ and not } B] \Longleftrightarrow [\text{not}(\text{not } A \text{ or } B)]$

Proof $\text{not}(\text{not } A \text{ or } B) \xleftrightarrow{\text{L-4}} [\text{not}(\text{not } A) \text{ and not } B]$
$\xleftrightarrow{\text{L-5}} [A \text{ and not } B]$ QED

L-4 $\text{not}(A \text{ or } B) \Longleftrightarrow$ $[\text{not } A \text{ and not } B]$

L-5 $\text{not}(\text{not } A) \Longleftrightarrow A$

Note that A, or any letter in a fundamental principle, represents any proposition. The A in L-4, DeMorgan's Laws, was represented in the first step of the above proof by the proposition *not A*.

1 $[(A \Longrightarrow B) \quad \text{and} \quad A] \Longrightarrow [A \text{ and } B]$
2 $[\text{not } A \Longrightarrow B] \Longrightarrow [A \text{ or } B]$
3 $[(A \text{ or } B) \quad \text{and} \quad \text{not } A] \Longrightarrow B$
4 $[A \Longrightarrow (B \quad \text{and} \quad \text{not } B)] \Longrightarrow \text{not } A$
5 $[\text{not}(\text{not } A \quad \text{or} \quad B) \Longrightarrow C] \Longrightarrow [C \quad \text{or} \quad \text{not}(A \quad \text{and} \quad \text{not } B)]$
6 $[\text{not}(A \quad \text{or} \quad \text{not } B)] \Longleftrightarrow [A \quad \text{and} \quad \text{not } B]$
7 $[\text{not}(\text{not } A \quad \text{or} \quad B)] \Longleftrightarrow [A \quad \text{and} \quad \text{not } B]$
8 $[A \quad \text{or} \quad B] \Longleftrightarrow [(\text{not } A \Longrightarrow B) \quad \text{or} \quad (A \text{ and } B)]$

Problem Set V ■ Just for Fun

1 Solve the following riddle: A boatman is to take a wolf, a goat, and a basket of cabbages across a river. His boat is so small that there is only room for himself and one of the three in the boat at a time. How can he accomplish this task without loss or damage to the property entrusted to him?

2 Suppose you have eight similar coins and a beam balance (without weights). One and only one of the coins is counterfeit and therefore is lighter. How can you identify the counterfeit coin using the balance *only twice?* (This problem appeared in the January 1945 issue of *American Mathematical Monthly*.)

1.6   Introduction to Sets

No mathematical innovation of the past century has had so profound and far-reaching an effect on modern mathematics as the theory of sets. Set theory is now found in almost all areas of mathematical interest, and it provides much of the language and symbolism of modern mathematics. Sets are so useful in formulating and analyzing many mathematical notions that they are now introduced at the most elementary level of one's formal education. Thus, several of the items discussed in this chapter may not be new. However, the presentation may well be more complete than in former exposures.

1.7   Sets and Subsets

A careful development of a mathematical system requires all terms and relations that characterize the system to be clearly defined, at least to the greatest extent possible. That is, a definition of one term or relation must necessarily contain other supposedly known terms or relations. The definitions of these, in turn, contain still other terms or relations and so on until eventually we find it necessary to use words whose definitions occurred earlier in this inevitably circular process. It is logically impossible to define everything. The foundation of a mathematical system must, therefore, contain certain *undefined* terms and relations from which all others are derived. In most cases our own experience and intuition supported by a few examples will render the meaning of these undefined forms quite clear and acceptable.

We begin set theory with two undefined terms, set and element, and one undefined relation, set membership. The term set is used to refer to any collection of objects or abstract entities, all having at least one property in common. The objects of a set can be anything: numbers, people, books, fish, sheep, etc. The objects comprising a set are called elements or members of the set. The undefined relation of set membership indicates whether or not an element belongs to a given set. A set is well-defined if it is always possible to determine its members. The following are examples of well-defined sets.

Example 1.7.1   (a) The collection of numbers 2, 4, 6, 8.
(b) All the words on this page.
(c) The vowels a, e, i, o, u.
(d) The capital of each state in the United States.
(e) All odd counting numbers.

Note that each set has at least one distinguishing characteristic by which you can determine its members. Examples of sets that are *not* well-defined would include the following:

Example 1.7.2   All books on the shelf.

Because we do not know which shelf is "the" shelf, it is not possible to identify the members of this set.

Example 1.7.3   All members of a club.

Here, we are not told which club and furthermore what constitutes club membership.

Two notational forms employed to denote sets and their members are in common use: one is the roster form and the other is the set-builder form. In the roster form elements are enclosed in brackets and separated by commas. Using this form to identify the sets of Example 1.7.1 (a), (c), and (e), we obtain, respectively,

$$\{2,4,6,8\} \qquad \{a,e,i,o,u\} \qquad and \qquad \{1,3,5,7, \ldots\}$$

If repeated reference to these sets were to be made, each could conveniently be labeled by a capital letter,● such as

$$A = \{2,4,6,8\} \qquad B = \{a,e,i,o,u\} \qquad and \qquad C = \{1,3,5,7, \ldots\}$$

● Although capital letters are also used to represent propositions, the context in which they are used should be sufficient to identify their role as sets or propositions.

The roster form is particularly well-suited to sets containing only a few members or, as in the case of C above, to sets whose unlisted members are clear from the pattern displayed by the few that are listed. On the other hand, the set-builder form is more adapted to sets whose elements are quite numerous. In this form, rather than listing the elements, we identify the common properties that the elements must satisfy in order to qualify as members. Using set-builder notation, Example 1.7.1 (b) may be stated

$$D = \{x \mid x \text{ is a word on this page}\}$$

which is read "*D* is the set *of all x such that x is a word on this page*." (The vertical bar $\mid$ is a symbol, usually limited to set-builder notation that means "such that" or "where.") The letter x here represents an *arbitrary* element of the set D. Utilizing this form, Examples 1.7.1 (c) and (e) can also be denoted as

$$B = \{x \mid x \text{ is a vowel}\}$$
$$C = \{x \mid x \text{ is an odd counting number}\}$$

Here, instead of listing the elements of B and C in the roster notation, we identify the elements by their classification.

If a set A contains a as one of its members, this relation is denoted

$$a \in A$$

and is read "*a is an element of A*." If, on the other hand, a is *not* a member of set A, then we write

$$a \notin A$$

Example 1.7.4 Let $A = \{1, 3, 4, 7\}$, then $1 \in A$ and $4 \in A$, whereas $2 \notin A$ and $24 \notin A$.

Example 1.7.5 Let $B = \{x \mid x \text{ is a state capital}\}$, then Austin $\in B$, whereas Chicago $\notin B$.

We shall now use our undefined terms of set, element, and set membership to define some important relations. We begin with the relation of identity — *equals*.

DEFINITION 1.7.1 Two sets, A and B, are equal, denoted $A = B$, if every element of A is an element of B and, conversely, every element of B is an element of A. That is, if A and B contain exactly the same members. Diagrammed, this definition becomes

$$[A = B] \iff [a \in A \iff a \in B]$$

Example 1.7.6 Let $A = \{1, 2, 3\}$, $B = \{1, 3\}$, $C = \{3, 1, 2\}$, and $D = \{1, 2, 2, 3\}$. Accordingly,

$$A \neq B$$

because $3 \in A$ and $3 \notin B$. However,

$$A = C$$

because A and C contain the same elements (sets are not changed if their elements are rearranged), and finally,

$$A = D$$

(sets are not changed if their elements are repeated).

DEFINITION 1.7.2 A is a subset of B, denoted $A \subseteq B$, if every element of A is also an element of B. Symbolically this definition becomes

$$[A \subseteq B] \Longleftrightarrow [a \in A \Longleftrightarrow a \in B]$$

Example 1.7.7 Let $A = \{a, b, c, d\}$, $B = \{b, d\}$, $C = \{a, b, c, d, e, f\}$, and $D = \{a, c, b, d\}$. Then the following relations hold:

$$A \subseteq C \qquad B \subseteq A \qquad A \subseteq D \qquad D \subseteq A$$

Remark 1.7.1 From the example above we note that $A = B$ is equivalent to $A \subseteq B$ and $B \subseteq A$. In fact, had Definition 1.7.2 preceded Definition 1.7.1, this equivalence could have been used as a definition for the relation equals.

Before presenting the next definition, two new symbols will be introduced. The first is "$\exists$" that is read "*there exist*," and the second "$\ni$" that is read "*such that*" or "*where*."

Example 1.7.8 The symbolic mathematical expression

$$\exists\, a \in A \ni a \notin B$$

is read "*there exists* an element a in set A *such that* a is not a member of set B."

Until, of course, the meaning of each symbol is well understood, these abbreviated mathematical forms may be less clear than their verbal counterparts. However, there are so few of these shorthand symbols to be introduced that you will quickly become familiar with them and thereafter often find the abbreviated forms not only simpler and more convenient, but you will find they actually render the proposition more clearly understandable.

DEFINITION 1.7.3 A is a proper subset of B, denoted $A \subset B$, if A is a subset of B and if *there exists* an element in B that is *not* in A. Diagrammed, this definition becomes

$$[A \subset B] \Longleftrightarrow [A \subseteq B \quad \text{and} \quad \exists \, b \in B \ni b \notin A]$$

Remark 1.7.2 Definitions 1.7.1, 1.7.2, and 1.7.3 are interrelated as the following two properties indicate:

$$[A \subseteq B] \Longleftrightarrow [A \subset B \quad or \quad A = B]$$

and

$$[A \subset B] \Longleftrightarrow [A \subseteq B \quad and \quad A \neq B]$$

Example 1.7.9 Let $A = \{1, 2, 5\}$, $B = \{2\}$, $C = \{5, 2, 1\}$, and $D = \{1, 2, 6\}$. Then

$$B \subset A$$

because $B \subseteq A$ and there exists the element 5 in A such that 5 is not in B, or, equivalently, $B \subseteq A$ and $B \neq A$; also

$$A \subseteq C$$

because $A = C$; and finally,

$$A \not\subseteq D$$

because there exists the element 5 in A that is not in D.

Frequently relationships among sets are illustrated by Venn diagrams (diagrams in which sets are represented as collections of points within plane enclosures such as circles or squares).

Example 1.7.10 Suppose A is a subset of B ($A \subseteq B$). This relationship may be described by any one of the Venn diagrams in Fig. 1.7.1.

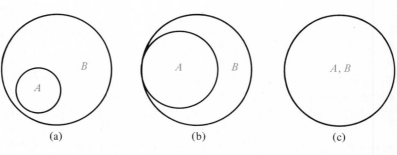

(a)   (b)   (c)

FIGURE 1.7.1

Remark 1.7.3 In Fig. 1.7.1 (a) and (b), A is a *proper* subset of B ($A \subset B$) because there are points in B not in A, whereas in (c) A is *equal* to B ($A = B$). These diagrams remind us that

$$[A \subseteq B] \Longleftrightarrow [A \subset B \quad \text{or} \quad A = B]$$

Example 1.7.11 Suppose A is not a proper subset of B ($A \not\subset B$). Venn diagrams illustrating this are shown in Fig. 1.7.2.

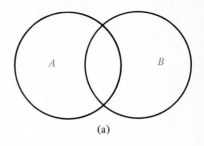

(a)

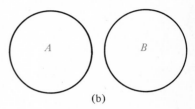

(b)

FIGURE 1.7.2

Remember that Venn diagrams represent only specific illustrations and not sets in general. Nevertheless, they can be a great help in developing a clearer understanding of relations such as these.

In most applications of set theory, all sets under consideration are subsets of a specific set called the universe of discourse or universal set. A universal set is the set of *all* elements that may be admitted as members of other sets considered within a given problem or study. This set is commonly denoted by the letter U.

Example 1.7.12 Let U be the set of all people living on the planet earth. Any subsequently given set must be a subset of U—say, the set of all citizens of Canada, or London, or Asia. Within this universe sets such as $\{1, 2, 4, -10\}$ or $\{$blue, green, black, orange$\}$ would not be admissible.

Example 1.7.13 Let U be the set of all *integers*. Then within this universe, the set $A = \{-2, 0, 4, 25\}$ would be acceptable but the set $B = \{-1, 3, \frac{1}{2}\}$ would not, because it contains an element, $\frac{1}{2}$, that does not belong to the universal set.

In direct contrast to the notion of a universal set is that of an empty or null set. The empty set is just what its name suggests—a set which has no members. This set is usually denoted by $\varnothing$ or $\{\ \}$.

Example 1.7.14 Let A be the set of all persons from the planet earth who are presently living on Mars. This set is, for now, *empty* and we write

$$A = \varnothing$$

Example 1.7.15 Let B be the set of all negative natural (counting) numbers. This set is also the *null* set because there are no natural numbers that are negative. Set B may be expressed, using set-builder notation, as

$$B = \{x \mid x \text{ is a negative natural number}\} = \varnothing$$

Remark 1.7.4 The empty set is a subset of every set (including itself). This can be reasoned by observing the consequence of supposing the empty set were *not* a subset of every set (arguing from the contrapositive approach). That is, suppose there existed a set, say D, for which the empty set is not a subset. This would imply that *there exists an element in* $\varnothing$, say a, such that $a \notin D$, which, of course, is a contradiction because $\varnothing$ has no members. Thus, by L-2 ● we are led to the conclusion that the empty set *is* a subset of every set. Note that a set containing the empty set as a member is *not* empty just as a box containing an empty box is not empty. Expressed symbolically, this comment becomes

$$A = \{\varnothing\} \neq \varnothing$$

We now observe an important property of sets—a transitive law for the relation subset.

● L-2 $[A \Longrightarrow B] \Longleftrightarrow$
$[\text{not } B \Longrightarrow \text{not } A]$

THEOREM 1.7.1   Let A, B, and C be three sets. Further, suppose A is a subset of B and B is a subset of C. Then A is a subset of C. Symbolically we have

$$[A \subseteq B \quad \text{and} \quad B \subseteq C] \Longrightarrow [A \subseteq C]$$

Proof

D1.7.2
$[A \subseteq B] \Longleftrightarrow [a \in A \Longrightarrow a \in B]$

$$[A \subseteq B \quad \text{and} \quad B \subseteq C] \overset{\textbf{D1.7.2}}{\Longleftrightarrow}$$

$$[(a \in A \Longrightarrow a \in B) \quad \text{and} \quad (a \in B \Longrightarrow a \in C)]$$

L-3 $[A \Longrightarrow B \quad \text{and} \quad B \Longrightarrow C]$
$\Longrightarrow [A \Longrightarrow C]$
(A, B, and C here are propositions represented in the proof of Theorem 1.7.1 by the propositions $a \in A$, $a \in B$. and $a \in C$, respectively.)

$$\overset{\textbf{L-3}}{\Longrightarrow} [a \in A \Longrightarrow a \in C] \overset{\textbf{D1.7.2}}{\Longleftrightarrow} [A \subseteq C]$$

Hence, again by the Transitive Law of Implication, L-3,

$$[A \subseteq B \quad \text{and} \quad B \subseteq C] \Longrightarrow [A \subseteq C] \qquad \text{QED}$$

Example 1.7.16   Let the universal set, U, be the set of points within a rectangle and sets A, B, and C be circles of points, satisfying the relations $A \subseteq B$ and $B \subseteq C$, as illustrated in Fig. 1.7.3. Note the obvious conclusion $A \subseteq C$, illustrating Theorem 1.7.1.

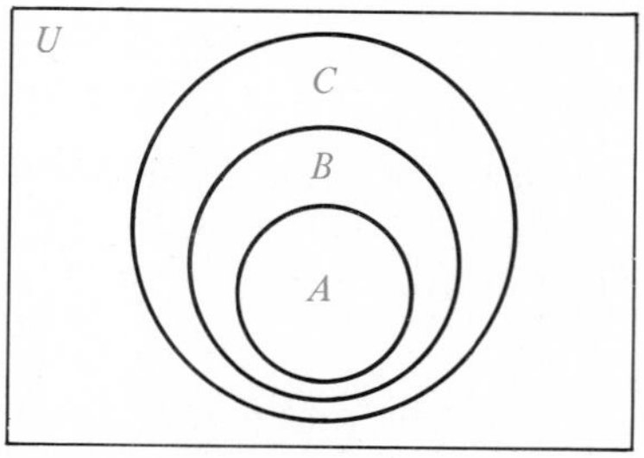

$$[A \subseteq B \quad \text{and} \quad B \subseteq C] \Longrightarrow [A \subseteq C]$$

FIGURE 1.7.3

Problem Set I ▪ Reading Comprehension

Determine whether statements 1 through 8 are true or false and give reasons for your responses.

1   Set theory is one of the oldest branches of mathematics.
2   It is logically impossible to define all terms and relations associated with a mathematical system.
3   One of the undefined relations in set theory is the relation of identity, called equals.
4   Suppose $A = \{3, 4, 7\}$, $B = \{7, 3, 4\}$, and $C = \{3, 4, 7, 7\}$. Then sets A and B are *equal*, but A and C are not equal because C has more members than A.
5   Sets A, B, and C in Problem 4 are expressed in the *roster form*.
6   The expression "a is a member of set A" is denoted $a \subset A$.
7   An equivalent definition of *subset* is $[A \subseteq B] \Longleftrightarrow [A \subset B$ and $A = B]$.
8   The relation *set membership* is undefined.
9   What constitutes a *well-defined* set?
10 For what kinds of sets is the *set-builder* notational form best suited?
11 Explain what a *Venn diagram* is and what its limitations are.
12 Is it possible for a *universal set* and the *null set* to be equal?

Problem Set II ▪ Skills Development

Express the sets defined in problems 1 through 5 in the *roster form*.

1   The set of even natural numbers between 1 and 10 inclusive.
2   The set of Presidents of the United States since 1950.
3   The set of all states of the United States that border on Mexico.
4   The set of all natural numbers less than 15 that are divisible by 4.
5   The set of all courses you are taking this year.

Express the sets of problems 6 through 11 in the *set-builder* form.

6   The set of all books in your home library.
7   The set of all integers between 1 and 100 that are multiples of 3.
8   The set of all positive integers that are perfect squares.
9   The set of all circles.
10 The set of all friends that are over 6 feet tall, have blue eyes, and sing well.
11 The set of all positive rational numbers.

Write statements 12 through 17 in symbolic form.

12 The element b does not belong to A.
13 B is a proper subset of C.
14 A is a subset of B, but B is not a subset of A, therefore, A is a proper subset of B.
15 D is not a subset of F. Hence, there must exist an element in D that is not in F.
16 The element a is a member of set A though it is not a member of set D.
17 If an element a is a member of B implies a is also a member of C and C is a proper subset of D, then there exists an element in D that is not a member of B.

Express the symbolic statements 18 through 23 in English.

18 $\{x \mid x \text{ is a Senator from Ohio or Iowa}\}$
19 $\{x \mid 2x = 8\}$
20 $\{3, 4, 5, 6\}$
21 $[A \subseteq B \text{ and } B \subset C] \Longrightarrow \exists\, c \in C \ni c \notin A$
22 $A \subseteq D \overset{\Longrightarrow}{\not\ } \exists\, d \in D \ni d \notin A$
23 $[A \subseteq B \text{ and } B \subseteq A] \Longleftrightarrow [A = B]$
24 If $A = \{0, 1, 3\}$, determine whether each of the four following statements is correct. If the statement is incorrect, give reasons.
  a $\{\ \} \subseteq A$ **b** $\{0\} \in A$
  c $3 \subset A$ **d** $3 \in A$
25 Determine which of the following sets are equal:
  (1) The letters of the word *able*
  (2) $\{x \mid x \text{ is a letter of the word } bale\}$
  (3) The letters of the word *label*
  (4) $\{e, a, l, b\}$
26 Determine which of the following sets are equal:
  (1) $\varnothing$
  (2) $\{0\}$
  (3) $\{\varnothing\}$
  (4) 0
  (5) $\{\ \}$
  (6) $\{x \mid 3x = 12 \text{ and } \sqrt{x} = 4\}$

1.9   Operations on Sets

Algebraic systems are composed not only of terms and relations, both defined and undefined, but also of *operations* on the terms. In fact the operations create the flexibility and utility of a given system. Take, for example, the rational numbers. How useful would this system be if it were not for the operations of addition, subtraction, multiplication and division? Nearly all applications of numbers in problem solving depend on the fact that two numbers can be combined under these operations to obtain a third number: $2 + 3 = 5$, $\frac{1}{3} - \frac{1}{4} = \frac{1}{12}$, $\frac{1}{5} \times 15 = 3$, etc.

● There are other operations on sets but these three will be sufficient for our purposes here.

To develop an algebra for sets, we define three operations — set union, intersection, and difference. ●

DEFINITION 1.9.1   The union of two sets A and B, denoted $A \cup B$, is the set of all elements belonging either to A *or* to B. Symbolically, this definition becomes

$$A \cup B = \{x \mid x \in A \quad or \quad x \in B\}$$

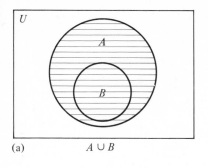

(a) $A \cup B$

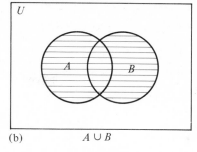

(b) $A \cup B$

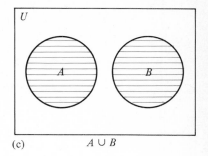

(c) $A \cup B$

FIGURE 1.9.1

Remark 1.9.1 Remember, particularly in the above definition, the inclusive meaning of the disjunction *or* in mathematics. Elements that are in *A or* in *B* include those that are in *both A* and *B*.

Example 1.9.1 Let $A = \{a, b, c, d\}$ and $B = \{b, d, e, f, g\}$. Then

$$A \cup B = \{a, b, c, d, e, f, g\}$$

Example 1.9.2 The shaded areas in the three Venn diagrams in Fig. 1.9.1 illustrate the *union* of point sets denoted in each case by disks A and B.

Remark 1.9.2 Part (a) of Fig. 1.9.1 illustrates the property

$$[B \subseteq A] \Longleftrightarrow [A \cup B = A]$$

DEFINITION 1.9.2 The intersection of two sets A and B, denoted $A \cap B$, is the set of all elements belonging to both A *and* B—the elements *common* to both. Symbolically, this definition is

$$A \cap B = \{x \mid x \in A \quad and \quad x \in B\}$$

Example 1.9.3 Let $A = \{1, 2, 3, 4\}$, $B = \{7, 8, 9\}$, and $C = \{3, 4, 5, 7\}$. Then,

$$A \cup C = \{1, 2, 3, 4, 5, 7\}$$
$$A \cap C = \{3, 4\}$$
$$A \cap B = \{ \ \} = \varnothing$$

Remark 1.9.3 Notice $A \cup C$ in Example 1.9.3. Although 3 and 4 were elements of both A and C, they were *not repeated* in the union of A and C. Usually elements are not repeated in set notation.

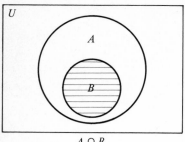

$A \cap B$

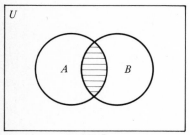

$A \cap B$

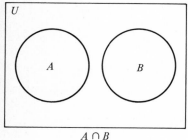

$A \cap B$

FIGURE 1.9.2

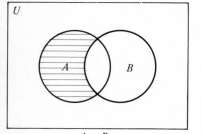

$A - B$

FIGURE 1.9.3

Example 1.9.4 The shaded areas in the three Venn diagrams of Fig. 1.9.2 illustrate the *intersection* of point sets A and B.

Remark 1.9.4 Part (a) of Fig. 1.9.2 illustrates the property

$$[B \subseteq A] \Longleftrightarrow [B \cap A = B]$$

Notice in Part (c) that $A \cap B = \varnothing$, the empty set.

DEFINITION 1.9.3 Two *nonempty* sets A and B are disjoint if their intersection is empty—they have *no* members in common. More briefly, this definition is written

$$[A \text{ and } B \text{ disjoint}] \Longleftrightarrow [A \cap B = \varnothing \text{ and } A, B \neq \varnothing]$$

Observe that the two sets A and B of Example 1.9.3 are *disjoint*.

DEFINITION 1.9.4 The difference of two sets A and B in the stated order, denoted $A - B$, is the set of *all* elements of A that are *not* elements of B. Diagrammed, this definition becomes

$$A - B = \{x \mid x \in A \text{ and } x \notin B\}$$

Example 1.9.5 Let $A = \{1, 2, 3, 4\}$, $B = \{7, 8, 9\}$, and $C = \{3, 4, 5, 7\}$. Then

$$A - C = \{1, 2\}$$

and

$$A - B = \{1, 2, 3, 4\} = A$$

Remark 1.9.5 If A and B are *disjoint* then $A - B = A$ and $B - A = B$.

Example 1.9.6 Let A and B be circles of points as defined in the Venn diagram in Fig. 1.9.3. The shaded area represents $A - B$.

DEFINITION 1.9.5 The complement of a set A, denoted A , is the *difference* $U - A$, where U is the universal set. Symbolically, this definition is

$$A' = U - A = \{x \mid x \in U \text{ and } x \notin A\}$$

Example 1.9.7 Let $U = \{1, 2, 3, 4, 5, 6, 7, 8, 9\}$ and $A = \{2, 4, 5\}$. Then

$A' = \{1, 3, 6, 7, 8, 9\}$

It is convenient in many applications to use more than one operation in a single expression. For example, we may have an expression of numbers such as

$[4 \times (2 - 3)] \div [5 - (2 + 4)]$

containing the four operations of addition, subtraction, multiplication, and division. This is also true in the algebra of sets. When more than one operation is present, the order in which operations are performed is indicated by parentheses, brackets, or braces. An expression of the form

$A \cap (B \cup C)$

means to intersect A with the union of B and C, whereas

$(A \cap B) \cup C$

means to find the union of C with the intersection of A and B.

We now examine several Venn diagrams illustrating various combinations of the set operations *union, intersection,* and *difference*. In Examples 1.9.8 through 1.9.11, U will represent a square of points and A, B, C, D, circles of points within U as indicated in Figure 1.9.4.

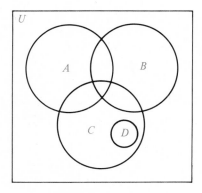

FIGURE 1.9.4

Example 1.9.8 In Fig. 1.9.5, the area shaded with horizontal lines represents $A \cap B$, whereas that shaded with vertical lines represents $B \cap C$. The area shaded with both horizontal and vertical lines is, of course, the intersection of these two sets: $(A \cap B) \cap (B \cap C)$.

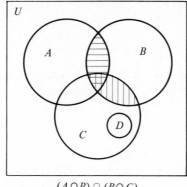

$(A \cap B) \cap (B \cap C)$

FIGURE 1.9.5

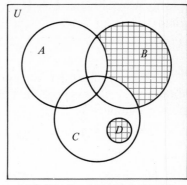

$(A' \cap B) \cup D = (A' \cup D) \cap (B \cup D)$

FIGURE 1.9.6

Example 1.9.9 In Figure 1.9.6, the horizontal lines represent $(A' \cap B) \cup D$ and the vertical lines $(A' \cup D) \cap (B \cup D)$. Notice that these sets are equal: $(A' \cap B) \cup D = (A' \cup D) \cap (B \cup D)$. This relation is not a coincidence but rather a fundamental property of these operations that will be proved in the next section.

Example 1.9.10 Figure 1.9.7 illustrates the two sets $(C - D) \cup (A \cap B)$ and $A - (C \cup B)$ with horizontal and vertical lines, respectively.

Example 1.9.11 Let the set $(A \cup B)'$ be represented in Fig. 1.9.8 by the area shaded with horizontal lines and the set $A' \cap B'$ with vertical lines. You will note that $(A \cup B)' = A' \cap B'$, which is true in general and is one of DeMorgan's laws for sets.

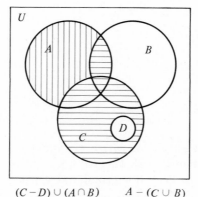

$(C - D) \cup (A \cap B)$ $A - (C \cup B)$

FIGURE 1.9.7

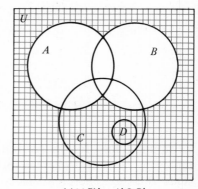

$(A \cup B)' = A' \cap B'$

FIGURE 1.9.8

It must be remembered, as mentioned earlier, that Venn diagrams are but specific illustrations of general principles or expressions. These diagrams are by no means unique. For example, if sets U, A, B, C, or D of Fig. 1.9.4 were defined with A and B *disjoint*, then the expression $A \cap B$ of Example 1.9.8 would be *empty*, and there would not be a shaded are in Fig. 1.9.5, which corresponds to $(A \cap B) \cap (B \cap C)$.

Problem Sets 1.10

Problem Set I ■ *Reading Comprehension*

Determine whether statements 1 through 4 are true or false and give reasons for your responses.

1 If $A = \{a, e, f, g\}$ and $B = \{e, g, h, i\}$, then A *union* B, denoted $A \cup B$, is the set $\{a, e, f, g, e, g, h, i\}$.
2 Two sets are *disjoint* if and only if their union is empty.
3 The *difference* of two sets is the set of all elements that exist in each but that are not common to both.
4 The *complement* of two sets is the set of all elements common to both sets.

Problem Set II ■ *Skills Development*

1 Let $U = \{1, 2, 3, 4, 5, 6, 7, 8, 9\}$, $A = \{1, 4\}$, $B = \{2, 3, 4, 5\}$, and $C = \{3, 8, 9\}$. Find
  a $A \cap B$ b $A \cup B$ c $(A \cup B) \cap C'$
  d $(B - C) \cup A$ e $B \cap U'$ f $(C - B) - A$
  g $(A \cap C)' \cap A$ h $(A \cup B) \cap (A \cup C)$
2 Complete each of the following, using Venn diagrams, if necessary, to assist in your decision.
  a $A \cup U =$ b $A \cap \varnothing =$ c $A \cup \varnothing =$
  d $A \cap A =$ e $\varnothing \cap \varnothing =$ f $A \cup A =$
  g $\varnothing \cup \varnothing =$ h $U \cap U =$ i $(A')' =$
  j $A \cap A' =$ k $A \cap U =$ l $A \cup A' =$
3 Let sets U, A, B, C, and D be defined as areas of points as indicated in Fig. 1.10.1. Reproduce this diagram twice on a sheet of paper. In the first diagram shade the set

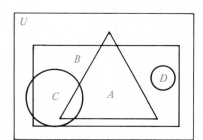

FIGURE 1.10.1

$C - (A \cup B')$ by horizontal lines and the set $(C \cap A) \cup D$ by vertical lines. In the second Venn diagram, shade the set $(C - A) \cup (A - C)$ by horizontal lines, and the set $[D' - (A \cup C)] \cap B$ by vertical lines.

Example

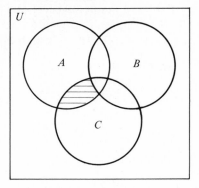

FIGURE 1.10.2

Answer $(A \cap C) - B$ or $(A \cap C) \cap B'$

4   Find an expression in operational notation for each of the shaded areas in the Venn diagrams of Figure 1.10.3. (Note first the example of Fig. 1.10.2).

5   Suppose that $U = \{x | x$ is a student at East High$\}$, that $A = \{x | x$ is a freshman$\}$, that $B = \{x | x$ is a student-body officer$\}$, that $C = \{x | x$ is a football player$\}$, and that $D = \{x | x$ has blue eyes$\}$. Using operational notation, describe the set of all freshmen who do not play football but who are either student-body officers or have blue eyes.

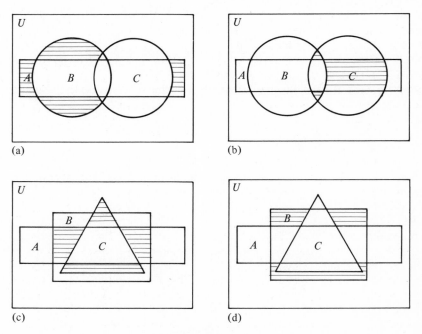

FIGURE 1.10.3

Problem Set III ■ Theoretical Developments

Determine the conditions under which the following statements are true.

Example $A \cup B = A$

Solution

$A \cup B = A \xLeftrightarrow{\text{D1.7.1}} [x \in (A \cup B) \Longleftrightarrow x \in A]$

$\xLeftrightarrow{\text{D1.9.1}} [(x \in A \quad \text{or} \quad x \in B) \Longleftrightarrow x \in A]$

$\xLeftrightarrow{\text{D1.4.1}} [x \in B \Longrightarrow x \in A]$

$\xLeftrightarrow{\text{D1.7.2}} B \subseteq A$

Therefore, by the Transitive Law of Implication, L-4, we have

$$[A \cup B = A] \Longleftrightarrow [B \subseteq A] \qquad \text{QED}$$

You may wish to begin your investigation of these problems by a Venn diagram; this will provide not only visual but also conceptual understanding.

1 $\quad A \cup \varnothing = \varnothing$	2 $\quad A \cap U = U$	3 $\quad A \cup \varnothing = U$
4 $\quad A \cap B = A$	5 $\quad A \cup B = \varnothing$	6 $\quad A \cap B = A \cup B$
7 $\quad A' \cap U = U$	8 $\quad A' \cup \varnothing = \varnothing$	9 $\quad A \cap B = B$

Prove each of the following propositions justifying every step in your proof by referring to a property permitting the step.

10 Let A and B be any two sets. Then
- a $\quad B \subseteq (A \cup B)$
- b $\quad (A \cap B) \subseteq A$
- c $\quad A - B = A \cap B'$
- d $\quad (A - B) \subseteq A$
- e $\quad (A - B) \cap B = \varnothing$

11 Prove the following implications where A and B are sets satisfying the given conditions (hypotheses).
- a $\quad A \subseteq B \Longrightarrow (A \cap B) = A$
- b $\quad A \subseteq B \Longrightarrow (A \cup B) = B$
- c $\quad A \subseteq B \Longrightarrow B' \subseteq A'$
- d $\quad A \cup B = A \Longrightarrow A' \cap B = A'$
- e $\quad A \cap B \neq \varnothing \Longrightarrow [A \neq \varnothing \quad \text{and} \quad B \neq \varnothing]$
- f $\quad A \cap B \neq \varnothing \Longrightarrow (A' \cup B') \neq U$

Sidebar:

D1.7.1
$[A = B] \Longleftrightarrow [A \in A \Longleftrightarrow a \in B]$

D1.9.1
$A \cup B = \{x \mid x \in A \quad \text{or} \quad x \in B\}$

D1.4.1
$[A \Longleftrightarrow B] \Longleftrightarrow$
$\qquad [A \Longrightarrow B \quad \text{and} \quad B \Longrightarrow A]$

D1.7.2
$A \subseteq B \Longleftrightarrow [a \in A \Longrightarrow a \in B]$

Problem Set IV ■ Just for Fun

1 What is the largest sum of money composed of pennies, nickels, dimes, quarters, and half dollars, that one can have without being able to give change for any one of these coins or for a dollar?

2 True or false: A candy dispenser is not functioning properly or you did not put in a dime, or it is really a slot machine. It is not the case that it is both not functioning and you did not put in a dime. If it is not a slot machine, then it is working and you did put in a dime. It is either a slot machine or it is not the case either that it is not functioning or that you did not put in a dime. Therefore, it is really a slot machine.

1.11 Summary of Definitions and Properties in Chapter 1

Definitions and Theorems

D1.7.1 $\quad [A = B] \Longleftrightarrow [a \in A \Longleftrightarrow a \in B]$ (p. 30)

D1.7.2 $\quad [A \subseteq B] \Longleftrightarrow [a \in A \Longrightarrow a \in B]$ (p. 31)

D1.7.3 $\quad [A \subset B] \Longleftrightarrow [A \subseteq B \quad \text{and} \quad \exists\, b \in B \ni b \notin A]$ (p. 32)

T1.7.1 $\quad [A \subseteq B \quad \text{and} \quad B \subseteq C] \Longrightarrow [A \subseteq C]$ (p. 34)

D1.9.1 $\quad A \cup B = \{x \mid x \in A \quad \text{or} \quad x \in B\}$ (p. 36)

D1.9.2 $\quad A \cap B = \{x \mid x \in A \quad \text{and} \quad x \in B\}$ (p. 37)

D1.9.3 $\quad [A \text{ and } B \text{ disjoint}] \Longleftrightarrow [A \cap B = \varnothing \quad \text{and} \quad A, B \neq \varnothing]$ (p. 38)

D1.9.4 $\quad A - B = \{x \mid x \in A \quad \text{and} \quad x \notin B\}$ (p. 38)

D1.9.5 $\quad A' = \{x \mid x \in U \quad \text{and} \quad x \notin A\}$ (p. 38)

Fundamental Properties

L-1 $\quad [A \Longrightarrow B]$ not equivalent to $[B \Longrightarrow A]$ (p. 17)

L-2 $\quad [A \Longrightarrow B]$ is equivalent to $[\text{not } B \Longrightarrow \text{not } A]$ (p. 18)

L-3 $\quad [A \Longrightarrow B \quad \text{and} \quad B \Longrightarrow C] \Longrightarrow [A \Longrightarrow C]$ (p. 19)

L-4 $\quad [\text{not}(A \text{ and } B)] \Longleftrightarrow [\text{not } A \quad \text{or} \quad \text{not } B]$ (p. 21)
$\quad\quad\ [\text{not}(A \text{ or } B)] \Longleftrightarrow [\text{not } A \quad \text{and} \quad \text{not } B]$

L-5 $\quad \text{not}(\text{not } A) \Longleftrightarrow A$ (p. 22)

Answers to Sample Problem Set 1.3 (page 15)

1 This is a proposition because it is a true statement.
2 This is a proposition because it is a false statement.
3 This is an open proposition because it contains the variable *his*.
4 This is not a proposition because it carries no notion of trueness nor falseness.
5 This is a proposition because it is false.
6 This is an open sentence.
7 An open sentence such as this, which is true for all values of the variable, is usually called an identity.
8 This is not a proposition because it is neither true nor false. If the expression were set equal to some number, forming an equation, then it would become an open sentence.
9 False. Antecedent and hypothesis are synonymous.
10 False. An implication is only true if the truth of the antecedent always leaves the truth of the consequent inescapable.
11 True.

"It hurts when I do square roots."

Courtesy of Irwin Caplan

Real Numbers

Introduction 2.1

We are all familiar with numbers. For most, these are limited to the natural (counting) numbers and fractions: "Jim weighs 173 pounds," or "there is only $\frac{1}{4}$ of a pie left." Man has, however, created many other numbers, all designed to help measure and analyze an almost endless array of physical objects or abstract notions. Just imagine how handicapped we would be if it were not for our concepts of numbers. How would a carpenter build a house, or an engineer construct a bridge? How would the astrophysicist analyze the movement and makeup of the stars, or the sociologist study the growth patterns of a large population? How would the zoologist study hereditary probabilities, or the businessman examine sales trends or pursue sound economic policies? How would the astronauts ever make it to the moon and back? As we have witnessed in the past, many of the successes of the future will depend upon man's knowledge of numbers and how well that knowledge is utilized.

2.2 Real Numbers

Let us briefly examine those numbers that as a collection are called real. We referred in the previous section to counting numbers (1, 2, 3, 4, . . .) and fractions ($\frac{1}{2}, \frac{3}{5}, \frac{16}{3}, . . .$), with which you are already well acquainted. These are all real numbers. There are, however, other important real numbers. These include the *negative* of each of the numbers above as well as those real numbers called irrational numbers.

Negative numbers can be quite descriptive in a variety of situations. For example, it is convenient to indicate an overdrawn bank account of, say 26 dollars, by showing a balance of −26 dollars. How often have we listened intently to a countdown prior to a rocket blastoff. "Ignition time is −13 minutes and counting."

One of the most common uses of numbers is in measuring distances: 34 miles to the next station, 24-inch zipper, or 6 feet 3 inches tall. Few who have studied geometry are surprised to learn that there are distances that cannot be measured by natural numbers or fractions. This fact was probably first discovered by the Pythagoreans—a group of Greek intellectuals founded by Pythagoras in the middle of the sixth century B.C. Prior to this discovery, they asserted that all distances could be measured by counting numbers and fractions. The Pythagoreans were familiar with the following property of right triangles (proved about 200 years later by Euclid): The sum of the squares of the lengths of the two shorter sides of a *right* triangle is equal to the square of the length of the hypotenuse. If a and b represent the lengths of the two shorter sides, and c, the length of the hypotenuse (Fig. 2.2.1), then the property above, called the Pythagorean theorem, may be expressed algebraically by the formula $a^2 + b^2 = c^2$.

The Pythagoreans discovered, with this formula, that a right triangle with two sides equal to a length of 1 unit yielded a hypotenuse whose length when squared was equal to 2 units (see Fig. 2.2.2). They further proved that a number whose square is equal to 2 cannot be a counting number nor a fraction. Yet they felt that corresponding to every distance there was or ought to be a number, and so they were faced with acceptance of numbers other than the counting numbers and fractions. Today the positive number whose square is 2 is called the square root of 2, usually denoted $\sqrt{2}$. The

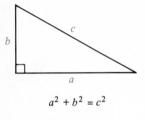

$$a^2 + b^2 = c^2$$

FIGURE 2.2.1

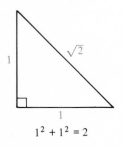

$$1^2 + 1^2 = 2$$

FIGURE 2.2.2

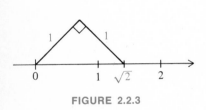

FIGURE 2.2.3

position of $\sqrt{2}$ on the number line is illustrated in Fig. 2.2.3. The square root of 2 is an *irrational* real number. Irrational numbers are those that, when expressed in decimal form, are *nonterminating* and *nonrepeating;* they have an infinite decimal expansion free of continually repeating blocks of numbers. The Greeks called the irrational numbers "unmentionables" or "unknowables." By combining the irrational numbers with the counting numbers and fractions, all distances can be represented by a number.

Subsets of the Set of Real Numbers 2.3

We call the collection of real numbers the *set of real numbers* and denote this set R. As noted in Section 2.2, there are several important subsets of R. We will single out and give names to some of these. First, the set of natural or counting numbers, denoted N, is

$$N = \{1, 2, 3, 4, 5, 6, \ldots\}$$

The set of integers, denoted J, is composed of the natural numbers, the negative of each natural number, and zero. Thus,

$$J = \{\ldots, -3, -2, -1, 0, 1, 2, 3, \ldots\}$$

Next we have the set of rational numbers, denoted Q. This set is composed of all numbers that can be expressed as the ratio of two integers, the denominator of which is never zero. Using set-builder notation, this definition becomes:

$$Q = \left\{ x \mid x = \frac{a}{b}, \quad a, b \in J \quad \text{and} \quad b \neq 0 \right\}$$

which is read "Q is the set of all numbers x such that x equals the quotient a/b where a and b are integers and b is not zero." Members of Q include -3 (which can be expressed as $-\frac{3}{1}$), $\frac{3}{5}$, $2\frac{1}{2}$ (which can be expressed as $\frac{5}{2}$), and $-\frac{7}{5}$.

The set of irrational real numbers, denoted H, is composed of all other real numbers. The set H may be expressed as

$$H = \{x \mid x \in R \quad \text{and} \quad x \notin Q\} = R - Q$$

that is, all real numbers that are not rational. Members of H include $\sqrt{2}$, $\sqrt{5}$, and π.

Some of the more obvious relations among these subsets are

$$N \subset J \subset Q \qquad Q \cap H = \varnothing \qquad Q \cup H = R$$

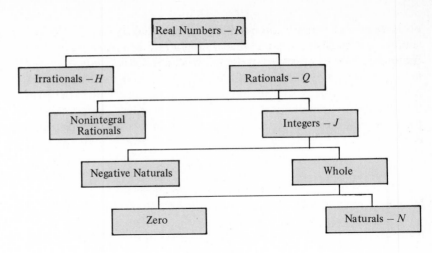

FIGURE 2.3.1

Figure 2.3.1 also helps to visualize the basic set structure of the real number system.

2.4   Numerals and Decimal Notation

Symbols used to denote numbers are called numerals. Those most commonly used today are the modern *Arabic* numerals: 0, 1, 2, 3, $\frac{1}{2}$, -12, 176, $\frac{16}{3}$, etc. Another group of symbols is the *Roman* numerals: I, II, III, IV, etc. Today we use Arabic numerals almost exclusively.

There are many different numerals (even within the Arabic system) that represent the *same* number. Take for example the numerals

$$\tfrac{8}{2}, 7 - 3, \sqrt{16}, 3 + 1, 2^2$$

Although each is a distinct numeral (symbol), they all represent the same number, whose simplest numeral form is 4. Numerals representing the *same* number are said to be *equal,* and we write

$$4 = \tfrac{8}{2} = 7 - 3 = \sqrt{16} = 3 + 1 = 2^2$$

It is a common practice to refer to the numeral for a number as if that numeral were the number. The *numeral* 2, for example, will frequently be referred to as the *number* 2. In most cases the meaning is clear from the context of the statement. Remember that the numeral is the symbol representing the abstraction, which is the number. We will write the numeral and speak of the number it represents.

Every real number can be represented in decimal form. In particular, nonintegral rational numbers are frequently expressed as decimals. You may be familiar with some of the following:

$$\tfrac{1}{4} = 0.25$$
$$\tfrac{1}{3} = 0.3333 \ldots$$
$$\tfrac{3}{8} = 0.375$$

and

$$\tfrac{3}{11} = 0.272727 \ldots$$

The decimal representation of a rational number is easily obtained by the process of long division, as shown in Example 2.4.1.

Example 2.4.1   Find the decimal form for $\tfrac{5}{8}$.

```
      0.625
  8) 5.000000
     4 8
      20
      16
       40
       40
        0
```

Thus

$$\tfrac{5}{8} = 0.625$$

Should the division terminate (obtain a zero remainder), as with Example 2.4.1, then we say the decimal is terminating, otherwise the decimal is nonterminating. If the decimal form of a rational number is nonterminating, as with $\tfrac{1}{3}$ and $\tfrac{3}{11}$ above, the expansion will always contain repeating blocks of digits. It is very important to remember this fact because it provides a characterization that distinguishes the rationals from the irrationals. (Recall that the decimal expansions of irrationals are nonterminating but do not have such repeating blocks of digits.)

Example 2.4.2   The decimal representation for $\tfrac{7}{13}$ is

$$\tfrac{7}{13} = 0.538461538461538461538461 \ldots$$

The repeating block is 538461.

When continually repeating blocks occur, we often place a *bar* above the first block and do not repeat it further. Thus, we may write

$$\tfrac{7}{13} = 0.\overline{538461}$$

and

$$\tfrac{3}{11} = 0.272727 \ldots = 0.\overline{27}$$

At times it is necessary to convert from decimal form to the rational. The next two examples will demonstrate a method to accomplish this.

Example 2.4.3 Find the fractional equivalent to $0.\overline{45}$. Let $x = 0.454545 \ldots$. Then

$$100x = 45.454545 \ldots$$

Subtracting the first expression from the second and solving for x we have

$$\begin{aligned} 100x &= 45.454545 \ldots \\ -x &= -0.454545 \ldots \\ \hline 99x &= 45 \end{aligned}$$

$$\implies \quad x = \frac{45}{99}$$

$$= \frac{5}{11}$$

Thus

$$\frac{5}{11} = 0.\overline{45}$$

Example 2.4.4 Find the fraction equivalent to $2.3121212 \ldots$. Let $x = 2.3121212 \ldots$. Then $10x = 23.121212 \ldots$ and $1000x = 2312.121212 \ldots$. Subtracting $10x$ from $1000x$ and solving for x we have

$$\begin{aligned} 1000x &= 2312.121212 \ldots \\ -10x &= -23.121212 \ldots \\ \hline 990x &= 2289 \end{aligned}$$

$$x = 2289/990$$

$$= 763/330$$

Thus

$$\frac{763}{330} = 2.3121212 \ldots$$

By methods similar to those of Examples 2.4.3 and 2.4.4 any nonterminating decimal with repeating blocks of digits can be converted to a fraction. Because irrational numbers cannot be expressed in finite form using modern Arabic numerals, we often resort to certain artificial symbols to represent some of the more commonly used of these numbers. For example, the principle square root of any nonnegative real number a is denoted $\sqrt{a}$, and in general, the principle n^{th} root is denoted by $\sqrt[n]{a}$. Unless a is a perfect n^{th} root (there exists a rational b such that $b^n = a$), $\sqrt[n]{a}$ is *irrational*.

Almost everyone is familiar with the irrational number called "pi" (the name of the Greek letter denoted π) that is equal to the ratio of the circumference of a circle to its diameter. This irrational number is used a great deal in both pure and applied mathematics. The number π is often approximated by the rational 3.1416. This approximation is sometimes written

$$\pi \cong 3.1416$$

Another very important irrational number is denoted by e, whose definition involves the notion of a limit that is beyond the scope of this text. However, a rational approximation for e is

$$e \cong 2.7182818284$$

The irrational number e will be used in Chapter 5 as the base of the natural system of logarithms.

The use of irrational numbers is made practical in applications by the fact that any irrational can be approximated to any degree of accuracy by a rational number.

Problem Sets 2.5

Problem Set I ■ *Reading Comprehension*

Determine whether statements 1 through 13 are true or false and give reasons for your responses:

1   One of the many advantages of fractions is that they provide a measurement for every distance.
2   The Pythagorean Theorem states that the sum of the squares of any two sides of a triangle is equal to the square of the third side.
3   It can be shown that a number whose square is 2 cannot be expressed as a fraction a/b where a and b are counting numbers.
4   A nonterminating decimal expression represents an irrational number.
5   Those numbers called "unmentionables" or "unknowables" by the early Greeks are today known as imaginary numbers.
6   Any number that can be expressed as the ratio of two other numbers is rational.
7   The irrational numbers are all those that are not rational.
8   The words numeral and number have the same meaning.
9   .285714 equals 2/7.
10   Irrational numbers can be characterized as nonterminating decimals having nonrepeating blocks of digits.
11   The irrational π is given by 3.1416.
12   Irrational numbers do not actually exist since they cannot be expressed in finite form using modern Arabic numerals.
13   Zero is not a natural number.

In Problems 1 through 12 let *N, J, Q, H,* and *R* be defined as in Section 2.3. Determine whether each statement is true or false and give reasons for your decision.

1 $-1 \in N$ 2 $\{3, -5, \frac{2}{3}\} \subset Q$ 3 $(Q \cup H) \cap R = H$

4 $2\pi \in H$ 5 $\frac{3}{4} \in J'$ 6 $\sqrt[3]{8} \in N$

7 $Q \subset H$ 8 $N \cap H \neq \varnothing$ 9 $\{\sqrt{4}, \pi, \sqrt{3}\} \subset H$

10 $-6 \in Q$ 11 $J \subset H'$ 12 $(J \cup H) = R$

13 Show geometrically that there exists a distance corresponding to the real numbers (a) $\sqrt{5}$, (b) $\sqrt{10}$, (c) $\sqrt{13}$. (HINT: Create some right triangles.)

14 Express some of the following numbers as nonterminating decimals: (a) $\frac{3}{8}$, (b) $\frac{2}{3}$, (c) $\frac{5}{11}$.

2.6 Real Numbers as Points on a Line

A very remarkable and extremely important property of real numbers is the fact that a one-to-one correspondence can be established between this set and the set of points on a line. This does not surprise us today, of course, because we have seen this property established, at least partially, on rulers, tapes, and various other measuring devices. This correspondence is usually constructed as shown in Fig. 2.6.1. An arbitrary point is selected and paired with the number zero; then another point is chosen, usually to the right of zero and paired with the number one. With a *unit* distance between 0 and 1 thus defined, there is a natural way to pair the remaining points and real numbers. When such a correspondence has been established the line is called the real line. Because of this correspondence, we may use the words *point* and *number* interchangeably—the *point* 3 is the point on the real line corresponding to the *real number* 3.

All numbers that have been paired with points to the *right* of zero are called positive real numbers and all those to the *left* of zero are called negative real numbers.

A very significant property of the real line is that it conveniently *orders* the real numbers. That is, given any two numbers *a* and *b* on the real line, one and only one of the following properties holds:

(1) *a* is to the *left* of *b*
(2) *a* is the *right* of *b*
(3) *a equals b*

FIGURE 2.6.1

If *a* is to the *left* of *b* on the real line, we say "*a* is less than *b*." Symbolically this is written $a < b$. If *a* is to the *right* of *b*, we say "*a* is greater than *b*" and write $a > b$. Using the relation symbols for "less than," "greater than," and "equals," these three properties can be written

(1) $a < b$
(2) $a > b$
(3) $a = b$

Example 2.6.1 Let points be identified by numbers and letters as in Figure 2.6.2. Accordingly, the following relations hold:

(a) $a < b$ (b) $d > c$ (c) $2 > -1$
(d) $b < c$ (e) $-3 < 1$ (f) $-3 < -1$

FIGURE 2.6.2

Expressions such as $a < b$ and $-2 < 3$ are called inequalities. A property of inequalities that follows immediately from the real line is the equivalence of the statements "*a* is less than *b*" and "*b* is greater than *a*." Diagrammed, this means

$$a < b \Longleftrightarrow b > a$$

(If *a* is to the left of *b*, then of course *b* is to the right of *a* and conversely.)

We denote the set of negative real numbers (those to the left of zero on the real line) by R_-, and the positive real numbers (those to the right of zero) by R_+. Thus

$$x \in R_+ \Longleftrightarrow x > 0$$

and

$$x \in R_- \Longleftrightarrow x < 0$$

Using inequalities and set-builder notation, we may identify specific line segments as sets of points. Take, for example, the following sets:

$$I_1 = \{x \mid -1 < x < 3\}$$
$$I_2 = \{x \mid -1 \le x < 3\}$$
$$I_3 = \{x \mid -1 < x \le 3\}$$
$$I_4 = \{x \mid -1 \le x \le 3\}$$

We note that I_2, is read "the set of all numbers *x* such that -1 is less than *or equal* to *x* and *x* is less than 3." We observe also that the two statements $-1 \le x$ and $x < 3$ are conveniently combined as

$-1 \le x < 3$. Sets such as these are called intervals. The interval I_1 above is called an open interval (does not include the end points -1 and 3) and is sometimes more briefly denoted $(-1,3)$. The fourth interval is a closed interval (does include the end points) and is denoted $[-1,3]$. The second and third intervals are half-open and half-closed intervals. The second is called closed–open, denoted $[-1,3)$, and the third, open–closed, denoted $(-1,3]$.

A representation of a set of numbers as a set of points is called a graph of the set. Graphs of the intervals I_1, I_2, I_3, I_4 are shown in Fig. 2.6.3.

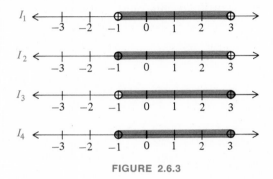

FIGURE 2.6.3

Now consider the sets:

$I_5 = \{x \mid x > 1\}$
$I_6 = \{x \mid x \le 2\}$
$I_7 = \{x \mid x \in \mathbf{R}\}$

These sets are called infinite intervals and are frequently denoted

$I_5 = (1, \infty)$
$I_6 = (-\infty, 2]$
$I_7 = (-\infty, \infty)$

The symbol ∞ is not a number but rather is used to indicate infinite in length. Since an interval such as $[1,\infty)$ cannot have a right-hand endpoint, the symbol) must be used on the right rather than]. Graphs of these intervals may be represented as shown in Fig. 2.6.4.

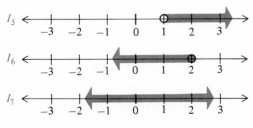

FIGURE 2.6.4

The relations "less than" and "greater than" of the previous section did not deal directly with the notion of magnitude. Absolute value, on the other hand, will deal directly with the magnitude of a number, as the following definition indicates.

DEFINITION 2.7.1   If x is a real number then the absolute value of x, denoted $|x|$, is equal to x, if x is positive or zero, and equal to $-x$, if x is negative. When diagrammed, this becomes

$$x \in \mathbf{R} \implies |x| = \begin{cases} x, \text{ if } x \geq 0 \\ \text{or} \\ -x, \text{ if } x < 0 \end{cases}$$

Observe that the absolute value of a number represents, geometrically, its distance from zero (in either direction) on the number line, as demonstrated in Fig. 2.7.1.

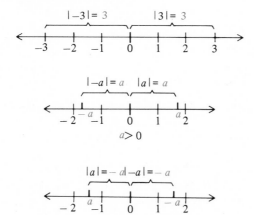

FIGURE 2.7.1

It would be helpful at this point to have the double minus property of real numbers, which will be proved later in this chapter.

THEOREM 2.7.1   If x is a real number, then minus a minus x, written $-(-x)$, is equal to x. Symbolically, this is

$$x \in \mathbf{R} \implies -(-x) = x$$

Example 2.7.1 $4 \in R \Longrightarrow -(-4) = 4$

Example 2.7.2 Determine the absolute value of $3, -2, 0$.

D2.7.1

$x \in R \Longrightarrow |x| = \begin{cases} x, & \text{if } x \geq 0 \\ -x, & \text{if } x < 0 \end{cases}$

$$3 \in R \xRightarrow{\text{D2.7.1}} |3| = \begin{cases} 3, & \text{if } 3 \geq 0 \\ -3, & \text{if } 3 < 0 \end{cases}$$

$$\Longrightarrow |3| = 3$$

$$-2 \in R \xRightarrow{\text{D2.7.1}} |-2| = \begin{cases} -2, & \text{if } -2 \geq 0 \\ -(-2), & \text{if } -2 < 0 \end{cases}$$

T2.7.1 $-(-a) = a$

$$\Longrightarrow |-2| = -(-2) \xRightarrow{\text{T2.7.1}} 2$$

$$0 \in R \xRightarrow{\text{D2.7.1}} |0| = \begin{cases} 0, & \text{if } 0 \geq 0 \\ -0, & \text{if } 0 < 0 \end{cases}$$

$$\Longrightarrow |0| = 0$$

A very important algebraic property is obtained by combining the relations of *absolute value* and *less than or equal*. This is presented in the next theorem, which will be proved later in this chapter.

THEOREM 2.7.2 If $b \geq 0$, then the following open sentences are equivalent.

$$|x| \leq b \Longleftrightarrow -b \leq x \leq b$$

The interval defined by these inequalities is bounded from above by b and from below by $-b$.

This important property is quite clear if you remember that the absolute value of a number, positive or negative, represents geometrically its distance from the origin. Thus $|1| = 1$ and $|-1| = 1$ and, in general, $|a| = a$, if $a \geq 0$, and $|a| = -a$, if $a < 0$. Let us examine Fig. 2.7.2. Note that if x lies between $-b$ and b, then $|x| < b$. Whereas, if $-x$ is less than $-b$ then $|-x|$, which is x, is greater than b, and we have $|x| > b$.

T2.7.2 $|x| \leq b \Longleftrightarrow -b \leq x \leq b$

Example 2.7.3 $|3| \leq 5 \xLeftrightarrow{\text{T2.7.2}} -5 \leq 3 \leq 5$

Example 2.7.4 $|-3| < 5 \xLeftrightarrow{\text{T2.7.2}} -5 < -3 < 5$

Example 2.7.5 $|x| \leq 5 \xLeftrightarrow{\text{T2.7.2}} -5 \leq x \leq 5$

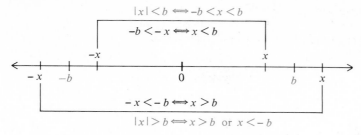

$$|x| < b \Longleftrightarrow -b < x < b$$

$$-b < -x \Longleftrightarrow x < b$$

$$-x < -b \Longleftrightarrow x > b$$

$$|x| > b \Longleftrightarrow x > b \ \text{ or } \ x < -b$$

The numbers x and $-x$ are equidistant from the origin.

FIGURE 2.7.2

Problem Sets 2.8

Problem Set I ■ *Reading Comprehension*

Determine whether statements 1 through 4 are true or false and give reasons for your responses.

1 The set of real numbers can be placed in a one-to-one correspondence with the set of points on a line.
2 If a is less than b on the real line, then a is to the right of b.
3 All intervals are either closed or open.
4 By definition of absolute value we have $[a \in \textbf{R} \Longrightarrow |-a| = a]$.

Problem Set II ■ *Skills Development*

1 Insert the proper relation symbol ($=$, $<$, or $>$) between each of the pairs of numbers below:
   (a) $2, 7$ (b) $-5, 2$ (c) $\frac{2}{3}, 0.6666 \ldots$
   (d) $\pi, 3.1416$ (e) $\frac{19}{32}, \frac{5}{8}$ (f) $0.347, 0.3469999 \ldots$
2 Express in set-builder notation and graph the intervals: (a) $[-1, 3]$, (b) $[-6, -2)$, (c) $(-\infty, 8]$, (d) $(1, \infty)$.
3 Determine the absolute value of the following expressions:

   Example $x - 3$

   Answer $|x - 3| = \begin{cases} x - 3, & \text{if } x - 3 \geq 0 \\ -(x - 3), & \text{if } x - 3 < 0 \end{cases}$

   $\qquad\qquad = \begin{cases} x - 3, & \text{if } x \geq 3 \\ -(x - 3), & \text{if } x < 3 \end{cases}$

   (a) 3 (b) -2 (c) $x + 1$
   (d) $4 - x$ (e) $-\frac{2}{3}$ (f) $-x$

4 Write the following statements in symbolic form:
   (a) a is greater than b (b) a is to the left of b
   (c) x is between $-a$ and b (d) x is to the right of a and to the left of b.

(e) The absolute value of x is less than or equal to 3.

(f) a is not less than b.　　　(g) x lies between -2 and 5.

(h) x is greater than or equal to 0 and less than 9.

5　On the real line sketch a graph of each of the sets below in a manner similar to the example. Then write the resultant set in interval notation.

Example　$\{x \mid -2 < x < 2\} \cap \{x \mid 0 < x < 3\}$.

The left-hand set will be represented above the real line and the right-hand set below the real line in Fig. 2.8.1. The resultant intersection of these two sets is clearly $(0, 2)$.

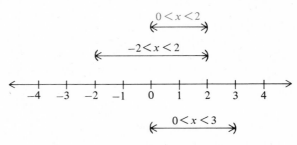

FIGURE 2.8.1

(a) $\{x \mid -2 < x < 3\} \cap \{x \mid 1 < x < 5\}$

(b) $\{x \mid x > 4\} \cap \{x \mid -1 < x < 6\}$

(c) $\{x \mid x < 3\} \cap \{x \mid x > -2\}$

Problem Set III ■ Theoretical Developments

1　Using intuitive reasoning show why the following properties are true for real numbers.

Example　If $a < b$ and $b < c$, then $a < c$.

Intuitive Argument　If $a < b$, then a is to the left of b on the real line; furthermore, if $b < c$ then b is to the left of c on the real line. If a is to the left of b and b is to the left of c, then certainly a must be to the left of c, which means $a < c$.

(a) If $a < b$, then $a + c < b + c$.

(b) If $a + c < b + c$, then $a < b$.

(c) $|x| < 4 \iff -4 < x < 4$.

(d) $a \leq b$ and $b \leq a \implies a = b$.

2　When would the intersection of two intervals be an interval?

Problem Set IV ■ Just For Fun

1　Two salesmen are selling similar hats. One is selling them at three for a dollar, whereas the other man sells his at two for a dollar. Business is not going very well, so the one selling at two for a dollar sells his remaining 30 hats to the other businessman, surprisingly enough, at two for a dollar. The remaining salesman now has 60 hats which he sells at 5 hats for $2, thus receiving $24 for the merchandise. Had they continued to sell separately, their separate earnings would have totaled $25. Because the combined price was essentially the same as the separate price, what happened to the missing dollar?

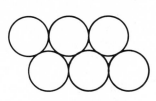

FIGURE 2.8.2

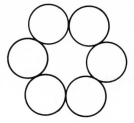

FIGURE 2.8.3

2 Take six coins of the same size and arrange them as in Fig. 2.8.2. Now see if you can rearrange them by sliding (not lifting them from the table) to match the configuration of Fig. 2.8.3 perfectly, where their centers would all lie on a common circle. The construction must be exact—not relying on something visually similar.

The Real Numbers as a Deductive System● 2.9

One of our most honored pursuits is our continuing quest for knowledge. Our sources of knowledge are many—recorded history, the five senses, experimentation, reasoning, intuition, and so forth. Although there is a great deal of overlapping among these, reasoning stands out as one of the most productive means of gaining knowledge. Reasoning is rather like performing mental experiments—the outcome of each step in the process becomes clear to the reasoner from previous experiences. Although there are several methods of reasoning, we will discuss only two here—reasoning by induction and reasoning by deduction.

The following statements are examples of reasoning by induction:

I'm sure John is at work this hour because he has been at work at this hour every Monday for the past year.

It appears that the price of gas has gone down; I checked many of the local stations this morning and all had reduced their prices.

There is a good chance of rain this afternoon; with a wind coming out of the southwest and the barometer dropping, we nearly always get rain.

The sun will rise tomorrow.

The process of *inductive reasoning* is an attempt to obtain a truth about all things by examining a limited number of those things. It tries to make generalizations or discover some regularity in situations through careful observations. Alfred North Whitehead calls it trying to "see what is general in what is particular."

Reasoning by induction is not an exact science; it is more probabilistic in nature. Inductive assertions may be true in general,

● Portions or all of this section and the succeeding section may be skipped without affecting seriously the continuity of the textual material. These sections are designed for the more advanced student or class where a more in-depth look at the real number system, its structure, and its properties is desired. However, a brief review of the basic properties of arithmetic and algebraic manipulation treated in these sections and presented in summary in Section 2.13 is advisable.

but there is no absolute assurance of their truth. Although inductive arguments do not produce results that are necessarily valid, it is true that such reasoning, for practical purposes, is quite effective in dealing with many of our everyday problems.

Deductive reasoning, on the other hand, is quite different from inductive reasoning. The following examples will illustrate this form of reasoning.

> No student can graduate from college without having taken at least one year of English. Carolyn has taken only one semester of English. *Therefore,* Carolyn cannot, as yet, graduate.

> The bill will pass Congress if it receives at least a two-thirds confirming vote from all members present. It was just learned that three-fourths of all Congressmen voted for the bill this morning. *Thus,* the bill has passed Congress and will now be sent to the President.

> If it is known that one-third an unknown quantity is four, *then* the unknown quantity must be twelve.

Notice in these examples explicit conditions were given under which certain results would inevitably follow. When those conditions were satisfied, the outcome was *inescapable.*

Although argument by deduction is often more difficult, if not impossible under certain circumstances, the results are absolutely conclusive. All mathematical *assertions* are obtained by deductive reasoning. Because of this, mathematics is often called "the exact science." This idea has, however, resulted in several misconceptions about mathematics. For one thing, people for centuries thought that mathematics contained only absolute and irrefutable truths—one could not argue about mathematical "truth." In a sense such a statement could not be further from the truth. Think about deductive reasoning for a moment. We start with certain statements or premises and reason that various conclusions must necessarily follow from these premises. Who is to say that the premises are correct? Two different sets of premises can yield contradictory results. We briefly discussed this very point in Chapter 0 when observing certain critical differences between Euclidean and non-Euclidean geometry. Conclusions are only valid or accepted if the premises are valid or accepted. No results can ever be obtained that are not already inherently contained within the premises. Because these postulated statements are generally our own creation, drawn from our experiences and knowledge of the world and people about us, they are certainly subject to error or misjudgment. Although mathematics is by no means perfect, it has served and continues to serve mankind in an amazingly effective and productive manner.

Another *misconception* is that mathematics is strictly deductive. It is true that all mathematical conclusions are, in the final analysis, proved by deductive reasoning, but the premises (surely the most important part) and the various techniques in reasoning deductively (the next most important part) are not acquired by deduction. Mathematics is never deductive in its creation. Morris Kline, commenting on this fact, says:

> Before mathematicians can obtain a body of knowledge that warrants the deductive organization which Euclid gave to geometry they must spend decades and even centuries in creating the material. And unlike the logical organization, the creative work does not proceed step by step from one argument to another, each supported by some axiom or previously established conclusion. What the creative process does involve is groping, blundering, conjecturing, and hypothesizing. Imagination, intuition, divination, insight, experimentation, chance association, luck, hard work, and immense patience are applied to grasp a key concept, to formulate a conjecture, and to find a proof. ●

● Morris Kline, "The Nature of Mathematics," in *Mathematics in the Modern World,* W. H. Freeman, San Francisco, 1968, p. 2.

Mathematician Herman Weyl remarked in an obituary of his colleague, David Hilbert,

> Mathematizing may well be a creative activity of man, like language or music, of primary originality, whose historical decisions defy complete objective rationalizations. ●

● Herman Weyl, ibid., p. 3.

A major objective of this section is to take you backstage and to have you experience some of the challenge of creating and developing a mathematical system deductively. It is hoped that this will give you a renewed appreciation for some of the more salient elements of mathematical discovery while reaffirming, as perhaps never before, your understanding of basic arithmetic and algebraic principles. We shall provide this experience by characterizing and analyzing the *real numbers* as a *deductive system.*

There are several advantages in a deductive approach to the real numbers. Almost everyone who studies numbers and their properties does so in order to more effectively apply them in problem-solving situations or in the investigation or creation of new mathematical systems. One's success in these endeavors may depend upon how well the fundamental properties of real numbers are understood and applied.

Another very significant advantage of the deductive approach (which will not be appreciated fully until Chapter 8) is that much of the work in our study of other number systems, such as the complex and matrix number systems, will already have been accomplished through our deductive development of the real system.

The final advantage we shall mention has to do with efficiency in our study. Mathematics has grown so rapidly that in order to obtain but a small understanding of the subject we must carefully organize, unify, and condense it. It has taken hundreds of years and thousands upon thousands of painstaking hours for man to create and organize the system of real numbers as we have it today. We do not have the time (and probably the inclination or patience) to re-create this system as it was done historically. Our partial redevelopment of the system deductively will speed up the learning and understanding processes a thousandfold, though, to be sure, we will lose many of the original creative thrills.

To develop a branch of mathematics deductively, we must begin with premises. These will include

(1) undefined terms;
(2) undefined relations;
(3) statements, called axioms or definitions, relating the undefined terms and undefined relations.

After a foundation is laid, we seek to find all the properties, called *theorems,* that follow logically from the axioms and definitions.

In our treatment of the real number system the undefined terms will be the *real numbers* themselves. That is, we shall not define the meaning of $1, -3, \frac{1}{4}, -\frac{1}{5}, \pi$, and so forth. This is not serious because almost everyone already has developed strong intuitive feelings for these numbers, except perhaps for the irrationals. The undefined relations are *addition* and *multiplication;* such relations as $2 + 3 = 5$ and $2 \times 3 = 6$ are well understood already and will not be defined.

Before we state the axioms and definitions that characterize the system of real numbers, we shall review and introduce certain notational forms. First, as presented earlier, the letter R will represent the set of *real numbers*. The notation

$a \in R$

means "*a is* a real number." When letters toward the beginning of the alphabet, such as a, b, c, d, are used to symbolize numbers, they represent *fixed though unspecified* numbers called constants. Letters toward the end of the alphabet, such as x, y, and z, will represent *unfixed* numbers called variables. (Variables and open sentences containing variables will be studied in Chapter 3.)

The conditional proposition

$a, b \in R \implies ab = ba$

is read "If a and b are real numbers, then the products ab and ba are equal (are numerals representing the same number)." Note that when no confusion exists, products will be written in the form ab rather than $a \cdot b$ or $a \times b$.

As noted earlier in this chapter, several English phrases occur so frequently in mathematical statements that we find it convenient to represent them by shorthand symbols. Here we shall introduce one new symbol and repeat two discussed earlier for convenience. They are:

$\forall$ meaning "for each" or "for every"

∃ meaning "there exists"
∋ meaning "such that" or "where"

Example 2.9.1 The statement, "*for each* real number a, *there exists* a real number, denoted $-a$, *such that* the sum of a and $-a$ is zero," can now be written in the abbreviated form:

$$\forall\ a \in \boldsymbol{R}\ \exists\ -a \in \boldsymbol{R}\ \ni\ a + (-a) = 0$$

To minimize any misunderstanding, we shall nearly always express propositions in both the verbal and symbolic forms.

We now introduce some definitions and axioms that characterize real numbers.

DEFINITION 2.9.1 If a and b are two numbers (not necessarily distinct), then the relation a equals b, denoted $a = b$, means a and b, though distinct letters, are symbols representing the *same* number. Equals is called the relation of identity, and the expression $a = b$ is called an equation.

It is important to understand that when letters are used to denote numbers, they represent those numbers in any one of an infinite variety of forms. For example, a may stand for 5 or any one of the following expressions, each equal to 5:

$$8 - 3,\ \frac{10}{2}, \frac{22 - 7}{3},\ \sqrt{25}$$

The fundamental properties of the relation *equals* are established by the following axioms (recall that axioms are accepted without proof):

E-1   REFLEXIVE LAW OF EQUALITY (RLE)

If a is a real number, then a equals a. Diagrammed, this becomes

$$a \in \boldsymbol{R} \Longrightarrow a = a$$

E-2   SYMMETRIC LAW OF EQUALITY (SLE)

If a and b are real numbers, then the two equations $a = b$ and $b = a$ are *equivalent*.

$$a, b \in \boldsymbol{R} \Longrightarrow [a = b \Longleftrightarrow b = a]$$

E-3 TRANSITIVE LAW OF EQUALITY (TLE)

If a, b, c, are real numbers where $a = b$ and $b = c$, then $a = c$.

$$a, b, c \in \mathbf{R} \implies [a = b \quad \text{and} \quad b = c \implies a = c]$$

This may also be written

$$[a = b = c] \implies [a = c]$$

Remark 2.9.1 A relation satisfying reflexive, symmetric, and transitive properties, such as the relation *equals* in E-1, E-2, and E-3, is called an equivalence relation.

E-4 SUBSTITUTION LAW OF EQUALITY (SubLE)

If $a = b$, then a can be replaced by b or b by a in any real mathematical statement (one composed of real numbers or real expressions) without changing the truth or falsity of that statement.

Remark 2.9.2 These four axioms for equality are by no means unique nor independent. Another author may choose a different set of axioms to provide the same results. Furthermore, E-1, E-2 and E-3 can be shown to follow from E-4. However, for our purposes here we shall accept all four.

Example 2.9.2 $\dfrac{8-2}{2} = \dfrac{6}{2}$ and $\dfrac{6}{2} = 3$

therefore, by the Transitive Law of Equality

$$\frac{8-2}{2} = 3$$

For notational convenience and simplicity, we shall often write the above as

$$\frac{8-2}{2} = \frac{6}{2} = 3 \xleftrightarrow{\text{TLE}} \frac{8-2}{2} = 3$$

In fact, any time we have a sequence of expressions joined by an unbroken chain of equalities, any two of the expressions, by the Transitive Law of Equality, can be set equal.

Example 2.9.3 By the Symmetric Law of Equality, $x = 4$ is equivalent to $4 = x$. Diagrammed, this becomes

$$x = 4 \xleftrightarrow{\text{SLE}} 4 = x$$

Example 2.9.4 If $a = 4$, $b = 3$, and $c = 2$, then by the Substitution Law of Equality,

$$a(b + c) = 4(3 + 2)$$

We may also write

$$a(b + c) \xeq{\text{SubLE}} 4(3 + 2) = 4 \cdot 5 = 20$$
$$\xRightarrow{\text{TLE}} a(b + c) = 20$$

Remark 2.9.3 Notice in the above examples that an equals symbol may be lengthened so that the abbreviated form of the justification for the equality may be placed directly above it. The reference is thereby placed at the very point where it is being applied, as has been done earlier with the implication symbol.

The next eleven axioms are called the field axioms. These will provide all the properties we shall need regarding the operations of *addition* and *multiplication*.

F-1   CLOSURE LAW OF ADDITION (Cl+)

If a and b are any two real numbers, then the sum of a and b, denoted $a + b$, is a real number.

$$a, b \in \textbf{R} \implies (a + b) \in \textbf{R}$$

Remark 2.9.4 This axiom asserts that the set of real numbers is *closed* under the operation of addition. A system is said to be closed under a given operation if any two members of the system will, under the operation, render a third member in the system, not necessarily distinct from either of the two.

F-2   ASSOCIATIVE LAW OF ADDITION (A+)

If a, b and c are any three real numbers, then the sum of a and b added to c is equal to a added to the sum of b and c.

$$a, b, c \in \textbf{R} \implies [(a + b) + c = a + (b + c)]$$

Example 2.9.5 $(2 + 5) + 3 = 7 + 3 = 10$

and

$$2 + (5 + 3) = 2 + 8 = 10$$
$$\xRightarrow{\text{TLE}} (2 + 5) + 3 = 2 + (5 + 3)$$

which illustrates the Associative Law of Addition for the real numbers 2, 5, and 3.

F-3 COMMUTATIVE LAW OF ADDITION (C+)

If a and b are any two real numbers, then the sum of a and b, in the order $a + b$, equals the sum in the *reverse* order, $b + a$.

$$a, b \in \mathbf{R} \Longrightarrow [a + b = b + a]$$

Example 2.9.6 $3, -4 \in \mathbf{R} \xrightarrow{\text{C+}} 3 + (-4) = -4 + 3$

F-4 ADDITIVE IDENTITY LAW (Id+)

There exists a real number called zero, denoted 0, where for every real number, say a, the sum of a and 0 equals a.

$$\exists\ 0 \in \mathbf{R}\ \ni\ \forall\ a \in \mathbf{R}\ [a + 0 = 0 + a = a]$$

Zero is called the additive identity element of the set of real numbers.

F-5 ADDITIVE INVERSE LAW (In+)

For each real number a, there exists an additive inverse,● denoted $-a$, which is real and where the sum of a and $-a$ equals 0.

$$\forall\ a \in \mathbf{R}\ \exists\ -a \in \mathbf{R}\ \ni\ [a + (-a) = (-a) + a = 0]$$

Example 2.9.7 $2 \in \mathbf{R} \xrightarrow{\text{In+}} \exists\ -2 \in \mathbf{R}\ \ni\ 2 + (-2) = 0$

and

$$-3 \in \mathbf{R} \xrightarrow{\text{In+}} \exists\ -(-3) \in \mathbf{R}\ \ni\ -3 + [-(-3)] = 0$$

Remark 2.9.5 The latter proposition of Example 2.9.7 draws attention to the important fact that the letter a represents *any* real number – positive or negative. You would *not* want to suppose that the symbol $-a$ necessarily denotes a negative real number.

F-6 CLOSURE LAW OF MULTIPLICATION (Cl·)

If a and b are any two real numbers, then the product of a and b, denoted $a \cdot b$ or just ab, is also a real number.

$$a, b \in \mathbf{R} \Longrightarrow (a \cdot b) \in \mathbf{R}$$

Thus, the set of real numbers is *closed* under the operation of multiplication as well as under addition.

F-7   ASSOCIATIVE LAW OF MULTIPLICATION $(A\cdot)$

If a, b and c are any three real numbers, then the product of a and b times c is equal to a times the product of b and c.

$$a, b, c \in \textbf{\textit{R}} \Longrightarrow [(a \cdot b) \cdot c = a \cdot (b \cdot c)]$$

Example 2.9.8 $(2 \cdot 3) \cdot 4 = 6 \cdot 4 = 24$

and

$$2 \cdot (3 \cdot 4) = 2 \cdot 12 = 24$$
$$\xRightarrow{\text{TLE}} (2 \cdot 3) \cdot 4 = 2 \cdot (3 \cdot 4)$$

which illustrates the Associative Law of Multiplication for the numbers 2, 3, and 4.

F-8   COMMUTATIVE LAW OF MULTIPLICATION $(C\cdot)$

If a and b are any two real numbers, then the product of a and b, in the order $a \cdot b$, equals the product in the *reverse* order, $b \cdot a$.

$$a, b \in \textbf{\textit{R}} \Longrightarrow [a \cdot b = b \cdot a]$$

Example 2.9.9 $2 \cdot 7 = 14$

and

$$7 \cdot 2 = 14$$
$$\xRightarrow{\text{TLE}} 2 \cdot 7 = 7 \cdot 2$$

which illustrates the Commutative Law of Multiplication for 2 and 7.

F-9   DISTRIBUTIVE LAW (D) (multiplication distributes over addition)

If a, b, and c are any three real numbers, then the product of a with the sum of b and c is equal to the sum of the products ab and ac.

$$a, b, c \in \textbf{\textit{R}} \Longrightarrow [a(b + c) = ab + ac]$$

Remark 2.9.6   The role of parentheses in expressions such as $a(b + c)$ is very significant. Here they indicate that the operation of addition is to be performed before that of multiplication. If the parentheses were not there, the expression would be $ab + c$, meaning the product is to be performed before the sum; that is, when no preference is dictated by parentheses, multiplication will always take precedence over addition.

Example 2.9.10 $2(4 + 6) = 2 \cdot 10 = 20$

and

$$2 \cdot 4 + 2 \cdot 6 = 8 + 12 = 20$$
$$\overset{\text{TLE}}{\Longrightarrow} 2(4 + 6) = 2 \cdot 4 + 2 \cdot 6$$

F-10 MULTIPLICATIVE IDENTITY LAW (Id·)

There exists a real number called one, denoted 1, such that for every real number a, the product of a and 1 equals a.

$$\exists\ 1 \in \mathbf{R} \ni \forall\ a \in \mathbf{R}\ [a \cdot 1 = 1 \cdot a = a]$$

The number 1 is called the multiplicative identity element of the set of real numbers.

F-11 MULTIPLICATIVE INVERSE LAW (In·)

For each real number a, except 0, there exists a multiplicative inverse,● denoted a^{-1}, that is real; the product of a and a^{-1} equals 1.

● Sometimes called a reciprocal.

$$\forall\ a \in \mathbf{R},\ a \neq 0,\ \exists\ a^{-1} \in \mathbf{R} \ni [a \cdot a^{-1} = a^{-1} \cdot a = 1]$$

It is very important to note here that every real number, *except 0,* has a multiplicative inverse.

Example 2.9.11 $3 \in \mathbf{R},\ 3 \neq 0, \Longrightarrow \exists\ 3^{-1} \in \mathbf{R} \ni 3 \cdot 3^{-1} = 1$

Remark 2.9.7 The fact $3^{-1} = \frac{1}{3}$, or in general $a^{-1} = 1/a$, will be proved later as an exercise. This latter form is the one more commonly used.

Before investigating some of the properties that follow from these axioms, we define two more operations, subtraction and division, familiar to everyone. Observe that the operations of subtraction and division are being *defined,* whereas addition and multiplication were presented as *undefined operations.*

DEFINITION 2.9.2 (DSub) If a and b are any two real numbers, then the difference of a and b in the stated order, denoted $a - b$, equals the *sum* of a and $-b$ (the additive inverse of b). The process of performing this difference is called subtraction and the statement $a - b$ is read "a subtract b" or "a minus b."

$$a, b \in \mathbf{R} \Longrightarrow [a - b = a + (-b)]$$

Remark 2.9.8 It is perhaps unfortunate that the symbol called minus, and denoted −, has been given three quite different meanings in mathematics. It is used, as indicated earlier, to formulate numerals representing negative real numbers such as $-2, -7, -\frac{1}{2}$. The minus symbol is also used to identify the additive inverse of a real number; -2 is the additive inverse of 2, and $-(-3)$ is the additive inverse of -3. Finally, it is used as in Definition 2.9.2 to symbolize the operation of subtraction. We must always be careful to distinguish among these three uses.

Example 2.9.12

$$4 - 7 \stackrel{\text{D2.9.2}}{=\!=\!=\!=} 4 + (-7) = -3$$

D2.9.2 $a - b = a + (-b)$

Example 2.9.13

$$-2 - (-5) \stackrel{\text{D2.9.2}}{=\!=\!=\!=} -2 + [-(-5)] = -2 + 5 = 3$$

The fact $-(-5) = 5$, or in general $-(-a) = a$ (which is Theorem 2.9.8), will be proved later as an exercise.

DEFINITION 2.9.3 (DDiv) If a and b are any two real numbers, $b \neq 0$, then the quotient of a and b in the stated order, denoted a/b, equals the product of a and b^{-1} (the multiplicative inverse of b). The process of performing this quotient is called division and the statement a/b is read "a divided by b."

$$a, b \in \mathbf{R}, \ b \neq 0 \Longrightarrow \left[\frac{a}{b} = a \cdot b^{-1} = b^{-1} \cdot a \right]$$

The eleven field axioms not only partially characterize the system of real numbers, but also several other significant number systems, one of which—the complex system—will be developed in Chapter 8. With this latter system, the only change in stating the field axioms will be in the choice of a universal set; instead of it being R (the set of real numbers), it would be C (the set of complex numbers), or in general simply U, which represents an arbitrary universal set. A number system for which there are two operations satisfying these eleven axioms is called a field. For this reason we call the real numbers a field.

We now wish to discover some of the more fundamental and useful properties of a field. It is very important to thoroughly understand these basic properties since about 95 percent of all arithmetic and algebraic calculations are but applications of them. Efficiency and effectiveness in mathematics requires a firm foundation in basic principles. Most of these properties you will already know. Hence, at first, this work may seem rather elementary; however, you will probably discover that the methods of analysis and techniques applied will not always be trivial. And these procedures are frequently the very tools one uses to solve applied problems, which is likely to be your main objective for studying mathematics.

To develop a system deductively, after a beginning has been made, each property, called a theorem, must be shown to be a logical consequence of those theorems, axioms, or definitions that have preceded it. Every step in a proof must be justified. Theorems are conditional propositions and hence have hypotheses and conclusions. We usually begin with that which is given — the hypotheses — and proceed step by step until we arrive at the conclusion. Of course discovering the steps is the main challenge. Several examples will now be given to illustrate how this may be accomplished.

THEOREM 2.9.1 ADDITION LAW OF EQUALITY (ALE)

If a, b, c are real numbers and $a = b$, then $a + c = b + c$.

$$a, b, c \in \mathbf{R} \Longrightarrow [a = b \Longrightarrow a + c = b + c]$$

Proof

CI+ $a, b \in \mathbf{R} \Longrightarrow (a + b) \in \mathbf{R}$

RLE $a \in \mathbf{R} \Longrightarrow a = a$

SubLE
Substitution Law of Equality

TLE $[a = b = c] \Longrightarrow [a = c]$

$$a, c \in \mathbf{R} \xrightarrow{\text{CI+}} (a + c) \in \mathbf{R}$$

$$\xrightarrow{\text{RLE}} a + c = a + c \xrightarrow[\text{SubLE}]{a = b} b + c$$

$$\xrightarrow{\text{TLE}} a + c = b + c \qquad \text{QED}$$

Notice that every step in the above proof is verified by identifying the property being used and at the very point where it applies. It is hoped that this will enable you to more easily follow the flow of the reasoning as the proof unfolds. Naturally this proof does not represent the original thinking, but rather the final form, clearly organized with every step justified so that anyone reading the proof may be equally convinced. (Read again the quotation of Morris Kline on page 61). *Original thinking* may have gone something like this:

"Knowing $a = b$, I wish to show $a + c = b + c$. Because $a = b$, I may, by the Law of Substitution, replace a by b or b by a in any mathematical statement. If I were to replace b by a in the statement I hope to discover, namely $a + c = b + c$, it would read $a + c = a + c$, which, by the Reflexive Law of Equality, holds if $a + c$ is a real number. But $a + c$ is real by the Closure Law of Addition, because a and c are given real." These thoughts would then be organized, beginning with the hypotheses until the conclusion is reached.

The Addition Law of Equality enables you to add the same real expression to both sides of an equation, thereby obtaining an equation that is equally valid, although generally different. Many equations are solved by this technique.

Example 2.9.14 Suppose $x + 2 = 6$, then

$$x + 2 = 6 \xrightarrow{\text{ALE}} (x + 2) + (-2) = 6 + (-2)$$
$$\xrightarrow{\text{A+}} x + [2 + (-2)] = 4$$
$$\xrightarrow{\text{In+}} x + 0 = 4$$
$$\xrightarrow{\text{Id+}} x = 4$$

ALE $[a = b \Longrightarrow a + c = b + c]$

A+ $(a + b) + c = a + (b + c)$

In+ $a + (-a) = 0$

Id+ $a + 0 = a$

Of course, these steps are so elementary that they are usually performed mentally.

THEOREM 2.9.2 MULTIPLICATION LAW OF EQUALITY (MLE)

If a, b, c are real numbers and $a = b$, then $ac = bc$.

$$a, b, c \in \mathbf{R} \Longrightarrow [a = b \Longrightarrow ac = bc]$$

Proof To be done as an exercise.

Remark 2.9.9 It is important to note that the converse of Theorem 2.9.2, namely

$$ac = bc \Longrightarrow a = b$$

is *not* true without a significant qualification. For example, take the true equation $4 \cdot 0 = 3 \cdot 0$; we would certainly not want to conclude from this equation that $4 = 3$. We will prove (Theorem 2.9.4), however, that

$$[ac = bc \quad \text{and} \quad c \neq 0] \Longrightarrow a = b$$

THEOREM 2.9.3 TERM CANCELLATION LAW OF EQUALITY (TCLE)

If a, b, c are real numbers and $a + c = b + c$, then the common term c may be cancelled, leaving $a = b$.

$$a, b, c \in \mathbf{R} \Longrightarrow [a + c = b + c \Longrightarrow a = b]$$

Proof

(1) $c \in \mathbf{R} \xrightarrow{\text{In+}} \exists\, -c \in \mathbf{R} \ni c + (-c) = 0$

(2) $a + c = b + c \xrightarrow{\text{ALE}} (a + c) + (-c) = (b + c) + (-c)$
$$\xrightarrow{\text{A+}} a + [c + (-c)] = b + [c + (-c)]$$
$$\xrightarrow{\text{SubLE}} a + 0 = b + 0$$
$$\xrightarrow{\text{Id+}} a = b \qquad \text{QED}$$

In+ $a + (-a) = 0$

ALE $a = b \Longrightarrow a + c = b + c$

A+ $(a + b) + c = a + (b + c)$

SubLE
Substitution Law of Equality

Id+ $a + 0 = a$

Original reasoning of the proof of Theorem 2.9.3 may have gone something like this: "Given the equation $a + c = b + c$, show the equation $a = b$ is equally valid. This can be achieved if the c's in the first equation can be eliminated. Now the only way to eliminate the *term c* is to add to it its additive inverse $-c$. Can I add $-c$ to both sides of the equation? Yes! This is precisely what is provided in the Addition Law of Equality—Theorem 2.9.1. Is $-c$ given in the hypotheses? No! Therefore, the first step in the proof will be to obtain $-c$ from the additive inverse axiom. The proof is now ready to be written up."

Remark 2.9.10 Combining the results of Theorem 2.9.1 and 2.9.3 we have the biconditional equivalence

$$a = b \Longleftrightarrow a + c = b + c$$

TCLE $\quad a + c = b + c \Longrightarrow a = b$

Example 2.9.15 $\quad x + 2 = 4 + 2 \xrightarrow{\text{TCLE}} x = 4$

THEOREM 2.9.4 FACTOR CANCELLATION LAW OF EQUALITY **(FCLE)**

If a, b, c are real numbers, $a \neq 0$, and $ab = ac$, then the common nonzero factor a may be cancelled leaving $b = c$.

$$a, b, c \in \textbf{\textit{R}} \Longrightarrow [ab = ac \quad \text{and} \quad a \neq 0 \Longrightarrow b = c]$$

Proof To be done as an exercise.

Example 2.9.16 $\quad 3x = 12 \Longrightarrow 3x = 3 \cdot 4 \xrightarrow{\text{FCLE}} x = 4$

THEOREM 2.9.5 If a and b are real numbers and $a + b = 0$, then $a = -b$ and $b = -a$.

$$a, b \in \textbf{\textit{R}} \Longrightarrow [a + b = 0 \Longrightarrow a = -b \quad \text{and} \quad b = -a]$$

Proof To be done as an exercise.

Remark 2.9.11 Theorem 2.9.5 identifies the *uniqueness* of additive inverses in the system of real numbers. By axiom F-5, we know $a + (-a) = 0$. Now by Theorem 2.9.5 we know the only number that when added to a yields 0 is its additive inverse $-a$. An important application of this property is that when a term is transferred to the opposite side of an equation its sign is always reversed—it is replaced by its additive inverse.

Example 2.9.17 $\quad x + 5 = 0 \xrightarrow{\text{T2.9.5}} x = -5$

THEOREM 2.9.6   If a is a real number, then the product of a and 0 is 0.

$$a \in \boldsymbol{R} \Longrightarrow a \cdot 0 = 0$$

Proof

$$0 \in \boldsymbol{R} \xrightarrow{\text{Id+}} 0 + 0 = 0$$

$$\xrightarrow{\text{MLE}} a \cdot (0 + 0) = a \cdot 0$$

$$\xrightarrow{\text{D}} a \cdot 0 + a \cdot 0 = a \cdot 0 + 0$$
$$\xrightarrow{\text{Id+}}$$

$$\xrightarrow{\text{TCLE}} a \cdot 0 = 0 \qquad \text{QED}$$

Id+	$a + 0 = a$
MLE	$a = b \Longrightarrow ac = bc$
D	$a(b + c) = ab + ac$
TCLE	$a + c = b + c \Longrightarrow a = b$

Example 2.9.18 $(27x - \pi) \in \boldsymbol{R} \xrightarrow{\text{T2.9.6}} (27x - \pi) \cdot 0 = 0$

The product of *any* real expression and 0 is always 0.

THEOREM 2.9.7   If a, b are real numbers and $ab = 0$, then either $a = 0$ or $b = 0$.

$$a, b \in \boldsymbol{R} \Longrightarrow [ab = 0 \Longrightarrow a = 0 \quad \text{or} \quad b = 0]$$

Proof   To be done as an exercise.

Example 2.9.19 $4(x + 3) = 0 \xrightarrow{\text{T2.9.7}} 4 = 0 \quad \text{or} \quad x + 3 = 0$

$$\xrightarrow{\text{or}} x + 3 = 0$$

$$\xrightarrow{\text{T2.9.5}} x = -3$$

T2.9.5	$a + b = 0 \Longrightarrow a = -b$

Remark 2.9.12   Theorem 2.9.7 is applied frequently in solving algebraic equations. If the left-hand side of an equation is factored and the right-hand side is zero, then at least one, perhaps both, of the left-hand factors is equal to zero, as illustrated in Example 2.9.19.

THEOREM 2.9.8   If a and b are real numbers, then

I $-(-a) = a$
II $-(a + b) = (-a) + (-b)$
III $(-a)b = -(ab)$
IV $(-a)(-b) = ab$

Proof   (Part I)

$$a \in \boldsymbol{R} \xrightarrow{\text{In+}} \exists\, -a \in \boldsymbol{R} \ni a + (-a) = 0$$

$$\xrightarrow{\text{T2.9.5}} a = -(-a) \qquad \text{QED}$$

In+	$a + (-a) = 0$
T2.9.5	$a + b = 0 \Longrightarrow a = -b$

Example 2.9.20

$$5(x - 4) = 0 \xrightarrow{\text{T2.9.7}} x - 4 = 0 \text{ (because } 5 \neq 0)$$

T2.9.7 $ab = 0 \Longrightarrow a = 0 \text{ or } b = 0$

$$\xrightarrow{\text{T2.9.5}} x = -(-4)$$

T2.9.5 $a + b = b \Longrightarrow a = -b$

$$\xrightarrow{\text{T2.9.8(I)}} x = 4$$

T2.9.8(I) $-(-a) = a$

Example 2.9.21

$$4 - (x + 2) = 0 \xrightarrow{\text{DSub}} 4 + [-(x + 2)] = 0$$

DSub $a - b = a + (-b)$

$$\xrightarrow{\text{T2.9.8(II)}} 4 + (-x) + (-2) = 0$$

T2.9.8(II) $-(a + b) = (-a) + (-b)$

$$\Longrightarrow 2 + (-x) = 0$$

$$\xrightarrow{\text{T2.9.5}} 2 = -(-x)$$

$$\xrightarrow{\text{T2.9.8(I)}} 2 = x$$

In practice, of course, you would not go through each of these steps to solve the equation $4 - (x + 2) = 0$ except, perhaps, mentally. Here, however, our *purpose* is to identify every step necessary to solve the equation. Shortcuts are acceptable only when they can be replaced by all the verifying intermediate steps, should you be challenged. The theorems themselves are important shortcuts; they help simplify calculations. Our mastery of the theorems, therefore, will determine how proficient we may be in performing many calculations, both arithmetic and algebraic.

THEOREM 2.9.9 If $a, b, c, d \in R$ and $b, d \neq 0$, then

● This theorem is sometimes called the cross-multiplication property for fractional equations.

$$\frac{a}{b} = \frac{c}{d} \Longleftrightarrow ad = bc \bullet$$

Proof To be done as an exercise.

FCLE $ac = bc, c \neq 0 \Longrightarrow a = b$

Example 2.9.22 $\dfrac{x}{4} = \dfrac{1}{2} \xleftarrow{\text{T2.9.9}} 2x = 4 \xrightarrow{\text{FCLE}} x = 2$

THEOREM 2.9.10 If $a, b, c \in R$ and $b, c \neq 0$, then

$$\frac{ac}{bc} = \frac{a}{b}$$

Proof

(1) $\dfrac{ac}{bc} = \dfrac{a}{b} \xleftarrow{\text{T2.9.9}} (ac)b = (bc)a$

$$\xLeftrightarrow{\text{A·}} a(cb) = (bc)a$$

$$\xLeftrightarrow{\text{C·}} (b \cdot c) \cdot a = (bc)a$$

Thus,

$$\frac{ac}{bc} = \frac{a}{c} \xLeftrightarrow{\qquad} (bc)a = (bc)a$$

(2) $a, b, c \in R \xRightarrow{\text{Cl·}} (bc)a \in R \xRightarrow{\text{RLE}} (bc)a = (bc)a$

$$\xRightarrow{\text{(1)}} \frac{ac}{bc} = \frac{a}{b} \qquad \text{QED}$$

A· $(ab)c = a(bc)$

C· $ab = ba$

Cl· $a, b \in R \Longrightarrow (a \cdot b) \in R$

RLE $a \in R \Longrightarrow a = a$

Remark 2.9.13 You may have been a bit perplexed by the above proof, realizing that a fundamental principle of logic is that one cannot use, *in a direct way,* the conclusion of a theorem in the proof of that very theorem. It may appear that this law is being violated in step (1) above. However, such is not the case, because step (1) is not the proof. Step (1) simply shows that the two equations $(ac)/(bc) = a/b$ and $(bc)a = (bc)a$ are *equivalent.* They are either both valid or both invalid, and step (1) does not indicate which. What the first step does accomplish is to show that the equation we are seeking is equivalent to one that can be readily verified, namely, $(bc)a = (bc)a$. Step (2) proceeds to verify this equation, which then leads to the conclusion of the theorem.

Notice the theorems are so ordered that frequently the proof of one theorem can be greatly simplified by applying the results of a previously proved theorem. You will want to take full advantage of this ordering. In contrast, however, you must remember that a valid proof cannot rely in any way on a theorem which *follows* the one in question, unless that theorem is proved independent of the one being proved.

THEOREM 2.9.11 If $a, b \in R$, but are not equal to zero, and $ab = 1$, then b is the multiplicative inverse of a and a is the multiplicative inverse of b.

$$a, b \in R, \ a, b \neq 0, \quad \text{and} \quad ab = 1 \Longrightarrow \begin{cases} a = b^{-1} \\ \text{and} \\ b = a^{-1} \end{cases}$$

Proof To be done as an exercise.

Theorem 2.9.11 states that the multiplicative inverse of a nonzero real number is *unique.*

THEOREM 2.9.12   If $a, b \in R$ and $a, b \neq 0$, then

$$\textbf{I}\quad \frac{a}{a} = 1 \qquad \textbf{II}\quad \frac{a}{1} = a \qquad \textbf{III}\quad \frac{1}{a} = a^{-1}$$

$$\textbf{IV}\quad \frac{1}{\dfrac{1}{a}} = (a^{-1})^{-1} = a \qquad \textbf{V}\quad \frac{-a}{b} = -\frac{a}{b} = \frac{a}{-b} \qquad \textbf{VI}\quad \frac{-a}{-b} = \frac{a}{b}$$

Proof   (Part III)

DDiv $\dfrac{a}{b} = a \cdot b^{-1}$

$$\frac{1}{a} \xrightarrow{\textbf{DDiv}} 1 \cdot a^{-1} \xrightarrow{\textbf{Id·}} a^{-1}$$

Id· $a \cdot 1 = a$

TLE $a = b = c \Longrightarrow a = c$

$$\xrightarrow{\textbf{TLE}} \frac{1}{a} = a^{-1}$$

(Proofs of Parts I, II, IV, V, and VI are to be done as exercises.)

Example 2.9.23 $\dfrac{2}{-3} \xrightarrow{\textbf{T2.9.12(III)}} -\dfrac{2}{3}$ and $\dfrac{-2}{-3} \xrightarrow{\textbf{T2.9.12(IV)}} \dfrac{2}{3}$

THEOREM 2.9.13   If $a, b, c, d \in R$ and $a, b, c, d \neq 0$, then

$$\textbf{I}\quad \frac{1}{a} \cdot \frac{1}{b} = \frac{1}{a \cdot b} \qquad \textbf{II}\quad \frac{a}{b} \cdot \frac{c}{d} = \frac{ac}{bd} \qquad \textbf{III}\quad \frac{a}{c} + \frac{b}{c} = \frac{a+b}{c}$$

$$\textbf{IV}\quad \frac{a}{b} + \frac{c}{d} = \frac{ad+bc}{bd} \qquad \textbf{V}\quad \frac{a}{b} - \frac{c}{d} = \frac{ad-bc}{bd} \qquad \textbf{VI}\quad \frac{1}{\dfrac{a}{b}} = \frac{b}{a}$$

$$\textbf{VII}\quad \frac{\dfrac{a}{b}}{\dfrac{c}{d}} = \frac{a}{b} \cdot \frac{d}{c}$$

Proof   (Part I)

Cl· $a, b \in R \Longrightarrow (ab) \in R$

(1) $a, b \in R \xrightarrow{\textbf{Cl·}} ab \in R \xrightarrow[\textbf{T2.9.7}]{\textbf{Hyp}} ab \neq 0$

T2.9.7
$ab = 0 \Longrightarrow a = 0$ or $b = 0$

(2) $\dfrac{1}{a} \cdot \dfrac{1}{b} = \dfrac{1}{a \cdot b} \xleftarrow[\textbf{(III)}]{\textbf{T2.9.12}} a^{-1} \cdot b^{-1} = (ab)^{-1}$

T2.9.12(III) $a^{-1} = \dfrac{1}{a}$

$$\xrightarrow{\textbf{MLE}}$$

MLE $a = b \Longrightarrow ac = bc$

$$\xleftarrow[\textbf{(1)}]{\textbf{FCLE}} \Bigg\} (ab)(a^{-1}b^{-1}) = (ab)(ab)^{-1}$$

FCLE $ac = bc, c \neq 0 \Longrightarrow a = b$

A· $(ab)c = a(bc)$

$$\xleftrightarrow[\textbf{C·}]{\textbf{A·}} (aa^{-1})(bb^{-1}) = (ab)(ab)^{-1}$$

C· $ab = ba$

$$\xleftrightarrow{\textbf{In·}} 1 \cdot 1 = 1$$

In· $a \cdot a^{-1} = 1$

$$\xleftrightarrow{\textbf{Id·}} 1 = 1$$

Id· $a \cdot 1 = a$

Thus,

$$\frac{1}{a} \cdot \frac{1}{b} = \frac{1}{a \cdot b} \Longleftrightarrow 1 = 1$$

(3) $\qquad 1 \in \boldsymbol{R} \xrightarrow{\text{RLE}} 1 = 1$

$\qquad\qquad \xrightarrow{\text{(2)}} \dfrac{1}{a} \cdot \dfrac{1}{b} = \dfrac{1}{a \cdot b}$ QED

RLE $\quad a \in \boldsymbol{R} \Longrightarrow a = a$

Remark 2.9.14 There are several observations that are important to note in the above proof. We began in step (2) with the conclusion; therefore, it was critical that every implication throughout step (2) be *two-way* (reversible). Technically the proof starts in step (3) with known information. That is, we must always remember that in developing a deductive system the final form of a proof must begin with known or given facts and proceed step by step — each verified by known properties — until the conclusion is reached. The flow from top to bottom is merely to set up the proof; the proof itself is initiated by $1 \in \boldsymbol{R}$ and the Reflexive Law of Equality and then proceeds upward to the final conclusion.

Note also the importance of step (1). Although it is given that a and b are not zero, nothing is mentioned in the hypotheses about the *product ab*. Because the multiplicative inverse of this product is used in the proof, it is essential that it be proved nonzero. Observe how this was accomplished by applying the hypotheses $a, b \neq 0$ and Theorem 2.9.7. Theorem 2.9.7 asserts that $[ab = 0 \Longrightarrow a = 0$ or $b = 0]$, which contradicts the hypotheses on a and b; hence, by the Second Fundamental Principle of Logic, $ab \neq 0$.

Note, finally, that the reasons for an implication holding in one direction are not always the same as those for the reverse direction. This occurred in the middle of step (2).

We now return to the proof of Theorem 2.9.13.

Proof (Part III)

$$\frac{a}{c} + \frac{b}{c} \xrightarrow{\text{DDiv}} a \cdot c^{-1} + b \cdot c^{-1}$$

$$\xrightarrow{\text{D}} (a + b) c^{-1}$$

$$\xrightarrow{\text{DDiv}} \frac{a + b}{c}$$

$$\xrightarrow{\text{TLE}} \frac{a}{c} + \frac{b}{c} = \frac{a + b}{c} \qquad \text{QED}$$

DDiv $\quad a/b = a \cdot b^{-1}$

D $\quad a(b + c) = ab + ac$

TLE $\quad a = b = c \Longrightarrow a = c$

This is the very important Addition Law of Fractions — you always obtain a common denominator when adding.

Example 2.9.24

T2.9.10 $ac/bc = a/b$

$$\frac{1}{3} + \frac{1}{4} \xlongequal{\textbf{T2.9.10}} \frac{1 \cdot 4}{3 \cdot 4} + \frac{1 \cdot 3}{4 \cdot 3}$$

T2.9.13(III) $a/c + b/c = (a + b)/c$

$$\xlongequal{\textbf{T2.9.13(III)}} \frac{4 + 3}{12}$$

$$= \frac{7}{12}$$

Example 2.9.25

T2.9.13(VII)
$(a/b)/(c/d) = (a/b) \cdot (d/c)$

$$\frac{-\dfrac{2}{3}}{\dfrac{8}{9}} \xlongequal{\textbf{T2.9.13(VII)}} \left(-\frac{2}{3}\right) \cdot \frac{9}{8}$$

T2.9.8(III) $(-a)b = -(ab)$

$$\xlongequal{\textbf{T2.9.8(III)}} -\left(\frac{2}{3} \cdot \frac{9}{8}\right)$$

T2.9.13(II) $(a/b)(c/d) = (ac/bd)$

$$\xlongequal{\textbf{T2.9.13(II)}} -\left(\frac{2 \cdot 9}{3 \cdot 8}\right)$$

$$= -\left(\frac{18}{24}\right)$$

$$= -\left(\frac{3 \cdot 6}{4 \cdot 6}\right)$$

T2.9.10 $ac/bc = a/b$

$$\xlongequal{\textbf{T2.9.10}} -\frac{3}{4}$$

These fundamental properties may seem quite elementary; yet there are few everyday calculations that cannot be performed by their application, and this includes the processes of solving algebraic equations, which we shall treat in Chapter 3. If you will master these basic properties, your ability to effectively use the system of real numbers in problem solving will be greatly enlarged.

It may appear in this section that we are overemphasizing the importance of justifying every step in a proof of a mere arithmetic simplification. However, such is the very purpose of this section — to gain a feeling or understanding of deductive thoroughness and completeness, and to express ourselves in the language of mathematics with increased precision and clarity. In most other sections of this book, we often skip certain obvious intermediate steps leading to a conclusion; even those steps that are included may not always be verified by a reference. Generally, only sufficient detail will be provided to preserve continuity of thought in the unfolding of a proof or the solving of a problem. Rigor overdone may become "rigor mortis," and it is not our intention that this occur.

Problem Set I ■ Reading Comprehension

Determine whether statements 1 through 6 are true or false and give reasons for your responses.

1 Reasoning by induction seeks to obtain a truth about all things or events of a specific kind by examining a limited number of them.
2 Reasoning by induction provides inevitable conclusions.
3 Deductive reasoning is usually easier to apply than is inductive reasoning.
4 Reasoning deductively seeks to make generalizations from specific incidents or experiments.
5 Mathematics employs deductive reasoning only; this is especially true in its creation.
6 In developing a branch of mathematics deductively, one must always begin with certain undefined terms and relations.
7 Under what conditions is a number system called a field?
8 How many different uses for the symbol '−' have been discussed in this chapter? Explain their differences.

Problem Set II ■ Skills Development

Show the following statements to be valid by inserting each intermediate step, including the source references that verify the statements.

Example $-2(3-4) = 2$

Solution $-2(3-4) \overset{\text{DSub}}{=\!=\!=} -2[3+(-4)]$

$$\overset{}{=\!=\!=} -2(-1)$$

$$\overset{\text{T2.9.8(IV)}}{=\!=\!=} 2 \cdot 1$$

$$\overset{\text{Id·}}{=\!=\!=} 2$$

DSub $a - b = a + (-b)$

T2.9.8(IV) $(-a)(-b) = ab$

Id· $a \cdot 1 = a$

(It is important to remember that often there is more than one way to simplify an expression — your steps may not be the same as someone else's.)

1 $\frac{1}{4} - \frac{1}{6} = \frac{1}{12}$
2 $\frac{1}{3} - \frac{5}{6} = -\frac{1}{2}$
3 $\frac{1}{2} + \frac{2}{3} = \frac{7}{6}$
4 $\frac{1}{4} \cdot \frac{2}{5} = \frac{1}{10}$
5 $\frac{2}{3} - \frac{8}{15} = \frac{2}{15}$
6 $(\frac{3}{4} + \frac{5}{12}) - (\frac{5}{12} - \frac{1}{4}) = 1$
7 $2\{-3 + 4(-2 - 5)\} = -62$
8 $(-2)\{-[-1 - 4]\} = -10$
9 $-\{-2[-3 + 2(-1 - 4) - 3] + 4\} = -36$
10 $(x - 3)(x - 1) = x^2 - 4x + 3$
11 $(x + 2)(x - 2) = x^2 - 4$
12 $(2x - 1)(x + 1) = 2x^2 + x - 1$
13 $(2x + 1)(3x - 2) = 6x^2 - x - 2$

Each statement 14 through 20 is an application of one of the equality or field axioms. Justify each statement by citing the appropriate axiom.

Example $x + 4 = y$ and $y = 6 \Longrightarrow x + 4 = 6$

Solution   E-3 (Transitive Law of Equality) or E-4 (Substitution Law of Equality).

14 $a + 3 = 4 \implies 4 = a + 3$ **15** $4 + b = 2 \implies b + 4 = 2$
16 $(3x + 4) + 6 = 3x + (4 + 6)$ **17** $-4 + 0 = -4$
18 $5 + (-5) = 0$ **19** $3(2 + (-4)) = 3 \cdot 2 + 3(-4)$ **20** $8 \cdot 8^{-1} = 1$

Each of the statements 21 through 26 is an application of one of the theorems. Justify each statement by citing the appropriate theorem.

21 $2x = 6 \implies \frac{1}{2}(2x) = \frac{1}{2} \cdot 6$ **22** $2x = 2 \cdot 3 \implies x = 3$
23 $x + 5 = 2 + 5 \implies x = 2$ **24** $3(x + 4) = 0 \implies x + 4 = 0$

25 $-2x + 5 = 0 \implies -2x = -5$ **26** $\dfrac{a}{2} = \dfrac{3}{4} \iff 4a = 2 \cdot 3$

Solve the following equations by finding the correct values of the variable x that render the equations valid. Justify each step by citing the appropriate definition, axiom, or theorem.

Example $x^2 - x - 2 = 0$

Solution $x^2 - x - 2 = 0 \iff (x - 2)(x + 1) = 0$

T2.9.7
$ab = 0 \implies a = 0$ or $b = 0$

                     T2.9.7
                     $\iff x - 2 = 0$ or $x + 1 = 0$

T2.9.5 $a + b = 0 \implies a = -b$

                     T2.9.5
                     $\iff x = -(-2)$ or $x = -1$

T2.9.8(I) $-(-a) = a$

                     T2.9.8(I)
                     $\iff x = 2$ or $x = -1$

REMEMBER: In normal practice all of the steps above are not considered; in fact, one can arrive at the answer merely by carefully examining the original equation. However, our purposes here are to acquaint ourselves thoroughly with the fundamental principles of arithmetic and algebraic manipulation and to illustrate how one can express and verify all steps leading to a conclusion, if challenged.

(The first step above is to factor the expression $x^2 - x - 2$, which itself requires several steps applying the distributive, associative, and commutative laws. These were verified by Problems 10 through 13.)

27 $x - 3 = 5$ **28** $2x + 4 = 8$ **29** $-2 + 4x = 7$
30 $3x - 4 = 6 - 2x$ **31** $4(x + 2) = 0$ **32** $(x - 1)(2x + 6) = 0$

33 $\dfrac{x - 4}{5} = \dfrac{2x + 1}{2}$ **34** $\dfrac{x + 2}{4} - \dfrac{6}{x - 2} = 0$

Problem Set III   ■   *Theoretical Development*

Which of the sets in Problems 1 through 5 are closed under the stated operation?

1 $N = \{x \mid x \text{ is a natural number}\}$, under subtraction.
2   {even integers}, under multiplication.
3 $Q = \{\text{rational numbers}\}$, under division.
4 $J = \{\text{integers}\}$, under subtraction.
5   {odd natural numbers}, under addition.
6   Is the operation of subtraction commutative in the field of real numbers?
7   Is the operation of division commutative in the field of real numbers?
8   Does there exist an identity element for division? That is, does there exist a real number, say p, such that for any real number a, we have $a/p = a$?

9 Does there exist an inverse element for subtraction for each real number? That is, for each $a \in R$, does there exist a $q \in R$ such that $a - q = 0$?

10 Argue that E-3 is a consequence of E-4 and therefore need not have been given as an axiom.

11 Argue that E-2 is a consequence of E-4.

Prove each of the following — include every step and cite the justification for each step:

12 Theorem 2.9.2 **13** Theorem 2.9.4 **14** Theorem 2.9.5

15 Theorem 2.9.7 **16** Theorem 2.9.8 **17** Theorem 2.9.9
          (a) Part II
          (b) Part III
          (c) Part IV

18 Theorem 2.9.11 **19** Theorem 2.9.12 **20** Theorem 2.9.13
                (a) Part I   **(a)** Part II
                (b) Part II   **(b)** Part IV
                (c) Part IV   **(c)** Part V
                (d) Part V   **(d)** Part VI
                (e) Part VI   **(e)** Part VII

21 Zero is unique — there is only one identity element for addition.

22 The number 1 is unique — there is only one identity element for multiplication.

23 $(a - b) + c = a - (b - c)$

24 $(a - b) - c = a - (b + c)$

25 $\dfrac{a}{b} = c \Longleftrightarrow a = bc \quad \text{and} \quad b \neq 0$

26 $-1 \cdot a = -a$

27 $a = b$ and $c \neq 0 \Longleftrightarrow \dfrac{a}{c} = \dfrac{b}{c}$

28 Identify the error in the following argument:

$$a = b + c$$

$$\Longrightarrow \quad a(a - b) = (b + c)(a - b)$$

$$\Longrightarrow \quad a^2 - ab = ab + ac - b^2 - bc$$

$$\Longrightarrow a^2 - ab - ac = ab - b^2 - bc$$

$$\Longrightarrow a(a - b - c) = b(a - b - c)$$

$$\Longrightarrow \qquad\qquad a = b$$

$$\overset{\text{TLE}}{\Longrightarrow} [a = b + c \Longrightarrow a = b]$$

The Real Numbers — An Ordered Field • 2.11

We briefly discussed an ordering for the real numbers in Section 2.6. Here we define more precisely the terms and relations associated with an ordering and proceed to discover several important properties.

• See sidenote page 59.

Recall from Section 2.6 the manner in which a one-to-one correspondence was established between the set of real numbers and the set of points on a straight line (see Fig. 2.11.1). We divide the real numbers into three mutually disjoint sets. The set of all negative numbers, denoted R_-, is the set of all numbers corresponding to points that lie to the *left* of 0 on the real line. The set of all positive numbers, denoted R_+, is the set of all points that lie to the *right* of 0. The third set, $\{0\}$, contains only zero, which, of course, is neither positive nor negative.

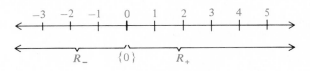

FIGURE 2.11.1

From this one-to-one correspondence, we observe the following two relations—denoted N_1 and N_2 for future reference:

(N_1) $a \in R_+ \Longleftrightarrow -a \in R_-$

(N_2) $a \in R_- \Longleftrightarrow -a \in R_+$

Remark 2.11.1 N_2 is particularly important to understand because, as noted earlier, we often make the mistake of assuming $-a$ is a negative number; all we know is that $-a$ is the additive inverse of a, which may or may not be negative.

Example 2.11.1 $\quad 3 \in R_+ \overset{N_1}{\Longleftrightarrow} -3 \in R_-$

$$-2 \in R_- \overset{N_2}{\Longleftrightarrow} -(-2) \in R_+$$

$$\overset{T2.9.8(I)}{\Longleftrightarrow} 2 \in R_+$$

T2.9.8(I) $\quad -(-a) = a$

The next definition and two axioms provide an ordering for the real numbers.

DEFINITION 2.11.1 If $a, b \in R$, then "a is less than b," denoted $a < b$, if $b - a$ is *positive*. The relation "b is greater than a," denoted $b > a$, is equivalent to $a < b$. Expressed symbolically this definition becomes

$$a, b \in R \Longrightarrow \begin{cases} a < b \Longleftrightarrow (b - a) \in R_+ \\ \text{and} \\ a < b \Longleftrightarrow b > a \end{cases}$$

O-1 If $a, b \in \mathbf{R}$, then one and only one of the following relations hold:

$$a < b$$
$$a = b$$
$$a > b$$

O-2 If $a, b \in \mathbf{R}_+$, then their sum and product are *positive*.

$$a, b \in \mathbf{R}_+ \implies (a + b), (ab) \in \mathbf{R}_+$$

that is, $\mathbf{R}_+$ is *closed* under addition and multiplication.

Whereas the field axioms provided properties associated with the relation equals, the two axioms above furnish the essential properties of the order relations "*less than*" and "*greater than.*"

It is important to observe that the relation $a < b$, interpreted *geometrically,* means a is to the *left* of b on the real line; this holds whether a and b are both positive, both negative, or a is negative and b positive.

Example 2.11.2 Given the set $\{-4, -1, 0, 2, 3\}$, then the following relations hold: $-4 < -1, -4 < 2, 0 < 2$, and $2 < 3$. Notice that the left-hand member of each relation is to the left of the right-hand member on the real line in Fig. 2.11.2.

FIGURE 2.11.2

It is sometimes desirable to express symbolically the relation "*a* is less than *or equal* to *b*"; this is done by writing $a \le b$. That is

$$a \le b \iff [a < b \quad \text{or} \quad a = b]$$

Thus it is mathematically correct to write $2 \le 2$ or $2 \le 5$, although when the exact relation is known, that relation alone is the preferred form.

We now turn our attention to some of the properties of an ordered field.

LEMMA 2.11.1 $a \in \mathbf{R} \implies \begin{cases} a > 0 \iff a \in \mathbf{R}_+ \\ \text{and} \\ a < 0 \iff a \in \mathbf{R}_- \end{cases}$

Proof (Part II)

$$a < 0 \xLeftrightarrow{\text{D2.11.1}} (0 - a) \in \boldsymbol{R}_+$$

$$\xLeftrightarrow{\text{Id+}} -a \in \boldsymbol{R}_+ \xLeftrightarrow[\text{T2.9.8(I)}]{\text{N}_2} a \in \boldsymbol{R}_-$$

$$\xRightarrow{\text{L-4}} [a < 0 \iff a \in \boldsymbol{R}_-] \qquad \text{QED}$$

(The proof of Part I is to be done as an exercise.)

D2.11.1 $a < b \iff (b - a) \in \boldsymbol{R}_+$

Id+ $a + 0 = a$

N$_2$ $a \in \boldsymbol{R}_- \iff -a \in \boldsymbol{R}_+$

T2.9.8(I) $-(-a) = a$

L-4
$[A \Longrightarrow B \Longrightarrow C] \Longrightarrow [A \Longrightarrow B]$

LEMMA 2.11.2 $a, b \in \boldsymbol{R} \Longrightarrow$

  I $a \leq a$
  II $[a \leq b \text{ and } b \leq a] \iff a = b$

Proof To be done as an exercise.

THEOREM 2.11.1 $a, b \in \boldsymbol{R} \Longrightarrow$

  I $a > 0 \text{ and } b < 0 \Longrightarrow ab < 0$
  II $a < 0 \text{ and } b < 0 \Longrightarrow ab > 0$
  III $a \in \boldsymbol{R} \Longrightarrow a \cdot a \geq 0$

Proof (Part I)

L2.11.1 $c > 0 \iff c \in \boldsymbol{R}_+$

O-2 $a, b \in \boldsymbol{R}_+ \Longrightarrow ab \in \boldsymbol{R}_+$

T2.9.8(III) $(-a)b = -(ab)$

N$_2$ $a \in \boldsymbol{R}_- \iff -a \in \boldsymbol{R}_+$

T2.9.8(I) $-(-a) = a$

L2.11.1(II) $c < 0 \iff c \in \boldsymbol{R}_-$

$$b < 0 \xRightarrow[\text{N}_2]{\text{L2.11.1}} -b > 0 \xRightarrow[\text{L2.11.1}]{\text{Hyp}} a, -b \in \boldsymbol{R}_+$$

$$\xRightarrow{\text{O-2}} a(-b) \in \boldsymbol{R}_+ \xRightarrow{\text{T2.9.8(III)}} -(ab) \in \boldsymbol{R}_+$$

$$\xRightarrow[\text{T2.9.8(I)}]{\text{N}_2} ab \in \boldsymbol{R}_- \xRightarrow[\text{(II)}]{\text{L2.11.1}} ab < 0 \qquad \text{QED}$$

(Proofs of Parts II and III are to be done as exercises.)

Example 2.11.3

$$2 > 0 \text{ and } -3 < 0 \xRightarrow{\text{T2.11.1(I)}} 2(-3) < 0 \iff -6 < 0$$

$$-2 < 0 \text{ and } -3 < 0 \xRightarrow{\text{T2.11.1(II)}} (-2)(-3) > 0 \iff 6 > 0$$

THEOREM 2.11.2 $a, b, c \in \boldsymbol{R} \Longrightarrow$

  I $[a < b \text{ and } b < c] \Longrightarrow a < c$
  II $a < b \iff a + c < b + c$
  III $a < b \text{ and } c > 0 \iff ac < bc \text{ and } c > 0$
  IV $a < b \text{ and } c < 0 \iff ac > bc \text{ and } c < 0$

Proof (Part III) (Proofs of Parts I, II, and IV are to be done as exercises.)

(1) $\quad a < b \overset{\textbf{D2.11.1}}{\Longleftrightarrow} (b-a) \in \textbf{\textit{R}}_+$

(2) $\quad c > 0 \overset{\textbf{L2.11.1}}{\Longleftrightarrow} c \in \textbf{\textit{R}}_+$

(3) (1) and (2) $\overset{\bullet}{\Longleftrightarrow} (b-a), c \in \textbf{\textit{R}}_+$

$\quad \overset{\textbf{O-2} \bullet}{\underset{\textbf{T2.11.1(I) + Hyp}}{\Longleftrightarrow}} [(b-a) \cdot c] \in \textbf{\textit{R}}_+$

$\quad \overset{\textbf{D}}{\Longleftrightarrow} (bc - ac) \in \textbf{\textit{R}}_+ \overset{\textbf{D2.11.1}}{\Longleftrightarrow} ac < bc \qquad \text{QED}$

Example 2.11.4 $\quad x - 2 < 0 \overset{\textbf{T2.11.2(II)}}{\Longrightarrow} (x-2) + 2 < 0 + 2$

$\quad \overset{\textbf{A+}}{\Longrightarrow} x + [-2+2] < 2$

$\quad \overset{\textbf{In+}}{\Longrightarrow} x + 0 < 2$

$\quad \overset{\textbf{Id+}}{\Longrightarrow} x < 2$

Example 2.11.5

$\quad 6 > -2x \overset{\textbf{T2.9.8(IV)}}{\Longleftrightarrow} (-2)(-3) > -2x$

$\quad \overset{\textbf{T2.11.2(IV)}}{\Longleftrightarrow} -3 < x \overset{\textbf{D2.11.1}}{\Longleftrightarrow} x > -3$

THEOREM 2.11.3 $\quad a, b \in \textbf{\textit{R}} \Longrightarrow$

I $\quad |a| < b \Longleftrightarrow -b < a < b$

II $\quad |a| > b \Longleftrightarrow a > b \quad \text{or} \quad a < -b$

Proof (Part I)

(1) $\quad |a| < b \overset{\textbf{D2.7.1}}{\Longleftrightarrow} \begin{cases} a < b, & \text{if } a \geq 0 \\ \text{or} \\ -a < b, & \text{if } a < 0 \end{cases}$

(2) $\quad a < b \overset{\textbf{T2.11.2(IV)}}{\longleftarrow}$

$\quad -a = (-1)a > (-1)b = -b$

$\quad \overset{\textbf{SubLE}}{\Longleftrightarrow} -a > -b \overset{\textbf{D2.11.1}}{\Longleftrightarrow} -b < -a$

$\quad \overset{a \geq 0}{\underset{a < b}{\Longleftrightarrow}} -b < -a \leq a < b \overset{\textbf{T2.11.2(I)}}{\Longleftrightarrow} -b < a < b$

(3) $\quad -a < b \overset{\textbf{T2.11.2(IV)}}{\longleftarrow}$

$\quad a = (-1)(-a) > (-1)b = -b$

$\quad \overset{\textbf{SubLE}}{\Longleftrightarrow} a > -b \overset{\textbf{D2.11.1}}{\Longleftrightarrow} -b < a$

$\quad \overset{a < 0}{\underset{-a < b}{\Longleftrightarrow}} -b < a \leq -a < b \overset{\textbf{T2.11.2(I)}}{\Longleftrightarrow} -b < a < b$

D2.11.1 $\quad a < b \Longleftrightarrow (b-a) \in \textbf{\textit{R}}_+$

L2.11.1 $\quad c > 0 \Longleftrightarrow c \in \textbf{\textit{R}}_+$

O-2 $\quad a, b \in \textbf{\textit{R}}_+ \Longrightarrow ab \in \textbf{\textit{R}}_+$

● O-2 justifies the implication "$\Longrightarrow$," whereas T2.11.1(I) and hypotheses justifies the reverse implication "$\Longleftarrow$." This is required because the theorem carries a *two-way* implication.

T2.11.1(I)
$a > 0 \quad \text{and} \quad b < 0 \Longrightarrow ab < 0$
D $\quad a(b+c) = ab + ac$

T2.11.2(II) $\quad a < b \Longleftrightarrow a+c < b+c$

A+ $\quad (a+b) + c = a + (b+c)$

In+ $\quad a + (-a) = 0$

Id+ $\quad a + 0 = a$

T2.9.8(IV) $\quad (-a)(-b) = ab$

T2.11.2(IV)
$a < b \quad \text{and} \quad c < 0 \Longleftrightarrow$
$\quad ac > bc, c < 0$

D2.11.1 $\quad a < b \Longleftrightarrow b > a$

D2.7.1 $\quad |a| = \begin{cases} a, & \text{if } a \geq 0 \\ -a, & \text{if } a < 0 \end{cases}$

SubLE
E-4 Substitution Law of Equality

T2.11.2(I)
$a < b \quad \text{and} \quad b < c \Longrightarrow a < c$

T2.11.2(IV)
$a < b \quad \text{and} \quad c < 0 \Longleftrightarrow ac > bc$

SubLE
E-4 Substitution Law of Equality

D2.11.1 $\quad a < b \Longleftrightarrow b > a$

T2.11.2(I)
$a < b \quad \text{and} \quad b < c \Longrightarrow a < c$

$$(4) \quad |a| < b \xleftrightarrow[\ (3)\ ]{(1)\ (2)}$$

O-1 $a, 0 \in \mathbf{R} \Longleftrightarrow a < 0, a = 0,$ or $a > 0$

$$\left. \begin{cases} -b < a < b, & \text{if } a \geq 0 \\ \text{or} & \\ -b < a < b, & \text{if } a < 0 \end{cases} \right\} \xleftrightarrow{\text{O-1}} -b < a < b \qquad \text{QED}$$

(Proof of Part II is to be done as an exercise.)

Example 2.11.6

T2.11.3(I) $|a| < b \Longleftrightarrow -b < a < b$

T2.11.2(II) $a < b \Longleftrightarrow a + c < b + c$

$$|x - 2| < 6 \xleftrightarrow{\text{T2.11.3(I)}} -6 < x - 2 < 6$$
$$\xleftrightarrow{\text{T2.11.2(II)}} -6 + 2 < (x - 2) + 2 < 6 + 2$$
$$\Longleftrightarrow -4 < x < 8$$

2.12 Problem Sets

Problem Set I ■ *Skills Development*

1 Write the following statements in symbolic form:
  a a is less than or equal to b.
  b a is not less than nor equal to b.
  c a is greater than b.
  d a is not less than b.
  e a is not greater than or equal to b.
  f a is not less than nor greater than b.

2 Insert the proper symbols (either $<$, $>$, or $=$) between the following pairs of numbers:
  a -2 and 1 b -5 and -2 c 3 and 1
  d $\sqrt{4}$ and 2 e 2 and -4 f $\frac{3}{5}$ and $-\frac{7}{10}$

3 Rewrite the following statements using mathematical notation:
  a a is to the right of b.
  b c is to the left of x.
  c x lies between -2 and 3.
  d The absolute value of x is greater than 4.
  e The absolute value of the quantity y minus 3 is less than or equal to 5.
  f z lies to the left of -3 or to the right of 6.

4 Rewrite the following expressions so that x is alone (solved) between the inequality symbols:
  a $-3 < x + 3 < 4$
  b $2 < 2x - 4 < 8$
  c $-1 < -3x - 4 < 5$

5 Rewrite the following without absolute value signs and solve for x:

Example $|4 - 2x| < 10$

Solution $|4 - 2x| < 10 \Longleftrightarrow -10 < 4 - 2x < 10$
$$\Longleftrightarrow -14 < -2x < 6$$
$$\Longleftrightarrow \frac{-14}{-2} > x > \frac{6}{2}$$
$$\Longleftrightarrow 7 > x > -3$$
$$\Longleftrightarrow -3 < x < 7$$

a $|x - 3| < 7$ b $|3 - 2x| > 4$

c $|4x + 5| \leq 0$ d $|-2 - 3x| \geq -5$

e $|2x + 3| < 4$ f $|-x - 4| < 6$

g $-|x - 3| > 4$ h $|3 - 2x| > 7$

Problem Set II ■ *Theoretical Developments*

1 Justify the following conditional propositions by citing the appropriate definition, lemma, or theorem.

a $x - 2 < 4 \Longleftrightarrow [4 - (x - 2)] \in \boldsymbol{R}_+$

b $[-2 < -1 \ \text{ and } \ -1 < 3] \Longrightarrow -2 < 3$

c $[x > 0 \ \text{ and } \ -2 < 0] \Longleftrightarrow -2x < 0$

d $x + 2 < 3 \Longleftrightarrow x < 1$

e $-3x > -6 \Longleftrightarrow x < 2$

f $x < 0 \Longleftrightarrow -x \in \boldsymbol{R}_+$

2 Prove the following, justifying every step by citing the appropriate known property:

a Lemma 2.11.1 Part I   b Lemma 2.11.2 Part I

c Lemma 2.11.2 Part II   d Theorem 2.11.1 Parts II and III

e Theorem 2.11.2 Part I   f Theorem 2.11.2 Part II

g Theorem 2.11.2 Part IV   h Theorem 2.11.3 Part II

i $a < b \ \text{ and } \ c < d \Longrightarrow a + c < b + d$

j $a < 0 \ \text{ and } \ b < 0 \Longrightarrow a + b < 0$

k $\dfrac{a}{b} < \dfrac{c}{d} \ \text{ and } \ bd > 0 \Longrightarrow ad < bc$

l $0 < a < 1 \Longrightarrow a^2 < a$ m $1 < a \Longrightarrow a^2 > a$

n $a < b \Longrightarrow a < \dfrac{a + b}{2} < b$ o $a > 0 \Longrightarrow a + \dfrac{1}{a} \geq 0$

Problem Set III ■ *Just For Fun*

1 Imagine that you have three boxes, one containing two black marbles, one containing two white marbles, and the third containing one black marble and one white marble. The boxes were labeled for their contents – BB, WW, and BW – but someone has switched the labels so that every box is now incorrectly labeled. You can determine the contents of all the boxes from the color of one marble that you draw from a particular box without looking inside. How? (HINT: Recall that all the boxes were wrongly labeled. Draw the marble from the box labeled BW and reason from that point.)

2 Consider the following trick: Pick any number. Multiply it by 3. Add 6 to that result. Divide this result by 3 and then subtract your original number. Your answer is always 2.

    a Perform the indicated operation using N for the number.

    b Explain how the trick works.

3 Suppose that you have before you two pitchers, one containing water and the other containing wine. Pour some of the water into the wine and return an equal amount of the diluted wine to the pitcher containing the water. Each pitcher thus contains its original amount of liquid. Is there now more wine in the water than water in the wine, is there less, or are the amounts the same? (Ignore the fact that a mixture of water and alcohol, in practice, occupies a little less volume than the sum of the volumes of the two liquids before they are mixed.) (HINT: Let x denote the amount of wine ultimately transferred to the water pitcher. The space this amount of wine originally occupied in the wine pitcher must now be filled with water.)

2.13   Summary of Definitions and Properties in Chapter 2

D2.7.1 $|x| = \begin{cases} x & \text{if } x \geq 0 \\ -x & \text{if } x < 0 \end{cases}$ (p. 55)

D2.9.1   Definition of equals   (p. 63)

E-1   (RLE) $a \in R \Longrightarrow a = a$ (p. 63)

E-2   (SLE) $a, b \in R \Longrightarrow [a = b \Longleftrightarrow b = a]$ (p. 63)

E-3   (TLE) $a, b, c \in R, [a = b \text{ and } b = c \Longrightarrow a = c]$ (p. 64)

E-4   (SubLE) Substitution Law of Equality   (p. 64)

F-1   (Cl+) $a, b \in R \Longrightarrow (a + b) \in R$ (p. 65)

F-2   (A+) $a, b, c \in R \Longrightarrow (a + b) + c = a + (b + c)$ (p. 65)

F-3   (C+) $a, b \in R \Longrightarrow a + b = b + a$ (p. 66)

F-4   (Id+) $\exists\ 0 \in R \ni \forall\ a \in R \quad a + 0 = 0 + a = a$ (p. 66)

F-5   (In+) $\forall\ a \in R \ \exists\ -a \in R \ni a + (-a) = (-a) + a = 0$ (p. 66)

F-6   (Cl·) $a, b \in R \Longrightarrow (a \cdot b) \in R$ (p. 66)

F-7   (A·) $a, b, c \in R \Longrightarrow (a \cdot b) \cdot c = a \cdot (b \cdot c)$ (p. 67)

F-8   (C·) $a, b \in R \Longrightarrow a \cdot b = b \cdot a$ (p. 67)

F-9   (D) $a, b, c \in R \Longrightarrow a(b + c) = ab + ac$ (p. 67)

F-10   (Id·) $\exists\ 1 \in R \ni \forall\ a \in R \quad a \cdot 1 = 1 \cdot a = a$ (p. 68)

F-11   (In·) $\forall\ a \in R, a \neq 0, \exists\ a^{-1} \in R \ni a \cdot a^{-1} = a^{-1}a = 1$ (p. 68)

D2.9.2   (DSub) $a, b \in R \Longrightarrow a - b = a + (-b)$ (p. 68)

D2.9.3   (DDiv) $a, b \in R, b \neq 0 \Longrightarrow \dfrac{a}{b} = a \cdot b^{-1}$ (p. 69)

T2.9.1   (ALE) $a, b, c \in R \ [a = b \Longrightarrow a + c = b + c]$ (p. 70)

T2.9.2 (MLE) $a, b, c \in \mathbf{R}$ $[a = b \Longrightarrow ac = bc]$ (p. 71)

T2.9.3 (TCLE) $a, b, c \in \mathbf{R}$ $[a + c = b + c \Longrightarrow a = b]$ (p. 71)

T2.9.4 (FCLE) $a, b, c \in \mathbf{R}$ $[ac = bc$ and $c \neq 0 \Longrightarrow a = b]$ (p. 72)

T2.9.5 $a, b \in \mathbf{R}$ and $a + b = 0 \Longrightarrow a = -b$ and $b = -a$ (p. 72)

T2.9.6 $a \in \mathbf{R} \Longrightarrow a \cdot 0 = 0$ (p. 73)

T2.9.7 $a, b \in \mathbf{R}$ and $ab = 0 \Longrightarrow a = 0$ or $b = 0$ (p. 73)

T2.9.8 $a, b \in \mathbf{R} \Longrightarrow$ (p. 73)

I $-(-a) = a$ **II** $-(a + b) = (-a) + (-b)$

III $(-a)b = -(ab)$ **IV** $(-a)(-b) = ab$

T2.9.9 $a, b, c, d \in \mathbf{R}$ and $b, d \neq 0 \Longrightarrow \left[\dfrac{a}{b} = \dfrac{c}{d} \Longleftrightarrow ad = bc \right]$ (p. 74)

T2.9.10 $a, b, c \in \mathbf{R}$ and $b, c \neq 0 \Longrightarrow \dfrac{ac}{bc} = \dfrac{a}{b}$ (p. 74)

T2.9.11 $a, b \in \mathbf{R}, a, b \neq 0$, and $ab = 1 \Longrightarrow \begin{cases} a = b^{-1} \\ \text{and} \\ b = a^{-1} \end{cases}$ (p. 75)

T2.9.12 $a, b \in \mathbf{R}$ and $a, b \neq 0 \Longrightarrow$ (p. 76)

I $\dfrac{a}{a} = 1$ **II** $\dfrac{a}{1} = a$ **III** $\dfrac{1}{a} = a^{-1}$

IV $\dfrac{1}{1/a} = a$ **V** $\dfrac{-a}{b} = -\dfrac{a}{b} = \dfrac{a}{-b}$ **VI** $\dfrac{-a}{-b} = \dfrac{a}{b}$

T2.9.13 $a, b, c, d \in \mathbf{R}$ and $a, b, c, d \neq 0 \Longrightarrow$ (p. 76)

I $\dfrac{1}{a} \cdot \dfrac{1}{b} = \dfrac{1}{a \cdot b}$ **II** $\dfrac{a}{b} \cdot \dfrac{c}{d} = \dfrac{ac}{bd}$

III $\dfrac{a}{c} + \dfrac{b}{c} = \dfrac{a + b}{c}$ **IV** $\dfrac{a}{b} + \dfrac{c}{d} = \dfrac{ad + bc}{bd}$

V $\dfrac{a}{b} - \dfrac{c}{d} = \dfrac{ad - bc}{bd}$ **VI** $\dfrac{1}{\dfrac{a}{b}} = \dfrac{b}{a}$

VII $\dfrac{\dfrac{a}{b}}{\dfrac{c}{d}} = \dfrac{a}{b} \cdot \dfrac{d}{c}$

N₁ $a \in \mathbf{R}_+ \Longleftrightarrow -a \in \mathbf{R}_-$ (p. 82)

N₂ $a \in \mathbf{R}_- \Longleftrightarrow -a \in \mathbf{R}_+$

D2.11.1 $a, b \in \mathbf{R} \Longrightarrow \begin{cases} a < b \Longleftrightarrow b - a > 0 \\ \text{and} \\ a < b \Longleftrightarrow b > a \end{cases}$ (p. 82)

O-1 $a, b \in \mathbf{R} \Longrightarrow [a < b, a = b,$ or $a > b]$ (Using the exclusive *or*) (p. 83)

O-2 $a, b \in \mathbf{R}_+ \Longrightarrow (a + b), (ab) \in \mathbf{R}_+$

L2.11.1 $a, b \in \mathbf{R} \Longrightarrow \begin{cases} a > 0 \Longleftrightarrow a \in \mathbf{R}_+ \\ \text{and} \\ a < 0 \Longleftrightarrow a \in \mathbf{R}_- \end{cases}$ (p. 83)

L2.11.2 $a, b \in \boldsymbol{R} \Longrightarrow$ I $a \leq a$ (p. 84)
II $[a \leq b \quad \text{and} \quad b \leq a \Longleftrightarrow a = b]$

T2.11.1 $a, b \in \boldsymbol{R} \Longrightarrow$ (p. 84)
I $a > 0 \quad \text{and} \quad b < 0 \Longrightarrow ab < 0$
II $a < 0 \quad \text{and} \quad b < 0 \Longrightarrow ab > 0$
III $a \cdot a \geq 0$

T2.11.2 $a, b, c \in \boldsymbol{R} \Longrightarrow$ (p. 84)
I $[a < b \quad \text{and} \quad b < c] \Longrightarrow a < c$
II $a < b \Longleftrightarrow a + c < b + c$
III $a < b \quad \text{and} \quad c > 0 \Longleftrightarrow ac < bc \quad \text{and} \quad c > 0$
IV $a < b \quad \text{and} \quad c < 0 \Longleftrightarrow ac > bc \quad \text{and} \quad c < 0$

T2.11.3 $a, b \in \boldsymbol{R} \Longrightarrow$ (p. 85)
I $|a| < b \Longleftrightarrow -b < a < b$
II $|a| > b \Longleftrightarrow a > b \quad \text{or} \quad a < -b$

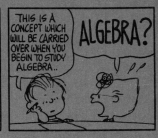
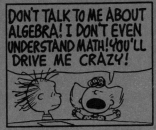

Open Sentences — Equations and Inequalities

3 CHAPTER

Introduction **3.1**

The mathematical representation or model of a problem usually takes the form of one or more equations or inequalities. These are, in fact, the most essential elements of the mathematical formulation.

To illustrate the need and utility of equations in problem-solving, we present the following two problems.

Problem 3.1.1 A large reservoir has three separate spillways. Let us name them A, B, and C. It is found that when A alone is open the reservoir will empty in 8 days, whereas if all three are open it takes only 2 days. It is known that spillway C has only half the capacity of B. Suppose it becomes necessary to drain the reservoir and spillways A and C are inoperative. How long will it take with only B open?

Problem 3.1.2 The owner of an apartment complex comprising 30 units has discovered he is able to keep all units occupied if he charges $200.00 rent per month for each apartment. Furthermore, he has found that for each $10.00 increase in monthly rent the number of units occupied decreases by 1. Because the service and maintenance costs of an unoccupied unit are significantly less than one being rented, he suspects he can increase his profit margin by raising the monthly rent and still collect the same amount as when all units were occupied renting at $200.00 per month. What should he charge per month to accomplish this, if it is possible, and how many units will be left vacant because of the higher cost?

We will not solve these problems at this time but rather we shall review and expand our knowledge of algebraic equations and inequalities. We will then return and easily solve the problems.

3.2   Open Sentences

Recall from Chapter 1 that a proposition is a statement that is either true or false but not both true and false. Observe the propositions

$$6 = 4 + 2 \quad \text{and} \quad 2 < -1$$

The first is a *true* statement, whereas the second is a *false* statement. When a proposition is true for all numbers comprising a given set, it is often stated using letters, each representing any one of the members of that set. In this manner, we wrote in Chapter 2 expressions such as

$$a, b, c \in \mathbf{R} \implies a(b + c) = ab + ac$$

and

$$a, b, c \in \mathbf{R} \implies [a = b \implies a + c = b + c]$$

meaning the propositions hold true for any three real numbers, here represented by the letters a, b, and c. It is important to understand that in such propositions the letters a, b, c represent *fixed* though unspecified numbers called constants. Remember also that these propositions are statements of fact (laws) in the systems they represent. The relation equals is used strictly according to Definition 2.9.1 — the left- and right-hand sides of such equations are literally equal in the sense that one can be replaced by the other in any mathematical statement without changing the truth or falsity of that statement. These equations are frequently called literal (always true) equations.

There are, however, other meanings and uses of the relation equals. Probably the most important of these are equations containing variables. These, you may recall from Chapter 1, we call open sentences, which we generally try to solve. Take for example the open sentence

$$3x = 12 \tag{1}$$

The solution equation equivalent to this is

$$x = 4$$

It is very important to understand the distinct meaning of the relation equals as used in equation (1). You would surely agree that of the many "equal" forms for the number 12, such as $2 \cdot 6$, 24/2, or $\sqrt{144}$, $3x$ is certainly not one of them; that is, $3x$ is not another way to write 12. Also, in general it would be *improper* to apply the Law of Substitution using such an equation: the truth or falsity of another statement containing $3x$ or 12 would likely be changed by substituting $3x$ for 12 or 12 for $3x$. Because such equations are not literal (not always true), we ask what kind are they?

To answer this question, we shall first review the concept of a variable. First, notationally a variable is a letter in a mathematical expression, usually chosen from those occurring toward the end of the alphabet, such as x, y, or z. This is done to help distinguish them from constants, which, as indicated earlier, are represented by letters usually chosen toward the beginning of the alphabet. As the name variable implies, it generally represents more than one number; in fact, it may stand for any one of the members of a general set called the replacement set. The replacement set is usually composed of all elements from the universal set for which the mathematical expressions comprising the equation are *defined*.

A replacement set for equation (1) above could be the entire set of real numbers, $\boldsymbol{R}$, since the product $3x$ is defined for all real x. If, however, we have an equation such as

$$\frac{16}{x - 4} = 36$$

then the replacement set could not include 4, because the quotient on the left is not defined when x is replaced by 4.

To distinguish between equations and inequalities that are literal (always true) and those that contain variables, we introduce the following definition.

DEFINITION 3.2.1 An equation or an inequality containing one or more variables is called an open proposition or an open sentence If the open sentence is an equation, it may be called an open equation; if it is an inequality, it may be called an open inequality. The variables represent any one of the members of a given replacement set.

In our brief treatment of open sentences in Chapter 1, we noted that when a variable is replaced by a specific element, the proposition is rendered either true or false.

Example 3.2.1   Let us examine the equation

$$2x = 16 \qquad\qquad (2)$$

over the replacement set $\{3, 7, 8, 9\}$. As the variable x varies over this set, equation (2) becomes $2 \cdot 3 = 16$, $2 \cdot 7 = 16$, $2 \cdot 8 = 16$, and $2 \cdot 9 = 16$. Each of these equations is false except, of course, the one in which x is replaced by 8. Thus, the mathematical statement (2) may be true or false, depending upon our choice of the variable.

Remark 3.2.1   Because an open sentence may be true or false depending upon one's choice for the variable, it is technically not a proposition (note Definition 1.1.1). For this reason we qualify them with the adjective open.

Example 3.2.2   Given the open equation in two variables

$$y = 2x - 4 \qquad\qquad (3)$$

over the replacement set R of real numbers. If x is replaced by 2 and y by 0, then equation (3) becomes

$$0 = 2 \cdot 2 - 4$$

which is a *true* proposition. If on the other hand, x is replaced by 3 and y by 5, then equation (3) becomes

$$5 = 2 \cdot 3 - 4$$

which is a *false* proposition.

The previous two examples lead us to our next definition.

DEFINITION 3.2.2   Each member or members of the replacement set associated with a given open sentence that reduces that open sentence to a true statement is called a solution or root of the open sentence. The set of all solutions of an open sentence is called the solution set. We will usually denote the solution set by *S*.

Example 3.2.3   Determine the solution set of the open equation

$$(x - 1)(x + 3) = 0 \qquad\qquad (4)$$

over the replacement set R of real numbers.

Solution   We now want to take full advantage of the properties of real numbers developed in Chapter 2. As these properties are being used, they will be identified and keyed in the outside margin of each page for easy reference. We proceed as follows:

$$(x-1)(x+3) = 0 \xrightarrow{\textbf{T2.9.7}} \begin{cases} x-1=0 \\ \text{or} \\ x+3=0 \end{cases}$$

T2.9.7
$ab = 0 \Longrightarrow a = 0$ or $b = 0$

$$\xrightarrow{\textbf{T2.9.5}} \begin{cases} x=1 \\ \text{or} \\ x=-3 \end{cases}$$

T2.9.5
$a+b = 0 \Longrightarrow a = -b$ or $b = -a$

We observe that if $x = 1$ or $x = -3$, the equation (4) is reduced to the true statements

$$(1-1)(1+3) = 0 \quad \text{or} \quad (-3-1)(-3+3) = 0,$$

respectively, and hence 1 and -3 are solutions. Because these are the only solutions, the solution set is

$$S = \{1, -3\}$$

Remark 3.2.2   It is important to understand, when applying the properties of real numbers, that the letters $a, b, c,$ or d present in the statements of these properties represent either real numbers or real expressions. That is, they may also represent algebraic expressions, such as $(x + 2)$, $x^2 - 3x + 5$, or $(x - 4)/(2x + 3)$, where the variable x is understood to vary over the real numbers only, thus preserving the realness of each expression and the validity of the properties.

Example 3.2.4   Solve the open inequality

$$|x - 2| < 4 \tag{5}$$

over $\boldsymbol{R}$.

Solution

$$|x-2| < 4 \xBiglongleftrightarrow{\textbf{T2.11.3(I)}} -4 < x - 2 < 4$$
$$\xBiglongleftarrow{\textbf{T2.11.2(II)}} -2 < x < 6$$
$$\xBiglongrightarrow{\textbf{D3.2.2}} S = \{x \mid -2 < x < 6\}$$

T2.11.3(I) $|a| < b \Longleftrightarrow -b < a < b$

T2.11.2(II) $a < b \Longleftrightarrow a + c < b + c$

D3.2.2   Definition of Solution Set

Notice that we were able to solve inequality (5) by converting it into a sequence of equivalent forms, each simpler than the former, until we arrived at an expression clearly identifying the solution set. Each step in the sequence was justified by one or more of the fundamental principles of real numbers.

Remark 3.2.3   In practice you usually do not identify the reasons for each step specifically. In fact, we will do this less frequently as you become increasingly familiar with the properties and their uses.

Example 3.2.5 Solve the open equation

$$x^2 = -4 \qquad (6)$$

over R.

● T2.11.1(III) $\quad a \in R \Longrightarrow a^2 \geq 0$

Solution We know by T2.11.1(III) ● that there does not exist a real number whose square is negative. Therefore, the solution set is empty, and we write

$$S = \{\ \} = \varnothing$$

Remark 3.2.4 There is nothing wrong or improper with open sentence (6). As mentioned earlier, the relation equals in an open sentence is not generally a literal (always true) one. In fact, as observed in the above example, the left side of that open equation is never equal to the right side for any choice of variable within the replacement set. We must remember when working with open sentences that we are, in essence, asking the *question*, "Do there exist solutions and, if so, what are they?" Bear in mind that an open sentence such as (6) may be an accurate mathematical model of a problem having no solution, although this fact may not be known initially. Hence the solution set being rendered empty provides the investigator with precisely the information he was seeking. We will find later that open sentence (6) over the replacement set of *complex numbers* will have a nonempty solution set. Whether or not an open sentence possesses solutions often depends upon the choice of replacement set.

Example 3.2.6 Solve the open equation

$$(x - 1)(x + 1) = x^2 - 1$$

over R.

Solution

$$(x - 1)(x + 1) \overset{\text{D}}{=\!=\!=} x^2 - x + x - 1 \overset{\text{In+}}{=\!=\!=} x^2 - 1$$

$$\overset{\text{TLE}}{=\!=\!=\!\Longrightarrow} (x - 1)(x + 1) = x^2 - 1 \ \forall \ x \in R$$

$$\overset{\text{D3.2.2}}{=\!=\!=\!\Longrightarrow} S = R$$

Examples 3.2.4, 3.2.5, and 3.2.6 illustrate that solution sets may range anywhere from the empty set to the replacement set itself.

An open sentence whose solution set includes all members of the replacement set for which the expressions comprising the open sentence are defined is called an unconditional open sentence or an identity. An open sentence is conditional if there exists at least one member of the replacement set that renders the open sentence false.

D $\quad a(b + c) = ab + ac$

In+ $\quad a \in R \Longrightarrow (-a) + a = 0$

TLE $\quad a = b = c \Longrightarrow a = c$

D3.2.2 Definition of Solution Set

$\forall$ "for every"

Example 3.2.7   Solve the open equation

$$\frac{x^2 - 1}{x + 1} = x - 1 \tag{7}$$

over **R**.

Solution   We first simplify the left-hand side of (7).

$$\frac{x^2 - 1}{x + 1} = \frac{(x - 1)(x + 1)}{x + 1} \xrightarrow[\substack{x \neq -1}]{\textbf{T2.9.10}} x - 1$$

$$\xrightarrow{\textbf{TLE}} \frac{x^2 - 1}{x + 1} = x - 1$$

for all $x \in \textbf{R}$ *except* $x = -1$

$$\xrightarrow{\textbf{D3.2.2}} S = \{x \mid x \in \textbf{R} \quad \text{but} \quad x \neq -1\}$$

T2.9.10
$$a, b, c \in \textbf{R}, c \neq 0 \Longrightarrow \frac{ac}{bc} = \frac{a}{b}$$

TLE $a = b = c \Longrightarrow a = c$

D3.2.2   Definition of Solution Set

Remark 3.2.5   Notice in the above example how important it is to exclude -1 from the solution set. The left side of (7) is not even defined for $x = -1$, let alone equal to the right side. Nevertheless, (7) does hold for all members of the replacement set for which the expressions comprising the equation *are defined*. Therefore, the open sentence is an identity

The presence of variables in an equation or an inequality easily identifies it as open. Thus, it is not always necessary to qualify it as being open. We do so here simply to emphasize the fact that these are not literal sentences (not always true), but those of a very important and special class called open sentences. We must distinguish between those that are literal equations (usually laws for which the Law of Substitution holds) and those which are open equations. Once a law occurring in the form of an equation has been proven true, or simply defined as being true, it is redundant to solve it. By contrast, the solving of an open sentence is the central issue because, before doing so, it is an *open* question as to whether it is ever valid, and if so, for what values.

The open sentences treated thus far have been quite simple to solve—with very little manipulation we have been able to recognize the solution sets. This, of course, is not typical of those encountered in practice. Often a very extensive and systematic analysis is required. Common in this analysis is the attempt to transform the open sentences into a sequence of increasingly simple forms until, eventually, we are able to clearly identify the solution set. In this process it is critical that each of these supposed simpler forms have the same solution set as the original, so that to solve any one of them is to solve the original. For this reason we introduce the following definition.

DEFINITION 3.2.3 Two open sentences are equivalent over a given replacement set if their solution sets are equal.

Example 3.2.8 Find the solution set of the open equation $\dfrac{x^2 + x - 6}{x + 3} = 0$.

Solution

MLE $\quad a = b \Longrightarrow ac = bc$

T2.9.10 $\quad \dfrac{ac}{bc} = \dfrac{a}{b}$ if $b, c \neq 0$

T2.9.6 $\quad a \in \mathbf{R} \Longrightarrow a \cdot 0 = 0$

D $\quad a(b + c) = ab + ac$

T2.9.7
$ab = 0 \Longrightarrow a = 0$ or $b = 0$

T2.9.5
$a + b = 0 \Longrightarrow a = -b$ and $b = -a$

$$\frac{x^2 + x - 6}{x + 3} = 0$$

$\xRightarrow{\text{MLE}} \dfrac{x^2 + x - 6}{x + 3}(x + 3) = 0 \cdot (x + 3)$

$\xRightarrow[\text{T2.9.6}]{\text{T2.9.10}} x^2 + x - 6 = 0$

$\xLeftrightarrow{\text{D}} (x - 2)(x + 3) = 0$

$\xRightarrow{\text{T2.9.7}} x - 2 = 0$ or $x + 3 = 0$

$\xRightarrow{\text{T2.9.5}} x = 2$ or $x = -3$

One might be led to believe from the foregoing that the solution set is composed of the elements 2 and -3. However, upon substituting -3 for the variable x in the *original* equation, you will find the denominator reduces to zero. Thus for this value the quotient does not even exist, let alone equal zero, as required. Where did we go wrong? The trouble is that all implications in the above analysis were not reversible; we did not maintain a sequence of equivalent equations. Notice the very first one where the Multiplication Law of Equality is applied. The reverse implication, the Factor Cancellation Law of Multiplication, only holds if $x + 3 \neq 0$, that is, $x \neq -3$. This fact is also overlooked in the next step where the Cancellation Law of Fractions is applied. Theorem 2.9.10 specifically indicates that the cancelled factor $x + 3$ cannot be zero; hence, again we should have $x \neq -3$. Thus, the two equations

$$\frac{x^2 + x - 6}{x + 3} = 0 \quad \text{and} \quad x^2 + x - 6 = 0$$

are *not* equivalent. Had we maintained equivalence by writing

$$\frac{x^2 + x - 6}{x + 3} = 0 \Longleftrightarrow x^2 + x - 6 = 0 \quad \text{and} \quad x + 3 \neq 0$$

we would not have been tempted to allow x to be -3.

False roots, such as -3 in Example 3.2.8, that are introduced by incorrect manipulation are called extraneous roots, although they are really not roots at all.

Example 3.2.9 Find the solution set of the open proposition $(x-1)(x+2) < 0$.

Solution

$$(x-1)(x+2) < 0$$

$$\overset{\text{O-2}}{\underset{\text{T2.11.1}}{\Longleftrightarrow}} \begin{cases} x-1 < 0 \quad \text{and} \quad x+2 > 0 \\ \text{or} \\ x-1 > 0 \quad \text{and} \quad x+2 < 0 \end{cases}$$

$$\Longleftrightarrow \begin{cases} x < 1 \quad \text{and} \quad x > -2 \\ \text{or} \\ x > 1 \quad \text{and} \quad x < -2 \end{cases}$$

$$\Longleftrightarrow -2 < x < 1$$

$$\Longrightarrow S = \{x \mid -2 < x < 1\}$$

O-2
$a > 0 \quad \text{and} \quad b > 0 \Longrightarrow ab > 0$

T2.11.1
$a > 0 \quad \text{and} \quad b < 0 \Longrightarrow ab < 0$

Remark 3.2.6 Notice the second to last of the equivalent forms above. Of the two alternatives only the first contributed members to the solution set. The second alternative, $x > 1$ and $x < -2$, admits no solution; there are no real numbers greater than 1 and *simultaneously* less than -2. This can be even more easily visualized by a graphical characterization. The first alternative may be represented by Fig. 3.2.1 and the second alternative by Fig. 3.2.2. Clearly, the only contribution to the solution set will be from the first.

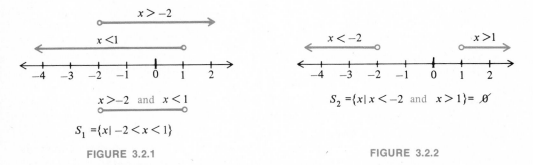

FIGURE 3.2.1   FIGURE 3.2.2

As we have observed in Examples 3.2.8 and 3.2.9, the process of solving open sentences usually involves converting a given proposition into a sequence of simpler yet equivalent ones until the solution set can be clearly identified. These examples illustrate the importance of maintaining equivalence, for when we do not, we may either lose solutions or gain extraneous solutions. Most of the information needed to maintain equivalence is contained within the field and order axioms and those properties (theorems) that follow from them. However, most of these axioms and properties were presented in Chapter 2 as one way implications, not necessarily equivalences. We can, notwithstanding, combine several of them forming equivalent relationships. Because these would be very helpful to us in solving open sentences, we shall state them as theorems for easy reference.

THEOREM 3.2.1 Let a, b, c represent real numbers or real expressions. Then

I $[a = b] \Longleftrightarrow [a + c = b + c]$

II $[a = b \quad \text{and} \quad c \neq 0] \Longleftrightarrow [ac = bc \quad \text{and} \quad c \neq 0]$

III $[a + b = c] \Longleftrightarrow [a = c - b \quad \text{or} \quad b = c - a]$

IV $[ab = 0] \Longleftrightarrow [a = 0 \quad \text{or} \quad b = 0]$

V $\left[\dfrac{a}{b} = \dfrac{c}{d} \quad \text{and} \quad b, d \neq 0\right] \Longleftrightarrow [ad = bc \quad \text{and} \quad b, d \neq 0]$

THEOREM 3.2.2 Let a, b, c represent real numbers or real expressions. Then

I $[a \leq b] \Longleftrightarrow [a + c \leq b + c]$

II $[a \leq b \quad \text{and} \quad c > 0] \Longleftrightarrow [ac \leq bc \quad \text{and} \quad c > 0]$

III $[a \leq b \quad \text{and} \quad c < 0] \Longleftrightarrow [ac \geq bc \quad \text{and} \quad c < 0]$

IV $\left[ab > 0 \quad \text{or} \quad \dfrac{a}{b} > 0\right] \Longleftrightarrow \begin{cases} a > 0 \quad \text{and} \quad b > 0 \\ \text{or} \\ a < 0 \quad \text{and} \quad b < 0 \end{cases}$

V $\left[ab < 0 \quad \text{or} \quad \dfrac{a}{b} < 0\right] \Longleftrightarrow \begin{cases} a < 0 \quad \text{and} \quad b > 0 \\ \text{or} \\ a > 0 \quad \text{and} \quad b < 0 \end{cases}$

The proof of each of the above equivalences follows immediately from the theorems of an ordered field presented in Chapter 2. For example, the implications from left to right of Theorem 3.2.1 Part I is Theorem 2.9.1, whereas the implication from right to left is Theorem 2.9.3, the two together forming the equivalence.

As noted earlier, it is important to understand that the letters a, b, c in the above equivalences may represent either real numbers or *real expressions*. By this we mean a may represent numbers such as 2, $-\frac{3}{5}$, and $\sqrt{7}$, or *expressions* such as $2x$, $(x - 4)$, and $(-3x + 2)/(7 - 2x + x^2)$, where the variable x varies over the real numbers only, thus preserving the realness of each expression and the validity of the above equivalences when containing such expressions.

Example 3.2.10 The equation $(x - 3)(x + 4) = 0$ has the form $ab = 0$ where $a = (x - 3)$ and $b = (x + 4)$. Thus, knowing that $[ab = 0 \Longleftrightarrow a = 0 \quad \text{or} \quad b = 0]$, we may conclude

$$[(x - 3)(x + 4) = 0] \Longleftrightarrow [x - 3 = 0 \quad \text{or} \quad x + 4 = 0]$$

Next we recall that $[a + b = 0 \Longleftrightarrow a = -b]$ and write

$$[x - 3 = 0 \quad \text{or} \quad x + 4 = 0] \Longleftrightarrow [x = 3 \quad \text{or} \quad x = -4]$$

At this point the solution set of the equation $(x - 3)(x + 4) = 0$ *is clearly* $\{3, -4\}$.

Before returning to the applied problems with which we began this chapter, let us review the steps one usually takes in solving applied or verbal problems.

Step 1 Make sure you understand the information given in the problem and what is to be discovered in solving it.

Step 2 If possible, draw an illustration describing the physical elements of the problem and then translate the given information into the language of mathematics using this illustration as a guide. It is usually essential in this step to discover one or more *equations* (or inequalities) that describe or satisfy conditions given in the problem and that contain *variables* (the *unknowns* to be discovered). These become the key elements of the mathematical model.

Step 3 Apply arithmetic and algebraic techniques to the equations, solving, if possible, for the unknowns. Find the solution set of the equations.

Step 4 Interpret correctly the elements of the solution set in light of the given information and the nature of the problem.

Now let us turn to our two applied problems:

Problem 3.1.1 A large reservoir has three separate spillways. Let us name them A, B, and C. It is found that when A alone is open the reservoir will empty in 8 days, whereas if all three are open it only takes 2 days. It is known that spillway C has only half the capacity of B. Suppose it becomes necessary to drain the reservoir and spillways A and C are inoperative. How long will it take with only B open?

Solution Let us follow steps 1 through 4 as suggested above.

Step 1 In this case the given information and the desired solution are clear.

Step 2 First let us draw a picture (Fig. 3.2.3) of the physical situation and then identify on the picture as much of the given information as possible in the language of mathematics. We let x represent the number of days required for B alone to empty the

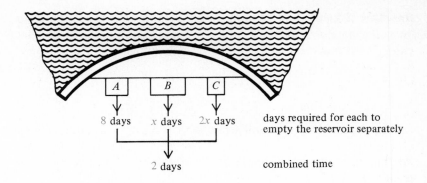

days required for each to empty the reservoir separately

combined time

FIGURE 3.2.3

reservoir. A fundamental principle of time and *rate* per unit time will be helpful to us in setting up an equation. If work is being performed at a constant rate and it takes x days to complete a job, then the rate at which work is being performed per day is $1/x$. For example, if it takes 12 days to complete a job, then $\frac{1}{12}$ of it is done in 1 day; that is, the job is being completed at the *rate* of $\frac{1}{12}$ per day.

Applying this principle to our problem above we find that A drains the reservoir at the rate of $\frac{1}{8}$ per day, B at the rate of $1/x$ per day, and C at the rate of $1/2x$ per day. Thus, the combined rate of all three draining simultaneously would be the *sum* of their individual rates, that is,

$$\frac{1}{8} + \frac{1}{x} + \frac{1}{2x}$$

But we happen to know that together they can empty the reservoir in 2 days, which is a rate of $\frac{1}{2}$ per day. Thus the *equation* we seek is

$$\frac{1}{8} + \frac{1}{x} + \frac{1}{2x} = \frac{1}{2}$$

and the most essential element in the mathematical model has been identified.

Step 3   Solve the equation. We proceed as follows:

T3.2.1(III) $a + b = c \Longleftrightarrow b = c - a$

T2.9.13 $a/b + c/d = (ad + bc)/bd$

T3.2.1(V)
$a/b = c/d \Longleftrightarrow ad = bc,\ b, d \neq 0$

T3.2.1(II)
$[ac = bc\ \text{ and }\ c \neq 0] \Longleftrightarrow [a = b]$

$$\frac{1}{8} + \frac{1}{x} + \frac{1}{2x} = \frac{1}{2} \xrightarrow{\text{T3.2.1(III)}} \frac{1}{x} + \frac{1}{2x} = \frac{1}{2} - \frac{1}{8} \xrightarrow{\text{T2.9.13}} \frac{3}{2x} = \frac{3}{8}$$

$$\xrightarrow{\text{T3.2.1(V)}} 6x = 24 \xrightarrow{\text{T3.2.1(II)}} x = 4$$

Step 4   Thus it will take 4 days to empty the reservoir using spillway B only.

Let us now consider the second of our applied problems.

Problem 3.1.2 The owner of an apartment complex comprising 30 units has discovered he is able to keep all units occupied if he charges $200.00 rent per month for each apartment. Furthermore, he has found that for each $10.00 increase in monthly rent the number of units occupied decreases by 1. Because the service and maintenance costs of an unoccupied unit are significantly less than one being rented, he suspects he can increase his profit margin by raising the monthly rent and still collect the same amount as when all units were occupied renting at $200.00 per month. What should he charge per month to accomplish this, if it is possible, and how many units will be left vacant because of the higher cost?

Solution   Here, too, we will follow the four steps suggested above.

   Step 1   We must understand here that though fewer apartments may be rented, the higher rent coming from the occupied ones may compensate for the loss resulting from the vacant apartments. Our problem is to find out if this is possible under the given conditions, and if so, to determine how much additional rent should be charged to thus come out even on the total rent received.

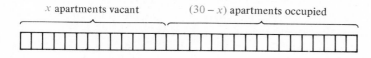

x apartments vacant $(30 - x)$ apartments occupied

FIGURE 3.2.4

   Step 2   We draw a picture (Fig. 3.2.4) illustrating the situation and translate into the language of mathematics as much of the given information as possible. We will let x represent the number of *vacant* apartments and y the *increase* in rent to be charged.

(1) $(200 + y)$ is the new rent charged per unit per month

(2) $(200 + y)(30 - x)$ is the total rent received from the occupied units.

(3) $(200)(30)$ is the total rent received when renting all units at $200.00 per unit per month.

(4) The problem states that the total rent received under the new and old programs is the *same*. We may thus set (2) and (3) *equal* thereby obtaining an equation. We write

$$(200 + y)(30 - x) = (200)(30)$$

(5) We are given that each vacant apartment represents an increase in rent of $10.00. Thus the total increase, which we are calling y, is given by $10x$. Thus we have another equation, namely,

$$y = 10x$$

   Step 3   From entries (4) and (5) above we have two equations with two unknowns. When the unknowns (or variables) of two equations are the *same unknowns,* then the two equations are said to be simultaneous equations. When solving a system of simultaneous equations the Law of Substitution does hold within the system. To solve the system we proceed as follows:

$$[(200 + y)(30 - x) = 6000 \quad \text{and} \quad y = 10x]$$

$$\xLeftrightarrow{\text{SubLE}} (200 + 10x)(30 - x) = 6000$$

$$\xLeftrightarrow{\text{D}} 6000 - 200x + 300x - 10x^2 = 6000 \Longleftrightarrow 100x - 10x^2 = 0$$

$$\xLeftrightarrow{\text{D}} 10x(10 - x) = 0 \xLeftrightarrow{\text{T3.2.1(IV)}} 10x = 0 \quad \text{or} \quad 10 - x = 0$$

$$\xLeftrightarrow[\text{T3.2.1(III)}]{\text{T3.2.1(IV)}} x = 0 \quad \text{or} \quad x = 10 \Longrightarrow S = \{0, 10\}$$

SubLE
Substitution Law of Equality

D $a(b + c) = ab + ac$

T3.2.1(IV)
$ab = 0 \Longleftrightarrow a = 0$ or $b = 0$

T3.2.1(III) $a + b = 0 \Longleftrightarrow a = -b$

Step 4   The solution 0 represents, of course, the original situation with no vacancies and no increase in rent. However, the solution 10 indicates that if we increase the rent by 10 times $10.00, that is, $100.00, we will still receive $6000.00 total rent per month from the occupied apartments even though 10 are vacant. This, you will recall, will give a greater overall margin of profit because the maintenance and service costs of the vacant units are less than for those that are occupied.

3.3   Problem Sets

Problem Set I   ■   Reading Comprehension

Determine whether problems 1 through 5 are true or false and give reasons for your responses.

1   A variable is a fixed, though unspecified, number.
2   The set over which a variable may vary is called the replacement set.
3   A solution or root of an open sentence is a specific number that, when substituted for the variable, reduces the open sentence to a true statement.
4   The solution set of an open sentence must contain at least one root (one element).
5   Certain kinds of roots are called extraneous roots.

Problem Set II   ■   Skills Development

Solve the following equations over the field R of real numbers.

1   $x - 3 = 4$ 2 $\dfrac{x}{2} = -3$ 3 $5x - 3 = 8$

4   $\dfrac{x}{2} - 4 = \dfrac{2x}{3}$ 5 $4(x + 3) = 3(2 - 3x) + 6x - 3$

6   $-2(x + 3) = 4(x + 3)$

Solve the following inequalities over the field R of real numbers.

7 $2x - 3 < 3x + 4$

8 $\dfrac{-x + 1}{x - 2} > 0$

9 $\dfrac{x - 2}{x + 3} < 0$

10 $x^2 - 7x + 12 > 0$

11 $-3x + 4 < x - 8$

12 $x^2 + 3x - 10 < 0$

Problem Set III ■ Theoretical Developments

Prove each of the following, justifying each step by citing the appropriate definition, axiom, or theorem.

1 Theorem 3.2.1 Part II
2 Theorem 3.2.2 Part III
3 Theorem 3.2.2 Part IV

Problem Set IV ■ Applications

1 A gasoline distributor has two pumps. The main pump can fill the tank of a delivery truck in 30 minutes, whereas the second, a smaller auxiliary unit, requires 45 minutes to fill the same truck. How long would it take if both pumps are used simultaneously?

2 Three earth-moving tractors are digging the basement for the construction of a large industrial building. It has been determined that they can complete the job in 30 days. The earth-moving capacity of the three tractors is not the same. The capacities of two of the tractors are double and three-fifths that of the third. Just before beginning work on the project, the tractor with the smallest capacity breaks down and so is withdrawn from the job. How long will it take the other two tractors to complete the work of digging the basement?

3 A builder can complete a job in 5 hours 15 minutes, whereas another can do the same job in 3 hours 30 minutes. How long would it take if they work together?

4 A businessman wishes to invest $20,000.00 in stocks in two different manufacturing companies. The stock of one of the companies is producing a profit of 8 percent annually, whereas the other produces only $5\frac{1}{2}$ percent annually. How much should he invest in each to realize a profit at year's end of $1400.00?

5 How many quarts of coolant must be drained out of a full 20-quart radiator and be replaced by antifreeze to increase the ratio of antifreeze to water from 1 part in 10 to 1 part in 3?

6 A boat, whose motor is running at a constant r.p.m., is able to travel upstream 50 miles in 8 hours. The return trip downstream takes only 5 hours. In still water what would be the speed of the boat (in miles per hour), and what is the rate of the stream's current?

Linear and Quadratic Open Sentences in One Variable 3.4

An algebraic expression in one variable of the form

$ax + b$

where a and b are constants, $a \neq 0$, is called linear. Thus, an open equation of the form

$ax + b = 0$

(or of one equivalent to this), is called a linear equation in one variable.

It is a simple matter to solve a linear equation. We have

$$ax + b = 0 \iff ax = -b \iff x = -\frac{b}{a} \implies S = \left\{-\frac{b}{a}\right\}$$

Example 3.4.1 Solve the equation $-2x + 6 = 0$ over $\mathbf{R}$.

Solution

$$-2x + 6 = 0 \iff x = \frac{-6}{-2} = 3 \implies S = \{3\}$$

Open inequalities of the form

$$ax + b < 0 \quad \text{or} \quad ax + b > 0$$

(or of forms equivalent to these), are called linear inequalities. These are also simple to solve.

Example 3.4.2 Solve the inequality $4x - 12 < 0$ over $\mathbf{R}$.

Solution

T3.2.2(I)
$[a \leq b] \iff [a + c \leq b + c]$

T3.2.2(II)
$[a \leq b \quad \text{and} \quad c > 0] \iff$
$\quad\quad [ac \leq bc \quad \text{and} \quad c > 0]$

$$4x - 12 < 0 \xleftrightarrow{\text{T3.2.2(I)}} 4x < 12$$

$$\xleftrightarrow{\text{T3.2.2(II)}} x < \frac{12}{4} \iff x < 3$$

$$\implies S = \{x \mid x < 3\}$$

The graph of S is represented in Fig. 3.4.1.

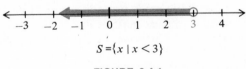

$$S = \{x \mid x < 3\}$$

FIGURE 3.4.1

An algebraic expression in one variable of the form

$$ax^2 + bx + c$$

where a, b, c are constants, $a \neq 0$, is called quadratic. Thus, an open equation of the form

$$ax^2 + bx + c = 0$$

(or of one equivalent to this), is called a quadratic equation in one variable.

There are two methods commonly used to solve a quadratic equation; one is by factoring and the other is by formula. We shall treat the factoring method first and then develop the formula.

To solve a quadratic equation *by factoring* you must first express

$$ax^2 + bx + c$$

as the *product* of two linear factors; that is, find constants d, e, f, g such that

$$ax^2 + bx + c = (dx + e)(fx + g)$$

Because you have studied factoring techniques earlier, we will not go into an extensive review here. However, if you feel a need to review, turn to Appendix A.

Example 3.4.3   Solve the quadratic equation $x^2 - x - 12 = 0$ over **R**.

Solution

$$x^2 - x - 12 = 0 \overset{\text{D}}{\Longleftrightarrow} (x - 4)(x + 3) = 0$$

$$\overset{\text{T3.2.1(IV)}}{\Longleftrightarrow} x - 4 = 0 \quad \text{or} \quad x + 3 = 0$$

$$\overset{\text{T3.2.1(III)}}{\Longleftrightarrow} x = 4 \quad \text{or} \quad x = -3$$

$$\Longrightarrow S \ \{4, -3\}$$

D $a(b + c) = ab + ac$

T3.2.1(IV)
$[ab = 0] \Longleftrightarrow [a = 0 \ \text{or} \ b = 0]$

T3.2.1(III)
$[a + b = c] \Longleftrightarrow$
$\quad [a = c - b \ \text{or} \ b = c - a]$

Example 3.4.4   Solve the quadratic equation $6x^2 - 13x + 6 = 0$ over **R**.

Solution

$$6x^2 - 13x + 6 = 0 \Longleftrightarrow (3x - 2)(2x - 3) = 0$$

$$\Longleftrightarrow 3x - 2 = 0 \quad \text{or} \quad 2x - 3 = 0$$

$$\Longleftrightarrow x = \tfrac{2}{3} \quad \text{or} \quad x = \tfrac{3}{2}$$

$$\Longrightarrow S = \{\tfrac{2}{3}, \tfrac{3}{2}\}$$

Example 3.4.5   Solve the quadratic equation $x^2 - 6x + 9 = 0$ over **R**.

Solution

$$x^2 - 6x + 9 = 0 \Longleftrightarrow (x - 3)(x - 3) = 0$$

$$\Longleftrightarrow x - 3 = 0 \quad \text{or} \quad x - 3 = 0$$

$$\Longleftrightarrow x = 3$$

$$\Longrightarrow S = \{3\}$$

This method of factoring is usually applied when the constant coefficients in the quadratic expression and in each of the linear factors are *integers*.

We next derive a formula that will yield the solutions of any quadratic equation. This will be accomplished by applying a method called completing the square. Because the method of completing the square is a popular technique applied in a variety of problem-solving situations, you will want to learn the steps involved. We proceed as follows:

D $a(b+c) = ab + ac$

$$ax^2 + bx + c \underset{\text{D}}{=\!=\!=} a\left[x^2 + \frac{b}{a}x + \frac{c}{a}\right]$$

In+ $a + (-a) = 0$

Id+ $a + 0 = a$

$$\underset{\text{Id+}}{\overset{\text{In+}}{=\!=\!=}} a\left[x^2 + \frac{b}{a}x + \left(\frac{b}{2a}\right)^2 - \left(\frac{b}{2a}\right)^2 + \frac{c}{a}\right]$$

$$= a\left[\left(x + \frac{b}{2a}\right)^2 - \frac{b^2}{4a^2} + \frac{c}{a}\right]$$

$$= a\left[\left(x + \frac{b}{2a}\right)^2 - \frac{b^2 - 4ac}{4a^2}\right]$$

Thus,

$$ax^2 + bx + c = 0 \iff a\left[\left(x + \frac{b}{2a}\right)^2 - \frac{b^2 - 4ac}{4a^2}\right] = 0$$

T3.2.1(IV)

$ab = 0 \iff a = 0$ or $b = 0$

$$\underset{a \neq 0}{\overset{\text{T3.2.1(IV)}}{\Longleftarrow\!=\!\Longrightarrow}} \left(x + \frac{b}{2a}\right)^2 - \frac{b^2 - 4ac}{4a^2} = 0$$

$$\iff \left(x + \frac{b}{2a}\right)^2 = \frac{b^2 - 4ac}{4a^2}$$

T3.8.1

$a^2 = b \iff a = \pm\sqrt{b}$ (proved on page 117)

$$\overset{\text{T3.8.1}}{\Longleftarrow\!=\!\Longrightarrow} x + \frac{b}{2a} = \pm\sqrt{\frac{b^2 - 4ac}{4a^2}}$$

$$\iff x = \frac{-b}{2a} \pm \frac{\sqrt{b^2 - 4ac}}{2a} \bullet$$

● It will be shown in Section 3.8, Theorem 3.8.2, that the principle square root of a^2, denoted $\sqrt{a^2}$, is $|a|$. However, with the plus or minus sign preceding the fraction here, the absolute value may be omitted.

$$\iff x = \frac{-b \pm \sqrt{b^2 - 4ac}}{2a}$$

This last equation in the above sequence of equivalences is called the quadratic formula. With this formula, solutions of the quadratic equation, if they exist, can always be identified. Moreover, whether solutions exist or not can also be determined by this formula. Notice the presence of the principle square root radical $\sqrt{b^2 - 4ac}$. Although a more detailed study of radicals will be presented in Section 3.8, you will probably recall that the radicand, $b^2 - 4ac$, which is called the discriminant, must be greater than or equal to zero for the radical to exist as a real number. That is, in the system of real numbers the square root of a negative number does not exist.

With this information we return to the quadratic formula

$$x = \frac{-b \pm \sqrt{b^2 - 4ac}}{2a}$$

and note that the roots may be characterized as follows:

real and unequal when $b^2 - 4ac > 0$
real and equal when $b^2 - 4ac = 0$
no real solutions exist when $b^2 - 4ac < 0$

The results of the preceding discussion regarding the quadratic equation are significant enough to be stated formally as a theorem.

THEOREM 3.4.1 The quadratic equation

$$ax^2 + bx + c = 0$$

where $a, b,$ and c are constants and $a \neq 0$, is equivalent to the equation

$$x = \frac{-b \pm \sqrt{b^2 - 4ac}}{2a}$$

If the constants a, b, c are real and the replacement set is the field *R* of real numbers, then the roots of the quadratic equation are

real and unequal if $b^2 - 4ac > 0$
real and equal if $b^2 - 4ac = 0$
no real solutions exist if $b^2 - 4ac < 0$

Example 3.4.6 Solve the equation $6x^2 + x - 2 = 0$ over the field *R* of real numbers.

Solution

$$6x^2 + 1 \cdot x - 2 = 0 \xLeftrightarrow{\textbf{T3.4.1}} x = \frac{-1 \pm \sqrt{(1)^2 - 4(6)(-2)}}{2 \cdot 6}$$

$$= \frac{-1 \pm \sqrt{49}}{12} = \frac{-1 \pm 7}{12}$$

$$\Longleftrightarrow x = \frac{1}{2} \quad \text{or} \quad x = -\frac{2}{3}$$

$$\Longrightarrow S = \left\{ \frac{1}{2}, -\frac{2}{3} \right\}$$

(two real roots: $b^2 - 4ac = 49$, greater than zero)

Example 3.4.7   Solve the equation $9x^2 + 6x + 1 = 0$ over $\boldsymbol{R}$.

Solution

T3.4.1
$ax^2 + bx + c = 0 \Longleftrightarrow$
$$x = \frac{-b \pm \sqrt{b^2 - 4ac}}{2a}$$

$$9x^2 + 6x + 1 = 0 \xrightarrow{\textbf{T3.4.1}} x = \frac{-6 \pm \sqrt{(6)^2 - 4(9)(1)}}{2 \cdot 9}$$

$$= \frac{-6 \pm \sqrt{0}}{18} = -\frac{1}{3}$$

$$\Longleftrightarrow x = -\frac{1}{3}$$

$$\Longrightarrow S = \left\{-\frac{1}{3}\right\}$$

(one real root: $b^2 - 4ac = 0$)

Example 3.4.8   Solve the equation $x^2 - 2x + 2 = 0$ over $\boldsymbol{R}$.

Solution

$$1 \cdot x^2 - 2x + 2 = 0 \xrightarrow{\textbf{T3.4.1}} x = \frac{-(-2) \pm \sqrt{(-2)^2 - 4(1)(2)}}{2 \cdot 1}$$

$$= \frac{2 \pm \sqrt{-4}}{2}$$

Because $\sqrt{-4}$ does not exist as a real number, we have

$$S = \varnothing \qquad \text{(no real roots: } b^2 - 4ac = -4)$$

Let us now apply our techniques to solving certain inequalities.

Example 3.4.9   Solve the inequality $\dfrac{x-2}{x+3} > 0$ over $\boldsymbol{R}$.

Solution

T3.2.2(IV)
$$\frac{a}{b} > 0 \Longleftrightarrow \begin{cases} a > 0 \quad \text{and} \quad b > 0 \\ \text{or} \\ a < 0 \quad \text{and} \quad b < 0 \end{cases}$$

$$\frac{x-2}{x+3} > 0 \xrightarrow{\textbf{T3.2.2(IV)}} \begin{cases} x - 2 > 0 \quad \text{and} \quad x + 3 > 0 \\ \text{or} \\ x - 2 < 0 \quad \text{and} \quad x + 3 < 0 \end{cases}$$

$$\Longleftrightarrow \begin{cases} x > 2 \quad \text{and} \quad x > -3 \qquad \text{(Fig. 3.4.2)} \\ \text{or} \\ x < 2 \quad \text{and} \quad x < -3 \qquad \text{(Fig. 3.4.3)} \end{cases}$$

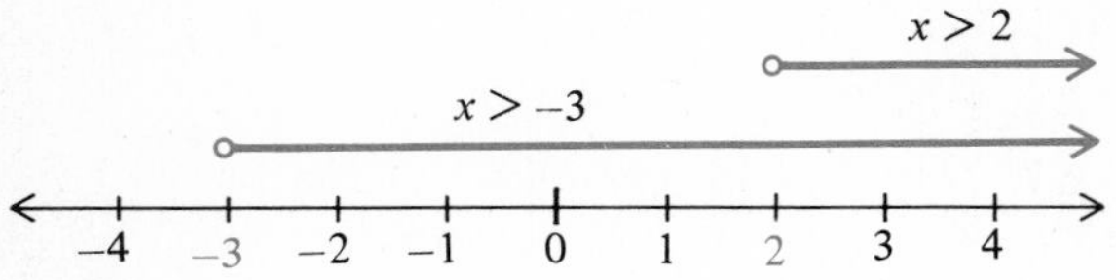

FIGURE 3.4.2

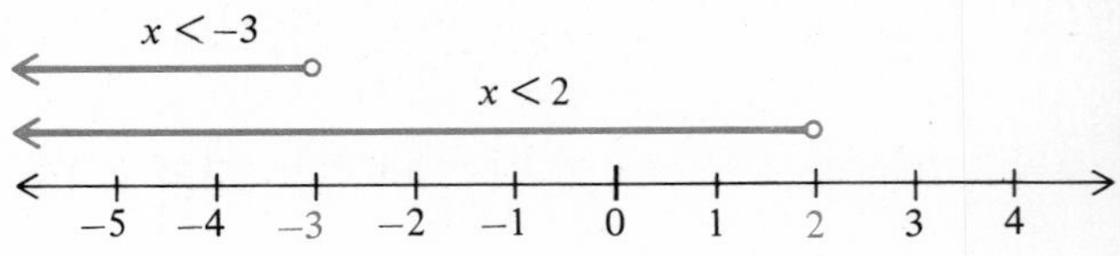

FIGURE 3.4.3

$$\xrightarrow{\text{and}} \begin{cases} x > 2 \\ \text{or} \\ x < -3 \end{cases}$$

$$\Longrightarrow S = \{x \mid x > 2 \quad \text{or} \quad x < -3\}$$

Example 3.4.10 Solve the inequality $x^2 - 6x + 8 < 0$ over R.

Solution

$$x^2 - 6x + 8 < 0 \Longleftrightarrow (x-2)(x-4) < 0$$

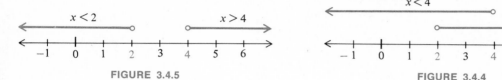

$$\xLeftrightarrow{\text{T3.2.2(V)}} \begin{cases} x - 2 > 0 \quad \text{and} \quad x - 4 < 0 \\ \text{or} \\ x - 2 < 0 \quad \text{and} \quad x - 4 > 0 \end{cases}$$

T3.2.2(V)
$$ab < 0 \Longleftrightarrow \begin{cases} a > 0 \quad \text{and} \quad b < 0 \\ \text{or} \\ a < 0 \quad \text{and} \quad b > 0 \end{cases}$$

$$\Longleftrightarrow \begin{cases} x > 2 \quad \text{and} \quad x < 4 \quad \text{(Fig. 3.4.4)} \\ \text{or} \\ x < 2 \quad \text{and} \quad x > 4 \quad \text{(Fig. 3.4.5)} \end{cases}$$

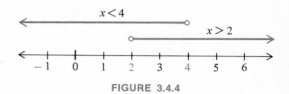

FIGURE 3.4.5

FIGURE 3.4.4

$$\xLeftrightarrow[\text{or}]{\text{and}} [x > 2 \quad \text{and} \quad x < 4]$$

$$\Longrightarrow S = \{x \mid 2 < x < 4\}$$

Sample Problem Set • 3.5

Find the solution set of each of the following open sentences:

1 $-2x + 4 = 8$

2 $-2x + 4 < 8$

3 $x^2 - x - 12 = 0$

4 $x^2 - x - 12 > 0$

5 $6x^2 + x - 2 = 0$

6 $x^2 + 2x + 3 = 0$

● Answers to Sample Problem Set questions are given at the end of the chapter.

Open Sentences in One Variable Containing Absolute Value   3.6

First we restate the definition of absolute value.

DEFINITION 3.6.1 If $a \in R$, then the absolute value of a, denoted $|a|$, is given by

$$|a| = \begin{cases} a & \text{if } a \geq 0 \\ -a & \text{if } a < 0 \end{cases}$$

The following theorem will be helpful to us in solving algebraic equations containing absolute value.

THEOREM 3.6.1 If $a, b \in R$, then

$$|a| = b \Longleftrightarrow \begin{cases} b = a & \text{if } a \geq 0 \\ \text{or} \\ b = -a & \text{if } a < 0 \end{cases}$$

Proof Follows immediately from Definition 3.6.1.

Example 3.6.1 Solve the equation $|x - 4| = 3x + 2$ over the field R of real numbers.

Solution $|x - 4| = 3x + 2$

T3.6.1

$$|a| = b \Longleftrightarrow \begin{cases} b = a & \text{if } a \geq 0 \\ \text{or} \\ b = -a & \text{if } a < 0 \end{cases}$$

$$\overset{\textbf{T3.6.1}}{\Longleftrightarrow} \begin{cases} 3x + 2 = x - 4 & \text{if } x - 4 \geq 0 \\ \text{or} \\ 3x + 2 = -(x - 4) & \text{if } x - 4 < 0 \end{cases}$$

$$\Longleftrightarrow \begin{cases} x = -3 & \text{if } x \geq 4 & \text{(Fig. 3.6.1)} \\ \text{or} \\ x = \tfrac{1}{2} & \text{if } x < 4 & \text{(Fig. 3.6.2)} \end{cases}$$

$$\Longleftrightarrow x = \tfrac{1}{2} \Longrightarrow S = \{\tfrac{1}{2}\}$$

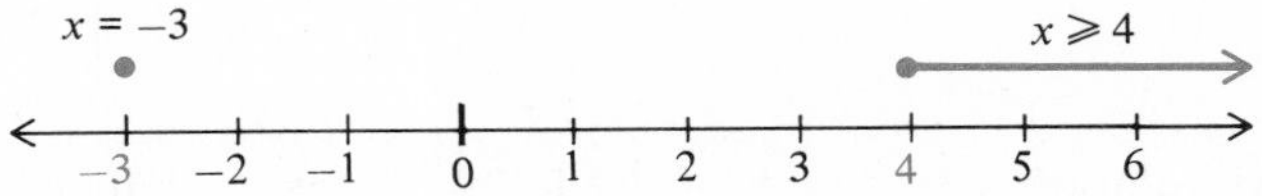

FIGURE 3.6.1

FIGURE 3.6.2

Example 3.6.2 Solve the equation $|x - 2| + |x + 4| = 6$

Solution $|x - 2| + |x + 4| = 6$

D3.6.1 $|a| = \begin{cases} a & \text{if } a \geq 0 \\ -a & \text{if } a < 0 \end{cases}$

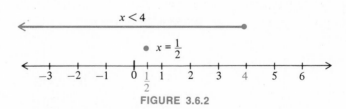

$$\overset{\textbf{D3.6.1}}{\Longleftrightarrow} \begin{cases} (x - 2) + (x + 4) = 6 & \text{if } \begin{cases} x - 2 \geq 0 \\ \text{and} \\ x + 4 \geq 0 \end{cases} \\ \\ -(x - 2) + (x + 4) = 6 & \text{if } \begin{cases} x - 2 < 0 \\ \text{and} \\ x + 4 \geq 0 \end{cases} \end{cases}$$

$$\begin{cases} (x-2)-(x+4)=6 & \text{if } \begin{cases} x-2 \ge 0 \\ \text{and} \\ x+4 < 0 \end{cases} \\ \\ -(x-2)-(x+4)=6 & \text{if } \begin{cases} x-2 < 0 \\ \text{and} \\ x+4 < 0 \end{cases} \end{cases}$$

$$\Longleftrightarrow \begin{cases} x=2 & \text{if } x \ge 2 \text{ and } x \ge -4 \quad \text{(Fig. 3.6.3)} \\ 6=6 & \text{if } x < 2 \text{ and } x \ge -4 \quad \text{(Fig. 3.6.4)} \\ -6=6 & \text{if } x \ge 2 \text{ and } x < -4 \quad \text{(Fig. 3.6.5)} \\ x=-4 & \text{if } x < 2 \text{ and } x < -4 \quad \text{(Fig. 3.6.6)} \end{cases}$$

$$\overset{\text{and}}{\Longleftrightarrow} \begin{cases} x=2 & \text{if } x \ge 2 \\ \text{or} \\ 6=6 & \text{if } x < 2 \text{ and } x \ge -4 \end{cases}$$

$$\Longrightarrow S = \{x \mid -4 \le x \le 2\}$$

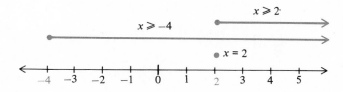

FIGURE 3.6.3

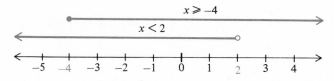

FIGURE 3.6.4

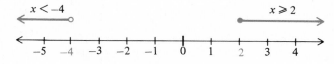

FIGURE 3.6.5

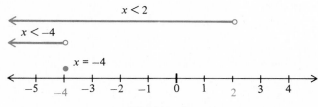

FIGURE 3.6.6

Example 3.6.3 Solve the inequality $|2x + 3| < 5$ over R.

T2.11.3(I) $|a| < b \Longleftrightarrow -b < a < b$

Solution $|2x + 3| < 5 \xrightarrow{\text{T2.11.3(I)}} -5 < 2x + 3 < 5$

$$\Longleftrightarrow -8 < 2x < 2$$

$$\Longleftrightarrow -4 < x < 1$$

$$\Longrightarrow S = \{x \mid -4 < x < 1\}$$

The graph of this solution set is represented in Fig. 3.6.7.

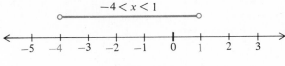

FIGURE 3.6.7

Example 3.6.4 Solve the inequality $\left|\dfrac{2x-4}{3}\right| \geq 2x$ over R.

Solution

T2.11.3(II)
$|a| \geq b \Longleftrightarrow a \geq b$ or $a \leq -b$

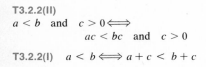

$$\left|\frac{2x-4}{3}\right| \geq 2x \xrightarrow{\text{T2.11.3(II)}} \begin{cases} \dfrac{2x-4}{3} \geq 2x \\ \text{or} \\ \dfrac{2x-4}{3} \leq -2x \end{cases}$$

T3.2.2(II)
$a < b$ and $c > 0 \Longleftrightarrow$
$\qquad ac < bc$ and $c > 0$

T3.2.2(I) $a < b \Longleftrightarrow a + c < b + c$

$$\xrightarrow{\text{T3.2.2(II)}} 2x - 4 \geq 6x \quad \text{or} \quad 2x - 4 \leq -6x$$

$$\xrightarrow{\text{T3.2.2(I)}} -1 \geq x \quad \text{or} \quad x \leq \tfrac{1}{2} \quad \text{(Fig. 3.6.8)}$$

$$\overset{\text{or}}{\Longleftrightarrow} x \leq \tfrac{1}{2} \Longrightarrow S = \{x \mid x \leq \tfrac{1}{2}\}$$

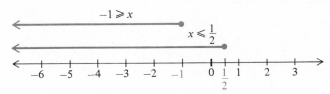

FIGURE 3.6.8

Example 3.6.5 Solve the inequality $|(x - 1)/(x + 2)| < 4$ over the field R. In treating increasingly difficult problems, we often must rely more upon previously defined or developed properties. In this example we shall be particularly careful not to overlook conditions that bear upon the solution.

Solution

T2.11.3(I)
$|a| < b \Longleftrightarrow -b < a < b$

$$(1). \left|\frac{x-1}{x+2}\right| < 4 \xrightarrow{\text{T2.11.3(I)}} -4 < \frac{x-1}{x+2} < 4$$

$$\Longleftrightarrow -4 < \frac{x-1}{x+2} \quad \text{and} \quad \frac{x-1}{x+2} < 4$$

We now treat each of these open inequalities separately in steps (2) and (3) that follow.

$$(2)\quad -4 < \frac{x-1}{x+2} \xLeftrightarrow[\text{(II), (III)}]{\text{T3.2.2}} \begin{cases} -4(x+2) < x-1 & \text{if } x+2 > 0 \\ \text{or} \\ -4(x+2) > x-1 & \text{if } x+2 < 0 \end{cases}$$

$$\Longleftrightarrow \begin{cases} -5x < 7 & \text{if } x > -2 \\ \text{or} \\ -5x > 7 & \text{if } x < -2 \end{cases}$$

$$\Longleftrightarrow \begin{cases} x > -\frac{7}{5} & \text{if } x > -2 \quad \text{(Fig. 3.6.9)} \\ \text{or} \\ x < -\frac{7}{5} & \text{if } x < -2 \quad \text{(Fig. 3.6.10)} \end{cases}$$

$$\Longrightarrow S_1 = \{x \mid x > -\tfrac{7}{5} \text{ or } x < -2\}$$

T3.2.2(II)
$$a < b \text{ and } c > 0 \Longleftrightarrow$$
$$ac < bc \text{ and } c > 0$$

T3.2.2(III)
$$a < b \text{ and } c < 0 \Longleftrightarrow$$
$$ac > bc \text{ and } c < 0$$

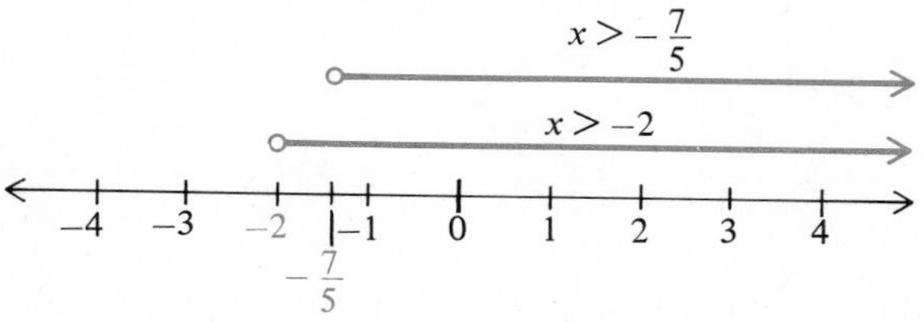

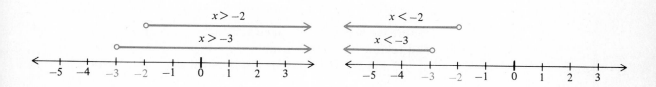

$x > -\frac{7}{5}$
$x > -2$

FIGURE 3.6.9

$x < -\frac{7}{5}$
$x < -2$

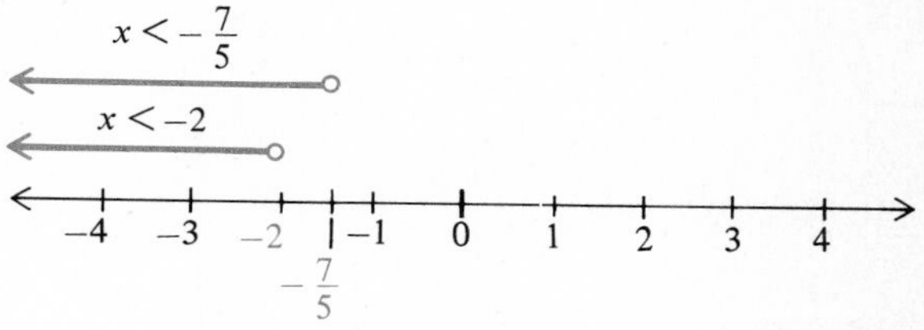

FIGURE 3.6.10

(Study the last equivalence above carefully.)

$$(3)\quad \frac{x-1}{x+2} < 4 \Longleftrightarrow \begin{cases} x-1 < 4(x+2) & \text{if } x+2 > 0 \\ \text{or} \\ x-1 > 4(x+2) & \text{if } x+2 < 0 \end{cases}$$

$$\Longleftrightarrow \begin{cases} -3x < 9 & \text{if } x > -2 \\ \text{or} \\ -3x > 9 & \text{if } x < -2 \end{cases}$$

$$\Longleftrightarrow \begin{cases} x > -3 & \text{if } x > -2 \quad \text{(Fig. 3.6.11)} \\ \text{or} \\ x < -3 & \text{if } x < -2 \quad \text{(Fig. 3.6.12)} \end{cases}$$

$$\Longrightarrow S_2 = \{x \mid x > -2 \text{ or } x < -3\}$$

$x > -2$
$x > -3$

FIGURE 3.6.11

$x < -2$
$x < -3$

FIGURE 3.6.12

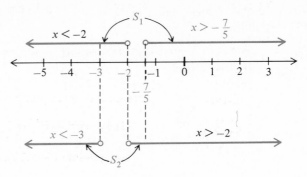

FIGURE 3.6.13

From step (1) we know the solution set of the original inequality is the *intersection* of the solution sets obtained in steps (2) and (3). Thus,

$$S = S_1 \cap S_2$$
$$= \{x \mid x > -\tfrac{7}{5} \quad \text{or} \quad x < -2\} \cap \{x \mid x > -2 \quad \text{or} \quad x < -3\}$$
$$= \{x \mid (x > -\tfrac{7}{5} \quad \text{or} \quad x < -2) \quad \text{and} \quad (x > -2 \quad \text{or} \quad x < -3)\}$$
$$= \{x \mid x < -3 \quad \text{or} \quad x > -\tfrac{7}{5}\}$$

The above steps may be more easily visualized by graphing sets S_1 and S_2 and thereby observing their intersection (see Fig. 3.6.13). Thus, again $S = \{x \mid x < -3 \quad \text{or} \quad x > -\tfrac{7}{5}\}$

3.7   Sample Problem Set

Find the solution set of each of the following open sentences.

1 $|2x - 3| = 5x + 6$ **2** $|x - 2| - |x + 4| = 4$ **3** $|-2x + 4| < 8$

4 $|3x + 2| > x + 4$ **5** $\left|\dfrac{x + 3}{x}\right| < 2$

3.8   Open Sentences in One Variable Containing Radicals

Before we begin solving open sentences containing radicals, we will first review briefly the meaning of a radical.

If $a \in R$ and $a^2 = b$, then a is called a square root of b.

Example 3.8.1 Because $(2)^2 = 4$, 2 is a square root of 4. But $(-2)^2 = 4$; thus, -2 is also a square root of 4.

It can be proven that every positive real number has two square roots — one positive and the other negative. To distinguish the positive square root from the negative square root, the radical symbol, denoted $\sqrt{}$, is employed. When the positive (*principal*) square root of a number, say b, is desired, this is denoted by $\sqrt{b}$. Hence, the positive square root of 4, written $\sqrt{4}$, equals 2. If you wish to identify the negative square root of 4, this can be accomplished by placing the negative symbol $-$ in front of the radical. Thus, the negative square root of 4 is denoted $-\sqrt{4}$ and equals -2. If both roots are desired, this is written $\pm\sqrt{4}$. In summary, we have the following:

DEFINITION 3.8.1 If $a, b \in R$, then

$$a^2 = b \quad \text{and} \quad a \geq 0 \Longleftrightarrow a = \sqrt{b}$$

The number a is called the principal square root of b.

THEOREM 3.8.1 If $a, b \in R$, then

$$a^2 = b \Longleftrightarrow a = \pm\sqrt{b}$$

Proof

(1) $a^2 = b \xLeftrightarrow{\text{T2.9.8(IV)}} (-a)^2 = b$

(2) $a^2 = b \xLeftrightarrow[\text{(1)}]{\text{D3.8.1}} \begin{cases} a = \sqrt{b} & \text{if } a \geq 0 \\ \text{or} \\ -a = \sqrt{b} & \text{if } a < 0 \end{cases}$

$\Longleftrightarrow a = \sqrt{b} \quad \text{or} \quad a = -\sqrt{b}$

$\Longleftrightarrow a = \pm\sqrt{b} \qquad$ QED

T2.9.8(IV) $(-a)(-b) = ab$

D3.8.1
$a^2 = b \quad \text{and} \quad a \geq 0 \Longleftrightarrow a = \sqrt{b}$

Notice the only difference between the left-hand side of Definition 3.8.1 and Theorem 3.8.1 is the absence of a ≥ 0 in the latter; the result of leaving $a \geq 0$ off is that we take both roots.

Remark 3.8.1 Note that b must be greater than or equal to zero in both Definition 3.8.1 and Theorem 3.8.1.● This comes as a result of Theorem 2.11.1(III) that states

$$a \in R \Longrightarrow a^2 \geq 0$$

● T3.8.1 $a^2 = b \Longleftrightarrow a = \pm\sqrt{b}$

That is, in the real number system there does not exist a number whose square is negative. Hence, the radical expression $\sqrt{b}$ only exists as a real number for $b \geq 0$.

THEOREM 3.8.2 If $a \in R$, then

$$\sqrt{a^2} = |a|$$

Proof To be done as an exercise.

Example 3.8.2

D3.6.1 $\quad |a| = \begin{cases} a & \text{if } a \geq 0 \\ -a & \text{if } a < 0 \end{cases}$

(a) $\sqrt{9} = \sqrt{(3)^2} \xrightarrow{\text{T3.8.2}} 3$

(b) $\sqrt{\frac{1}{4}} = \sqrt{(\frac{1}{2})^2} \xrightarrow{\text{T3.8.2}} \frac{1}{2}$

(c) $\sqrt{(-2)^2} \xrightarrow{\text{T3.8.2}} |-2| \xrightarrow{\text{D3.6.1}} 2$

The square root of a real number that is not a perfect square is irrational, and for this reason it is usually left in radical form. Thus, such radicals as $\sqrt{3}$, $\sqrt{6}$, or $\sqrt{10}$ are not simplified further unless a rational approximation is desired.

THEOREM 3.8.3 If $a, b \in R$ and $a, b \geq 0$, then

$$\sqrt{ab} = \sqrt{a} \cdot \sqrt{b}$$

Proof To be done as an exercise.

With the use of Theorem 3.8.3, the square root of some real numbers that are not perfect squares can be simplified, if among their integral factors there exist perfect squares.

Example 3.8.3 $\quad \sqrt{12} = \sqrt{2^2 \cdot 3} = \sqrt{2^2} \cdot \sqrt{3} = 2\sqrt{3}$
and

$$\sqrt{450} = \sqrt{3^2 \cdot 5^2 \cdot 2} = \sqrt{3^2} \cdot \sqrt{5^2} \cdot \sqrt{2} = 3 \cdot 5\sqrt{2} = 15\sqrt{2}$$

Let us now apply our knowledge of radicals to solving certain open sentences containing radicals.

Example 3.8.4 Solve the equation $x - 1 = \sqrt{x + 1}$ over the field *R* of real numbers.

Solution $\quad x - 1 = \sqrt{x + 1}$

$\xLeftrightarrow{\text{D3.8.1}} (x - 1)^2 = x + 1 \quad \text{and} \quad x - 1 \geq 0$

$\Longleftrightarrow x^2 - 2x + 1 = x + 1 \quad \text{and} \quad x \geq 1$

$\Longleftrightarrow x^2 - 3x = 0 \quad \text{and} \quad x \geq 1$

$$\Longleftrightarrow x(x-3) = 0 \quad \text{and} \quad x \geq 1$$

$$\Longleftrightarrow [x = 0 \quad \text{or} \quad x - 3 = 0] \quad \text{and} \quad x \geq 1$$

$$\Longleftrightarrow [x = 0 \quad \text{or} \quad x = 3] \quad \text{and} \quad x \geq 1$$

$$\Longrightarrow S = \{3\}$$

Remark 3.8.2 The above example is another illustrating the importance of maintaining equivalent open sentences when determining a solution set. Had the inequality $x - 1 \geq 0$ been omitted in the first step, as often happens, then one would have been tempted to include 0 in the solution set, which is incorrect.

THEOREM 3.8.4   If $a, b \in R$ and $a, b \geq 0$, then

 I $a \leq b \Longleftrightarrow a^2 \leq b^2$
 II $a \leq b \Longleftrightarrow \sqrt{a} \leq \sqrt{b}$

Proof   Part II (Part I is to be done as an exercise.)

$$\sqrt{a} \leq \sqrt{b} \xleftrightarrow{\textbf{T3.8.4(I)}} (\sqrt{a})^2 \leq (\sqrt{b})^2$$

$$\xleftrightarrow{\textbf{D3.8.1}} a \leq b \quad \text{and} \quad a, b \geq 0 \qquad \text{QED}$$

D3.8.1
$a^2 = b$ and $a \geq 0 \Longleftrightarrow a = \sqrt{b}$

Example 3.8.5   Solve the inequality $\sqrt{2x+3} \leq \sqrt{x+4}$ over R.

Solution $\sqrt{2x+3} \leq \sqrt{x+4}$

$$\xleftrightarrow{\textbf{T3.8.4(II)}} 2x + 3 \leq x + 4 \quad \text{and} \quad \begin{cases} 2x + 3 \geq 0 \\ \text{and} \\ x + 4 \geq 0 \end{cases}$$

$$\Longleftrightarrow x \leq 1 \quad \text{and} \quad \begin{cases} x \geq -\frac{2}{3} \\ \text{and} \\ x \geq -4 \end{cases} \qquad \text{(Fig. 3.8.1)}$$

$$\Longleftrightarrow -\tfrac{2}{3} \leq x \leq 1$$

$$\Longrightarrow S = \{x \mid -\tfrac{2}{3} \leq x \leq 1\}$$

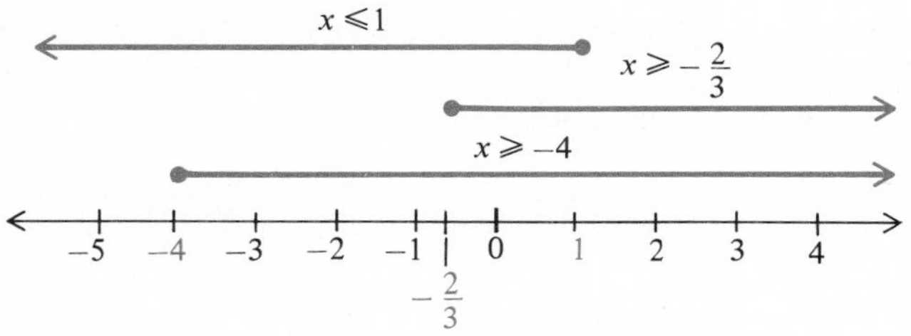

FIGURE 3.8.1

Example 3.8.6 Solve the equation $\sqrt{x+2}+4=x$ over $\boldsymbol{R}$.

Solution　$\sqrt{x+2}+4=x \Longleftrightarrow \sqrt{x+2}=x-4$

D3.8.1
$a^2 = b$ and $a \geq 0 \Longleftrightarrow a = \sqrt{b}$

$$\overset{\text{D3.8.1}}{\Longleftrightarrow} x+2=(x-4)^2 \quad \text{and} \quad x-4 \geq 0$$

$$\Longleftrightarrow x+2=x^2-8x+16 \quad \text{and} \quad x \geq 4$$

$$\Longleftrightarrow x^2-9x+14=0 \quad \text{and} \quad x \geq 4$$

$$\Longleftrightarrow (x-7)(x-2)=0 \quad \text{and} \quad x \geq 4$$

$$\Longleftrightarrow (x=7 \quad \text{or} \quad x=2) \quad \text{and} \quad x \geq 4$$

$$\Longleftrightarrow x=7$$

$$\Longrightarrow S=\{7\}$$

Example 3.8.7 Solve the equation $\sqrt{5+\sqrt{x}}=\sqrt{x}-1$ over $\boldsymbol{R}$.

Solution　$\sqrt{5+\sqrt{x}}=\sqrt{x}-1$

D3.8.1
$a^2 = b$ and $a \geq 0 \Longleftrightarrow a = \sqrt{b}$

$$\overset{\text{D3.8.1}}{\Longleftrightarrow} 5+\sqrt{x}=(\sqrt{x}-1)^2 \quad \text{and} \quad \sqrt{x}-1 \geq 0$$

$$\Longleftrightarrow 5+\sqrt{x}=x-2\sqrt{x}+1 \quad \text{and} \quad \sqrt{x} \geq 1$$

T3.8.4(I)
$a, b \geq 0 \Longrightarrow [a \leq b \Longleftrightarrow a^2 \leq b^2]$

$$\overset{\text{T3.8.4(I)}}{\Longleftrightarrow} 3\sqrt{x}=x-4 \quad \text{and} \quad x \geq 1$$

$$\overset{\text{T3.8.4(I)}}{\Longleftrightarrow} [(3\sqrt{x})^2=(x-4)^2 \quad \text{and} \quad x-4 \geq 0] \quad \text{and} \quad x \geq 1$$

$$\Longleftrightarrow [9x=x^2-8x+16 \quad \text{and} \quad x \geq 4] \quad \text{and} \quad x \geq 1$$

$$\Longleftrightarrow x^2-17x+16=0 \quad \text{and} \quad x \geq 4$$

$$\Longleftrightarrow (x-16)(x-1)=0 \quad \text{and} \quad x \geq 4$$

$$\Longleftrightarrow (x=16 \quad \text{or} \quad x=1) \quad \text{and} \quad x \geq 4$$

$$\Longleftrightarrow x=16 \Longrightarrow S=\{16\}$$

3.9　Problem Sets

Problem Set I　■　*Reading Comprehension*

Determine whether problems 1 through 5 are true or false and give reasons for your responses.

1　A linear open sentence is one that describes a line.
2　$|-a| = a$
3　$\sqrt{4} = \pm 2$
4　The principal square root of a negative number is always negative.
5　$a^2 = b^2 \Longleftrightarrow a = b$

Problem Set II   ■   Skills Development

Solve the following equations over the field R of real numbers.

1   $3x + 4 = x - 5$

2   $5(x - 1) = 2x + 4$

3   $3x + [2x - (4x - 4)] = 2 - \dfrac{x}{3}$

4   $\dfrac{1}{2} + \dfrac{2}{3} = \dfrac{1}{x}$

5   $\dfrac{4}{x - 3} = \dfrac{3}{x - 4}$

6   $\dfrac{2}{x - 1} - \dfrac{3}{x - 2} = \dfrac{4}{x - 3}$

7   $x^2 - x - 6 = 0$

8   $x^2 + 2x - 8 = 0$

9   $6x^2 + 11x - 10 = 0$

10   $12x^2 + 11x - 15 = 0$

11   $x^4 - 12x^2 + 32 = 0$ (HINT: $x^4 = (x^2)^2$)

12   $x^2 + 9 = 0$

13   $x^2 - 3x - 10 = 0$

14   $4x^2 - 12x + 9 = 0$

15   $\sqrt{2}x^2 - 5x + \sqrt{18} = 0$

16   $\sqrt{3}x^2 - 2x - \sqrt{12} = 0$

17   $x^2 - 3x + 12 = 0$

18   $x^2 + 3x + 4 = 0$

19   $\dfrac{x}{x - 1} = \dfrac{2}{x - 4}$

20   $2x = \dfrac{7x - 1}{x + 1}$

21   $|x - 3| = 4$

22   $|2 - 4x| = 6$

23   $|x + 2| + |x - 3| = 5$

24   $|3 - x| - |x + 1| = 7$

25   $\sqrt{x + 3} = 2$

26   $\sqrt{x - 4} = -2$

27   $\sqrt{x - 1} = 3 - x$

28   $\sqrt{4x + 6} = 1 + \sqrt{2x + 7}$

29   $\sqrt{13 + \sqrt{x}} = \sqrt{x} + 1$

30   $(\sqrt{x + 3})(\sqrt{x - 9}) = 8$

31   $(\sqrt{x - 3})(\sqrt{x - 4}) = \sqrt{6}$

Solve the following inequalities over the field R of real numbers and graph the solution set.

32   $4 - 5x > 2x - 10$

33   $x^2 + 5x - 6 < 0$

34   $\dfrac{4 - x}{x + 1} > 0$

35   $|2x + 3| \leq 5$

36   $|2 - 5x| < 8$

37   $|-2x + 3| > 7$

38   $|2x^2 - 4x - 5| > -3$

39   $\dfrac{x + 3}{2} \leq \dfrac{2 - x}{-2}$

40   $(x - 1)(x + 2)(x - 3) < 0$

Problem Set III   ■   Applications

1   A rectangular plot of ground is 20 feet wide and 30 feet long. Across one of the shorter ends it is necessary to put in a 4-foot walk. How much must the shorter dimension be increased in order to maintain the original area?

2   A man wishes to borrow a total of $6000.00 from two different sources. One charges 4 percent interest, whereas the other charges only 3 percent. If he wishes the total interest from both loans to be $225.00, determine how much he should borrow from each of the two sources.

3   A man is able to invest a portion of his $10,000.00 savings at $4\frac{1}{4}$ percent annual interest and the remaining amount at 6 percent interest. If his total earnings on the first year is $530.00, how much of the $10,000.00 was invested at 6 percent?

4   If one man can do a job alone in 7 hours and another can do it in 9 hours, how long will the job take them if they work together?

5   A freight plane leaves New York for Los Angeles flying at a rate of 300 miles per hour. Two hours later a passenger flight flying at a rate of 500 miles per hour also leaves New York for Los Angeles following the same flight plan. How long will it take the passenger flight to overtake the first plane?

6   A plane travels 630 miles in 3 hours with the wind and makes the return trip against the wind in $4\frac{1}{2}$ hours. What would the speed of the plane have been without the wind? Also, determine the speed of the wind.

7   How many quarts of antifreeze must be added to 15 quarts of a 30 percent solution to bring it up to a 45 percent solution?

8   If a radiator is filled with a 40 percent antifreeze solution, how much must be drained off and replaced by pure antifreeze in order to obtain a concentration of 60 percent, assuming the radiator holds 20 quarts when full?

9   Suppose a rectangular plot of ground 10 yards long and 5 yards wide is to be made into a garden. A path of uniform width will be necessary around the outer edge of this plot so as to leave an area of 24 square yards. How wide should the path be?

10 A park, 100 yards by 100 yards, is to be designed having a road around the entire inside perimeter. How wide should the road be so as to preserve 6400 square yards of area for the park?

11 A train travels 180 miles. A second train, moving 10 miles per hour faster than the first, can make the same trip in $1\frac{1}{2}$ hours less time than the first. Find the speed of each.

12 If the formula for the distance in feet that a freely falling body travels in t seconds is $s = 16t^2$, how long will it take a stone dropped from an altitude of 400 feet to reach the ground?

13 A stone is dropped from a vertical cliff overlooking a lake; 9 seconds later the sound of the splash is heard from the point where it was dropped. If the speed of sound is 1024 feet per second, find the height of the cliff. (HINT: Note the formula of the previous problem.)

14 If $3000.00 is invested at a certain interest rate compounded annually and earns $276.07 in two years, find the interest rate.

Problem Set IV   ▓   *Theoretical Developments*

1   If $ax^2 + bx + c = 0$, where a, b, c are integers and b is even, prove that an alternative form for the quadratic formula is

$$x = \frac{-b' \pm \sqrt{(b')^2 - ac}}{a}$$

where $b' = \dfrac{b}{2}$.

2   Prove that if x_1 and x_2 are solutions of $ax^2 + bx + c = 0$, then

$$x_1 + x_2 = -\frac{b}{a} \quad \text{and} \quad (x_1)(x_2) = \frac{c}{a}$$

Prove each of the following, justifying each step by citing the appropriate definition, axiom, or theorem.

3   Theorem 3.8.2

4   Theorem 3.8.3

5   Theorem 3.8.4 Part I

1 In H. G. Wells's novel *The First Men in the Moon* our natural satellite is found to be inhabited by intelligent insect creatures that live in caverns below the surface. These creatures, let us assume, have a unit of distance that we shall call a "lunar." It was adopted because the moon's surface area, if expressed in square lunars, exactly equals the moon's volume in cubic lunars. The moon's diameter is 2160 miles. How many miles long is a lunar?

2 An airplane flies in a straight line from airport A to airport B, then back in a straight line from B to A. It travels with a constant engine speed and there is no wind. Will its travel time for the same round trip be greater, less, or the same if, throughout both flights, at the same engine speed, a constant wind blows from A to B?

3 Smith drove at a steady clip along the highway, his wife beside him. "Have you noticed," he said, "that those annoying signs for Flatz beer seem to be regularly spaced along the road? I wonder how far apart they are."

Mrs. Smith glanced at her wrist watch, then counted the number of Flatz beer signs they passed in one minute.

"What an odd coincidence!" exclaimed Smith. "When you multiply that number by ten, it exactly equals the speed of our car in miles per hour."

Assuming that the car's speed is constant, that the signs are equally spaced and that Mrs. Smith's minute began and ended with the car midway between two signs, how far is it between one sign and the next?

4 A collection of nickels, dimes, and quarters totals $4.20. If there are twice as many quarters as nickels, and the total number of coins is 27, find how many nickels, how many dimes, and how many quarters there are in this collection.

5 At what time between 8 and 9 o'clock does the minute hand on a clock coincide with the hour hand? At what time between 9 and 10 o'clock do both hands of a clock coincide?

6 A man is $\frac{1}{4}$ of the way across a railway bridge when he sees a train from behind one bridge-length away. Which way should he run?

Summary of Definitions and Properties in Chapter 3   3.10

D3.2.1 Open Sentence   (p. 93)

D3.2.2 Solution, Root, and Solution Set   (p. 94)

D3.2.3 Equivalent Open Sentences   (p. 98)

T3.2.1   (p. 100)

  I $[a = b] \Longleftrightarrow [a + c = b + c]$

  II $[a = b \quad \text{and} \quad c \neq 0] \Longleftrightarrow [ac = bc \quad \text{and} \quad c \neq 0]$

 III $[a + b = c] \Longleftrightarrow [a = c - b \quad \text{or} \quad b = c - a]$

 IV $[ab = 0] \Longleftrightarrow [a = 0 \quad \text{or} \quad b = 0]$

  V $\left[\dfrac{a}{b} = \dfrac{c}{d} \quad \text{and} \quad b, d \neq 0\right] \Longleftrightarrow [ad = bc \quad \text{and} \quad b, d \neq 0]$

T3.2.2   (p. 100)

  I $[a \leq b] \Longleftrightarrow [a + c \leq b + c]$

  II $[a \leq b \quad \text{and} \quad c > 0] \Longleftrightarrow [ac \leq bc \quad \text{and} \quad c > 0]$

III $\quad [a \leq b \quad \text{and} \quad c < 0] \Longleftrightarrow [ac \geq bc \quad \text{and} \quad c < 0]$

IV $\quad \left[ab > 0 \quad \text{or} \quad \dfrac{a}{b} > 0 \right] \Longleftrightarrow \begin{cases} a > 0 \quad \text{and} \quad b > 0 \\ \text{or} \\ a < 0 \quad \text{and} \quad b < 0 \end{cases}$

V $\quad \left[ab < 0 \quad \text{or} \quad \dfrac{a}{b} < 0 \right] \Longleftrightarrow \begin{cases} a < 0 \quad \text{and} \quad b > 0 \\ \text{or} \\ a > 0 \quad \text{and} \quad b < 0 \end{cases}$

T3.4.1 $\quad ax^2 + bx + c = 0 \Longleftrightarrow x = \dfrac{-b \pm \sqrt{b^2 - 4ac}}{2a}$ (p. 109)

T3.6.1 $\quad |a| = b \Longleftrightarrow \begin{cases} b = a \qquad \text{if } a \geq 0 \\ \text{or} \\ b = -a \qquad \text{if } a < 0 \end{cases}$ (p. 112)

D3.6.1 $\quad |a| = \begin{cases} a \qquad \text{if } a \geq 0 \\ -a \qquad \text{if } a < 0 \end{cases}$ (p. 111)

D3.8.1 $\quad a^2 = b \quad \text{and} \quad a \geq 0 \Longleftrightarrow a = \sqrt{b}$ (p. 117)

T3.8.1 $\quad a^2 = b \Longleftrightarrow a = \pm\sqrt{b}$ (p. 117)

T3.8.2 $\quad \sqrt{a^2} = |a|$ (p. 118)

T3.8.3 $\quad a, b \geq 0 \Longrightarrow \sqrt{ab} = \sqrt{a} \cdot \sqrt{b}$ (p. 118)

T3.8.4 $\quad a, b \geq 0 \Longrightarrow$ (p. 119)
I $\quad a \leq b \Longleftrightarrow a^2 \leq b^2$
II $\quad a \leq b \Longleftrightarrow \sqrt{a} \leq \sqrt{b}$

Answers to Sample Problem Set 3.5 (p. 111)

1 $\;S = \{-2\}$ **2** $\;S = \{x \mid x > -2\}$ **3** $\;S = \{4, -3\}$
4 $\;S = \{x \mid x < -3 \quad \text{or} \quad x > 4\}$ **5** $\;S = \{\frac{1}{2}, -\frac{2}{3}\}$ **6** $\;S = \varnothing$

Answers to Sample Problem Set 3.7 (p. 116)

1 $\;S = \{-\frac{3}{7}\}$ **2** $\;S = \{-3\}$ **3** $\;S = \{x \mid -2 < x < 6\}$
4 $\;S = \{x \mid x < -\frac{3}{2} \quad \text{or} \quad x > 1\}$ **5** $\;S = \{x \mid x < -1 \quad \text{or} \quad x > 3\}$

Relations and Functions

4 CHAPTER

Introduction 4.1

Our everyday experiences are full of relations of various sorts. How often do we hear such expressions as "Jack is older than Tom," "Jim is married to Helen," "Ten is greater than four," or "My car is faster than yours." The word *relation* suggests some kind of association or correspondence that exists between objects, both physical and abstract, which is characterized by some property that they possess. In mathematics the notion of a relation is used in a similar, although more precise, manner.

Closely related to the concept of a relation is that of a function, which is unquestionably the single most important notion in all mathematical analysis and is fundamental to the study of nearly all branches of mathematics. Our study here will be limited to but a small class of functions; these, however, will find application in a generously wide variety of significant problems.

4.2 Relations

DEFINITION 4.2.1 Suppose each element of a set A is assigned in some manner to one or more elements of a set B. We call such an *assignment* a relation of A into B. If this relation is denoted by the letter f (any letter would do), then the relation may be symbolically represented as

$$[f\colon A \longrightarrow B] \quad \text{or} \quad [A \xrightarrow{\ f\ } B]$$

Furthermore, if b is an element of B to which an element a in A is assigned by f, then b is called an image of a under f and denoted $f(a)$ (which is read "f of a"). The set A is called the domain of f and is denoted $D(f)$, and the set of all images in B is called the range of f and is denoted $R(f)$.

Example 4.2.1 Let $A = \{1, 2, 3, 4\}$ and $B = \{2, 4, 5, 7, 9\}$. Suppose f is a relation of A into B defined by Fig. 4.2.1. From this relation we may write

$$1 \in A \Longrightarrow f(1) = 2$$
$$2 \in A \Longrightarrow f(2) = 2$$
$$3 \in A \Longrightarrow f(3) = 4 \quad \text{and} \quad f(3) = 5$$
$$4 \in A \Longrightarrow f(4) = 9$$

The *domain* of f is

$$D(f) = \{1, 2, 3, 4\} = A$$

and the *range* of f, composed of the image elements $f(a)$ in B, is

$$R(f) = \{2, 4, 5, 9\}$$

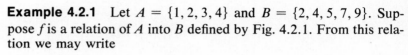

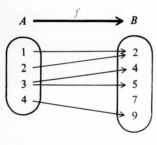

FIGURE 4.2.1

Remark 4.2.1 Notice that the *range* of a relation f of A into B is always a *subset* of B. Observe also that such a relation pairs each element of A with one *or more* elements of B.

When two elements are paired in such a manner that one, say a, is designated the first and the other, say b, is designated second, such a pairing is called an ordered pair and is denoted (a, b). The left-hand member of an ordered pair is called the first coordinate or abscissa and the right-hand member is called the second coordinate or ordinate.

Example 4.2.1 illustrates that relations define ordered pairs. If we agree that the left-hand coordinate of an ordered pair defined by a relation is a member of the domain of f and the right-hand coordinate is a member of the range of f, then we have the following equivalences.

$$1 \rightarrow 2 \iff (1,2)$$
$$2 \rightarrow 2 \iff (2,2)$$
$$3 \rightarrow 4 \iff (3,4)$$
$$3 \rightarrow 5 \iff (3,5)$$
$$4 \rightarrow 9 \iff (4,9)$$

Because this collection of ordered pairs completely characterizes the relation f of A into B, a frequently employed alternative notation for *the relation f* is this set of defining ordered pairs; we write

$$f = \{(1,2), (2,2), (3,4), (3,5), (4,9)\}$$

We shall often define relations as sets of ordered pairs; this manner of representation is usually simpler and more compact.

Example 4.2.2 It is well known that commodities can often be purchased at reduced rates if bought in larger quantities. Suppose that laundry detergent can be purchased according to the scale in Table 4.2.1. This table, of course, defines a relation. Let us call it the relation g, with the quantities of 25-lb boxes the domain A, and the costs per box the range B. The relation g may be diagrammed as shown in Fig. 4.2.2. The relation g may also be represented as the following set of ordered pairs:

$$g = \{(1, 4.75), (5, 4.15), (10, 4.00), (20, 3.50)\}$$

where the first coordinate represents quantity of boxes, and the second the cost per box. The *domain* of g is

$$D(g) = \{1, 5, 10, 20\}$$

and the *range* of g is

$$R(g) = \{4.75, 4.15, 4.00, 3.50\}$$

One of the most convenient and practical means of defining a relation is by use of open sentences in two variables. Take, for example,

$$y = x^2$$

where the replacement set for both x and y is the field R of real numbers. To reduce this open equation to a true or false proposition both x and y must be replaced by members of the replacement set. More specifically, if $x = 2$ and $y = 4$, then the equation becomes

$$4 = 2^2$$

Cost per box	Number of 25-lb boxes
4.75	1
4.15	5
4.00	10
3.50	20

Table 4.2.1

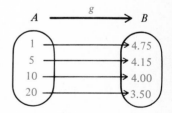

FIGURE 4.2.2

which, of course, is a *true* proposition, whereas if $x = 3$ and $y = 5$, the equation is reduced to the *false* proposition

$$5 = 3^2$$

Thus we observe that a pair of numbers is required to transform an open equation in two variables to a true or false proposition. In fact, you will note, the pair must be an *ordered* pair; one member represents our choice for x and the other our choice for y, and, in general, these choices cannot be interchanged. We noted above that $x = 2$ and $y = 4$ form a solution pair for the equation $y = x^2$. If we *interchange* these choices, that is, take $x = 4$ and $y = 2$, then the equation becomes

$$2 = 4^2$$

which, of course, is false.

The accepted standard in forming an ordered pair from the expressions $x = 2$ and $y = 4$ is to choose the first coordinate as the x-coordinate and the second coordinate as the y-coordinate. Thus the solution $x = 2$ and $y = 4$ of $y = x^2$ may be denoted by the ordered pair $(2, 4)$. The solution set of an open sentence in two variables will be the set of *all* ordered pairs that satisfy the open sentence, reducing it to a true proposition. It is clear that the equation $y = x^2$ over R has infinitely many solutions (any real number, when squared, will yield another real number by the closure axiom for multiplication of the real number system). We will often, therefore, use set-builder notation to denote such a solution set. Hence, the solution set of $y = x^2$ may be expressed as

$$\{(x, y) \mid y = x^2\}$$

which is read "the set of all ordered pairs (x, y) such that $y = x^2$." Because a set of ordered pairs defines a relation, we may call this solution set the relation h (any letter would do) and write

$$h = \{(x, y) \mid y = x^2\}$$

We now summarize the above observations in the following definition:

DEFINITION 4.2.2 (An *algebraic* characterization of a relation.) The nonempty *solution set* of an open sentence in two variables defines a relation. Let f represent the relation. Then

$$f = \{(x, y) \mid p(x, y) = 0 \text{ is true over } U\}$$

where $p(x, y) = 0$ is the open sentence over the replacement set U. An ordered pair must satisfy $p(x, y) = 0$ in order to qualify as a member of f.

Example 4.2.3 Given the relation f defined by Fig. 4.2.3. This may also be described by

$$1 \to 3 \iff (1,3)$$
$$2 \to 5 \iff (2,5)$$
$$3 \to 7 \iff (3,7)$$
$$4 \to 9 \iff (4,9)$$
$$\vdots \qquad \vdots$$
$$n \to 2n+1 \iff (n, 2n+1)$$
$$\vdots \qquad \vdots$$

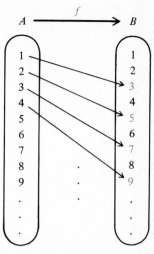

FIGURE 4.2.3

For each natural number n in the domain of this relation, we have a formula for its image in the range given by

$$f(n) = 2n + 1$$

When a formula for $f(n)$ can be found, such as $2n + 1$ in the relation above, then the relation may be conveniently denoted in set-builder form. We have

$$f = \{(x, y) \mid y = 2x + 1 \quad \text{and} \quad x \in N\}$$

where the replacement set N is the set of natural numbers. The domain of f is

$$D(f) = N = \{1, 2, 3, 4, \ldots\}$$

and the range

$$R(f) = \{1, 3, 5, 7, 9, \ldots\}$$

Associated with each relation is another relation called its inverse. This relation is easily defined using ordered pair notation.

DEFINITION 4.2.3 If f is a relation represented by a set of ordered pairs of numbers, then the inverse relation of f, denoted f^{-1}, is the set of all ordered pairs of the relation f in the reverse order. Symbolically this may be written

$$f^{-1} = \{(b, a) \mid (a, b) \in f\}$$

Example 4.2.4 Let f be a relation of A into B, defined by Fig. 4.2.4 or, as a collection of ordered pairs,

$$f = \{(2, -2), (4, 0), (6, 0), (8, 2)\}$$

The *inverse* relation of f is

$$f^{-1} = \{(-2, 2), (0, 4), (0, 6), (2, 8)\}$$

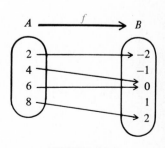

FIGURE 4.2.4

Notice that

$$D(f) = \{2, 4, 6, 8\} \quad \text{and} \quad R(f) = \{-2, 0, 2\}$$

whereas

$$D(f^{-1}) = \{-2, 0, 2\} \quad \text{and} \quad R(f^{-1}) = \{2, 4, 6, 8\}$$

that is,

$$D(f) = R(f^{-1}) \quad \text{and} \quad R(f) = D(f^{-1})$$

which, of course, should not surprise us because an inverse relation is obtained by reversing the order of each ordered pair. The inverse relation f^{-1} is also shown in Fig. 4.2.5.

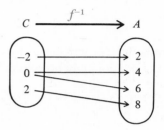

FIGURE 4.2.5

Remark 4.2.2 Notice that the elements -1 and 1 of B, which are not images of any elements in A, are not included in the domain C of f^{-1}. The range of f, which is a subset of B, is the domain of f^{-1}.

Example 4.2.5 Define a relation g as

$$g = \{(x, y) \mid y = \sqrt{4 - x^2} \quad \text{and} \quad x, y \in R\}$$

where the replacement set R is the field of real numbers. Members of this relation include

$$(-2, 0), (0, 2), (1, \sqrt{3})$$

These relationships may be expressed

$$g(-2) = 0, \ g(0) = 2, \ g(1) = \sqrt{3}$$

and, in general, for each real x for which $\sqrt{4 - x^2}$ is real, we have

$$g(x) = \sqrt{4 - x^2}$$

When a relation is defined by an open equation, as g, the members of the domain and range are not as easily identified as when the ordered pairs are explicitly expressed. To identify the domain and range sets one must carefully examine the defining equation. Any choice of x from the replacement set which, when substituted into the equation, yields a choice for y in the replacement set is a member of the domain. In this example, we have $y = \sqrt{4 - x^2}$, and, therefore, any $x \in R$ for which $\sqrt{4 - x^2} \in R$ will be in the domain of g.

To determine these, we proceed as follows:

$$\sqrt{4 - x^2} \in R \xrightarrow[]{\text{R3.8.1}} 4 - x^2 \geq 0 \Longleftrightarrow 4 \geq x^2$$

$$\Longleftrightarrow x^2 \leq 4 \xrightarrow[]{\text{T3.8.4(II)}} \sqrt{x^2} \leq \sqrt{4}$$

$$\xrightarrow[]{\text{T3.8.2}} |x| \leq 2 \xrightarrow[]{\text{T2.11.3(I)}} -2 \leq x \leq 2$$

$$\Longrightarrow D(g) = \{x \mid -2 \leq x \leq 2\}$$

We will determine the range of g by solving the equation $y = \sqrt{4 - x^2}$ for x.

$$y = \sqrt{4 - x^2} \xrightarrow[]{\text{D3.8.1}} y^2 = 4 - x^2 \quad \text{and} \quad y \geq 0$$

$$\Longleftrightarrow x^2 = 4 - y^2 \quad \text{and} \quad y \geq 0$$

$$\xrightarrow[]{\text{T3.8.1}} x = \pm\sqrt{4 - y^2} \quad \text{and} \quad y \geq 0$$

Thus, the conditions on y that will render a real x are

$$4 - y^2 \geq 0 \quad \text{and} \quad y \geq 0$$

By the same steps taken to obtain the domain, these latter expressions become equivalent to

$$-2 \leq y \leq 2 \quad \text{and} \quad y \geq 0 \Longleftrightarrow 0 \leq y \leq 2$$

Therefore

$$R(g) = \{y \mid 0 \leq y \leq 2\}$$

Let us now determine the inverse of g. Recall that the inverse of a relation is a relation whose ordered pairs are those of the original relation in the reverse order. When a relation is defined by an open sentence in two variables x and y, the simplest way to define the corresponding inverse relation is to interchange the roles of x and y in the defining open sentence. The inverse of the relation g, therefore, may be expressed

$$g^{-1} = \{(x, y) \mid x = \sqrt{4 - y^2} \quad \text{and} \quad x, y \in R\}$$

$$= \{(x, y) \mid y = \pm\sqrt{4 - x^2} \quad \text{and} \quad x \geq 0 \text{ and } x, y \in R\}$$

Note that whereas the ordered pair $(-2, 0)$ satisfies the equation $y = \sqrt{4 - x^2}$ that defines g, the pair $(0, -2)$ (obtained by reversing the order of the pair) satisfies the equation $x = \sqrt{4 - y^2}$ that defines g^{-1}. Remembering, finally, that the domain of a relation is equal to the range of its inverse and the range of a relation is equal to the domain of its inverse, we have

$$D(g^{-1}) = \{x \mid 0 \leq x \leq 2\}$$

and

$$R(g^{-1}) = \{y \mid -2 \leq y \leq 2\}$$

R3.8.1
$a = \sqrt{b}$ and $a \in R \Longleftrightarrow b \geq 0$
(R3.8.1 refers to Remark 3.8.1)

T3.8.4(II)
$a, b \geq 0$ and $a \leq b \Longleftrightarrow \sqrt{a} \leq \sqrt{b}$

T3.8.2 $\sqrt{a^2} = |a|$

T2.11.3(I)
$|a| \leq b \Longleftrightarrow -b \leq a \leq b$

D3.8.1
$a = \sqrt{b}$ and $a \geq 0 \Longleftrightarrow$
$\qquad a^2 = b$ and $a \geq 0$

T3.8.1 $a^2 = b \Longleftrightarrow a = \pm\sqrt{b}$

4.3 Sample Problem Set

Determine whether the following are relations and, if so, find the domain, the range, and the inverse of each.

1

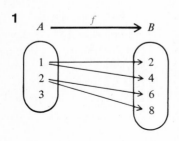

FIGURE 4.3.1

2

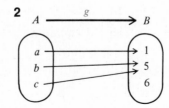

FIGURE 4.3.2

3

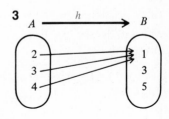

FIGURE 4.3.3

4 $f = \{(1, 4), (2, -3), (2, 1)\}$

5 $f = \left\{(x, y) \mid y = \dfrac{2x - 1}{x + 2} \quad \text{and} \quad x, y \in R\right\}$

6 $h = \{(x, y) \mid x^2 + y^2 = -1 \quad \text{and} \quad x, y \in R\}$

7 $f = \{(x, y) \mid y = \sqrt{x^2 - 9} \quad \text{and} \quad x, y \in R\}$

4.4 Functions

We are now prepared to define a function. It was mentioned in Section 4.1 that a function is a special kind of relation. This becomes clear in the following definition.

DEFINITION 4.4.1 If f is a relation of A into B, which assigns to each element of A a *unique* (exactly one) element of B, then f is called a function (sometimes a mapping) of A into B.

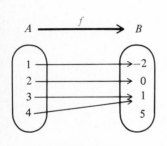

FIGURE 4.4.1

Example 4.4.1 Let $A = \{1, 2, 3, 4\}$ and $B = \{-2, 0, 1, 5\}$. Suppose f is a function of A into B defined as shown in Fig. 4.4.1 or, expressed in order pair notation,

$$f = \{(1, -2), (2, 0), (3, 1), (4, 1)\}$$

The inverse relation of the function f is most easily obtained from the ordered pair notation by simply reversing the order of each pair in f. Hence,

$$f^{-1} = \{(-2, 1),\ (0, 2),\ (1, 3),\ (1, 4)\}$$

This relation may be diagrammed as shown in Fig. 4.4.2. Notice that f^{-1} does not assign 1 to a *unique* element in the range, but rather assigns 1 to both 3 and 4. Thus, the relation f^{-1} is *not* a function. You will find that it is not infrequent that the inverse of a function is not a function.

It is often important in both pure and applied mathematics to know if a given relation or its inverse are functions. The following definitions and examples will assist in determining when the inverse of a function remains a function.

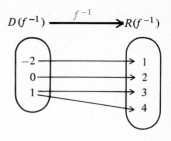

FIGURE 4.4.2

DEFINITION 4.4.2 Let f be a function of A into B. Then f is called a one-to-one (abbreviated 1-1) function, if each image element in B corresponds to a *unique* element in A — that is, it defines a 1-1 correspondence between its domain and range sets.

Example 4.4.2 Let f and g be two functions of A into B defined by Fig. 4.4.3. The map f is clearly *not* 1-1 because c and d are both mapped into 4; that is, $f(c) = 4$ and $f(d) = 4$. The map g is, however, 1-1; each image in B under g is coupled with a unique element of the domain A.

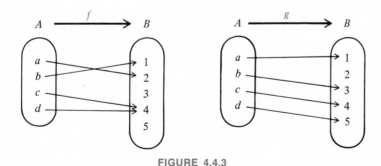

FIGURE 4.4.3

DEFINITION 4.4.3 Let f be a relation of A into B. If the range of f is equal to B, then f is said to be a relation of A onto B — every element of B is the image of at least one element of A.

Remark 4.4.1 Recall that the definition of a relation of A into B requires every element of A be assigned at least one element of B. An onto relation requires the same of B.

Example 4.4.3 Let f be a function of A into B defined by Fig. 4.4.4. Because every element of B is the image of at least one element of A, f is a function of A *onto* B. Notice, however, that f is not 1-1. Let us examine the inverse relation of f as diagrammed in Fig. 4.4.5. Observe that f^{-1} is a relation of B onto A but is *not* a function: $f^{-1}(d) = 3$ and $f^{-1}(d) = 4$. It is significant to note here that f^{-1} is not a function because f is *not* 1-1.

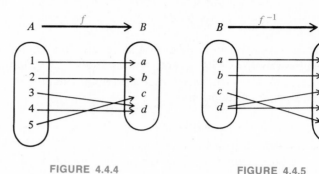

FIGURE 4.4.4   FIGURE 4.4.5

These observations are verified in the following theorem.

THEOREM 4.4.1 f is a 1-1 function of A onto B if and only if f^{-1} is a 1-1 function of B onto A.

Example 4.4.4 Let f and g be two functions defined as follows:

$$f = \{(x, y) \mid y = x^2 \quad \text{and} \quad x, y \in R\}$$
$$g = \{(x, y) \mid y = x^3 \quad \text{and} \quad x, y \in R\}$$

● T2.11.1(III) $x \in R \Longrightarrow x^2 \geq 0$

The domain of f is R; the range of f is $\{y \mid y \geq 0$ and $y \in R\}$,● a *proper* subset of R. Thus, f is not a function of R *onto* R. Furthermore, $f(-2) = f(2) = 4$ so that -2 and 2 are mapped into the same image element. Hence, f is not 1-1 either. The function g, on the other hand, is both 1-1 and onto.

Let us now examine the inverses of these two functions. The form for these is easily obtained because each is defined by an equation in two variables x and y; all that is necessary is to interchange the roles of x and y in the defining equations. We obtain

$$f^{-1} = \{(x, y) \mid x = y^2 \quad \text{and} \quad x, y \in R\}$$

and

$$g^{-1} = \{(x, y) \mid x = y^3 \quad \text{and} \quad x, y \in R\}$$

Because of the equivalences

$$x = y^2 \Longleftrightarrow y = \pm\sqrt{x}$$

and

$$x = y^3 \Longleftrightarrow y = \sqrt[3]{x} \bullet \qquad\qquad \bullet\ a^n = b \Longleftrightarrow a = \sqrt[n]{b} \quad \text{if } n \text{ is odd.}$$

these inverses may also be expressed in the form

$$f^{-1} = \{(x, y) \mid y = \pm\sqrt{x} \quad \text{and} \quad x, y \in R\}$$

and

$$g^{-1} = \{(x, y) \mid y = \sqrt[3]{x} \quad \text{and} \quad x, y \in R\}$$

Because the range of f is $\{y \mid y \geq 0 \quad \text{and} \quad y \in R\}$, the domain of f^{-1} is

$$D(f^{-1}) = \{x \mid x \geq 0 \quad \text{and} \quad x \in R\}$$

The defining equation for f^{-1} clearly provides two distinct choices of y for each positive choice of x; thus the relation f^{-1} is *not* a function. In contrast, g^{-1} is a 1-1 *function* of R onto R, just as g is.

We now define two special functions—both simple, yet quite significant.

DEFINITION 4.4.4 Let f be a function of A into A defined by the equation $f(x) = x$, that is, f maps x into itself. Then f is called the identity function on A and is denoted I_A. The identity function on A may be expressed as

$$I_A: A \xrightarrow{\ y = x\ } A$$

or, in ordered pair notation:

$$I_A = \{(x, y) \mid y = x \quad \text{and} \quad x, y \in A\}$$

The identity function is clearly 1-1 and onto.

Example 4.4.5 Let f and g be two functions of A into A defined as shown in Fig. 4.4.6. Here f is not the identity function on A, whereas g clearly is such a function.

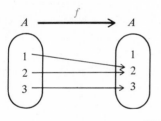 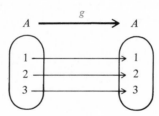

FIGURE 4.4.6

DEFINITION 4.4.5 Let f be a function of A into B where every element of A is mapped into one and the same element in B (where the range of f consists of a *single* element). Then f is called a con-stant function.

Example 4.4.6 Let f and g be functions of A into B defined as shown in Fig. 4.4.7. In this example, f is not a constant function because we have $f(1) \neq f(2)$. However, the function g is constant, because for each $a \in A$ we have only one image in B, that is $g(a) = c$ for each $a \in A$.

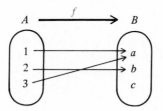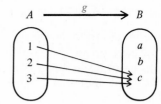

FIGURE 4.4.7

Example 4.4.7 Let f be a constant function defined by

$$f = \{(x, y) \mid y = f(x) = 3 \quad \text{and} \quad x, y \in R\}$$

Here, for each $x \in R$, we have the image $f(x) = 3$; the image is a constant — always 3. Every ordered pair in this collection of the form $(x, 3)$ for each $x \in R$.

4.5 Sample Problem Set

Determine if each of the following is 1-1, onto, a function, and has an inverse that is a function.

1

2

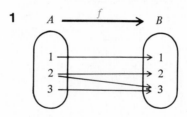

FIGURE 4.5.1   FIGURE 4.5.2

3
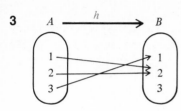

FIGURE 4.5.3

4
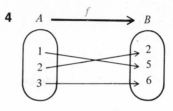

FIGURE 4.5.4

5 $f = \{(x, y) \mid y = \sqrt{25 - x^2}$ and $x, y \in R\}$ where $D(f) = \{x \mid -5 \le x \le 5\}$ and $R(f) = \{y \mid 0 \le y \le 5\}$

6 $g = \{(x, y) \mid y = \sqrt{4 - x^2}, x \ge 0,$ and $x, y \in R\}$ where $D(f) = \{x \mid 0 \le x \le 2\}$ and $R(f) = \{y \mid 0 \le y \le 2\}$

A Graph of Relations and Functions   4.6

Relations can be represented geometrically in a very natural way. We have already characterized sets of real numbers as points on a line. This was brought about by observing the existence of a one-to-one correspondence between the set R of real numbers and the set of all points on a two-way infinite line, as illustrated in Fig. 4.6.1. This is called a linear or one-dimensional coordinate system. A single number is all that is necessary to identify a point on this line.

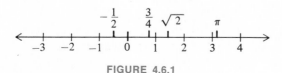

FIGURE 4.6.1

There are several *two*-dimensional coordinate systems in common use. The rectangular (also called Cartesian) coordinate system is most frequently used. You are probably already familiar with this system. It is formed by taking two real lines, one oriented horizontally and the other vertically, and joining them at their zero point, as illustrated in Fig. 4.6.2. Notice on the vertical line that positive real numbers correspond to points above the horizontal line, whereas negative real numbers correspond with points below this line. The horizontal and vertical real lines are called the axes of the coordinate system. These two lines define a plane surface area. It is possible to identify uniquely a point on this surface by an ordered pair of real numbers. This is done as follows: Given the

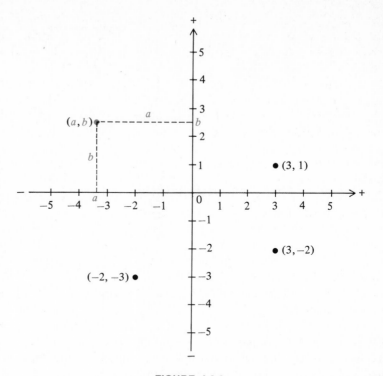

FIGURE 4.6.2

ordered pair (a, b), we determine the position of the point that it represents by first locating the point corresponding to a on the horizontal axis and then locating the same for b on the vertical axis. The perpendiculars through these two points intersect at the desired point (see Fig. 4.6.2). By reversing this process you can see that given any point there exists a unique ordered pair corresponding to it. Thus, in a rectangular coordinate system, there exists a one-to-one correspondence between the set of all points and the set of all ordered pairs of real numbers.

It is now obvious how we represent a relation geometrically. Because a relation may be expressed as a set of ordered pairs of real numbers and each such ordered pair represents a point in a rectangular coordinate system, we designate the set of all such points as the geometric representation of the relation. More specifically, this set of points is called the graph of the relation.

Example 4.6.1 Graph the relations

$$f = \{(-1, 2), (0, 3), (1, 4)\}$$

and

$$g = \{(2, 1), (3, 1), (3, 2), (4, 2)\}$$

Solution We plot the three points corresponding to the three ordered pairs comprising f and call this geometric collection the graph of f (Fig. 4.6.3(a)). Similarly we identify the graph of g (Fig. 4.6.3(b)).

Recall that a function is a relation that associates with each element in its domain a *unique* element in its range. This means, when a function is expressed as a set of ordered pairs, that no two distinct ordered pairs can have the *same first* coordinate. Notice in g the two pairs $(3, 1)$ and $(3, 2)$. Although distinct ordered pairs, these do have the same first coordinate; thus, g is *not* a function. Distinct ordered pairs with the same first coordinate can be quickly identified geometrically because they will lie one directly above the other—a vertical line passing through one of them will also pass through the other (see Figure 4.6.3 (b)). It is for this reason that you may easily recognize a relation as a function from its graph. If nowhere on the graph there exist two points, one directly above the other, then the relation is a function; otherwise the relation is not a function. From Fig. 4.6.3 it is easy to see that f is a function, whereas g is not.

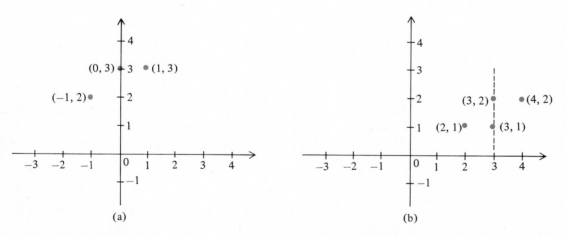

(a)   (b)

FIGURE 4.6.3

Example 4.6.2 Graph the function h defined as

$$h = \{(x, y) \mid y = x^2 \quad \text{and} \quad x, y \in R\}$$

Solution When a function such as h has infinitely many members, it is not possible, of course, to plot individually each of its members. Rather, one usually plots sufficiently many ordered pairs from h to suggest a geometric pattern that the others may likely follow. You then fill in the gaps following the suggested pattern. We will proceed in this manner to graph h.

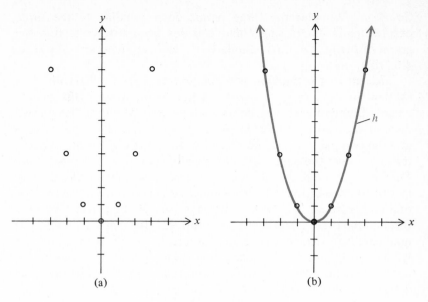

(a)   (b)

FIGURE 4.6.4

We first obtain a few members of the function h. The pairs $(-3, 9)$, $(-2, 4)$, $(-1, 1)$, $(0, 0)$, $(1, 1)$, $(2, 4)$, $(3, 9)$ all belong to h, because they satisfy the equation $y = x^2$. These pairs are plotted as illustrated in Figure 4.6.4 (a).

Remark 4.6.1 Because the first coordinate of the pair (x, y), namely x, is located on the horizontal axis, and second coordinate, y, on the vertical axis, these axes will be called, respectively, the x-axis, and the y-axis (Fig. 4.6.4).

From the pattern suggested in Fig. 4.6.4 (a), we join the points by a smooth curve as illustrated in part (b). *Arrow points* are placed at the end of the portion drawn indicating that the curve continues indefinitely in the suggested direction.

Let us now graph the inverse relation h^{-1}. Because the ordered pairs $(-3, 9)$, $(-2, 4)$, $(-1, 1)$, $(0, 0)$, $(1, 1)$, $(2, 4)$, $(3, 9)$ were all elements of h, we have but to reverse the order of each to obtain a corresponding set belonging to h^{-1}. Thus $(9, -3)$, $(4, -2)$, $(1, -1)$, $(0, 0)$, $(1, 1)$, $(4, 2)$, $(9, 3)$ all belong to h^{-1}. After plotting each and sketching the suggested curve, we obtain Fig. 4.6.5. This graph clearly indicates that h^{-1} is *not* a function, for there are, in fact, infinitely many cases where a vertical line may intersect two points on the curve. The fact that h^{-1} is not a function can also be evidenced by noting that

$$(4, -2), (4, 2) \in h^{-1}$$

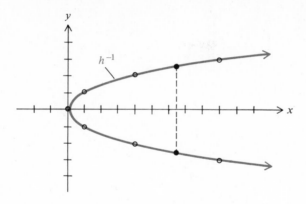

FIGURE 4.6.5

that is, two distinct pairs having the same first coordinate. This would also be apparent from the set-builder form for h^{-1}, which is

$$h^{-1} = \{(x, y) \mid y = \pm\sqrt{x} \quad \text{and} \quad x, y \in R\}$$

The defining open sentence $y = \pm\sqrt{x}$ makes it clear that for each positive choice of x, there are two choices for y, namely, $\sqrt{x}$ and $-\sqrt{x}$.

There is a significant geometric relationship between a relation and its inverse that we shall now observe. Let us take the relations h and h^{-1} of Example 4.6.2 and superimpose their graphs one upon the other as indicated in Fig. 4.6.6. Observe that h^{-1} is the mirror

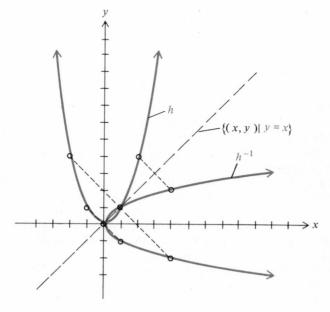

FIGURE 4.6.6

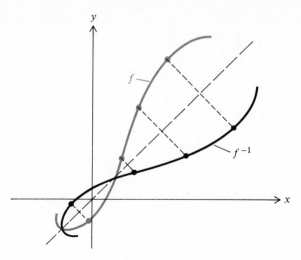

FIGURE 4.6.7

image or reflection of h through the line defined by $y = x$. This geometric relationship between a relation and its inverse holds true in general when employing a rectangular coordinate system. We illustrate this property further in Fig. 4.6.7.

4.7   Problem Sets

Problem Set I   ■   Reading Comprehension

Determine whether statements 1 through 13 are true or false and give reasons for your responses.

1   If f is a relation of A into B then every element in A has its image in B.

2   If f is a relation of A into B then the domain of f may be a proper subset of A.

3   If f is a relation defined by the open equation $y = 2x + 1$ where $x, y \in R$, then the inverse of f, denoted f^{-1}, is defined by $y = 1/(2x + 1)$; putting it another way, the inverse of f is the reciprocal of f, i.e., $f^{-1} = 1/f$.

4   If f is a relation on R defined by an open equation of the form $y = f(x)$ and if for each real x the equation $y = f(x)$ yields only one real choice for y, then the relation f is a function.

5   If f is a function of A into B, then f^{-1} is necessarily a function of B into A.

6   Two functions f and g are one-to-one if they have the same number of elements.

7   If g is a map of A onto B, then g^{-1} is a relation of B onto A.

8   If h is an identity function on A, then there must exist at least one element in A that is mapped into itself.

9   A constant function has no variables.

10 The graph of a function is a geometric characterization of the function.

11 If a function is represented as a collection of ordered pairs, then no two distinct members of the collection can have the same second coordinate.

12 The graph of a function cannot have two points that lie on the same horizontal line.

13 Geometrically f^{-1} is the mirror image of f through the x-axis.

Problem Set II ■ Skills Development

1 For each relation below determine the domain, range, and inverse. Also determine if each relation and its inverse are functions and whether they are 1-1, onto, constant, or identity relations.

a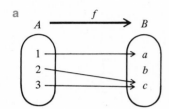

FIGURE 4.7.1

b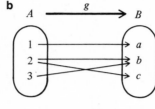

FIGURE 4.7.2

c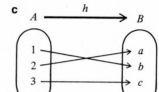

FIGURE 4.7.3

d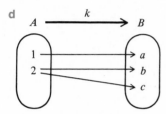

FIGURE 4.7.4

e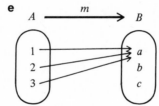

FIGURE 4.7.5

f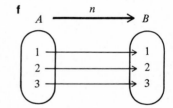

FIGURE 4.7.6

g $n = \{(-2, 2), (-1, -1), (0, 3), (3, 3), (3, -1)\}$

h $p = \{(x, y) \mid y = (x + 1)^2 \text{ and } x, y \in R\}$

i $q = \{(x, y) \mid y = \sqrt{x + 4} \text{ and } x, y \in R\}$

j $r = \{(x, y) \mid y = \sqrt{x^2 - 9} \text{ and } x, y \in R\}$

k $s = \{(x, y) \mid x = y^2 + 4 \text{ and } x, y \in R\}$

l $t = \left\{(x, y) \mid y = \dfrac{2}{x - 1} \text{ and } x, y \in R\right\}$

m $u = \{(x, y) \mid x^2 + y^2 = 4 \text{ and } y \geq 0 \text{ and } x, y \in R\}$

2 Let $A = \{1\}$ and $B = \{-1, 0\}$.

  a Identify all possible functions from A into B.

  b Identify all possible 1-1 functions from A into B.

  c Identify all possible onto functions from A onto B.

  d Identify all possible 1-1 and onto functions from A onto B.

3   Which of the following graphs represents a function?

a

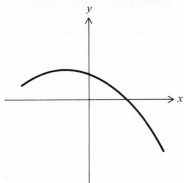

FIGURE 4.7.7

b

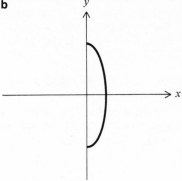

FIGURE 4.7.8

c

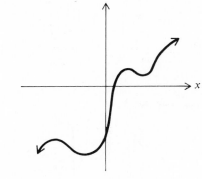

FIGURE 4.7.9

d

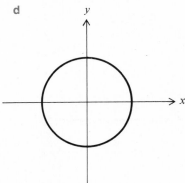

FIGURE 4.7.10

4   Which of the following represents the graph of a relation (solid) and its inverse (dotted)?

a

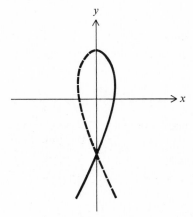

FIGURE 4.7.11

b

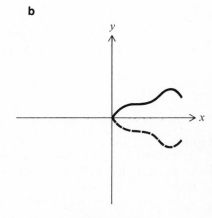

FIGURE 4.7.12

c

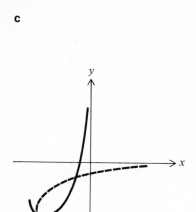

FIGURE 4.7.13

d

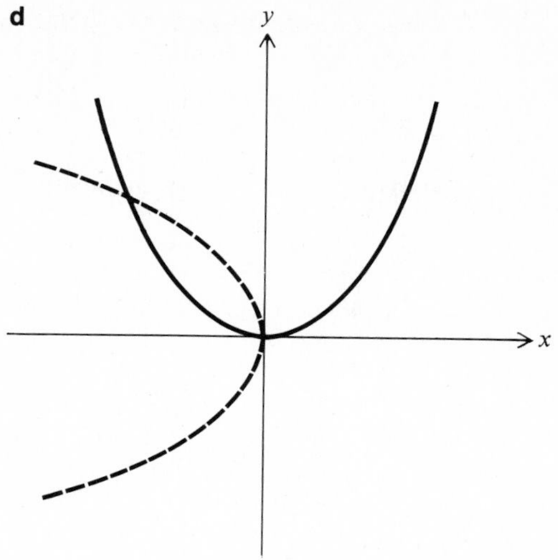

FIGURE 4.7.14

5 If $A = \{-3, -2, -1, 0, 1, 2, 3\}$ and f is a function of A into R defined by $y = x^2 - 4$, find the range of f.

6 Graph the relation $f = \{(-3, 2), (-2, -1), (0, 4), (1, 2), (2, -1)\}$.

Graph the relations defined by the following equations where $x, y \in R$.

7 $y = \sqrt{4 - x^2}$ **8** $y = -2x + 3$ **9** $y = |x - 2|$ **10** $y = x^2 - 2x$

11 $y = \sqrt{4 + x^2}$ **12** $y = \pm\sqrt{4 - x^2}$ **13** $y = |x + 3| + 1$

14 $y = 2 - |x - 1|$ **15** $y = \dfrac{1}{x}$ **16** $y = \dfrac{1}{x^2}$ **17** $y = x^2 - 5x + 6$

Problem Set III ■ Theoretical Developments

1 If $f = \{(x, y) \mid y = x^2 \text{ and } x, y \in R\}$ and $g = \{(p, q) \mid q = p^2 \text{ and } p, q \in R\}$, does $f = g$? Explain.

2 If f is a 1-1 function of A onto B, prove the proposition $[f(x) = f(y) \iff x = y]$.

3 If $f(x) = 4 - x^2$, find a domain interval in R that will render f one-to-one.

4 Under what conditions, if any, will a constant function be (a) 1-1? (b) onto?

Problem Set IV ■ Applications

In problems 1–4 express the relations as a collection of ordered pairs of numbers using set-builder notation. Also, find the domain, range, and inverse of each and determine whether each relation and its inverse are functions, and finally, graph the relation.

Example Find the function describing the relation between the radius of a circle and its area.

Solution Let g represent the relation. The formula for the area of a circle is, of course, πr^2, where r is the radius. Because the radius must be greater than zero for a

circle to exist, we have for the domain $A = R_+$ and for the range $B = R_+$. In set-builder notation this relation becomes
$$g = \{(r, A) \mid A = \pi r^2 \quad \text{and} \quad r, A \in R_+\}$$
or, in x and y coordinates,
$$g = \{(x, y) \mid y = \pi x^2 \quad \text{and} \quad x, y \in R_+\}$$
As noted earlier, the *domain* and *range* are R_+. Finally, the inverse of g is
$$g^{-1} = \{(x, y) \mid x = \pi y^2 \quad \text{and} \quad x, y \in R_+\}$$
or
$$g^{-1} = \left\{(x, y) \mid y = \sqrt{\frac{x}{\pi}} \quad \text{and} \quad x, y \in R_+\right\}$$

Note: Although $[x = \pi y^2 \iff y = \pm\sqrt{x/\pi}]$, $-\sqrt{x/\pi}$ is eliminated because the range of $g^{-1} = R_+$. Thus, both g and g^{-1} are *functions*. To graph g we first determine a few points in g. Remembering $\pi \cong 3.14$, we have the values listed in Table 4.7.1. Thus, $(1, 3.14)$, $(2, 12.56)$, $(3, 28.26)$ are elements of g and the graph of g may be sketched as shown in Fig. 4.7.15. Notice, for convenience, the scale on the A-axis is much finer than on the r-axis. The graph of g^{-1} is also included.

Table 4.7.1

r	$A = \pi r^2$	A
1	$A = \pi(1)^2 \cong$	3.14
2	$= \pi(2)^2 \cong$	12.56
3	$= \pi(3)^2 \cong$	28.26

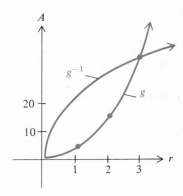

FIGURE 4.7.15

1 The relation between the circumference of a circle and the radius.
2 The relation between the perimeter and the area of a square.
3 The relation between the circumference and the area of a circle.
4 The relation between inches and yards.
5 If a car is traveling at a speed of 60 miles per hour, find the relation between time and distance.
6 Suppose 10 books are purchased. The first three are $12.00 each, whereas the price of the remaining books, because of quantity discounts, decreases $1.00 for each additional book purchased. Determine the relation between the number of books and price.
7 Suppose a state imposes an income tax according to the following schedule: For net earnings from $0 to $600.00, no tax; from $600.00 to $1200.00, the tax is $20.00; $1200.00 to $2400.00, the tax is $40.00; $2400.00 to $4800.00, the tax is $80.00; and so forth. Determine the relation between income and taxes.
8 The relationship between the distance (in feet) traveled by a freely falling object and time (in seconds) is given by
$$d = \tfrac{1}{2}t^2$$
Graph this function.

We begin this section with an applied problem for motivation.

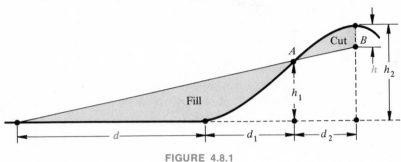

FIGURE 4.8.1

Problem 4.8.1 Suppose a road construction company is designing a highway to pass over a hill. They are limited to a maximum grade (steepness) of 4 percent. Because of various soil conditions on the hill they have determined that a straight and flat road rising at the maximum grade should intersect the hillside at a Point A (see Fig. 4.8.1). They have also determined the height h_1 of point A above the valley floor, h_2, the height of the hill, and distances d_1 and d_2. They wish to determine how much of the top of the hill should be cut away and also how far from the foot of the hill the fill should begin; that is, they wish to determine distances h and d, respectively, of Figure 4.8.1.

Before solving this problem, we will, as usual, develop the mathematical tools and techniques that will greatly simplify the resolving of this and other problems.

DEFINITION 4.8.1 A linear function is one defined by an open equation equivalent to

$$y = mx + b$$

where $m, b \in R$. If we denote such a function by $L_{m,b}$, then we may write

$$L_{m,b} = \{(x, y) \mid y = mx + b \quad \text{and} \quad x, y \in R\}$$

Example 4.8.1 Graph the linear function

$$L_{-2,3} = \{(x, y) \mid y = -2x + 3 \quad \text{and} \quad x, y \in R\}$$

Solution First we obtain a few ordered pairs in L (see Table 4.8.1).

Table 4.8.1

x	$y = -2x + 3$	y	(x, y)
-1	$y = -2(-1) + 3 =$	5	$(-1, 5)$
0	$= -2(0) \ \ + 3 =$	3	$(0, 3)$
2	$= -2(2) \ \ + 3 =$	-1	$(2, -1)$
5	$= -2(5) \ \ + 3 =$	-7	$(5, -7)$

Plotting these points and joining them by the suggested curve pattern, we obtain the graph shown in Fig. 4.8.2. As you may have suspected or already know, it was no coincidence that the graph appears to be that of a straight line.

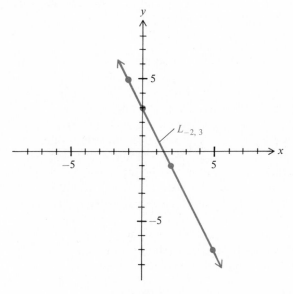

FIGURE 4.8.2

We would like to determine if and why open equations of the form $y = mx + b$ always define straight lines. Let us begin by observing certain properties or characteristics of a line. We know geometrically, for example, that two distinct points determine a line. Also important in the characterization of a line is its slope or inclination—its steepness. We all have strong intuitive feelings as to what slope is. However, to apply the notion mathematically requires a precise definition.

To motivate this definition we observe the following. There are three similar triangles in Fig. 4.8.3. Because these triangles are similar, the following "rise over run" ratios are equal:

$$\frac{BC}{AC} = \frac{B'C'}{AC'} = \frac{B''C''}{AC''}$$

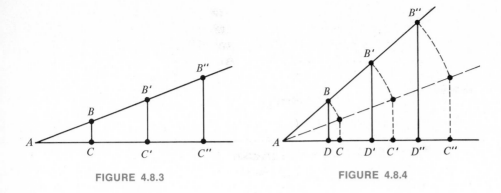

FIGURE 4.8.3   FIGURE 4.8.4

Furthermore, if the inclination of the line AB'' is increased, as illustrated in Figure 4.8.4, we obtain a similar, although distinct, set of ratios. We have

$$\frac{AB}{AD} = \frac{AB'}{AD'} = \frac{AB''}{AD''}$$

Important to observe here are two significant properties. First it does not matter which two points on a line we choose, that is, A and B, A and B', or A and B''; the above "rise over run" ratios remain constant. Second, the slightest change in the inclination of such a line brings about an immediate change in these ratios. Thus such a ratio becomes an ideal way to define the amount of inclination or slope of a line.

With this introduction, let us take two points denoted (x_1, y_1) and (x_2, y_2) and determine the "rise over run" ratio for the line passing through them. Geometrically we have the situation pictured in Fig. 4.8.5. In terms of the coordinates of these two points, our slope ratio becomes

$$\frac{\text{rise}}{\text{run}} = \frac{y_2 - y_1}{x_2 - x_1}$$

This leads us to our definition.

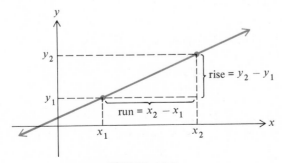

FIGURE 4.8.5

DEFINITION 4.8.2 The slope or inclination of the line passing through two distinct points (x_1, y_1) and (x_2, y_2), where $x_1 \neq x_2$, is

$$m = \frac{y_2 - y_1}{x_2 - x_1}$$

Remark 4.8.1 Notice that

$$\frac{y_2 - y_1}{x_2 - x_1} = \frac{-(y_1 - y_2)}{-(x_1 - x_2)} = \frac{y_1 - y_2}{x_1 - x_2}$$

Thus, it makes no difference which of the two points is represented by (x_1, y_1) and which by (x_2, y_2) in the formula for slope; the ratio remains unchanged.

Example 4.8.2 Determine the slope of the line passing through the points $(-2, 1)$ and $(4, 3)$.

Solution Let $(x_1, y_1) = (-2, 1)$ and $(x_2, y_2) = (4, 3)$. Then $x_1 = -2$, $y_1 = 1$, $x_2 = 4$, and $y_2 = 3$ and we have

$$m = \frac{y_2 - y_1}{x_2 - x_1} = \frac{3 - 1}{4 - (-2)} = \frac{2}{6} = \frac{1}{3}$$

If, on the other hand, we let $(x_1, y_1) = (4, 3)$ and $(x_2, y_2) = (-2, 1)$, we have

$$m = \frac{1 - 3}{-2 - 4} = \frac{-2}{-6} = \frac{1}{3}$$

Either way, the result is the same.

The graph of this line is illustrated in Fig. 4.8.6.

Now let us examine with even greater care the full geometric meaning of slope as we have defined it. We refer to the line in Fig.

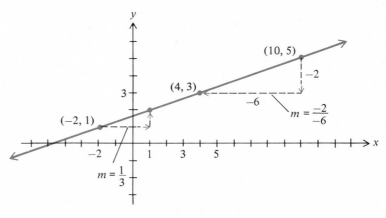

FIGURE 4.8.6

4.8.6. Knowing the slope of this line to be $\frac{1}{3}$, we can, from a given point on the line, say $(-2, 1)$, move 3 units to the right and 1 unit upward to arrive at another point on this line. Similarly, because $\frac{1}{3} = -2/-6$, we can, from the point $(10, 5)$ on the line, move 2 units downward and 6 units to the left and thereby reach another point on the same line. In fact, the following is true in general. When the slope of a line is expressed as a ratio, say a/b, then from a given point, the denominator b represents the amount of *horizontal* movement (to the right if positive and to the left if negative), and the numerator a represents the amount of *vertical* movement (upward if positive and downward if negative) that will lead to a second point on the line.

Finally, note that a line with positive slope will always be inclined upward (from left to right), and a line with negative slope will be inclined downward (from left to right).

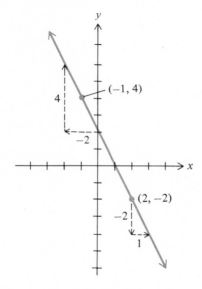

FIGURE 4.8.7

Example 4.8.3 Determine the slope of the line passing through the points $(-1, 4)$ and $(2, -2)$.

Solution Let $(x_1, y_1) = (-1, 4)$ and $(x_2, y_2) = (2, -2)$, then

$$m = \frac{y_2 - y_1}{x_2 - x_1} = \frac{-2 - 4}{2 - (-1)} = \frac{-6}{3} = -2$$

The graph of this line is illustrated in Fig. 4.8.7. Observe here that the slope -2 can be represented as the ratio $-2/1$ or $4/-2$, or, of course, any other ratio that equals -2. All you need to remember is that the numerator of the ratio represents the vertical movement and the denominator the horizontal movement, which leads to another point on the line.

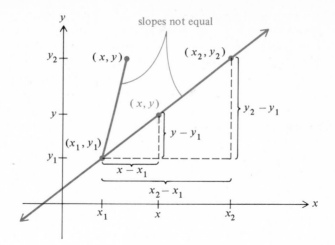

FIGURE 4.8.8

We are now prepared to determine the algebraic form (open equation) of a line determined by two points. Let (x_1, y_1) and (x_2, y_2) be two distinct points that define a line. Furthermore, for geometric motivation, let the graph of the line be represented as in Fig. 4.8.8. We wish to determine an equation in two variables x and y whose solution set, when expressed geometrically, coincides with the line passing through the points (x_1, y_1) and (x_2, y_2). To accomplish this, we take a variable point (x, y) and assume that it is constrained to vary only along the given line. In Fig. 4.8.8, this variable point has been placed between (x_1, y_1) and (x_2, y_2); however, it may be placed anywhere on the line. We now determine *algebraic conditions* that will assure the variable point be constrained to movement along the given line only. The notion of slope will be the key to the conditions we need. Notice that the *slopes* of the lines joining the two pairs of points (x_1, y_1), (x_2, y_2) and (x_1, y_1), (x, y) can only be *equal* if the variable point (x, y) remains on the line (see Fig. 4.8.8). Hence, using the definition of slope, we obtain

$$\frac{y - y_1}{x - x_1} = \frac{y_2 - y_1}{x_2 - x_1} \tag{1}$$

which is an equation in two variables x and y whose *solutions* must lie on the line joining the points (x_1, y_1), (x_2, y_2).

Equation (1) may be expressed in several different yet equivalent forms. The first of these forms is obtained by multiplying both sides of the equation by $(x - x_1)$. This yields

$$y - y_1 = \frac{y_2 - y_1}{x_2 - x_1}(x - x_1) \tag{2}$$

which is called the two-point form of a line.

As indicated earlier, the slope of a line passing through two points (x_1, y_1), (x_2, y_2) is $(y_2 - y_1)/(x_2 - x_1)$. Because the slope of a line is frequently denoted by the letter m, equation (2) may be further simplified notationally by substituting m for the slope. We obtain

$$y - y_1 = m(x - x_1) \tag{3}$$

which is called the point–slope form of a line. If one point on a line, say (x_1, y_1), is known along with the slope m of that line, then these may be substituted into equation (3), obtaining an open equation whose solution set, geometrically, is this line.

An even simpler form is obtained as follows, beginning with equation (3):

$$y - y_1 = m(x - x_1) \iff y - y_1 = mx - mx_1$$
$$\iff y = mx + (y_1 - mx_1)$$
$$\iff y = mx + b \tag{4}$$

where b is equal to $y_1 - mx_1$.

This equation is called the slope–intercept form of a line. Notice that the point $(0, b)$ satisfies equation (4), and further, that this point lies on the y-axis; this is the point where the graph of the line intercepts the y-axis. In addition, from (4) you will notice that the coefficient of x is the slope m of the line. Thus the slope–intercept form of a line provides both the y-intercept, $(0, b)$, and the slope, m, of the line. With this information the line is readily sketched.

Now let us convince ourselves that *any* equation of the form $y = mx + b$ is an equation defining a straight line with $(0, b)$ as the y-intercept and m as the slope. We proceed as follows:

$$y = mx + b \iff y - b = mx$$
$$\iff \frac{y - b}{x} = m \qquad \text{if } x \neq 0$$
$$\iff \frac{y - b}{x - 0} = m \qquad \text{if } x \neq 0$$

But the ratio $(y - b)/(x - 0)$ is, by definition, the slope of the line passing through the two points (x, y) and $(0, b)$. We observed earlier (Figure 4.8.8) that if the slope of a line through a fixed point, $(0, b)$, and a variable point, (x, y), is held *constant,* as it is here by m, then the variable point could only vary along a straight line that passes through the fixed point with the given slope. Hence, we conclude that the equation $y = mx + b$ does, indeed, define a straight line.

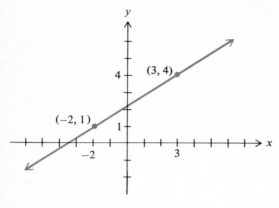

FIGURE 4.8.9

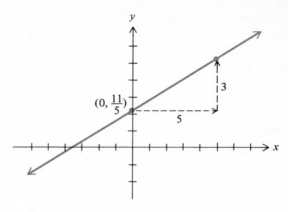

FIGURE 4.8.10

Example 4.8.4 Determine both algebraically and geometrically the line passing through the points $(-2, 1)$ and $(3, 4)$.

Solution Let $(x_1, y_1) = (-2, 1)$ and $(x_2, y_2) = (3, 4)$. By Definition 4.8.1, the slope of this line is

$$m = \frac{y_2 - y_1}{x_2 - x_1} = \frac{4 - (1)}{3 - (-2)} = \frac{4 - 1}{3 + 2} = \frac{3}{5}$$

● (3) $y - y_1 = m(x - x_1)$

Substituting the coordinates of the point $(3, 4)$ and the slope $\frac{3}{5}$ into the point–slope ● form of a line, we obtain

$$y - 4 = \tfrac{3}{5}(x - 3) \Longleftrightarrow y - 4 = \tfrac{3}{5}x - \tfrac{9}{5} \Longleftrightarrow y = \tfrac{3}{5}x + (4 - \tfrac{9}{5})$$

$$\Longleftrightarrow y = \tfrac{3}{5}x + \tfrac{11}{5}$$

This latter equation is in slope–intercept form. This then becomes an *algebraic* form of the line. More precisely, we may write

$$L = \{(x, y) \mid y = (\tfrac{3}{5})x + \tfrac{11}{5}\}$$

although, when there is no confusion, we will just let the equation itself represent the function.

Now for the graph of the line. This we shall obtain in two ways. The first will be by plotting the two points $(-2, 1)$, $(3, 4)$ and then drawing the straight line passing through them. (See Fig. 4.8.9.)

● (4) $y = mx + b$

The second approach is by using the slope–intercept form ● of the line that we have just obtained; namely, $y = \frac{3}{5}x + \frac{11}{5}$. From this equation we identify the slope to be $\frac{3}{5}$ and the y-intercept to be $(0, \frac{11}{5})$. With this information the line is drawn as in Fig. 4.8.10. Of course the line in both graphs is the same.

Example 4.8.5 Determine both algebraically and geometrically the line passing through the point $(-3, 4)$ and *parallel* to the line defined by $2x + 3y = 12$.

Solution Because the line we seek is parallel (meaning same slope) to the one defined by $2x + 3y = 12$, we will express $2x + 3y = 12$ in *slope-intercept* form (so as to identify the slope). We proceed as follows:

$$2x + 3y = 12 \Longleftrightarrow 3y = -2x + 12 \Longleftrightarrow y = -\tfrac{2}{3}x + 4 \Longrightarrow m = -\tfrac{2}{3}$$

We now have the slope of the line we are seeking. With this slope, the point $(-3, 4)$ through which the line passes, and the point–slope form ● of a line, we obtain

● (3) $y - y_1 = m(x - x_1)$

$$y - 4 = -\tfrac{2}{3}(x - [-3]) \Longleftrightarrow y = -\tfrac{2}{3}x + (4 - 2)$$
$$\Longleftrightarrow y = -\tfrac{2}{3}x + 2$$

which represents the algebraic form of the line. We graph this line by taking advantage of the equation in slope–intercept form just obtained. We have

$$y = -\tfrac{2}{3}x + 2 \quad\nearrow\; m = -\tfrac{2}{3}$$
$$\searrow\; y\text{-intercept} = (0, 2)$$

Knowing the y-intercept $(0, 2)$ and the slope $-2/3$ we sketch the line as illustrated in Fig. 4.8.11, showing also the line that was given as being parallel to it.

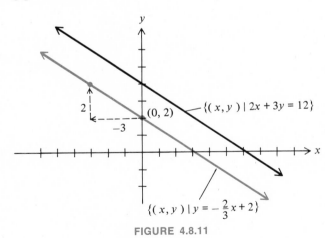

FIGURE 4.8.11

Example 4.8.6 Describe algebraically and geometrically the line passing through the points $(-3, 2)$ and $(2, 2)$.

Solution We first determine the slope. Let $(x_1, y_1) = (2, 2)$ and $(x_2, y_2) = (-3, 2)$, then

$$m = \frac{y_2 - y_1}{x_2 - x_1} = \frac{2 - 2}{-3 - 2} = \frac{0}{-5} = 0$$

● (3) $y - y_1 = m(x - x_1)$

Using the point $(2, 2)$ and the slope 0 in the point–slope form ● of a line, we obtain

$$y - 2 = 0(x - 2) \iff y - 2 = 0 \iff y = 2$$

To describe the linear function defined by $y = 2$, let us use the ordered pair notation. That is

$$L_{0,2} = \{(x, y) \mid y = 2\}$$

Because y is always 2, these ordered pairs take the form $(x, 2)$, where x can be *any* real number (there are no restrictions on x). Note that the *range* of this function is $\{2\}$. Recall that when the range of a function contains but a single real number, the relation is a constant function. The graph of this function is illustrated in Fig. 4.8.12. The graph of a constant function in the plane in rectangular coordinates is always a horizontal line passing through the y-axis at the constant value, in this case at 2.

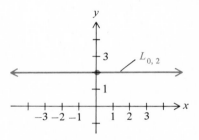

FIGURE 4.8.12

Example 4.8.7 Determine both algebraically and geometrically the line passing through the points $(2, 3)$ and $(2, -1)$.

Solution Let $(x_1, y_1) = (2, 3)$ and $(x_2, y_2) = (2, -1)$, then the slope would appear to be

$$m = \frac{-1 - 3}{2 - 2} = \frac{-4}{0}$$

But this expression, of course, is undefined. Thus, not one of the three algebraic forms of a line (two-point, point–slope, slope–intercept) is applicable here; they all require a defined slope. Of course there is no problem geometrically; because we have the two points, the line is easily drawn as indicated in Fig. 4.8.13. As you can see, the line is vertical. Vertical lines do *not* have a slope defined. Observe that the *domain* of this linear relation contains only one element, namely 2; that is, the variable x has only one admissible value. The *range,* on the other hand, is unrestricted and hence is R. This relation may be expressed

$$\{(x, y) \mid x = 2\}$$

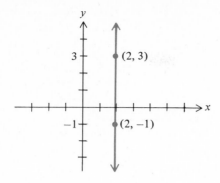

FIGURE 4.8.13

Notice that this relation is *not* a function. In fact, every pair of points lies one directly above the other.

Our final development of this section will be to derive a formula for the distance between two points. Let the points be denoted (x_1, y_1) and (x_2, y_2). Applying the Pythagorean theorem to the right triangle in Fig. 4.8.14, we obtain

$$d^2 = (x_2 - x_1)^2 + (y_2 - y_1)^2 \implies d = \sqrt{(x_2 - x_1)^2 + (y_2 - y_1)^2}$$

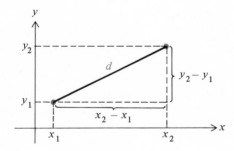

FIGURE 4.8.14

Example 4.8.8 Find the distance between the points $(2, -3)$ and $(-1, 1)$.

Solution Let $(x_1, y_1) = (2, -3)$ and $(x_2, y_2) = (-1, 1)$. Then

$$d = \sqrt{(-1 - 2)^2 + (1 - [-3])^2} = \sqrt{(-3)^2 + 4^2} = \sqrt{25} = 5$$
$$\implies d = 5$$

Let us now return to our applied problem with which we introduced this section.

Suppose a road construction company is designing a highway to pass over a hill. They are limited to a maximum grade (steepness) of 4 percent. Because of various soil conditions on the hill they have determined that a straight and flat road rising at the maximum grade

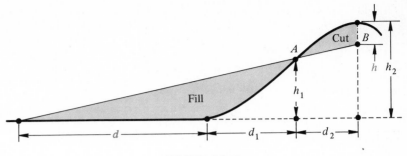

FIGURE 4.8.15

should intersect the hillside at a point A (see Fig. 4.8.15). They have also determined the height h_1 of point A above the valley floor along with h_2, the height of the hill, and in addition the horizontal distances d_1 and d_2. They wish to determine how much of the top of the hill should be cut away and also how far from the foot of the hill the fill should begin; that is, they wish to determine distances h and d, respectively, of Fig. 4.8.15.

Solution   First we set up our diagram in a rectangular coordinate system as illustrated in Fig. 4.8.16. Note that the fixed point A is now on the y-axis so that h_1 becomes the y-intercept of the line (the road) we wish to describe algebraically. We are given that the grade or slope of the line is 4 percent or $\frac{4}{100}$, which simplifies to $\frac{1}{25}$. Thus, using the *slope–intercept* form of a line we have

$$y = \tfrac{1}{25}x + h_1$$

Because the point where the fill begins occurs where the line crosses the x-axis, we set $y = 0$ and solve for x. We have

$$0 = \tfrac{1}{25}x + h_1 \iff x = -25h_1$$

From Fig. 4.8.16, observe that this value for x is also given by $x = d + d_1$. Setting these equal we have

$$d + d_1 = -25h_1 \iff d = -25h_1 - d_1$$

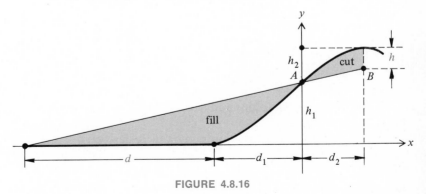

FIGURE 4.8.16

To find h we first let $x = d_2$. This will give the y-coordinate of the point B. Thus

$$y = \tfrac{1}{25}(d_2) + h_1$$

However, we also know from Fig. 4.8.16 that this value for y may be expressed $y = h_2 - h$. Setting these equal we have

$$\tfrac{1}{25}d_2 + h_1 = h_2 - h \Longleftrightarrow h = h_2 - h_1 - \tfrac{1}{25}d_2$$

With d and h now determined, we have solved the problem.

Problem Sets 4.9

Problem Set I ■ *Reading Comprehension*

Determine which of the problems 1 through 4 are true or false and state the reasons for your responses.

1 The slope m of a line passing through two distinct points (x_1, y_1) and (x_2, y_2) may be expressed

$$m = \frac{y_2 - y_1}{x_1 - x_2}$$

2 According to the definition of slope, a vertical line has no slope.
3 A horizontal line has no slope because the slope is zero.
4 All lines in the plane represent functions.

Problem Set II ■ *Skills Development*

1 Determine the algebraic form of each of the following lines and then graph them.
   a The line passing through $(2, -1)$ with slope $-\tfrac{1}{3}$.
   b The line passing through $(-3, -2)$ with slope $\tfrac{2}{5}$.
   c The line passing through $(-1, 4)$ with slope 4.
   d The line passing through $(0, -2)$ and $(3, 0)$.
   e The line passing through $(-2, 3)$ and $(-2, -1)$.
   f The line passing through $(-3, 2)$ and $(2, 2)$.
   g The line passing through $(0, 3)$ and parallel to the line defined by $2x - y = 3$.
   h The line passing through $(-2, -3)$ and parallel to the line defined by $3y - 2x = 6$.
   i The line passing through $(3, -1)$ and parallel to the line defined by $x - 2y = 4$.
   j The line whose x-intercept is 2 and y-intercept -3.
   k The line whose x-intercept is -1 and y-intercept 3.
   l The line whose x-intercept is 4 and y-intercept 2.
2 Determine algebraically which of the following sets of points lie on the same line:
   a $(-4, 4)$, $(-1, 2)$, $(5, -2)$
   b $(4, 4)$, $(2, 0)$, $(0, -4)$
   c $(-3, -1)$, $(0, 0)$, $(3, 2)$
3 Consider the linear function $f = \{(x, y) \mid y = f(x)\}$. If $(-1, 1)$, $(2, -3) \in f$, find $f(x)$.

4 If $g = \{(x, y) \mid y = -x + 2\}$, graph g and g^{-1} on the same set of axes.

5 If $h = \{(x, y) \mid y = \frac{1}{3}x - 1\}$, graph h and h^{-1} on the same set of axes.

6 If $f = \{(x, y) \mid y = 3\}$, graph f and f^{-1} on the same set of axes.

Problem Set III ■ *Theoretical Developments*

1 Show that the coordinates of the *midpoint* of the line segment joining the points (x_1, y_1) and (x_2, y_2) are given by

$$x = \frac{x_1 + x_2}{2} \quad \text{and} \quad y = \frac{y_1 + y_2}{2}$$

(HINT: Use properties of similar triangles or the distance formula.)

2 Show that the formula for the point (x, y) that divides the line segment joining the points (x_1, y_1) and (x_2, y_2) into two segments whose ratio is a/b is given by

$$x = \frac{bx_1 + ax_2}{a + b} \quad \text{and} \quad y = \frac{by_1 + ay_2}{a + b}$$

Show also that these reduce to the formulas of problem 1 above when $a = b$.

3 Show that two lines defined by $y = m_1x + b_1$ and $y = m_2x + b_2$, where $m_1 \neq m_2$, will intersect at the point whose coordinates are

$$x = \frac{b_2 - b_1}{m_1 - m_2} \quad \text{and} \quad y = \frac{m_1b_2 - b_1m_2}{m_1 - m_2}$$

4 Show that the three lines issuing from the vertices of a triangle and intersecting the opposite side at their midpoint will all intersect at a common point. (See Fig. 4.9.1.)

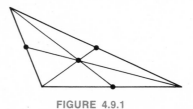

FIGURE 4.9.1

4.10 Quadratic Functions and Relations

FIGURE 4.10.1

We begin with a problem.

Problem 4.10.1 An airlines running passenger service between a large city and a ski resort has a special round-trip charter flight for skiers. The fare is $60.00 per person if at least 100 take the special. As a bonus, however, they will reduce the fare by the continuous ratio of $2.00 for each 10 passengers over the 100. From experience it has been established that the extra cost to the airlines for each passenger over 100 is $10.00. Determine the fare and number of passengers that will bring a *maximum* net revenue to the airlines from this charter flight.

The mathematical tools we will develop in this section will assist us in analyzing certain maximum and minimum problems such as this one as well as many other types of problems involving quadratic functions.

An open equation equivalent to one of the form

$$ax^2 + bxy + cy^2 + dx + ey + f = 0$$

where $a, b, c, d, e, f \in R$ and a, b, c are not all zero is said to be quadratic over R. Relations defined by quadratic open equations are called conic sections or simply conics. Here we will restrict our brief study of the conics to those defined by equations equivalent to

$$y = ax^2 + bx + c \qquad (1)$$

where $a, b, c \in R$ and $a \neq 0$.● The graphical representation of a relation defined by an equation of the above form is called a parabola.

● A more complete treatment of the conics is found in Chapter 12 on coordinate geometry.

Example 4.10.1   Graph the parabola defined by

$$g = \{(x, y) \mid y = -x^2 + 4\}$$

Solution   First we set up a table of points satisfying the defining equation. Plotting these points, we sketch the graph as illustrated in Fig. 4.10.2.

x	$y = -x^2 + 4$	y	(x, y)
0	$y = -0^2 + 4$	4	$(0, 4)$
1	$= -1^2 + 4$	3	$(1, 3)$
-1	$= -(-1)^2 + 4$	3	$(-1, 3)$
2	$= -(2)^2 + 4$	0	$(2, 0)$
-2	$= -(-2)^2 + 4$	0	$(-2, 0)$
3	$= -(3)^2 + 4$	-5	$(3, -5)$
-3	$= -(-3)^2 + 4$	-5	$(-3, -5)$

Table 4.10.1

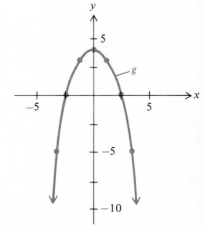

FIGURE 4.10.2

The parabola in Fig. 4.10.2 is said to be cup-down. All parabolas defined by equations of the form (1) are either cup-down or cup-up, depending upon the sign of the coefficient a (see Fig. 4.10.3). Of particular significance is the peak or maximum point for a cup-down parabola and the lowest or minimum point for a cup-up parabola. Also important are the zeros of a parabola; these are the x-coordinates of the points where the parabola intercepts the x-axis, when

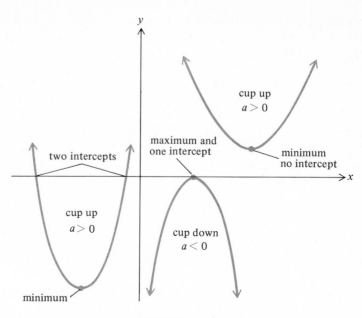

cup up
a > 0

maximum and
one intercept

two intercepts

minimum
no intercept

cup up
a > 0

cup down
a < 0

minimum

FIGURE 4.10.3

they exist. These special points are identified for several cases in Fig. 4.10.3.

To identify the maximum and minimum points, we employ a technique introduced in the previous chapter—the method of completing the square. By completing the square we obtain the following equivalence.

$$y = ax^2 + bx + c \iff y = a\left(x + \frac{b}{2a}\right)^2 - \frac{b^2 - 4ac}{4a}$$

It is from this latter equivalent form, that we identify the maximum or minimum. We reason as follows: The only term on the right side of the complete square equation containing the variable x is $a(x + b/2a)^2$. Because $(x + b/2a)^2 \geq 0$ for all x,● the term

● T2.11.1(III) $a \in R \Longrightarrow a^2 \geq 0$

$$a\left(x + \frac{b}{2a}\right)^2 \geq 0 \qquad \text{if } a > 0$$

and

$$a\left(x + \frac{b}{2a}\right)^2 \leq 0 \qquad \text{if } a < 0$$

This means, if $a > 0$, the expression

$$a\left(x + \frac{b}{2a}\right)^2$$

becomes increasingly large in the positive direction as $|x|$ becomes large (which allows x to take on increasingly large values in either the positive or negative directions). But because

$$y = a\left(x + \frac{b}{2a}\right)^2 - \frac{b^2 - 4ac}{4a}$$

we may conclude that y also becomes increasingly large as $|x|$ increases. Therefore, if $a > 0$, the parabola must exhibit a *cup-up* configuration. Reasoning similarly, if $a < 0$, y must become increasingly large negatively as $|x|$ increases, and hence such a parabola would possess a *cup-down* orientation.

To identify an *extremum* (a maximum or a minimum), we will first take the case where $a > 0$, which as analyzed above is a cup-up parabola. Hence, for this case we wish to find the minimum — lowest point. Because the only term in the expression for y, which varies with x, is $a(x + b/2a)^2$ and because this term is *greater than* or *equal* to zero, y will be minimized when this term is minimized. This is achieved by setting $x = -b/2a$, which reduces the term, to zero. For any other choice of x, the term $a(x + b/2a)^2$ would be greater than zero and hence the value for y would be increased. This substitution yields

$$y = a\left(-\frac{b}{2a} + \frac{b}{2a}\right)^2 - \frac{b^2 - 4ac}{4a} = -\frac{b^2 - 4ac}{4a}$$

Thus, the coordinates of the minimum point are

$$\left(-\frac{b}{2a}, \frac{4ac - b^2}{4a}\right)$$

We now examine the other case where $a < 0$, that is, a cup-down parabola possessing a maximum point. Again we refer to the completed square form for y, that is,

$$y = a\left(x + \frac{b}{2a}\right)^2 - \frac{b^2 - 4ac}{4a}$$

The only term containing the variable x in this expression is $a(x + b/2a)^2$, which for $a < 0$ is *less than* or *equal* to zero. Thus, to maximize y we must again minimize the effect of this term. As before, this is accomplished by setting $x = -b/2a$, which reduces the term to zero, and we obtain as the coordinates of the maximum point the exact entries obtained for the minimum, namely,

$$\left(-\frac{b}{2a}, \frac{4ac - b^2}{4a}\right)$$

This formula, therefore, will *always* identify the extremum for a parabola, whether cup-up or cup-down.

Although the above *formula* for the extreme values of a parabola may be used, it is important to become familiar with the "method of completing the square" because this is a mathematical tool frequently used in solving other mathematical problems, particularly some encountered in calculus.

The other special points we wish to identify are the *x-intercepts* (zeros of the function) when they exist. The x-intercepts are, of course, those points of the parabola lying on the x-axis, and hence those points for which the *y* coordinate is *zero*. For such a point, the equation defining the parabola, namely, $y = ax^2 + bx + c$, becomes

$$ax^2 + bx + c = 0 \qquad\qquad (2)$$

and this you will recall from Chapter 3 is called the quadratic equation. Furthermore, from Chapter 3 we proved the equivalence

$$ax^2 + bx + c = 0 \Longleftrightarrow x = \frac{-b \pm \sqrt{b^2 - 4ac}}{2a}$$

the latter equation being called the quadratic formula. It was from this formula that we observed the following characterization for the zeros of a quadratic equation using the discriminant $b^2 - 4ac$.

$b^2 - 4ac > 0 \Longrightarrow$ two real and unequal zeros (crosses twice)

$b^2 - 4ac = 0 \Longrightarrow$ one real zero (just touches once)

$b^2 - 4ac < 0 \Longrightarrow$ no real zeros (does not cross anywhere)

For the case $b^2 - 4ac = 0$, the x-intercept is also the extremum— the maximum if $a < 0$ and the minimum if $a > 0$.

One final observation before looking at some examples. Whenever a function possesses the property

$$f(a + x) = f(a - x)$$

for some real constant a and for all real x, then we are assured geometric symmetry about the vertical line defined by $x = a$. (See Fig. 4.10.4.) Let

$$f(x) = a\left(x + \frac{b}{2a}\right)^2 - \frac{b^2 - 4ac}{4a},$$

then

$$f\left(\frac{-b}{2a} + x\right) = a\left[\left(\frac{-b}{2a} + x\right) + \frac{b}{2a}\right]^2 - \frac{b^2 - 4ac}{4a} = ax^2 - \frac{b^2 - 4ac}{4a}$$

and

$$f\left(\frac{-b}{2a} - x\right) = a\left[\left(\frac{-b}{2a} - x\right) + \frac{b}{2a}\right]^2 - \frac{b^2 - 4ac}{4a} = ax^2 - \frac{b^2 - 4ac}{4a}$$

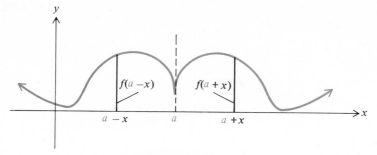

FIGURE 4.10.4

Therefore, $f(-b/2a + x) = f(-b/2a - x)$ and we conclude that the graph of a parabola is symmetric about the vertical line passing through $x = -b/2a$. But this is the x coordinate of the *extremum* point. Thus a parabola always is symmetric about the vertical line passing through its maximum or minimum point.

Example 4.10.2   Let $f = \{(x, y) \mid y = 2x^2 + 4x - 6\}$. Find the maximum or minimum point and the x-intercepts; then sketch the graph.

Solution   Applying the method of completing the square, we obtain

$$y = 2x^2 + 4x - 6 = 2[x^2 + 2x - 3] = 2[x^2 + 2x + (1)^2 - (1)^2 - 3]$$
$$= 2[(x + 1)^2 - 4] = 2(x + 1)^2 - 8$$

Because the leading coefficient 2 is positive, the parabola is *cup-up;* thus, we wish to find the minimum. Because

$$y = 2(x + 1)^2 - 8$$

it is clear the minimum value for y is -8, which is achieved when $x = -1$. Hence, the minimum point is $(-1, -8)$.

To determine the x-intercepts, we take the defining equation and substitute 0 for y. This yields the quadratic equation

$$2x^2 + 4x - 6 = 0 \Longleftrightarrow 2(x^2 + 2x - 3) = 0$$
$$\Longleftrightarrow x^2 + 2x - 3 = 0$$
$$\Longleftrightarrow (x + 3)(x - 1) = 0$$
$$\Longleftrightarrow x + 2 = 0 \quad \text{or} \quad x - 1 = 0$$
$$\Longleftrightarrow x = -3 \quad \text{or} \quad x = 1$$

To graph this parabolic function we plot the minimum point $(-1, -8)$ and the two x-intercepts $(1, 0)$ and $(-3, 0)$. Knowing basically the form of a parabola, we join these points by a smooth curve to obtain a reasonably accurate graph, as illustrated in Fig. 4.10.5.

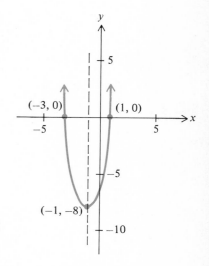

FIGURE 4.10.5

Example 4.10.3 Let $g = \{(x, y) \mid y = -x^2 + 3x - 3\}$. Find the maximum or minimum point, the x-intercepts, and then graph.

Solution By the method of completing the square, we have

$$y = -x^2 + 3x - 3 = -[x^2 - 3x + 3]$$

$$= -\left[x^2 - 3x + \left(\frac{-3}{2}\right)^2 - \left(\frac{-3}{2}\right)^2 + 3\right]$$

$$= -\left[\left(x - \frac{3}{2}\right)^2 + \frac{3}{4}\right] = -\left(x - \frac{3}{2}\right)^2 - \frac{3}{4}$$

Because the leading coefficient is -1, the parabola is *cup-down*. From the completed square form of the equation, that is,

$$y = -\left(x - \frac{3}{2}\right)^2 - \frac{3}{4}$$

it is clear that y will attain a *maximum* of $-\frac{3}{4}$ when $x = \frac{3}{2}$. Hence, the maximum point is $\left(\frac{3}{2}, -\frac{3}{4}\right)$.

Notice that this *maximum* point is *below* the x-axis; therefore, there are *no x-intercepts*. Nevertheless, let us verify this by checking the sign of the discriminant $b^2 - 4ac$ of the quadratic equation $-x^2 + 3x - 3 = 0$. Here $a = -1$, $b = 3$, and $c = -3$. Substituting these values into the discriminant yields

$$b^2 - 4ac = 3^2 - 4(-1)(-3) = 9 - 12 = -3 < 0$$

As we observed earlier, a *negative* discriminant implies that the parabola has no x-intercepts.

To obtain a reasonable graph of g, where no x-intercepts exist, we must determine a few other points in addition to the extremum. These can be obtained, of course, from the defining equation

$$y = -x^2 + 3x - 3$$

For example, if $x = 3$, then

$$y = -(3)^2 + 3(3) - 3 = -9 + 9 - 3 = -3$$

and further, if $x = 0$, then

$$y = -(0)^2 + 3(0) - 3 = -3$$

Thus, the points $(3, -3)$ and $(0, -3)$ are elements of g. The graph of g can now be sketched as in Fig. 4.10.6.

Remark 4.10.1 One point you can always quickly obtain is the y-intercept. Because $x = 0$ at this point, we have

$$y = a(0)^2 + b(0) + c \Longleftrightarrow y = c$$

Therefore the point $(0, c)$ will always lie on the graph.

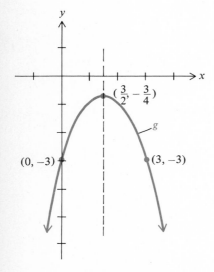

FIGURE 4.10.6

Now let us return to our problem concerning that airlines' skiers charter flight.

Problem 4.10.1 An airline running passenger service between a large city and a ski resort has a special round-trip charter flight for skiers. The fare is $60.00 per person if at least 100 people take the special. As a bonus, however, they will reduce the fare by a continuous ratio of $2.00 for each 10 passengers over the 100. From experience it has been established that the extra cost to the airline for each passenger over 100 is $10.00. Determine the fare and number of passengers that will bring a *maximum* net revenue to the airline from this charter flight.

Solution Let R represent the net revenue, x the number of passengers on the flight, and f the fare charged each passenger. Then

$$R = \overbrace{f \cdot x}^{\text{revenue from fare}} - \overbrace{10(x - 100)}^{\left\{\begin{array}{l}\text{minus the extra cost to the airline} \\ \text{for the passengers in excess of 100}\end{array}\right.}$$

But

$$f = \left[60 - \left(\frac{x - 100}{10}\right)2\right]$$

Therefore

$$R = \left[60 - \left(\frac{x - 100}{10}\right)2\right]x - 10(x - 100)$$

$$= \left[\frac{300 - x + 100}{5}\right]x - 10x + 1000$$

$$= -\tfrac{1}{5}[-350x + x^2 - 5000]$$

$$= -\tfrac{1}{5}[x^2 - 350x + (175)^2 - (175)^2 - 5000] \qquad \text{(completing the square)}$$

$$= -\tfrac{1}{5}[(x - 175)^2 - 35625]$$

$$= -\tfrac{1}{5}(x - 175)^2 + 7125$$

Thus, with the leading coefficient negative we can see that we will obtain a *maximum* of $7125.00 when the number of passengers is 175. To find the air fare charged each passenger, we first add to the net revenue the extra cost to the airlines for the passengers over 100, which would be $750.00, and then divide this amount by the number of passengers. Performing these operations, we obtain

$$\frac{(7125 + 750)}{175} = 45$$

Hence, the fare paid by each passenger is $45.00 instead of the original $60.00.

4.11 Problem Sets

Problem Set I ■ *Reading Comprehension*

Determine whether problems 1 through 3 are true or false and state the reasons for your responses.

1 If a parabola is defined by an equation of the form $y = ax^2 + bx + c$, where $a, b, c \in R$ and $a > 0$, then the parabola must possess a minimum.
2 Every quadratic function has two zeros.
3 A function defined by an equation of the form $y = ax^2 + bx + c$, where $a, b, c \in R$ and $a \neq 0$, may not have an extremum.

Problem Set II ■ *Skills Development*

1 For each of the following parabolic functions find the maximum or minimum point by completing the square, find the x-intercepts if they exist, and graph the curve.
   a $g = \{(x, y) \mid y = x^2 - 3x + 2 \quad \text{over } R\}$
   b $h = \{(x, y) \mid y = -x^2 + 4x - 2 \quad \text{over } R\}$
   c $k = \{(x, y) \mid y = -\frac{1}{2}x^2 + 2x \quad \text{over } R\}$
   d $m = \{(x, y) \mid y = 2x^2 - 5x \quad \text{over } R\}$
   e $n = \{(x, y) \mid y = 2x^2 - 14x + 24 \quad \text{over } R\}$
   f $p = \{(x, y) \mid y = -2x^2 - 6x + 36 \quad \text{over } R\}$
2 Let $f = \{(x, y) \mid y = 4x^2 - 12x + 9\}$. Graph f and f^{-1} on the same set of axes.
3 By means of the discriminant, determine whether the graphs defined by the following equations cross the x-axis twice, once, or not at all.
   a $y = x^2 - 4x - 5$ b $y = -2x^2 + 3x - 2$ c $y = x^2 - 2x + 1$
   d $y = -x^2 + 5x - 4$ e $y = 4x^2 - 12x + 9$ f $y = x^2 + x + 1$
4 Determine the values of k for which the graphs defined by the following equations will have no x-intercepts, one x-intercept, or two x-intercepts.
   a $y = kx^2 - 3x + 1$ b $y = x^2 + kx + 2$ c $y = 2x^2 - 4x + k$

Problem Set III ■ *Theoretical Developments*

1 On a single set of axes, sketch the family of parabolas defined by $y = -x^2 + k$ where $k \in \{-2, 0, 1, 2\}$. How do the members of the family vary as k varies?
2 On a single set of axes, sketch the family of parabolas defined by $y = kx^2 + 2$ where $k \in \{-2, -1, 1, 2\}$. How do the members of the family vary as k varies?

Problem Set IV ■ *Applications*

1 Determine the dimensions of a rectangle whose perimeter is 40 inches if its area is to be a maximum.
2 Find the dimensions of a rectangle whose perimeter is 30 feet and such that the square of its diagonal is a minimum.
3 It is a well-known fact that if a body is thrown vertically upward with an initial velocity

of 60 feet per second (neglecting the effects of air resistance), then the distance it travels in t seconds is given by the formula

$$s = 60t - 16t^2$$

After how many seconds will the body reach its maximum height? After how many seconds will it hit the ground? How far will it have traveled when it hits the ground? Graph the function

$$d = \{(t, s) \mid s = 60t - 16t^2\}$$

4 Suppose after careful study it is determined that the number of sales of a certain product varies as the cost per unit varies in the following manner: 500 are sold in one day if the price per unit is 25 cents, 450 if the price is 30 cents, 400 if the price is 35 cents, and so forth. That is, if the price is increased by 5 cents, the number of sales decreases by 50 units. Determine the cost per unit that will yield a maximum of gross receipts in one day.

5 Suppose a magazine company has determined that it can obtain 500 new subscriptions each month if the yearly rate is $5.00, and 50 more subscriptions for each 10-cent decrease in the yearly rate. What yearly rate will yield a maximum yearly gross income from new subscribers and what is the amount?

6 A farmer wishes to build a rectangular enclosure with 40 yards of fencing; one side of the enclosure will be a barn and hence will need no fencing. What should the dimensions be in order to obtain a maximum area?

7 A school wishes to make a rectangular enclosure to be used as a playground by the first-grade children only. They have 30 yards of fencing available. One side of the enclosure will be the school itself and hence will need no fencing. What should be the dimensions of the enclosure in order to obtain a maximum area?

8 Suppose a manufacturer of furniture loses $100.00 an hour when the plant is idle, whereas if operating at peak capacity, which is 500 units per hour, its profit earnings per hour amount to $600.00. Assume the relation between units produced per hour and profit per hour is a linear function. Determine an algebraic model of this function and graph. How many units per hour must be produced to "break even?"

Inequalities 4.12

Thus far in this chapter we have limited ourselves to open sentences that are equations. Recall that an *open sentence* is an equation or an *inequality* containing one or more variables. We will now treat certain special inequalities in two variables. Let us begin by presenting a problem that will involve an inequality.

It is not uncommon in business to have the cost C of the production of an item be a linear function of the number of items produced. Similarly, the revenue R received from the sale of those items may also be a linear function of the number of items sold. For the business to be a success, it is critical, of course, for the revenue R from sales to exceed the costs C of production, that is, it must experience a profit. When there is no loss nor any profit, we have what is referred to as the *break-even point*. Consider the following problem.

Problem 4.12.1 A bakery has a daily overhead from salaries and building cost of $2000.00. The cost of ingredients for a certain loaf of bread is $0.20 and it is sold for $0.50. How many loaves of bread must be made and sold each day in order to guarantee a profit?

Before solving this problem we introduce the technique commonly used to graph inequalities by analyzing the following example.

Example 4.12.1 Graph the relation

$$g = \{(x, y) \mid y < -\tfrac{1}{2}x + 3\}$$

Solution Recall the first order axiom.

O-1 $\quad a, b \in R \implies a < b, a = b, \quad$ or $a > b$

Because the variables x and y in the relation g are real, we have, applying O-1, one and only one of the following three possibilities for each point (x, y).

$$y < -\tfrac{1}{2}x + 3$$

$$y = -\tfrac{1}{2}x + 3$$

or

$$y > -\tfrac{1}{2}x + 3$$

We will now show that the curve defined by the equality

$$y = -\tfrac{1}{2}x + 3$$

separates those points satisfying

$$y > -\tfrac{1}{2}x + 3 \quad \text{and} \quad y < -\tfrac{1}{2}x + 3$$

We graph the set of points satisfying the equation

$$y = -\tfrac{1}{2}x + 3$$

Note that this equation is in the slope–intercept form of a line – the *slope* is $-\tfrac{1}{2}$ and the *y-intercept* is 3. The graph of this line is dotted *in red* in Fig. 4.12.1. To determine the location of those points satisfying $y < -\tfrac{1}{2}x + 3$, draw a vertical line through an arbitrary point on the x-axis, say x_1. Let the point where this vertical line intercepts the one defined by $y = (-\tfrac{1}{2})x + 3$ be (x_1, y_1). We have for this point, of course,

$$y_1 = -\tfrac{1}{2}x_1 + 3$$

Now let the point (x_1, y_2) be any point on the vertical line above (x_1, y_1), and (x_1, y_3) any point below (x_1, y_1). Recall, on the y-axis if y_2 is above y_1, and y_1 above y_3 then

$$y_3 < y_1 < y_2$$

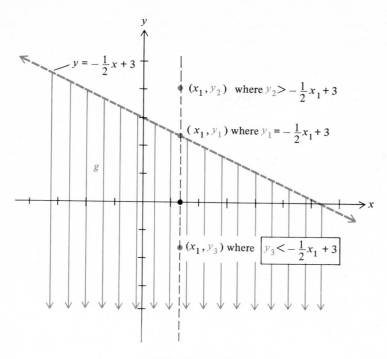

$y = -\frac{1}{2}x + 3$

(x_1, y_2) where $y_2 > -\frac{1}{2}x_1 + 3$

(x_1, y_1) where $y_1 = -\frac{1}{2}x_1 + 3$

g

(x_1, y_3) where $\boxed{y_3 < -\frac{1}{2}x_1 + 3}$

FIGURE 4.12.1

But, because $y_1 = -\frac{1}{2}x_1 + 3$, this inequality may also be expressed as

$$y_3 < -\tfrac{1}{2}x_1 + 3 < y_2$$

Because the vertical line intercepted the x-axis at an *arbitrary* point x_1, the above analysis holds no matter where the coordinate x_1 is positioned. Therefore, we conclude that *all* points *below* the line defined by $y = -\frac{1}{2}x + 3$ satisfy

$$y < -\tfrac{1}{2}x + 3$$

whereas *all* points *above* satisfy

$$y > -\tfrac{1}{2}x + 3$$

Hence, the graph of g is as illustrated in Fig. 4.12.1. Note that because the points satisfying $y = (-\frac{1}{2})x + 3$ are not in g, the line so defined is only dotted in, to suggest it is not included in g, although it is an upper boundary for g.

Example 4.12.2 Graph the relation

$$h = \left\{ (x, y) \mid y \geq \frac{x^2 - 9}{6} \right\}$$

Solution We proceed exactly as in the previous example by first graphing those points satisfying

$$y = \frac{x^2 - 9}{6}$$

This graph is a *cup-up parabola*. Because the square is already completed, the *minimum* point is readily determined as $(0, -\frac{3}{2})$ and the *x*-intercepts are $(-3, 0)$, $(3, 0)$. The graph of this parabola is drawn in Fig. 4.12.2. Again, to motivate our analysis, we draw a vertical line through an arbitrary point x_1 on the *x*-axis. Let (x_1, y_1) be the point where this line intercepts the parabola, that is, the point for which

$$y_1 = \frac{x_1{}^2 - 9}{6}$$

Further, let (x_1, y_2) be any point *below* (x_1, y_1) on the vertical line and (x_1, y_3) any point *above* (x_1, y_1). An obvious relationship between y_1, y_2, and y_3 is

$$y_2 < y_1 < y_3$$

Hence,

$$y_2 < \frac{x_1{}^2 - 9}{6} < y_3$$

That is

$$y_3 > \frac{x_1{}^2 - 9}{6}$$

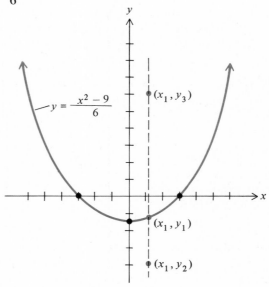

FIGURE 4.12.2

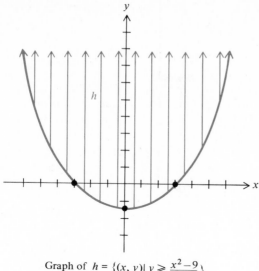

Graph of $h = \{(x, y)\mid y \geqslant \dfrac{x^2 - 9}{6}\}$

FIGURE 4.12.3

Thus, *all* points *on or above* the parabola satisfy the open inequality

$$y \geq \frac{x^2 - 9}{6}$$

and therefore all such points comprise the relation h. The graph of h is illustrated in Figure 4.12.3. In this case the boundary is a part of h and hence is not dotted as was the boundary in Example 4.12.1.

Example 4.12.3 Graph the relation

$$f = \{(x, y) \mid 3y - 2x + 6 \geq 0 \quad \text{and} \quad y < -x^2 + 4x\}$$

Solution We shall determine the points satisfying the two open inequalities separately first and then take the intersection of these two sets to obtain f. We begin with the linear inequality $3y - 2x + 6 \geq 0$. First, solve for y, obtaining

$$3y - 2x + 6 \geq 0 \Longleftrightarrow y \geq \tfrac{2}{3}x - 2$$

Then graph the line defined by $y = \tfrac{2}{3}x - 2$. Being in slope–intercept form we have slope $\tfrac{2}{3}$ and y-intercept $(0, -2)$. All points *on or above* this line satisfy

$$y \geq \tfrac{2}{3}x - 2$$

(See Fig. 4.12.4). Next we take the quadratic inequality $y < -x^2 + 4x$. Because we must graph the related equality, we shall first complete the square on the equation

$$y = -x^2 + 4x$$

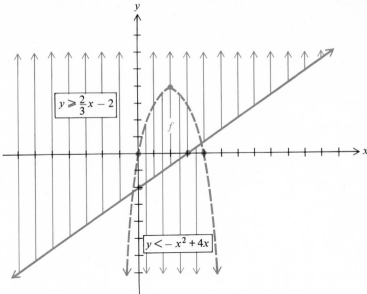

Graph of $f = \{(x, y) \mid y \geqslant (\frac{2}{3})x - 2$ and $y < -x^2 + 4x\}$

FIGURE 4.12.4

to determine the maximum point. Thus

$$y = -x^2 + 4x = -[x^2 - 4x] = -[x^2 - 4x + 4 - 4]$$

$$= -[(x - 2)^2 - 4]$$

$$= -(x - 2)^2 + 4$$

Thus, the maximum value of y is 4, which will be attained when $x = 2$; so the *maximum point* is $(2, 4)$. The x-intercepts must satisfy

$$-x^2 + 4x = 0 \Longleftrightarrow x(x - 4) = 0$$

$$\Longleftrightarrow x = 0 \quad \text{or} \quad x = 4$$

Hence, the intercepts are $(0, 0)$ and $(4, 0)$. With the maximum point at $(2, 4)$ and x-intercepts at $(0, 0)$ and $(4, 0)$, we dot in the graph as illustrated in Fig. 4.12.4. Because the points we desire must satisfy the inequality

$$y < -x^2 + 4x$$

they would lie *below* this cup-down parabola. The relation f now becomes the intersection of these two plane areas as indicated by the doubly shaded area of Fig. 4.12.4.

We return now to our applied problem.

Problem 4.12.1 A bakery has a daily overhead from salaries and building cost of $2000.00. The cost of ingredients for a certain loaf of bread is $0.20 and it is sold for $0.50. How many loaves of bread must be made and sold each day in order to guarantee a profit?

Solution   We first determine a formula for daily costs as follows:

$$C = 2000 + 0.20x$$

where x represents the number of loaves produced and sold. The revenue received is given by

$$R = 0.50x$$

These two equations may also be written

$$C = \tfrac{1}{5}x + 2000 \quad \text{and} \quad R = \tfrac{1}{2}x$$

Our analysis of the problem will be determined from the graphs of these two linear functions (see Figure 4.12.5). It is clear that the *break-even point* is 6667. For a profit they must sell more than 6667 loaves, and the problem is solved.

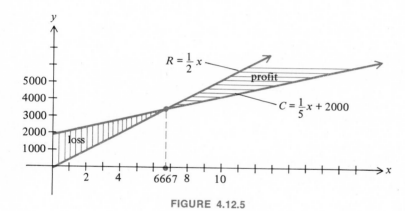

FIGURE 4.12.5

Problem Set I   ▨   *Reading Comprehension*

Determine whether problems 1 through 3 are true or false and state the reasons for your responses.

1   An inequality in two variables x and y does not represent an open sentence because functions are defined by open sentences and an inequality does not define a function.
2   The graph of a relation defined by an inequality in two variables is usually obtained by first working with an equation in two variables.
3   The plane area defined by the open sentence $2x - y < 4$ will lie below the line defined by $2x - y = 4$.

Problem Set II ■ Skills Development

Graph the following relations over R:

1 $\{(x, y) \mid y < x^2 - 3x + 2\}$
2 $\{(x, y) \mid y \geq -x^2 + 4x - 2\}$
3 $\{(x, y) \mid y \leq (-\frac{1}{2})x + 2\}$
4 $\{(x, y) \mid y > (\frac{2}{3})x - 1\}$
5 $\{(x, y) \mid y \geq (\frac{1}{2})x - 2 \quad \text{and} \quad y < x^2 + 4\}$
6 $\{(x, y) \mid y < -2x + 3 \quad \text{and} \quad y > x^2 - 2x - 1\}$

Problem Set III ■ Applications

1 A manufacturer produces two products, A and B. The only outlet for these products can receive up to 12 units of A and 8 units of B per day. Determine the relation that exists showing the possible combinations of the two products that can be shipped daily and then graph this relation.

2 A flour mill produces a high quality mix by combining two ingredients, A and B. Ingredient A contains 4 ounces of protein per pound and B 1 ounce per pound. Law requires that each 10-pound bag contain at least 1 pound of protein. Find the relation that determines the combinations of ingredients A and B satisfying this requirement and then graph it.

3 Suppose a manufacturing firm produces two styles of bicycles, which we call A and B. Further, suppose each unit of product A requires 6 man-hours to produce and each unit of B requires 10 man-hours to produce; finally, suppose the firm has a maximum of 400 man-hours available each week. If the only restraint upon weekly production is the maximal number of man-hours available, determine an algebraic relation that identifies the combinations of the two products that the firm is capable of producing weekly, and then represent the relation graphically.

4 A candy factory has a daily overhead from salaries and building costs of $1500.00. The cost of the ingredients to produce a pound of candy is $1.00; the candy is sold for $2.50 a pound. How much candy must be produced and sold for the business to make a profit? If the selling price is increased to $3.00 a pound, what is the *break-even point?* If it is known that at least 2000 pounds can be produced daily, what should be charged per pound to guarantee no loss?

5 A manufacturer produces bicycles at a daily cost of $50.00 per unit and sells them for $75.00 a piece. His operational overhead is $300.00 a day. How many bicycles must he produce a day to make a profit? If the manufacturer is able to reduce his cost per item to $40.00 a unit but with an increase of operational overhead from $300.00 to $350.00, is it to his advantage to do so?

Problem Set IV ■ Just For Fun

1 The following problem is an old one, believed to have been proposed by Metrodorus in 310 A.D.: Demochares has lived one-fourth of his life as a boy, one-fifth as a youth, one-third as a man, and has spent thirteen years in old age. How old is he?

2 An *unlimited supply* of gasoline is available at one edge of a desert 800 miles wide, but there is no source on the desert itself. A truck can carry enough gasoline to go 500

miles (this will be called one "load"), and it can build up its own refueling stations at any spot along the way. These caches may be any size, and it is assumed that there is no evaporation loss.

What is the minimum amount (in loads) of gasoline the truck will require in order to cross the desert? Is there a limit to the width of a desert the truck can cross?

Summary of Definitions and Properties in Chapter 4   4.14

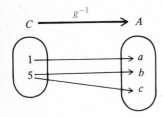

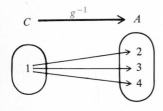

1 This is not a relation of A into B because an element in A, namely 3, is not paired with an element in B. (Note Definition 4.2.1.)

2 This is a relation with domain $\{a, b, c\}$ and range $\{1, 5\}$. The inverse of g can be represented as shown in the top figure or

$$g^{-1} = \{(1, a), (5, b), (5, c)\}$$

3 This is a relation with domain $\{2, 3, 4\}$ and range $\{1\}$. The inverse of g can be represented as shown in the bottom figure or

$$g^{-1} = \{(1, 2), (1, 3), (1, 4)\}$$

4 This is a relation with domain $\{1, 2\}$ and range $\{4, -3, 1\}$. The inverse is $f^{-1} = \{(4, 1), (-3, 2), (1, 2)\}$.

5 This is a relation with domain $\{x \mid x \neq 2 \quad$ and $\quad y \in R\}$. The inverse can be represented as

$$g^{-1} = \left\{(x, y) \mid x = \frac{2y - 1}{y + 2} \quad \text{and} \quad x, y \in R\right\}$$

or

$$g^{-1} = \left\{(x, y) \mid y = -\frac{2x + 1}{x - 2} \quad \text{and} \quad x, y \in R\right\}$$

6 This is not a relation because the set is empty. (See Definition 4.2.2.)

7 This is a relation with domain $\{x \mid x \geq 3 \quad$ or $\quad x \leq -3\}$ and range $\{y \mid y \geq 0\}$. The inverse can be represented as

$$f^{-1} = \{(x, y) \mid x = \sqrt{y^2 - 9} \quad \text{and} \quad x, y \in R\}$$

or

$$f^{-1} = \{(x, y) \mid y = \pm\sqrt{x^2 + 9}, x \geq 0 \quad \text{and} \quad x, y \in R\}$$

1 This relation is not 1-1, is onto, is not a function, and the inverse relation is not a function.

2 This relation is 1-1, is not onto, is a function, and the inverse relation is a function.

3 This relation is not 1-1, is not onto, is a function, and the inverse relation is not a function.

4 This relation is 1-1, is onto, is a function, and the inverse relation is a function.

5 This relation is not 1-1 $((3, 4), (-3, 4) \in f)$, is onto, is a function, and the inverse relation is not a function.

6 This relation is 1-1, is onto, is a function, and the inverse relation is a function.

"You've got to hand it to those computers."
Drawing by Al Kaufman; © 1968 Saturday Review, Inc.

Exponential and Logarithmic Functions

5 CHAPTER

Introduction 5.1

Exponential and logarithmic functions (which we find later are just inverses of each other) are probably the two most common and significant classes of functions in all mathematical analysis. In calculus (the mathematics of motion and change), for example, we discover certain exponential functions to be the simplest and most natural with which to work. In applications, it is surprising how frequently problems associated with natural phenomena (whether physical, sociological, economical, or otherwise) are analyzed and solved by means of exponential or logarithmic functions. Examples include the relationship between atmospheric pressure and altitude, problems of heat conduction, motion against friction, population growth, radioactive decay, and current patterns in electrical circuits.

Thus, you can appreciate the importance of a careful study of these significant functions. Our presentation here will be, of course, merely introductory, but one from which you may easily launch into a more in-depth study.

5.2 Exponents and Roots

We frequently encounter products where the factors are all the same. For example,

$$3 \cdot 3 \cdot 3 \cdot 3 \cdot 3 \cdot 3 \cdot 3 \cdot 3$$

If we had to repeat this expression several times within the context of some problem, we would find it rather cumbersome and tedious to work with. In fact if these expressions occurred very often you would probably be looking for a shorter, more compact way to express them. The way this is commonly done is to write the factor 3 just once but attach a superscript 8 to it signifying that the factor 3 is repeated 8 times. That is, we write

$$3^8 = 3 \cdot 3 \cdot 3 \cdot 3 \cdot 3 \cdot 3 \cdot 3 \cdot 3$$

In general, we have

DEFINITION 5.2.1 If $a \in R$ and $n \in N$, then

$$a^n = \overbrace{a \cdot a \cdot a \ldots a}^{n \text{ factors of } a}$$

The number a is called the base of the exponential expression and the positive integer n is called the exponent of a. The expression a^n is read "a to the nth power" or "the nth power of a."

Example 5.2.1

(a) $\quad (-2)^3 = (-2)(-2)(-2) = -8$

(b) $(-2)^3(-2)^2 = [(-2)(-2)(-2)][(-2)(-2)]$
$$= (-2)^5 = -32$$

(c) $\quad [(-2)^3]^2 = [(-2)(-2)(-2)]^2$
$$= [(-2)(-2)(-2)][(-2)(-2)(-2)]$$
$$= (-2)^6 = 64$$

(d) $[(-2) \cdot 3]^2 = [(-2) \cdot 3][(-2) \cdot 3] = [(-2)(-2)][3 \cdot 3]$
$$= (-2)^2 \cdot 3^2 = 36$$

(e) $\quad [(-2)/3]^2 = (-2)^2/3^2 = \frac{4}{9}$

Parts of (b), (c), (d), and (e) of Example 5.2.1 illustrate four significant properties of exponents. These are formally stated in the following theorem.

THEOREM 5.2.1 Let $a, b \in R$ and $m, n \in N$. Then

I $a^m \cdot a^n = a^{m+n}$

II $[a^m]^n = a^{m \cdot n}$

III $[a \cdot b]^n = a^n \cdot b^n$

IV $\left(\dfrac{a}{b}\right)^n = \dfrac{a^n}{b^n}, \; b \neq 0$

Proof (Part I)

$$a^m \cdot a^n \stackrel{\text{D5.2.1}}{=\!=\!=} \overbrace{[a \cdot a \ldots a]}^{m \text{ factors of } a} \cdot \overbrace{[a \cdot a \ldots a]}^{n \text{ factors } a}$$

$$= \overbrace{[a \cdot a \cdot a \ldots a]}^{m + n \text{ factors of } a} \stackrel{\text{D5.2.1}}{=\!=\!=} a^{m+n}$$

$$\stackrel{\text{TLE}}{=\!=\!\Longrightarrow} a^m \cdot a^n = a^{m+n} \qquad \text{QED}$$

The proofs of Parts II, III, and IV are to be done as exercises.

If we were to restrict exponents to the positive integers only, we would be too severely limited for many applications. We wish, therefore, to give exponential meaning to other real numbers. First we expand the meaning of exponents to include the integers. It is clear that Definition 5.2.1 alone is not sufficient to handle negative integers; for, if $n \in N$, the statement

$$a^{-n} = \overbrace{a \cdot a \cdot a \ldots a}^{-n \text{ factors of } a}$$

does not even make sense. The following definitions for negative integers and zero will, on the other hand, be meaningful, particularly because the properties of Theorem 5.2.1 will be preserved.

DEFINITION 5.2.2 If $a \in R$, $a \neq 0$, and $n \in N$, then

$$a^{-n} = \frac{1}{a^n}$$

Example 5.2.2

(a) $(-3)^{-2} = \dfrac{1}{(-3)^2} = \dfrac{1}{9}$

(b) $(\frac{1}{2})^{-3} = \dfrac{1}{(\frac{1}{2})^3} = \dfrac{1}{\frac{1}{8}} = 8$

(c) $(-\frac{1}{2})^{-3} = \dfrac{1}{(-\frac{1}{2})^3} = \dfrac{1}{-\frac{1}{8}} = -8$

(d) $\dfrac{2^3}{2^5} = \dfrac{2 \cdot 2 \cdot 2}{2 \cdot 2 \cdot 2 \cdot 2 \cdot 2} = \dfrac{1}{2 \cdot 2} = \dfrac{1}{2^2} = 2^{-2}$

In addition to the four basic properties of Theorem 5.2.1, Definition 5.2.2 provides us with a useful property for quotients, namely,

$$\frac{a^m}{a^n} = a^{m-n} \tag{1}$$

where $m, n, (m - n) \neq 0$. (This latter restriction will be lifted following Definition 5.2.3.)

Example 5.2.3 $\quad \dfrac{2^4}{2^3} \overset{(1)}{=\!=\!=} 2^{4-3} = 2$

To complete our definitions for integral exponents we next give meaning to zero exponents. Our definition will be motivated by property (1) above. Notice what would result if we allowed m to equal n.

$$1 = \frac{a^n}{a^n} = a^{n-n} = a^0$$

Because we wish the exponential properties determined thus far to hold for *all* integral exponents, including zero, we follow the lead suggested and formulate the following definition:

DEFINITION 5.2.3 If $a \in R$ and $a \neq 0$, then

$a^0 = 1$

Remark 5.2.1 It is important to note that Definition 5.2.3 restricts the base a from being zero. That is, we are *not* defining $0^0 = 1$. This could lead to several inconsistencies in our system. For example

$$1 = \frac{1}{1} = \frac{2^0}{0^0} = \left(\frac{2}{0}\right)^0$$

but $\frac{2}{0}$ does not exist.

Example 5.2.4   If $x, y \in R$, then

$$\left[2x^4 + \frac{4y^2 + 6}{15}\right]^0 = 1$$

Remark 5.2.2   Keep in mind, exponential properties hold for products and quotients — *not* for sums nor differences. That is

$$\frac{a^4}{a^2 \cdot b^3} = \frac{a^{-2} \cdot a^4}{b^3} = \frac{a^2}{b^3}$$

but

$$\frac{a^4}{a^2 + b^3} \neq \frac{a^{-2} + a^4}{b^3}$$

or equally incorrect

$$\frac{a^4}{a^2 + b^3} \neq \frac{a^2}{b^3}$$

Another to keep in mind is $(a + b)^n \neq a^n + b^n$.

We shall now extend exponents to include the rational numbers. To accomplish this, however, we first define the *principal nth root* of a number.

DEFINITION 5.2.4   The principal *n*th root of a real number b, denoted $\sqrt[n]{b}$, is a real number a, when it exists, such that $b = a^n$. Furthermore, if n is even, a must be nonnegative. Symbolically, this becomes

$$\sqrt[n]{b} = a \Longleftrightarrow \begin{cases} b = a^n \\ \text{and} \\ a \geq 0 \qquad \text{if } n \text{ is even} \end{cases}$$

where $a, b \in R$ and $n \in N$. The symbol $\sqrt{}$ is called a *radical*, the number b the *radicand*, and n the *index* of the radical.

Remark 5.2.3   It is important to observe from Definition 5.2.4 that if n is even, the radicand b *must* be nonnegative. This follows because

$$b = a^n$$

and in the real number system any number raised to an even power is nonnegative. For this reason there does not exist in the real number system an even root of a negative number. We will find in Chapter 8 that such roots will exist in the complex number system.

When $n = 2$, it is common to write $\sqrt{a}$ instead of $\sqrt[2]{a}$ and to call this the *square root of a*. Also, when $n = 3$, $\sqrt[3]{a}$ is called the *cube root of a*.

Example 5.2.3

(a) $\sqrt[3]{-8} = x \iff -8 = x^3$

$\iff x^3 = (-2)^3 \iff x = -2$

(b) $\sqrt[4]{16} = x \iff 16 = x^4 \quad \text{and} \quad x \geq 0$

$\iff (\pm 2)^4 = x^4 \quad \text{and} \quad x \geq 0$

$\iff x = 2$

(c) $\sqrt[4]{-16} = x \iff -16 = x^4$

But there does not exist a *real* x satisfying that condition. Therefor the solution set is *empty*.

With principal *n*th roots defined, we are prepared to give meaning to fractional exponents. We begin with the following definition:

DEFINITION 5.2.5   If $a \in R$ and $n \in N$, then

$$a^{1/n} = \sqrt[n]{a}$$

where $a \geq 0$ if *n* is even. Also

$$a^{-1/n} = \frac{1}{a^{1/n}}$$

where $a \neq 0$.

Example 5.2.4

(a) $8^{1/3} = \sqrt[3]{8} = 2$

(b) $(-32)^{1/5} = \sqrt[5]{-32} = -2$

(c) $(-27)^{-1/3} = \dfrac{1}{(-27)^{1/3}} = \dfrac{1}{\sqrt[3]{-27}} = \dfrac{1}{-3} = -\dfrac{1}{3}$

The next definition will give meaning to rational exponents in general.

DEFINITION 5.2.6   If $a \in R$ and $m, n \in J$, $n > 0$, then

$$a^{m/n} = (a^{1/n})^m = (a^m)^{1/n}$$

That is,

$$a^{m/n} = (\sqrt[n]{a})^m = \sqrt[n]{a^m}$$

where $a \geq 0$ if *n* is even and $a \neq 0$ if $m < 0$.

From Definition 5.2.6 we understand that when a number is raised to a rational exponent, the numerator of the exponent represents a power operation, whereas the denominator represents a root operation, and that these two operations can be performed in any order—first root and then power, or first power and then root. It is, however, to your advantage to take the root first if $|a| > 1$.

Example 5.2.5 $27^{2/3} = [\sqrt[3]{27}]^2 = 3^2 = 9$

or

$$27^{2/3} = \sqrt[3]{(27)^2} = \sqrt[3]{729} = 9$$

As you can see, taking the root first, as in the first part, generally results in simpler computations.

The properties of rational exponents correspond with those of integral exponents. We summarize these in the next theorem.

THEOREM 5.2.2 If $a, b \in R$, $b \neq 0$, and $m, n \in Q$, then

I $a^m \cdot a^n = a^{m+n}$

II $[a^m]^n = a^{m \cdot n}$

III $(ab)^n = a^n \cdot b^n$

IV $\left(\dfrac{a}{b}\right)^n = \dfrac{a^n}{b^n}$

V $\dfrac{b^m}{b^n} = b^{m-n} = \dfrac{1}{b^{n-m}}$

VI $a^{-n} = 1/a^n$

VII $\left(\dfrac{a}{b}\right)^{-n} = \left(\dfrac{b}{a}\right)^n$

In each of these properties it is understood that $a \geq 0$ and $b > 0$ if an even root is involved; furthermore, $a \neq 0$ if raised to a negative exponent.

Example 5.2.6

(a) $\dfrac{32^{2/5}}{32^{3/5}} \xundertext{T5.2.2(V)} 32^{(2/5)-(3/5)} = 32^{-1/5}$

$\xundertext{T5.2.2(VI)} \dfrac{1}{32^{1/5}} = \dfrac{1}{\sqrt[5]{32}} = \dfrac{1}{2}$

T5.2.2(V) $b^n/b^m = b^{n-m} = 1/b^{m-n}$

T5.2.2(VI) $a^{-n} = 1/a^n$

(b) $\left(-\dfrac{8}{27}\right)^{-2/3} \xundertext{D5.2.6} \left(\sqrt[3]{-\dfrac{8}{27}}\right)^{-2} = \left(-\dfrac{2}{3}\right)^{-2}$

D5.2.6 $b^{n/m} = (\sqrt[m]{b})^n$

$\xundertext{T5.2.2(VII)} \left(-\dfrac{3}{2}\right)^2 = \dfrac{9}{4}$

T5.2.2(VII) $(a/b)^{-n} = (b/a)^n$

T5.2.2(V) $b^m/b^m = 1/b^{n-m}$

(c) $\left(\dfrac{x^3 y^5}{y^{17}}\right)^{-1/6}$ $\underset{\text{T5.2.2(V)}}{=\!=\!=}$ $\left(\dfrac{x^3}{y^{17-5}}\right)^{-1/6}$

$$= \left(\dfrac{x^3}{y^{12}}\right)^{-1/6}$$

T5.2.2(VII) $(a/b)^{-n} = (b/a)^n$

$\underset{\text{T5.2.2(VII)}}{=\!=\!=}$ $\left(\dfrac{y^{12}}{x^3}\right)^{1/6}$

T5.2.2(IV) $(a/b)^n = a^n/b^n$

$\underset{\text{T5.2.2(IV)}}{=\!=\!=}$ $\dfrac{y^{12(1/6)}}{x^{3(1/6)}} = \dfrac{y^2}{x^{1/2}} = \dfrac{y^2}{\sqrt{x}}$

D5.2.6 $b^{n/m} = \sqrt{b^n}$

(d) $(-4)^{-3/2}$ $\underset{\text{D5.2.6}}{=\!=\!=}$ $[\sqrt{-4}]^{-3}$

which does not exist in R, for we have an even root of a negative number.

D5.2.6 $b^{n/m} = (\sqrt[m]{b})^n$

(e) $\dfrac{\sqrt{12xy^2}}{\sqrt[3]{y^2}}$ $\underset{\text{D5.2.6}}{=\!=\!=}$ $\dfrac{(3 \cdot 2^2 \cdot xy^2)^{1/2}}{(y)^{2/3}}$

T5.2.2(III) $(ab)^n = a^n \cdot b^n$

$\underset{\text{T5.2.2(III)}}{=\!=\!=}$ $\dfrac{3^{1/2} \cdot 2 \cdot x^{1/2} \cdot y}{y^{2/3}}$

T5.2.2(V) $b^n/b^m = b^{n-m} = 1/b^{m-n}$

$\underset{\text{T5.2.2(V)}}{=\!=\!=}$ $2 \cdot 3^{1/2} \cdot x^{1/2} \cdot y^{1-2/3}$

$$= 2 \cdot (3 \cdot x)^{1/2} y^{1/3}$$

D5.2.6 $b^{n/m} = \sqrt[m]{b^n}$

$\underset{\text{D5.2.6}}{=\!=\!=}$ $2\sqrt{3x} \cdot \sqrt[3]{y}$

With exponents now expanded to include rational numbers, the next logical step would be to push forward, if possible, to the irrational numbers. Here, however, we encounter our first serious obstacle. Every definition thus far, except Definition 5.2.3 (of a zero exponent), depended directly or indirectly upon our initial definition of a^n, where n is a *natural* number. Recall, for example, rational exponents: If $m, n \in N$ then

$$a^{-m/n} = \frac{1}{a^{m/n}} = \frac{1}{[\sqrt[n]{a}]^m}$$

where $\sqrt[n]{a} = b \Longleftrightarrow a = b^n$. In the end, positive integral exponents, m and n, provided the essential meaning; that is, we always relied eventually upon Definition 5.2.1. Irrational numbers, on the other hand, cannot be expressed in finite form by means of natural numbers exclusively. Our only hope here lies in the fact that irrational numbers can be approximated to any degree of accuracy by rational numbers. But even this, when defined carefully and rigorously, requires the notion of a limit, which is beyond the scope of this book. Thus, we will not attempt a definition of irrational exponents at this time, but merely state that irrational exponents can be defined to satisfy the properties of Theorem 5.2.2.

Problem Set I   ■   *Reading Comprehension*

Determine whether statements 1 through 8 are true or false and give reasons for your responses.

1   The expression a^x means a times itself x times.
2   One of the important properties of exponents is

$$a^{-n} = \frac{1}{(-a)^n}$$

3   Any real number raised to the zero power is defined to be one.
4   The principal fourth root of 16, denoted $\sqrt[4]{16}$, is ± 2; that is, we should not forget -2 in addition to the more obvious 2.
5   The nth root of a negative number does not exist in the real number system.
6   In the expression $a^{m/n}$, where $a, m, n \in J_+$, m represents the power to which a is to be raised, whereas n determines the root to be taken; these two processes may be administered in either order with the same result.
7   The properties of exponents, in general, include the following:

I $a^n \cdot b^m = (a \cdot b)^{n+m}$

II $\dfrac{b^n}{b^m} = \dfrac{1}{b^{n-m}}$

III $(a + b)^n = a^n + b^n$

8   Because all the various definitions of exponents include the positive integers, either directly or indirectly, and further, because the irrational numbers cannot be expressed in terms of positive integers, we do not define irrational exponents here.

Problem Set II   ■   *Skills Development*

Express each of the following as a product or quotient of powers in which each integer is prime and each variable, along with the prime integers, occurs but once. Furthermore, leave all exponents positive.

1   $(-32)^{1/5}$ 2 $(27)^{1/3}$ 3 $(-81)^{-3/4}$ 4 $(81)^{3/4}$ 5 $\left(-\dfrac{1}{8}\right)^{-5/3}$

6   $\left(\dfrac{4}{9}\right)^{-3/2}$ 7 $x^{1/3} \cdot x^{2/5}$ 8 $x^{3/4} \cdot x^{1/2}$ 9 $x^{1/2} \cdot x^{2/3} \cdot x^{-1/5}$

10   $\left(\dfrac{x^6}{y^2}\right)^{-1/2}$ 11 $\left(\dfrac{27x^{-6}y}{125x^3y^3}\right)^{-1/3}$ 12 $\dfrac{(x^2)^{-n/2} \cdot (y^n)^{2/n}}{(x^{n/2})^2(y^{-2})^n}$ 13 $\dfrac{x^{3n} \cdot y^{2n-1}}{(x^n \cdot y^{2n})^{1/2}}$

14   $\left(\dfrac{a^n}{b}\right)^{2/3} \cdot \left(\dfrac{a^n}{b^3}\right)^{-1/2}$ 15 $\left(\dfrac{a}{2b}\right)^{-3}\left(\dfrac{2b}{3a}\right)^3$ 16 $\dfrac{(x^{-3} \cdot y^{-5})^{-2}}{(x^{-1} \cdot y)^{-3}}$ 17 $\left(\dfrac{a^{n+1}}{a^n}\right)^{-2}$

18   $\left(\dfrac{-3a^2b^3}{2ab^{-2}}\right)^3 \cdot \left(\dfrac{3a^{-2} \cdot b}{4ab^{-3}}\right)^{-2} \cdot \left(\dfrac{2a^4 \cdot b^{-3}}{a \cdot b^{1/2}}\right)^0$

Write the following radicals in exponential form and simplify:

Example $\sqrt[3]{x^2} \cdot \sqrt[4]{x^3}$

Solution $\sqrt[3]{x^2} \cdot \sqrt[4]{x^3} = x^{2/3} \cdot x^{3/4} = x^{2/3+3/4}$
$$= x^{8/12+9/12} = x^{17/12}$$

19 $\sqrt[4]{a^3}$ **20** $\sqrt[3]{27a^6b^9}$ **21** $\sqrt[4]{16a^4b^0}$ **22** $\sqrt[3]{\dfrac{-27}{x^3y^6}}$ **23** $\sqrt[4]{\dfrac{81x^4}{y^{12}z^8}}$

24 $\sqrt{50x^2y^3}$ **25** $\sqrt[3]{\dfrac{-125x^3z^6}{8y^6}}$ **26** $\sqrt[3]{\sqrt{128}}$ **27** $\sqrt{\sqrt[3]{xy^2}\,\sqrt[4]{x^3y^2}}$

Problem Set III   ■   *Theoretical Developments*

Prove each of the following, justifying each step by referring to the appropriate property:

1 Theorem 5.2.1-
    a Part II   **b** Part III
2 Let $a, b \in R$, $b \neq 0$, and $n \in N$ (if n is even, then $a, b \in R_+$). Then

    I $\sqrt[n]{a^n} = a$

    II $\dfrac{\sqrt[n]{a}}{\sqrt[n]{b}} = \sqrt[n]{\dfrac{a}{b}}$

    III $\sqrt[m]{\sqrt[n]{a}} = \sqrt[m \cdot n]{a}$

3 Theorem 5.2.2
    a Part I   **b** Part II   **c** Part III   **d** Part IV   **e** Part V
    f Part VI   **g** Part VII

5.4 Exponential Functions

We introduce this section with an application concerning radioactive decay.●

● When a radioactive substance such as radium or uranium disintegrates, it emits particles each having a definite mass, however small. This process continually diminishes the amount of the substance present and is called *radioactive decay.*

Problem 5.4.1 It is known that the rate of radioactive decay of a substance is proportional, at any instant, to the amount of the substance present. Suppose it is found that 0.5 percent of 4 grams of radium has disappeared in 12 years. With the above information it can be shown (using techniques of calculus) that a relationship between time and amount of radium present is given by

$$A = 4e^{-0.0004t}$$

where A is the amount in grams, t is the time in years, and e is that very important irrational number approximated by 2.718218. From the above equation determine graphically the relationship between time and amount of radium present. From the graph estimate the *half-life* of radium (the time required for 50 percent of the substance to disappear).

Before solving this problem we will familiarize ourselves with exponential functions and some of their more common properties.

The groundwork for defining exponential functions was laid in Section 5.2. Recall that if b is a real number, then the exponential expression b^x is also real for all real choices of the variable x *except* those leading to even roots or division by zero. If our choice for x introduces an even root, then b must be greater than or equal to zero for the result to be real. Because we wish complete freedom for the variable x within R, we choose the constant b greater than zero. Next we set the exponential expression b^x equal to y, obtaining the open equation

$$y = b^x$$

This equation in two variables has as its solution set a set of ordered pairs of real numbers that characterizes one of the most important classes of functions — the exponential functions.

DEFINITION 5.4.1   If $b \in R_+$ and $b \neq 1$, then the function, denoted E_b, defined by

$$E_b = \{ (x, y) \mid y = b^x \quad \text{and} \quad x, y \in R \}$$

is called the exponential function to the base b.

Remark 5.4.1   You might question the assertion of Definition 5.4.1 that the equation $y = b^x$ defines a *function*, that is, for each $x \in R$, b^x is unique (single valued). When defining exponents we were careful to disallow multiple-valued exponential expressions. Recall when roots were introduced we defined $\sqrt[n]{a}$ to be the *principal* nth root (only the *positive* one of the two accepted when n is *even* and $a \geq 0$). And then we defined $a^{1/n}$ to be $\sqrt[n]{a}$. Had $a^{1/n}$ not been defined as the *principal* nth root, the equation $y = b^x$ would not have defined a function.

Another observation concerns the reason for the restriction $b \neq 1$. There is nothing wrong with raising 1 to a real exponent, except that all you ever obtain is 1; that is, 1 raised to any real power is still 1. So we eliminate 1 as a base for an exponential function because of lack of interest — it would remain a constant function.

Let us now look at the graphical representation of certain exponential functions.

Table 5.4.1

x	$\left(\frac{2}{3}\right)^x$
-4	$\frac{81}{16}$
-3	$\frac{27}{8}$
-2	$\frac{9}{4}$
-1	$\frac{3}{2}$
0	1
1	$\frac{2}{3}$
2	$\frac{4}{9}$
3	$\frac{8}{27}$
4	$\frac{16}{81}$

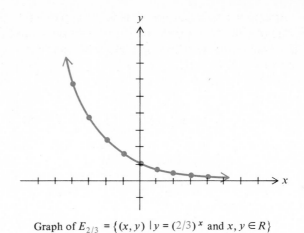

Graph of $E_{2/3} = \{(x, y) \mid y = (2/3)^x \text{ and } x, y \in R\}$

FIGURE 5.4.1

Example 5.4.1 Graph the exponential function

$$E_{2/3} = \{(x, y) \mid y = \left(\tfrac{2}{3}\right)^x\}$$

Solution We first determine a table of points in $E_{2/3}$ as shown in Table 5.4.1. This implies that $(-4, \frac{81}{16})$, $(-3, \frac{27}{8})$, $(-2, \frac{9}{4})$, $(-1, \frac{3}{2})$, $(0, 1)$, $(1, \frac{2}{3})$, $(2, \frac{4}{9})$, $(3, \frac{8}{27})$, $(4, \frac{16}{81})$ are points all belonging to $E_{2/3}$. Plotting these points and joining them by a smooth curve results in the graph shown in Fig. 5.4.1. This example illustrates the fact that for $0 < b < 1$, the graph of the function defined by $y = b^x$ is *strictly decreasing* (drops continuously as x increases from left to right).

Example 5.4.2 Graph the exponential function

$$E_{3/2} = \{(x, y) \mid y = \left(\tfrac{3}{2}\right)^x\}$$

Table 5.4.2

x	$\left(\frac{3}{2}\right)^x$
-4	$\frac{16}{81}$
-3	$\frac{8}{27}$
-2	$\frac{4}{9}$
-1	$\frac{2}{3}$
0	1
1	$\frac{3}{2}$
2	$\frac{9}{4}$
3	$\frac{27}{8}$
4	$\frac{81}{16}$

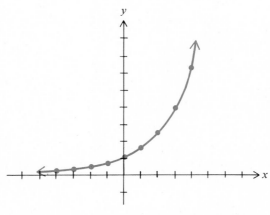

Graph of $E_{3/2} = \{(x, y) \mid y = (3/2)^x \text{ and } x, y \in R\}$

FIGURE 5.4.2

Solution Again, we first set up a table of points in $E_{3/2}$ (Table 5.4.2). Next, we plot these points and join them by a smooth curve (Fig. 5.4.2). This example illustrates the fact that for $b > 1$, the graph of the function defined by $y = b^x$ is *strictly increasing* (rises continuously as x increases from left to right).

Example 5.4.3 Graph the exponential function defined by $y = 2^{-x}$.

Solution Set up a table of points (Table 5.4.3). Plotting these points and connecting them by a smooth curve yields Fig. 5.4.3. Observe that although the base is greater than 1, the function is nevertheless strictly *decreasing*. This occurs because of the form of the exponent; rather than being x, it is $-x$. That is, graphing $y = 2^{-x}$ is the same as graphing $y = (\frac{1}{2})^x$ and as noted before this gives a decreasing function.

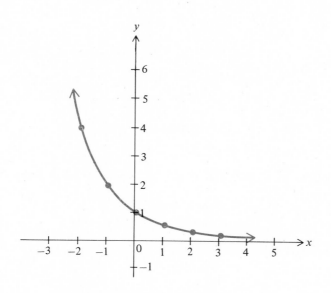

Graph of $\{(x, y) \mid y = 2^{-x} \text{ and } x, y \in R\}$

FIGURE 5.4.3

Table 5.4.3

x	2^{-x}
-2	4
-1	2
0	1
1	$\frac{1}{2}$
2	$\frac{1}{4}$
3	$\frac{1}{8}$

Let us now return to our problem about radioactive decay. The equation relating the amount of radium present in grams with time measured in years was

$$A = 4e^{-0.0004t}$$

We are to graph this exponential function and thereby determine the half-life of radium.

Solution   We first set up a table of values (Table 5.4.4) obtained from Table II found on page 534 of the Appendix. Plotting these points and joining them by a smooth curve yields the graph shown in Fig. 5.4.4. We can see that the half-life of radium (when only $\frac{1}{2}$ the original sample remains) is approximately 1660 years.

Table 5.4.4

t	A
0	4
1000	2.68
2000	1.8
4000	0.8
8000	0.16

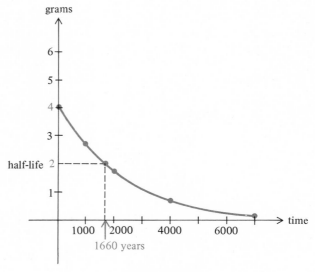

FIGURE 5.4.4

5.5   Problem Sets

Problem Set I   ■   *Reading Comprehension*

Determine whether statements 1 through 5 are true or false and give reasons for your responses.

1   Exponential functions are defined by open sentences of the form $y = b^x$ where $b \in R_+$ and $b \neq 1$.

2   The restriction that the base b for exponential functions be greater than zero is a matter of convention; we could have allowed $b < 0$ without affecting the domain or range of the function.

3   Exponential functions are either strictly increasing or strictly decreasing.

4   One of the lesser known but equally useful exponential functions is the one where the base b is chosen as 1.

5   The exponential function defined by $y = 3^{-x}$ is strictly increasing because the base 3 is greater than 1.

Problem Set II ■ Skills Development

In each of the following, find the missing coordinates so that the ordered pairs satisfy the given equation.

1   $y = 2^x$; $(0, \ )$, $(-2, \ )$, $(2, \ )$.
2   $y = -3^x$; $(-2, \ )$, $(2, \ )$, $(0, \ )$, $(-3, \ )$.
3   $y = (\frac{1}{2})^x$; $(-3, \ )$, $(3, \ )$, $(-1, \ )$, $(1, \ )$.
4   $y = 10^x$; $(-1, \ )$, $(0, \ )$, $(1, \ )$, $(2, \ )$.
5   $y = (-2)^x$; $(0, \ )$, $(-1, \ )$, $(2, \ )$, $(\frac{1}{2}, \ )$.
6   $y = (\frac{1}{3})^{-x}$; $(-2, \ )$, $(2, \ )$, $(3, \ )$, $(-4, \ )$.

Graph the function defined by each of the following equations:

7   $y = 2^x$ 8 $y = (\frac{1}{2})^x$ 9 $y = (\frac{1}{4})^x$ 10 $y = 10^x$ 11 $y = 2^{-x}$
12   $y = 2^{-x^2}$ 13 $y = 2 \cdot (\frac{1}{2})^x$ 14 $y = (\frac{1}{3})^x$ 15 $y = 5^x$

Simplify the following exponentials using Table II, p. 534:

16   $e^{2.2}$ 17 $e^{-4.3}$ 18 e^3 19 $e^{-3.4}$ 20 e^5 21 $e^{-0.5}$

Simplify the following squares and roots using Table IV, p. 538:

22   89^2 23 $\sqrt{6700}$ 24 $\sqrt{0.0048}$ 25 $\sqrt{0.000027}$ 26 37^2
27   $\sqrt{970000}$

Problem Set III ■ Applications

1   A tank is initially filled with 100 gallons of a salt solution containing 1 lb of salt per gallon. Another salt solution containing 2 lb of salt per gallon runs into the tank at a rate of 5 gal/min, and the mixture, kept uniform by stirring, runs out at the same rate. It is found that the amount of salt S in the tank at any time t is given by the formula
$$S = 200 - 100e^{-t/20}$$
Find the amount of salt in the tank at the end of (a) 10 minutes, (b) 20 minutes, and (c) 40 minutes.

2   Under certain conditions it is observed that the rate at which atmospheric pressure changes with altitude is proportional to the pressure. It is further observed that the pressure at sea level is 14.7 lb/in.², whereas at 18,000 ft it has dropped to half this amount. From this information it is found that a formula relating pressure and altitude is given by
$$P = 14.7e^{-0.0000385h}$$
where P is pressure measured in pounds per square inch and h is altitude measured in feet. Find the pressure in pounds per square inch at an altitude of 80,519.74 ft.

3   According to Newton's law of cooling, the rate at which the temperature of a body decreases is proportional to the difference between the instantaneous temperature of the body and the temperature of the surrounding medium. A body whose temperature initially is 100°C is allowed to cool in air that remains at a constant temperature of 20°C. It is observed after 10 minutes that the body has cooled to 60°C. From this information it is found that the temperature T of the body is given by the formula
$$T = 20 + 80e^{-0.0693t}$$
where t is time measured in minutes. What will the temperature be after 20.20 minutes?

5.6 Logarithmic Functions

Let us begin this section with the following computational problem.

Problem 5.6.1 Simplify the expression

$$\sqrt[5]{\frac{(281)^8 \cdot \sqrt[3]{0.0047}}{(7840)^6}}$$

With only the mathematical tools we have developed up to this point, the simplifying of the above expression would be extremely difficult. Fortunately, with certain tables and properties of logarithms, this problem can be reduced to a relatively simple one involving nothing more than elementary arithmetic. But first we must become acquainted with logarithms.

Because we shall define logarithmic functions by means of exponential functions, ● let us restate our definition of the exponential function E_b.

● Logarithmic functions are not always defined in this manner. In calculus, logarithmic functions are frequently defined by means of an integral, after which exponential functions are defined using logarithmic functions.

$$E_b = \{(x, y) \mid y = b^x, \, b > 0, \, b \neq 1, \text{ and } x, y \in R\}$$

In Chapter 4 we learned that when a relation is defined by an open sentence in two variables, the corresponding *inverse* relation could be defined using the same open sentence but with the roles of the two variables reversed — replacing x by y and y by x. We may, therefore, express the *inverse of the exponential function E_b*, which for the present we denote E^{-1}, as follows:

$$E_b^{-1} = \{(x, y) \mid x = b^y, \, b > 0, \, b \neq 1, \text{ and } x, y \in R\}$$

Although in general the inverse relation of a function is not necessarily a function, it is true that the *inverse* of an exponential function is a *function*. The reason is that the exponential functions are one-to-one functions. As observed earlier, they are either strictly increasing or strictly decreasing, which means for each real x the exponential expression b^x admits one and only one real y and, conversely, for each $y > 0$ there exists one and only one real x such that $y = b^x$.

The inverse exponential functions are so important in their own right that they are given a special name, the logarithmic functions.

When possible, it is desirable to express open equations containing two variables in the form $y = f(x)$, that is, y expressed explicitly in terms of x. Frequently, when the proposition is not in this form, we simply solve for y. For example,

$$x = \sqrt{4 - y^2} \quad \text{and} \quad y \geq 0 \Longleftrightarrow x^2 = 4 - y^2 \quad \text{and} \quad x, y \geq 0$$

$$\Longleftrightarrow y^2 = 4 - x^2 \quad \text{and} \quad x, y \geq 0$$

$$\Longleftrightarrow y = \sqrt{4 - x^2} \quad \text{and} \quad x \geq 0$$

If, however, we were to attempt to solve for y in the equation $x = b^y$, we would find the task impossible. No manner of applying the properties of addition, subtraction, multiplication, and division to this equation would remove the y as an exponent. We shall, therefore, take the liberty of creating a solution form. ● This is a mathematical way of renaming y.

● Remember all mathematical symbols and expressions are mere human creations to assist in describing various ideas or phenomena both physical and abstract.

DEFINITION 5.6.1   Let $b \in R_+$ and $b \neq 1$. Then

$$x = b^y \iff y = \log_b x$$

The expression $\log_b x$ is read "logarithm to the base b of x," or just "log to the base b of x."

Example 5.6.1   Note the following equivalences:

(a) $2 = \log_{10} 100 \iff 10^2 = 100$

(b) $8^{-1/3} = \frac{1}{2} \iff \log_8 \frac{1}{2} = -\frac{1}{3}$

(c) $3^4 = 81 \iff \log_3 81 = 4$

Remark 5.6.1   It is important to note here that there is nothing strange or mysterious about the logarithmic equation $y = \log_b x$. It is just another way of expressing that b to the y power equals x, which we also write $x = b^y$.

It is natural now to define logarithmic functions by the equation $y = \log_b x$.

DEFINITION 5.6.2   If $b \in R_+$ and $b \neq 1$, then the function, denoted L_b, defined by

$$L_b = \{(x, y) \mid y = \log_b x \text{ and } x, y \in R\}$$

is called the logarithmic function to the base b.

Example 5.6.2   Graph the logarithmic function

$$L_{3/2} = \{(x, y) \mid y = \log_{3/2} x\}$$

Solution   Because we are more familiar with the exponential equation equivalent to $y = \log_{3/2} x$, that is, $x = \left(\frac{3}{2}\right)^y$, we shall use this latter form to identify a few of the ordered pair solutions in $L_{3/2}$. Using the equation $x = \left(\frac{3}{2}\right)^y$, we construct a table of solution points (Table 5.6.1). Thus, the points $\left(\frac{16}{81}, -4\right)$, $\left(\frac{8}{27}, -3\right)$, $\left(\frac{4}{9}, -2\right)$, $\left(\frac{2}{3}, -1\right)$, $(1, 0)$, $\left(\frac{3}{2}, 1\right)$, $\left(\frac{9}{4}, 2\right)$, $\left(\frac{27}{8}, 3\right)$, $\left(\frac{81}{16}, 4\right)$ all belong to $L_{3/2}$. Figure

Table 5.6.1

x	$\left(\frac{3}{2}\right)^y$
-4	$\frac{16}{81}$
-3	$\frac{8}{27}$
-2	$\frac{4}{9}$
-1	$\frac{2}{3}$
0	1
1	$\frac{3}{2}$
2	$\frac{9}{4}$
3	$\frac{27}{8}$
4	$\frac{81}{16}$

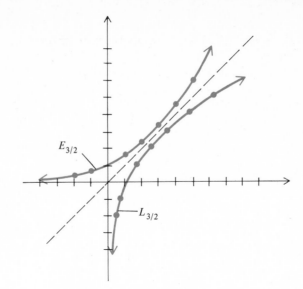

FIGURE 5.6.1

5.6.1 shows how these points look when plotted and when joined by a smooth curve.

Since the function $L_{3/2}$ is the inverse of $E_{3/2}$, the graph of $L_{3/2}$ is just the reflection of $E_{3/2}$ through the line defined by $y = x$, as illustrated in Figure 5.6.1.

Note that the domain and range for $L_{3/2}$ are R_+ and R, respectively, which is just the reverse of those for $E_{3/2}$. This reversal of domain and range is, of course, true in general for inverse functions.

Just as the exponential expression b^x possesses several significant and useful properties (Theorem 5.2.2), so it is with the logarithmic expression $\log_b x$, as the following theorem will illustrate.

THEOREM 5.6.1   Let $a, b, c \in R_+$, $b \neq 1$, $m \in J$, and $n \in J_+$. Then

    I $\log_b {}^b = 1$

    II $b^{\log_b a} = a$

    III $\log_b (a \cdot c) = \log_b a + \log_b c$

    IV $\log_b \left(\dfrac{a}{c}\right) = \log_b a - \log_b c$

    V $\log_b (a)^m = m \cdot \log_b a$

    VI $\log_b \sqrt[n]{a^m} = \dfrac{m}{n} \cdot \log_b a$

Proof   Part I

Let $\log_b b = x$, then

$$\log_b b = x \xrightarrow{\quad \text{D5.6.1} \quad} b^x = b \implies x = 1$$

The latter implication follows because the expression b^x is *single valued.*   QED

Proof   Part III

Let $\log_b a = x$ and $\log_b c = y$, then

(1) $\log_b a = x \xLeftrightarrow{\text{D5.6.1}} a = b^x$

and

(2) $\log_b c = y \xLeftrightarrow{\text{D5.6.1}} c = b^y$

$\xRightarrow[\text{(2)}]{\text{(1)}} ac = b^x \cdot b^y$

$\xLeftarrow{\text{T5.2.2(I)}} ac = b^{x+y}$

$\xLeftarrow{\text{D5.6.1}} \log_b (ac) = x + y$

$\xLeftarrow[\substack{x = \log_b a \\ y = \log_b c}]{} \log_b (ac) = \log_b a + \log_b c$ QED

D5.6.1 $y = \log_b x \Longleftrightarrow x = b^y$

T5.2.2(I) $a^m \cdot a^n = a^{m+n}$

(The proofs of Parts II, IV, V, and VI are to be done as exercises.)

Example 5.6.3

(a)   Express $\log_b 18$ in terms of $\log_b 2$ and $\log_b 3$.

Solution

$\log_b 18 = \log_b 2 \cdot 3^2$

$\xequal{\text{T5.6.1(III)}} \log_b 2 + \log_b 3^2$

$\xequal{\text{T5.6.1(V)}} \log_b 2 + 2 \log_b 3$

T5.6.1(III)
$\log_b (a \cdot c) = \log_b a + \log_b c$

T5.6.1(V) $\log_b (a)^m = m \cdot \log_b a$

(b) Express $\log_b \sqrt[3]{\dfrac{x^2 y}{z^4}}$ in terms of $\log_b x$, $\log_b y$, and $\log_b z$

Solution

$\log_b \sqrt[3]{\dfrac{x^2 y}{z^4}} \xequal{\text{D5.2.5}} \log_b \left(\dfrac{x^2 y}{z^4}\right)^{1/3}$

$\xequal{\text{T5.6.1(V)}} \dfrac{1}{3} \log_b \left(\dfrac{x^2 y}{z^4}\right)$

$\xequal{\text{T5.6.1(IV)}} \dfrac{1}{3}[\log_b x^2 y - \log_b z^4]$

$\xequal{\text{T5.6.1(III)}} \dfrac{1}{3}[\log_b x^2 + \log_b y - \log_b z^4]$

$\xequal{\text{T5.6.1(V)}} \dfrac{1}{3}[2 \log_b x + \log_b y - 4 \log_b z]$

D5.2.5 $a^{1/n} = \sqrt[n]{a}$

T5.6.1(V) $\log_b (a)^m = m \cdot \log_b a$

T5.6.1(IV) $\log_b \left(\dfrac{a}{c}\right) = \log_b a - \log_b c$

T5.6.1(III)
$\log_b (a \cdot c) = \log_b a + \log_b c$

This latter example illustrates how certain rather complicated expressions involving products, quotients, powers, or roots may be greatly simplified by using logarithms.

5.7 Sample Problem Set

Determine whether the following statements are true or false and give reasons for your responses.

1 The logarithmic functions are inverses of exponential functions.

2 If you algebraically solve $x = b^y$ for y, you will obtain the equation $y = \log_b x$.

3 An equivalent way to express $9^{-1/2} = \frac{1}{3}$ is $\log_9(-\frac{1}{2}) = \frac{1}{3}$.

4 An equivalent way to express $(-8)^{1/3} = -2$ is $\log_{-8}(-2) = \frac{1}{3}$.

5 The domain and range of the function defined by $y = \log_2 x$ are, respectively, R and R_+.

6 $\log_2(-4) = -2$.

7 $3^{\log_3 \sqrt{5}} = \sqrt{5}$

8 $\dfrac{\log_5 3}{\log_5 7} = \log_5 3 - \log_5 7$

5.8 Logarithms to the Base Ten

There are two bases that are in common use today — the base 10 and the base e (you will recall that e is the symbol for the irrational number 2.718218 . . .). Logarithms to the base 10 are called common • logarithms, whereas those to the base e are called natural • logarithms. Logarithms to the base e and other bases will be studied in Section 5.10. The logarithmic function to the base 10 is defined, of course, by the equation

$$y = \log_{10} x$$

- Also Briggs logarithms after the English mathematician Henry Briggs (1556–1630).

- Also Napierian logarithms after the Scottish mathematician John Napier (1550–1617).

Logarithms to the base 10 are especially well-suited to simplifying computations involving our decimal number system, which is also to the base 10. For example, notice how easy it is to take logarithms, to the base 10, of powers of 10.

T5.6.1(V) $\log_b (a)^m = m \cdot \log_b a$

T5.6.1(I) $\log_b b = 1$

$$\log_{10} 1000 = \log_{10} 10^3 \underline{\overset{\text{T5.6.1(V)}}{}} 3 \log_{10} 10$$
$$\underline{\overset{\text{T5.6.1(I)}}{}} 3 \cdot 1 = 3$$

$$\log_{10} 100 = \log_{10} 10^2 = 2 \log_{10} 10 = 2$$

$$\log_{10} 10^1 = 1$$

$$\log_{10} 1 = \log_{10} 10^0 = 0 \log_{10} 10 = 0$$

$$\log_{10} 0.1 = \log_{10} 10^{-1} = -1 \log_{10} 10 = -1$$

$$\log_{10} 0.01 = \log_{10} 10^{-2} = -2 \log_{10} 10 = -2$$

and, in general,

$$\log_{10} 10^x = x \qquad\qquad (1)$$

The above properties of logarithms to the base 10 produce a very important principle that we shall illustrate by the following example.

Suppose you wish to determine the logarithm of 38600 and 0.00386. One number is very large compared to the other, yet their only difference is a factor of 10^7, that is, the position of their decimal points varies by 7 places. Observe that

$$\log_{10} 38600 = \log_{10} 3.86 \cdot 10^4 \xlongequal{\text{T5.6.1(III)}} \log_{10} 3.86 + \log_{10} 10^4$$
$$\xlongequal{(1)} \log_{10} 3.86 + 4$$

whereas

$$\log_{10} 0.00386 = \log_{10} 3.86 \cdot 10^{-3} \xlongequal{\text{T5.6.1(III)}} \log_{10} 3.86 + \log_{10} 10^{-3}$$
$$\xlongequal{(1)} \log_{10} 3.86 + (-3)$$

T5.6.1(III)
$$\log_b (a \cdot c) = \log_b a + \log_b c$$

(1)　$\log_{10} 10^x = x$

Hence, the *logarithm* of 38600 and the *logarithm* of 0.00386 only differ by 7. More significantly, the above analysis demonstrates that the logarithm of any number, however large or small, may be expressed as the sum of the logarithm of a number between 1 and 10 plus an integer. This integer represents the number of decimal places the decimal point is moved over to obtain a number between 1 and 10; the number is positive if the point is moved to the left and negative if moved to the right, as the above example illustrates. Thus, it is only necessary to have logarithmic tables of numbers between 1 and 10. Instead of requiring virtually endless volumes of tables, we are able with just two pages of tables to provide sufficient entries to approximate any number to within one-hundredth, and the logarithm of each such number to within one-thousandth. A portion of this is reproduced in Table 5.8.1. Notice in this table the units and tenths digits of a number are represented in the left-hand column, whereas the hundredths digits from 0 to 9 are located across the top as the heads of nine columns, with the appropriate logarithmic entries listed below. Thus, as indicated in the table, $\log 3.86 = 0.5866$. Because the exact numbers for these logs are generally irrational, the entries tabulated are only approximations, exact through three decimal places and "rounded off" at the fourth.

Table 5.8.1

log x	0	1	2	3	4	5	6	7	8	9
log 3.7	.5682	.5694	.5705	.5717	.5729	.5740	.5752	.5763	.5775	.5786
log 3.8	.5798	.5809	.5821	.5832	.5843	.5855	.5866	.5877	.5888	.5899
log 3.9	.5911	.5922	.5933	.5944	.5955	.5966	.5977	.5988	.5999	.6010
log 4.0	.6021	.6031	.6042	.6053	.6064	.6075	.6085	.6096	.6107	.6117
log 4.1	.6128	.6138	.6149	.6160	.6170	.6180	.6191	.6201	.6212	.6222

$\searrow$log 3.86

Using this table we can complete our determination of $\log_{10} 38600$ and $\log_{10} 0.00386$.

$$\log_{10} 38600 = \log_{10} 3.86 + 4 = 0.5866 + 4 = 4.5866$$

and

$$\log_{10} 0.00386 = \log_{10} 3.86 + (-3) = 0.5866 + (-3) = -2.4134$$

So that we do not lose sight of the meaning of these logarithmic expressions, let us recall the following equivalences:

D5.6.1 $y = \log_b x \Longleftrightarrow x = b^y$

$$\log_{10} 38600 = 4.5866 \underset{\textbf{D5.6.1}}{\Longleftrightarrow} 10^{4.5866} = 38600$$

and

$$\log_{10} 0.00386 = -2.4134 \underset{\textbf{D5.6.1}}{\Longleftrightarrow} 10^{-2.4134} = 0.00386$$

That is, the logarithm (to the base of 10) of 38600 is the *power* to which 10 must be raised in order to obtain 38600. Because most logarithms in this text will be to the base 10, we will, unless otherwise specified, generally write $\log 38600$ rather than $\log_{10} 38600$ — the base will be omitted and understood to be 10.

Example 5.8.1 Determine the following logarithms:

● The value 0.5632 for log 3.70 is obtained from Table 5.8.1.

(a) $\log 370 = \log 3.70 \cdot 10^2 = \log 3.70 + \log 10^2 = \log 3.70 + 2$ ●

$$= 0.5682 + 2 = 2.5682$$

Thus,

$$\log 370 = 2.5682 \Longleftrightarrow 10^{2.5682} = 370$$

(b) $\log 0.00037 = \log 3.7 \cdot 10^{-4} = \log 3.7 + \log 10^{-4}$

$$= \log 3.7 + (-4)$$

$$= 0.5682 + (-4) = -3.4318$$

Hence,

$$\log 0.00037 = -3.4318 \iff 10^{-3.4318} = 0.00037$$

In summary, if you wish to take the logarithm of, say 38600, you first express the log of this number in two parts, as indicated below.

$$\log 38600 = \log 3.86 + 4$$

the first term on the right (always the logarithm of a number between 1 and 10) is called the mantissa and is usually evaluated using the tables of common logarithms. The second term (always an integer representing the number of places the decimal point is moved to obtain a number between 1 and 10) is called the characteristic. As noted earlier, the characteristic for numbers greater than 10 is a *positive* integer, whereas the characteristic for those less than 1 is a *negative* integer.

Although our table provides entries for the logarithms of numbers between 1 and 10 up to two decimal places, we may approximate one more place by linear interpolation. Before describing linear interpolation, let us briefly review a geometric property that provides the principle upon which the approximating process is based. Consider the geometric figure in Fig. 5.8.1. Triangles $AB'C'$ and ABC are said to be *similar* (all corresponding interior angles are equal). An important property of similar triangles is that corresponding sides are *proportional*. That is, we may write

$$\frac{B'C'}{BC} = \frac{AC'}{AC} \quad \text{or} \quad \frac{AB'}{AB} = \frac{AC'}{AC}$$

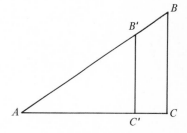

FIGURE 5.8.1

where AB represents the length of the line segment joining A and B and similarly for the other entries.

Returning now to linear interpolation, suppose we wish to find $\log 3.864$. From the table of logarithms we obtain

$$\log 3.86 = 0.4866$$

and

$$\log 3.87 = 0.5877$$

Because 3.864 lies between 3.86 and 3.87 and because logarithmic functions are strictly increasing, we know $\log 3.864$ lies somewhere between 0.5866 and 0.5877. To suggest this fact we write

$$\log 3.860 = 0.5866$$

$$\log 3.864 = a$$

$$\log 3.870 = 0.5877$$

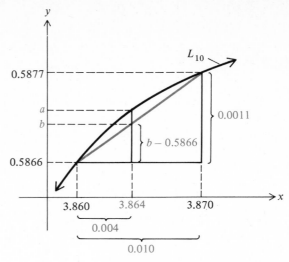

FIGURE 5.8.2

In order to more fully understand the reason for the next step, let us examine a portion of the graph of L_{10} as illustrated in Fig. 5.8.2. ● Observe that the two points (3.86, 0.5866) and (3.87, 0.5877), which are members of L_{10}, are joined by a straight line. The second coordinate of the point $(3.864, a)$ is, of course, the number we are seeking. However, we will not determine a exactly but rather approximate a by b. We find the number b by employing the property of similar triangles noted above. This is accomplished as follows:

● This segment of the curve is exaggerated slightly out of proportion to more clearly describe the process.

$$\frac{b - 0.5866}{0.0011} = \frac{0.004}{0.010}$$
$$\Longleftrightarrow \frac{b - 0.5866}{0.0011} = \frac{4}{10} \qquad \Longleftrightarrow b - 0.5866 = \frac{4(0.0011)}{10}$$
$$\Longleftrightarrow b - 0.5866 = 0.00044 \Longleftrightarrow b = 0.58704 \cong 0.5870$$

rounding off to four places. Thus,

$$\log 3.864 \cong 0.5870$$

Remark 5.8.1 The symbol $\cong$ reminds us here that b is only an approximation for a. However, because nearly all entries in the table are themselves approximations, we generally disregard employing the approximation symbol and simply use equals. In general, linear interpolation for logs gives accuracy to one more place than in the tables.

Example 5.8.2 Determine $\log 3.727$ by linear interpolation.

Solution From the tables we obtain

$$0.010 \begin{bmatrix} 0.007 \begin{bmatrix} \log 3.720 = 0.5705 \\ \log 3.727 = b \end{bmatrix} b - 0.5705 \\ \log 3.730 = 0.5717 \end{bmatrix} 0.0012$$

As in the preceding example, the following proportions (ratios) are equal:

$$\frac{b - 0.5705}{0.0012} = \frac{0.007}{0.010}$$

$$\Longleftrightarrow \frac{b - 0.5705}{0.0012} = \frac{7}{10} \Longleftrightarrow b - 0.5705 = (0.0012) \cdot \frac{7}{10} = 0.00084$$

$$\Longleftrightarrow b = 0.57134$$

Hence, rounding off to four places, we have

$$\log 3.727 = 0.5713$$

It is common practice when interpolating to simplify somewhat the mechanics of this process as follows. Set up the proportions, disregarding for the moment the actual decimal differences.

$$10 \left[7 \left[\begin{array}{l} \log 3.720 = 0.5705 \\ \log 3.727 = \\ \log 3.730 = 0.5717 \end{array} \right] x \right] 12$$

Form the proportions

$$\frac{x}{12} = \frac{7}{10}$$

solve for x

$$x = \tfrac{84}{10} = 8.4 \cong 8$$

and then add 0.0008 to 0.5705, obtaining 0.5713 and finally writing

$$\log 3.727 = 0.5713$$

As another example, take $\log 3.24684$. Because we are interpolating at the fourth significant digit, we round this off to $\log 3.247$. Then

$$10 \left[7 \left[\begin{array}{l} \log 3.240 = 0.5105 \\ \log 3.247 = \\ \log 3.250 = 0.5119 \end{array} \right] x \right] 14$$

Form the proportion

$$\frac{7}{10} = \frac{x}{14}$$

and solve for x

$$x = 9.8 \cong 10$$

Then add 0.0010 to 0.5105 obtaining 0.5115, and write

$$\log 3.247 = 0.5115$$

When simplifying calculations by means of logarithms, the process of finding logarithmic values by means of the tables is just half the battle. The reverse procedure is equally important. That is, if we employ logarithmic properties and tables to simplify an expression, we must reverse this process to obtain our final answer. The reverse procedure is sometimes called finding antilogarithms. More formally, if $y = \log x$ then $x = \text{antilog } y$. Thus, antilog y really means 10^y (recall $y = \log x \iff x = 10^y$). To find the antilog of y is to find a number whose log is y. For example, we know from Example 5.8.2 that

$$\log 3.727 = 0.5713$$

therefore

$$3.727 = \text{antilog } 0.5713 = 10^{0.5713}$$

It is important to note in this regard that our tables contain the logarithms of numbers between 1 and 10 only; therefore the entries within the tables, representing these logarithms, lie between 0 and 1. This means we can only take the antilogarithm of a number between 0 and 1 from the tables. This poses no serious difficulty because the role of the *characteristic* works equally well in reverse, as the following examples will illustrate.

Example 5.8.3   Find the antilogarithm of 0.5955.

Solution   Notice in column 4 opposite $\log 3.9$ in Table 5.8.1, we find 0.5955. Thus

$$\log 3.94 = 0.5955;$$

hence, the antilog of 0.5955 is 3.94.

This first example of finding an antilogarithm was, of course, oversimplified. First, the number 0.5955 was between 0 and 1, and second, 0.5955 was a specific entry in the table. Normally, one would not be so lucky.

Let us determine how to use the table when either or both of the above conditions is not met. Suppose the number, of which we wish to take the antilog, is between 0 and 1 but not an entry in the table. Here we apply linear interpolation as explained earlier. Take, for example, the antilogarithm of 0.5962. Observe in the table that

0.5962 lies between 0.5955 and 0.5966, which are, respectively, log 3.94 and log 3.95. Therefore,

$$10\left[\begin{array}{c}x\left[\begin{array}{l}\log 3.940 = 0.5955 \\ \log a \quad = 0.5962\end{array}\right]7 \\ \log 3.950 = 0.5966\end{array}\right]11$$

Hence, our proportions are

$$\frac{x}{10} = \frac{7}{11} \Longleftrightarrow x = \frac{70}{11} \cong 6$$

We add 0.006 to 3.940 to obtain 3.946. Thus,

$$\log 3.946 = 0.5962,$$

or equivalently

$$3.946 = \text{antilog } 0.5962$$

Now let us find the antilogarithm of a number greater than 1, say 4.6085. The technique here is to express this as the *sum* of a mantissa (a number between 0 and 1) and a characteristic (always an integer); that is, let

$$4.6085 = 0.6085 + 4$$

First, from the table of logarithms (Table 5.8.1), we find that

$$\log 4.06 = 0.6085$$

Then we proceed as follows:

$$4.6085 = 0.6085 + 4 = \log 4.06 + 4 = \log 4.06 + 4 \cdot \log_{10} 10 \; \bullet$$
$$= \log 4.07 + \log_{10} 10^4 = \log (4.06) \cdot (10^4) = \log 40600$$

● These steps are usually skipped in practice, but are retained here to illustrate the characteristic working in reverse.

Thus,

$$\text{antilog } 4.6095 = 40600$$

which also means

$$10^{4.6085} = 40600$$

Now let us take the antilog of a *negative* number.

Example 5.8.4   Find the antilog of -3.4067.

Solution   We must express -3.4067 as the sum of a mantissa (between 0 and 1) and a characteristic (any integer — positive or negative). This can be accomplished by adding *and* subtracting 4 to -3.4067. ● The effect of adding 4 to -3.4067 provides us with a

● Remember $-3.4067 \neq -3 + 0.4067$, but rather $-3.4067 =$
$$(-3) + (-0.4067).$$

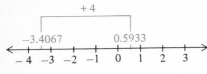

FIGURE 5.8.3

mantissa lying between 0 and 1, as illustrated on the number line in Fig. 5.8.3. Continuing as suggested, we obtain

$$-3.4067 = 4 + (-3.4067) + (-4) = 0.5933 + (-4)$$
$$= \log 3.92 + (-4) = \log (0.000392)$$

Hence, 0.000392 is the antilogarithm of -3.4067. That is,

$$10^{-3.4067} = 0.000392$$

We shall now take full advantage of the properties of logarithms to simplify certain rather complicated numerical calculations. Critical to our success will be our reliance on the following theorem that results from the logarithmic function being strictly increasing and, hence, one-to-one. It is also because of this theorem that the process of taking antilogarithms is meaningful, that is, it admits *unique* results.

THEOREM 5.8.1   If $a, b \in R_+$, then

$$\log a = \log b \iff a = b$$

Because of this equivalence, our approach to simplify an expression represented, say by a, that is complicated by roots, powers, products, and quotients will be to take the logarithm of a, thereby admitting the use of those properties that may eliminate some or all the roots, powers, products, and quotients. After performing the more simple numerical calculations resulting at this point, we then take the antilogarithm and obtain a number, say b, whose logarithm equals that of a. Hence, by Theorem 5.8.1, we have the complicated expression a equal to the simplified version b. The next example will illustrate this process.

Example 5.8.5   Simplify the expression

$$\sqrt[5]{\frac{(574)^2}{(0.76)^3}}$$

Solution   First we take the logarithm of the expression so as to take advantage of several logarithmic properties.

D5.2.5 $\sqrt[n]{a} = a^{1/n}$

T5.6.1(V) $\log_b (a)^n = n \cdot \log_b a$

T5.6.1(IV) $\log_b \left(\dfrac{a}{c}\right) = \log_b a - \log_b c$

$$\log \sqrt[5]{\frac{(574)^2}{(0.76)^3}} \xlongequal{\text{D5.2.5}} \log \left(\frac{(574)^2}{(0.76)^3}\right)^{1/5} \xlongequal{\text{T5.6.1(V)}} \frac{1}{5} \log \frac{(574)^2}{(0.76)^3}$$

$$\xlongequal{\text{T5.6.1(IV)}} \tfrac{1}{5}[\log (574)^2 - \log (0.76)^3]$$

$$\xlongequal{\text{T5.6.1(V)}} \tfrac{1}{5}[2 \log 574 - 3 \log (0.76)]$$

$$\overline{} = \tfrac{1}{5}[2(\log 5.74 + 2) - 3(\log 7.6 - 1)]$$

$$\overline{} = \tfrac{1}{5}[2(0.7589 + 2) - 3(0.8808 - 1)]$$
$$\text{(from Table I, p. 534)}$$

$$\overline{} = \tfrac{1}{5}[1.5178 + 4 - 2.6424 + 3]$$

$$\overline{} = \tfrac{1}{5}[8.5178 - 2.6424] = \tfrac{1}{5}[5.8754]$$

$$\overline{} = 1.17508 \cong 1.1751 = 0.1751 + 1$$

$$\overset{\bullet}{\overline{}} = \log 1.497 + 1 = \log 14.97$$

$$\overset{\text{TLE}}{\Longrightarrow} \log \sqrt[5]{\frac{(574)^2}{(0.76)^3}} = \log 14.97$$

$$\overset{\text{T5.8.1}}{\Longleftrightarrow} \sqrt[5]{\frac{(574)^2}{(0.76)^3}} = 14.97$$

$$\bullet \quad 10\begin{bmatrix} x\begin{bmatrix} \log 1.490 = 0.1732 \\ \log\ b\ = 0.1751 \\ \log 1.500 = 0.1761 \end{bmatrix}19 \end{bmatrix}29$$

$$\Longrightarrow \frac{x}{10} = \frac{19}{29} \Longleftrightarrow x \cong 7 \Longleftrightarrow b = 1.497$$

T5.8.1
$$\log a = \log b \Longleftrightarrow a = b,\ a, b \in R_+$$

Example 5.8.6 Simplify the expression

$$\sqrt[3]{\frac{(0.036)^5}{\sqrt{78.9}}}$$

Solution Again we take the logarithm of this expression and simplify as follows:

$$\log \sqrt[3]{\frac{(0.036)^5}{\sqrt{78.9}}} = \log\left(\frac{(0.036)^5}{\sqrt{78.9}}\right)^{1/3} = \frac{1}{3}\log\left(\frac{(0.036)^5}{\sqrt{78.9}}\right)$$

$$\overline{} = \tfrac{1}{3}\left[\log (0.036)^5 - \log (78.9)^{1/2}\right]$$

$$\overline{} = \tfrac{1}{3}\left[5 \log (0.036) - \tfrac{1}{2}\log 78.9\right]$$

$$\overline{} = \tfrac{1}{3}\left[5(\log 3.6 - 2) - \tfrac{1}{2}(\log 7.89 + 1)\right]$$

$$\overline{} = \tfrac{1}{3}\left[5(0.5563 - 2) - \tfrac{1}{2}(0.8971 + 1)\right] \quad \text{(from Table I, p. 534)}$$

$$\overline{} = \tfrac{1}{3}\left[2.7815 - 10 - \tfrac{1}{2}(1.8971)\right] = \tfrac{1}{3}\left[2.7815 - 10 - 0.9485\right]$$

$$\overline{} = \tfrac{1}{3}\left[2.7815 - 10.9485\right] = \tfrac{1}{3}\left[-8.1670\right] \cong -2.7223$$

$$\overline{} = 3 + (-2.7223) + (-3) \quad \text{(to obtain a mantissa}$$
$$\text{between 0 and 1)}$$

$$\overline{} = 0.2777 + (-3) \overset{\bullet}{\overline{}} = \log 1.895 + (-3) = \log (0.001895)$$

$$\Longrightarrow \log \sqrt[3]{\frac{(0.036)^5}{\sqrt{78.9}}} = \log (0.001895)$$

$$\overset{\text{T5.8.1}}{\Longrightarrow} \sqrt[3]{\frac{(0.036)^5}{\sqrt{78.9}}} = 0.008195$$

$$\bullet \quad 10\begin{bmatrix} x\begin{bmatrix} \log 1.890 = 0.2765 \\ \log\ b\ = 0.2777 \\ \log 1.900 = 0.2788 \end{bmatrix}12 \end{bmatrix}23$$

$$\Longrightarrow \frac{x}{10} = \frac{12}{23} \Longrightarrow x \cong 5 \Longrightarrow b = 1.895$$

T5.8.1 $\log a = \log b \Longleftrightarrow a = b$

As our final example, we return to Problem 5.6.1 with which we initiated our study of logarithms.

Problem 5.6.1 Simplify the expression

$$\sqrt[5]{\frac{(281)^8 \cdot \sqrt[3]{0.0047}}{(7840)^6}}$$

Solution We take the logarithm of the above expression and simplify as follows:

$$\log \sqrt[5]{\frac{(281)^8 \cdot \sqrt[3]{0.0047}}{(7840)^6}} = \frac{1}{5} \log\left[\frac{(281)^8 \cdot \sqrt[3]{0.0047}}{(7840)^6}\right]$$

$$= \tfrac{1}{5}[\log (281)^8 + \log (0.0047)^{1/3} - \log (7840)^6]$$

$$= \tfrac{1}{5}[8 \log (281) + \tfrac{1}{3} \log (0.0047) - 6 \log (7840)]$$

$$= \tfrac{1}{5}[8 \, (\log 2.81 + 2) + \tfrac{1}{3} \, (\log 4.7 - 3) - 6 \, (\log 7.84 + 3)]$$

$$= \tfrac{1}{5}[8(0.4487 + 2) + \tfrac{1}{3}(0.6721 - 3) - 6(0.8943 + 3)]$$

$$= \tfrac{1}{5}[3.5896 + 16 + 0.2240 - 1 - 5.3658 - 18]$$

$$= \tfrac{1}{5}[-4.5522] = -0.9104$$

$$= 1 + (-0.9104) + (-1) \quad \text{(to obtain mantissa between 0 and 1)}$$

$$= 0.0896 - 1$$

$$\overset{\bullet}{=} \log 1.229 - 1$$

$$= \log 0.1229$$

$$\underset{\text{T5.8.1}}{\Longrightarrow} \sqrt[5]{\frac{(281)^8 \cdot \sqrt[3]{0.0047}}{(7840)^6}} = 0.1229$$

$$
\begin{array}{l}
\bullet \\
10 \left[\begin{array}{l} x\left[\begin{array}{l}\log 1.220 = 0.0864 \\ \log \quad b \;\; = 0.0896\end{array}\right]32 \\ \log 1.230 = 0.0899 \end{array}\right]35
\end{array}
$$

$$\Longrightarrow \frac{x}{10} = \frac{32}{35} \Longrightarrow x \approx 9 \Longrightarrow b = 1.229$$

T5.8.1 $\log a = \log b \Longleftrightarrow a = b$

5.9 Problem Sets

Problem Set I ■ *Reading Comprehension*

Determine whether statements 1 through 15 are true or false and give reasons for your responses.

1 Logarithmic functions are just inverses of exponential functions.
2 The inverse of a strictly increasing exponential function is a logarithmic function that is strictly decreasing.
3 The logarithm of a negative number does not exist.
4 The function f defined as $f = \{(x, y) \mid x = 3^y\}$ is an exponential function.
5 The range of the function defined by $y = \log x$ is R, the set of real numbers.
6 A fundamental property of logarithms is
$$\frac{\log a}{\log b} = \log a - \log b$$

7 $\log (3.640 - 2) = \log (0.03640)$.

8 $-3.6421 = 0.6421 - 3$.

9 All entries within the table of common logarithms represent the logarithm of numbers between 1 and 10 only.

10 In the expression $\log 4.76 + 2$, the number 4.76 is called the mantissa.

11 $\log 3.84 + 3 = \log 6.84$

12 Linear interpolation is a method of approximating the logarithm or antilogarithm of numbers not found explicitly represented in the tables.

13 The antilogarithm of a number can be written as an exponential expression.

14 $[a, b \in R$ and $a = b] \Longrightarrow [\log a = \log b]$

15 $y = \text{antilog}_b\, x \Longleftrightarrow x = b^y$

Problem Set II ■ Skills Development

Express in logarithmic notation:

1 $2^3 = 8$	2 $3^2 = 9$	3 $16^{1/4} = 2$	4 $9^{1/2} = 3$	5 $8^{-1/3} = \frac{1}{2}$
6 $27^{-2/3} = \frac{1}{9}$	7 $10^{-2} = 0.01$	8 $10^5 = 100{,}000$	9 $10^0 = 1$	

10 $2^2 \cdot 2^3 = 32$ 11 $\dfrac{3^2}{3^4} = \dfrac{1}{9}$ 12 $(2 \cdot 3)^3 = 216$

Express in exponential notation:

13 $\log_{10} 1000 = 3$ 14 $\log_2 8 = 3$ 15 $\log_3 \frac{1}{9} = -2$ 16 $\log_{1/2} 4 = -2$

17 $\log_{1/3} 27 = -3$ 18 $\log_{1/4} 64 = -3$ 19 $\log_{10} 10 = 1$ 20 $\log_{10} 1 = 0$

21 $\log_5 25 = 2$

Determine x in each of the following:

22 $\log_{1/2} x = -3$ 23 $\log_{1/3} x = 2$ 24 $\log_{10} x = -3$ 25 $\log_4 \frac{1}{2} = x$

26 $\log_2 \frac{1}{8} = x$ 27 $\log_3 \left(-\frac{1}{9}\right) = x$ 28 $\log_x \frac{1}{8} = -3$ 29 $\log_x 16 = -2$

30 $\log_x \sqrt{3} = \frac{1}{2}$ 31 $\log_{10} 0.01 = x$ 32 $\log_5 \frac{1}{5} = x$ 33 $\log_2 \frac{1}{32} = x$

34 $\log_a a = x$ 35 $\log_b 1 = x$ 36 $\log_x 1 = 0$

Graph the logarithmic functions defined by the following equations:

37 $y = \log_2 x$ 38 $y = \log_{1/2} x$ 39 $y = \log_{1/3} x$ 40 $y = \log_3 x$

41 $y = \log_{10} x$ 42 $y = \log_5 x$

Express as the sum or difference of simpler logarithmic quantities:

Example $\log_b \left(\dfrac{x^2 \cdot y}{\sqrt{z}}\right)$

Solution $\log_b \left(\dfrac{x^2 y}{\sqrt{z}}\right) = \log_b x^2 y - \log_b z^{1/2}$

$\qquad\qquad\quad = \log_b x^2 + \log_b y - \log_b z^{1/2} = 2 \log_b x + \log_b y - \frac{1}{2} \log_b z$

43 $\log \sqrt[3]{x^2 z}$ 44 $\log \sqrt[4]{\dfrac{a \cdot b^3}{\sqrt[3]{c}}}$ 45 $\log \left(\dfrac{\sqrt{x}}{y \cdot z^2}\right)^3$ 46 $\log \left(2\pi \cdot \sqrt{\dfrac{L}{g}}\right)$

47 $\log \sqrt{\dfrac{2\sqrt{x}}{y^2 z}}$ 48 $\log \sqrt[3]{\dfrac{(0.036)^2}{\sqrt{943}}}$

Express the following as a single logarithm:

Example $2 \log (x + 1) - \frac{1}{2} \log y + \log z$

Solution $2 \log (x + 1) - \frac{1}{2} \log y + \log z$
$$= \log (x + 1)^2 - \log y^{1/2} + \log z$$
$$= \log \frac{(x + 1)^2}{\sqrt{y}} + \log z = \log \frac{z(x + 1)^2}{\sqrt{y}}$$

49   $\log x + 2 \log y - \log z$ 50 $\frac{1}{2} \log a - 3 \log (b + 1) + \frac{1}{3} \log c$
51   $\frac{1}{2} [\log x - 2 \log y - 3 \log z]$ 52 $\frac{1}{3} [\log 2 - 2 \log 3 + \frac{1}{2} \log 5]$
53   $\log 2 - 3 \log 7 + 1$ 54 $-2 \log 3 + \log 2 - 2$

Using Table I, determine the following logarithms:

55   $\log 467$ 56 $\log 37400$ 57 $\log 0.016$ 58 $\log 0.00439$
59   $\log 0.3692$ 60 $\log 0.003428$ 61 $\log 2763$ 62 $\log 4837$

Find the antilogarithm of the following numbers, using Table I.

63   0.4871 64 3.2923 65 -2.4345 66 -1.6328 67 4.4129
68   -10.1314

Complete the following calculations using logarithmic techniques (use linear interpolation as needed):

69   $\sqrt[3]{468}$ 70 $\sqrt{0.0684}$ 71 $\sqrt[5]{0.1634}$ 72 $(-0.132)^6$

73   $(0.0245)^{-5}$ 74 $(24.5)^{-4}$ 75 $\sqrt[4]{\dfrac{0.0623}{489}}$ 76 $\sqrt[3]{\dfrac{(64)^2}{(123)^3}}$

77   $\dfrac{(62.9)^2 \sqrt{493}}{94}$ 78 $\dfrac{(0.063)^3 \cdot (284)}{\sqrt{768}}$ 79 $\dfrac{\sqrt{292} \cdot \sqrt[3]{486}}{\sqrt[4]{56.2}}$

80   $\sqrt{\dfrac{283(0.062)^3}{\sqrt{768}}}$

Problem Set III   ■   *Theoretical Developments*

1   If $b, x \in R_+$ and $b \neq 1$, prove $b^{\log_b x} = x$.
2   If $b \in R_+$ and $b \neq 1$, prove $\log_b b^x = x$.
3   If $a, b \in R_+$ and $a, b \neq 1$, prove $\log_b a = 1/\log_a b$.
4   Prove the following theorems:
     Theorem 5.6.1   **a**   Part IV   **b**   Part V   **c**   Part VI

Problem Set IV   ■   *Applications*

1   The formula for the characteristic impedance of an open-wire transmission line is

$$Z = 276 \log \frac{a}{b}$$

where Z is the impedance of the line in ohms, a is the distance between the centers of two parallel conductors (wires), and b the radius of the conductors (the units of a and b must be the same). Suppose we wish to determine the distance necessary between the centers of two parallel conductors so that the impedance of the line be

250 ohms, using No. 12 AWG copper wire as the conductors (radius of this wire is 0.0404 inches). Find that distance.

2 If sound is increased or decreased in intensity, the response to this change by the human ear is approximately proportional to the logarithm of the quotient of the acoustical powers involved. Mathematically this is expressed as

$$N_{dB} = 10 \log \frac{P_1}{P_2}$$

where, in this case, N_{dB} is the number of decibels of acoustical difference between the two sounds (a change of 1 decibel is just barely audible by one with acute hearing ability), and P_1 and P_2 are power levels of two sounds. Find the decibel difference between an audio amplifier with an output of 20 watts compared with one with an audio output of 40 watts.

3 Using the formula of Problem 2, find the decibel gain of an electrical circuit if the input power is 3 watts and the output power is 7 watts.

4 The range of sound volume in speech may extend from as high as 250 to 1. What is the range in decibels?

Logarithms to Bases Other than Ten   5.10

The most frequently used base in mathematical analysis is not 10 but rather the irrational number e ($e \cong 2.718218$). As several of the applied problems in the exercises will demonstrate, a wide variety of natural phenomena is described mathematically by exponential or logarithmic expressions to the base e.

Logarithms to the base e are called natural logarithms. ● Because natural logarithms are so widely used, a separate table is provided for them. This table, however, is not as complete nor as easily managed as that for logarithms to the base 10. This is due, of course, to our number system being to the base 10, thus lending itself so readily to base 10 logarithmic tables. However, Table III of natural logarithms (also with a limited range) can be utilized even for numbers outside its range by employing a procedure similar to that used with the table of common logarithms, as the following examples illustrate.

● In some treatments $\log_e a$ is written $\ln a$ where the natural base e is understood.

Example 5.10.1 Determine $\log_e 5600$.

Solution Because the natural logarithmic table contains entries for numbers between 0.1 and 190, inclusive, we must convert $\log_e 5600$ to one within the range indicated. Thus

$$
\begin{aligned}
\log_e 5600 &= \log_e (5.6 \cdot 10^3) \\
&= \log_e 5.6 + \log_e 10^3 \\
&= \log_e 5.6 + 3 \log_e 10 \\
&= 1.7228 + 3(2.3026) \\
&= 8.6306
\end{aligned}
$$

Hence, the only difference in using tables of natural logarithms is that one must add to the logarithm of a number between 1 and 10 an *integral multiple of* $\log_e 10$ rather than just an integer, as with common logarithms where $\log_{10} 10 = 1$.

Example 5.10.2 Evaluate $\log_e 0.0082$.

Solution $\log_e 0.0082 = \log_e (8.2 \cdot 10^{-3}) = \log_e 8.2 + \log_e 10^{-3}$
$$= \log_e 8.2 - 3 \log_e 10 = 2.1041 - 3(2.3026)$$
$$= -4.8037$$

which means, as we have noted several times earlier,

$$e^{-4.8037} = 0.0082$$

There are problems or situations when a base different from 10 or e is desirable. It is to our advantage, therefore, to have a formula that will convert from one base to any other. The following theorem provides such a formula.

THEOREM 5.10.1   Let $a, b, c \in R_+$ and $b, d \neq 1$. then

$$\log_d a = \frac{\log_b a}{\log_b d}$$

Proof   Let $\log_d a = c$, then

D5.6.1 $y = \log_b x \Longleftrightarrow x = b^y$

T5.8.1
$a = b, \ a, b, c \in R_+ \Longleftrightarrow$
$\qquad\qquad \log_c a = \log_c b$

T5.6.1(V) $\log_b a^n = n \log_b a$

$$\log_d a = c \underset{\text{D5.6.1}}{\Longleftrightarrow} a = d^c \underset{\text{T5.8.1}}{\Longleftrightarrow} \log_b a = \log_b d^c$$
$$\underset{\text{T5.6.1(V)}}{\Longleftrightarrow} \log_b a = c \log_b d$$
$$\Longleftrightarrow c = \frac{\log_b a}{\log_b d}$$

($\log_b d \neq 0$, because $d \neq 1$)

But because $c = \log_d a$, we obtain the result

$$\log_d a = \frac{\log_b a}{\log_b d} \qquad \text{QED}$$

Example 5.10.3   Evaluate the following logarithms:

    (a) $\log_3 4.63$

Solution $\log_3 4.63 \underset{\text{T5.10.1}}{=\!=\!=\!=} \dfrac{\log_{10} 4.63}{\log_{10} 3} = \dfrac{0.6656}{0.4771} \cong 1.395$

    (b) $\log_7 0.0148$

Solution $\log_7 0.0148 = \dfrac{\log_{10} 0.0148}{\log_{10} 7} = \dfrac{\log 1.48 - 2}{\log 7} = \dfrac{0.1703 - 2}{0.8451}$
$$= \dfrac{-1.8297}{0.8451} = -2.165$$

A wide variety of equations containing exponential or logarithmic expressions can be solved taking advantage of the many properties of these functions along with the very important equivalence

$$y = b^x \Longleftrightarrow x = \log_b y$$

The following examples will illustrate some useful techniques in solving certain equations.

Example 5.11.1   Solve the following exponential equations:

(a) $3^x = 5$

Solution $3^x = 5 \xleftrightarrow{\textbf{D5.6.1}} x = \log_3 5$

$$\xupoverline{\textbf{T5.10.1}} \frac{\log 5}{\log 3} = \frac{0.6990}{0.4771}$$

$$=\!=\!= 1.465$$

$\textbf{D5.6.1} \quad y = \log_b x \Longleftrightarrow x = b^y$

$\textbf{T5.10.1} \quad \log_a a = \dfrac{\log_b a}{\log_b d}$

(b) $2^{3x-4} = 6$

Solution $2^{3x-4} = 6 \xleftrightarrow{\textbf{D5.6.1}} 3x - 4 = \log_2 6$

$$\Longleftrightarrow x = \frac{\log_2 6 + 4}{3}$$

(We will generally not simplify further expressions such as $\log_2 6$ in the above answer because the evaluation of these forms using Theorem 5.10.1 and tables is routine.)

(c) $3^{x^2-2x} = 4^{3x}$

Solution $3^{x^2-2x} = 4^{3x} \xrightarrow{\textbf{T5.8.1}} \log 3^{x^2-2x} = \log 4^{3x}$

$$\xleftarrow{\textbf{T5.6.1(V)}} (x^2 - 2x)\,\log 3 = 3x\,\log 4$$

$$\Longleftrightarrow x[x \log 3 - 2 \log 3 - 3 \log 4] = 0$$

$$\xleftarrow{\textbf{T2.9.7}} x = 0 \quad \text{or} \quad x = \frac{2 \log 3 + 3 \log 4}{\log 3}$$

$\textbf{T5.8.1}$
$a, b \in R_+ \quad \text{and} \quad a = b \Longleftrightarrow$
$\qquad\qquad\qquad\qquad \log a = \log b$

$\textbf{T5.6.1(V)} \quad \log a^m = m \log a$

$\textbf{T2.9.7}$
$ab = 0 \Longleftrightarrow [a = 0 \quad \text{or} \quad b = 0]$

(d) $2 \cdot 3^x = 5^{2x}$

Solution $2 \cdot 3^x = 5^{2x} \xrightarrow{\textbf{T5.8.1}} \log (2 \cdot 3^x) = \log 5^{2x}$

$$\xleftarrow{\textbf{T5.6.1(III)}} \log 2 + \log 3^x = \log 5^{2x}$$

$$\xleftarrow{\textbf{T5.6.1(V)}} \log 2 + x \log 3 = 2x \log 5$$

$$\Longleftrightarrow x(\log 3 - 2 \log 5) = -\log 2$$

$$\Longleftrightarrow x = \frac{-\log 2}{\log 3 - 2 \log 5}$$

$\textbf{T5.8.1}$
$a, b \in R_+ \quad \text{and} \quad a = b \Longleftrightarrow$
$\qquad\qquad\qquad\qquad \log a = \log b$

$\textbf{T5.6.1(III)}$
$\log (a \cdot b) = \log a + \log b$

$\textbf{T5.6.1(V)} \quad \log a^m = m \log a$

Example 5.11.2 Solve the following logarithmic equations:

(a) $\log (3x + 2) = \log (x - 4) + 1$

Solution

$$\log (3x+2) = \log (x-4) + 1 \Longleftrightarrow \log (3x+2) - \log (x-4) = 1$$

T5.6.1(IV) $\log \frac{a}{b} = \log a - \log b$

$$\xleftarrow{\text{T5.6.1(IV)}} \log \left(\frac{3x+2}{x-4}\right) = 1, \quad 3x+2 > 0 \quad \text{and} \quad x-4 > 0$$

D5.6.1 $y = \log_b x \Longleftrightarrow x = b^y$

$$\xleftarrow{\text{D5.6.1}} \frac{3x+2}{x-4} = 10^1 \quad \text{and} \quad x > 4$$

$$\Longleftrightarrow 3x + 2 = 10x - 40 \quad \text{and} \quad x > 4$$

$$\Longleftrightarrow 7x - 42 = 0 \quad \text{and} \quad x > 4 \Longleftrightarrow x = 6 \quad \text{and} \quad x > 4$$

$$\Longleftrightarrow x = 6$$

(b) $\log x + \log (x - 3) = 1$

Solution $\log x + \log (x - 3) = 1$

T5.6.1(III) $\log (ab) = \log a + \log b$

$$\xleftarrow{\text{T5.6.1(III)}} \log [x(x-3)] = 1, \; x > 0 \quad \text{and} \quad x - 3 > 0$$

$$\Longleftrightarrow x^2 - 3x - 10 = 0 \quad \text{and} \quad x > 3$$

$$\Longleftrightarrow (x-5)(x+2) = 0 \quad \text{and} \quad x > 3$$

$$\Longleftrightarrow [x = 5 \quad \text{or} \quad x = -2] \quad \text{and} \quad x > 3 \Longleftrightarrow x = 5$$

It is important to note here that if $x > 0$ and $x - 3 > 0$ had not been included in the first equivalence (which is required by Theorem 5.6.1), you may have been tempted to allow $x = -2$ as one of the solutions. Remember that the domain of the logarithm function contains *positive* numbers only.

(c) $2 \log (x + 2) - \log (x - 2) = 1$

Solution $2 \log (x + 2) - \log (x - 2) = 1$

T5.6.1(V) $\log a^m = m \log a$

$$\xleftarrow{\text{T5.6.1(V)}} \log (x+2)^2 - \log (x-2) = 1$$

$$\text{and} \quad x + 2 > 0 \quad \text{and} \quad x - 2 > 0$$

T5.6.1(IV) $\log \frac{a}{b} = \log a - \log b$

$$\xleftarrow{\text{T5.6.1(IV)}} \log \left[\frac{(x+2)^2}{x-2}\right] = 1, \quad x+2 > 0 \quad \text{and} \quad x-2 > 0$$

D5.6.1 $y = \log_b x \Longleftrightarrow x = b^y$

$$\xleftarrow{\text{D5.6.1}} \frac{(x+2)^2}{x-2} = 10 \quad \text{and} \quad x > 2$$

$$\Longleftrightarrow x^2 + 4x + 4 = 10x - 20 \quad \text{and} \quad x > 2$$

$$\Longleftrightarrow x^2 - 6x + 24 = 0 \quad \text{and} \quad x > 2$$

$$\Longleftrightarrow x = \frac{6 \pm \sqrt{36 - 96}}{2} \quad \text{and} \quad x > 2$$

$$\Longleftrightarrow x = \frac{6 \pm \sqrt{-60}}{2} \quad \text{and} \quad x > 2$$

$$\Longrightarrow S = \varnothing$$

The solution set here is empty because we are only admitting real solutions and, of course, $\sqrt{-60}$ is not real.

Problem Set I   ■   Reading Comprehension

Determine whether statements 1 through 5 are true or false and give reasons for your responses.

1 Because we only have tables of logarithms to the base 10 and e, these are the only ones we can actually evaluate.
2 Logarithms to the base e are called natural logarithms.
3 $\log_d a = \dfrac{\log_b a}{\log_b d} = \log_b a - \log_b d$
4 In mathematical analysis the logarithms most commonly used are the common logarithms to the base 10.
5 $\log x + \log (x + 1) = \log x[x + 1]$ for all x.

Problem Set II   ■   Skills Development

Evaluate the following logarithms using Table II.

1 $\log_3 84$ 2 $\log_2 0.013$ 3 $\log_5 846$ 4 $\log_7 0.086$
5 $\log_{23} 46$ 6 $\log_{15} 115$

Solve the following equations for x:

7 $2^x = 6$ 8 $3^{2x} = 4$ 9 $5^{2x} = 3 \cdot 2^{-3x}$ 10 $10^x = 3 \cdot 5^{-2x}$
11 $6^{x^2} = 4$ 12 $12^{-x^2} = 2 \cdot 3^{2x}$ 13 $\log x + 2 \log 3 = 3$
14 $\log (x + 2) - \log 5 = 2$ 15 $\log (x + 3) + \log x = 1$
16 $\log x + \log (x + 21) = 2$ 17 $\log x + \log (x - 3) = 1$
18 $\log (x + 3) - \log (x + 1) = 1$ 19 $\log_2 x = \log_3 5$ 20 $|\log_5 x| = 2$
21 $\log |x - 3| = 2$ 22 $|\log_x 9| = 2$

Problem Set III   ■   Applications

1 A tank is initially filled with 100 gallons of a salt solution containing 1 lb of salt per gallon. Another salt solution containing 2 lb of salt per gallon runs into the tank at a rate of 5 gal/min, and the mixture, kept uniform by stirring, runs out at the same rate.

It is found that the amount of salt S in the tank at any time t is given by the formula
$$S = 200 - 100e^{-t/20}$$
How long will it take for 150 lb of salt to accumulate in the tank?

2   Under certain conditions it is observed that the rate at which atmospheric pressure changes with altitude is proportional to the pressure. It is further observed that the pressure at sea level is 14.7 lb/in.², whereas at 18,000 ft it has dropped to half this amount. From this information it is found that a formula relating pressure and altitude is given by
$$P = 14.7e^{-0.0000385h}$$
where P is pressure measured in pounds per square inch and h is altitude measured in feet. At what altitude will the pressure be 1 lb/in.²?

3   According to Newton's law of cooling, the rate at which the temperature of a body decreases is proportional to the difference between the instantaneous temperature of the body and the temperature of the surrounding medium. A body whose temperature initially is 100°C is allowed to cool in air that remains at a constant temperature of 20°C. It is observed after 10 minutes that the body has cooled to 60°C. From this information it is found that the temperature T of the body is given by the formula
$$T = 20 + 80e^{-0.0693t}$$
where t is time measured in minutes. How long will it take the body to cool to a temperature of 25°C?

4   When a switch is closed in a circuit containing a resistance R, an inductance L, and a battery which supplies a constant voltage E, the current i builds up at a certain rate given by the formula
$$LR_i + R \cdot i = E$$
where R_i represents the rate at which the current changes with time. From this it is found that the current i is related to time t by the formula
$$i = \frac{E}{R} (1 - e^{-Rt/L})$$
How long will it take i to reach half its final value, that is, $\frac{1}{2} \frac{E}{R}$?

Problem Set IV   ■   **Just For Fun**

1   Determine which is larger, e^{π} or π^{e}.

5.13   Summary of Definitions and Properties in Chapter 5

n factors

D5.2.1 $a \in R, n \in N \Longrightarrow a^n = \overbrace{a \cdot a \ldots a}$ (p. 180)

D5.2.2 $a \in R, a \neq 0$, and $n \in N \Longrightarrow a^{-n} = \dfrac{1}{a^n}$ (p. 181)

D5.2.3 $a \in R, a \neq 0 \Longrightarrow a^0 = 1$ (p. 182)

D5.2.4 $\sqrt[n]{a} = b \Longleftrightarrow \begin{cases} a = b^n \\ \\ b \geq 0 \quad \text{if } n \text{ is even} \end{cases}$ (p. 183)

where $a, b \in R$ and $n \in N$

D5.2.5 $a^{1/n} = \sqrt[n]{a}$ where $a \geq 0$ if n is even (p. 184)

T5.2.2 $a, b \in R,\ b \neq 0,$ and $m, n \in Q \Longrightarrow$ (p. 185)

I $a^m \cdot a^n = a^{m+n}$

II $[a^m]^n = a^{mn}$

III $[ab]^n = a^n \cdot b^n$

IV $\left(\dfrac{a}{b}\right)^n = \dfrac{a^n}{b^n}$

V $\dfrac{b^m}{b^n} = b^{m-n} = \dfrac{1}{b^{n-m}}$

VI $a^{-n} = \dfrac{1}{a^n}$

VII $\left(\dfrac{a}{b}\right)^{-n} = \left(\dfrac{b}{a}\right)^n$

D5.4.1 $b \in R_+,\ b \neq 1 \Longrightarrow E_b = \{(x, y) \mid y = b^x\}$ (p. 189)
where E_b is called the exponential function to the base b.

D5.6.1 $b \in R_+$ and $b \neq 1 \Longrightarrow [y = \log_b x \Longleftrightarrow x = b^y]$ (p. 195)

D5.6.2 $b \in R_+$ and $b \neq 1 \Longrightarrow L_b = \{(x, y) \mid y = \log_b x\}$ (p. 195)
where L_b is called the logarithmic function to the base b.

T5.6.1 $a, b, c \in R_+,\ b \neq 1,\ m \in J,$ and $n \in J_+ \Longrightarrow$ (p. 196)

I $\log_b b = 1$

II $b^{\log_b a} = a$

III $\log_b (ac) = \log_b a + \log_b c$

IV $\log_b \dfrac{a}{c} = \log_b a - \log_b c$

V $\log_b a^n = n \log_b a$

VI $\log_b \sqrt[n]{a^m} = \dfrac{m}{n} \log_b a$

T5.10.1 $a, b, d \in R_+,\ b, d \neq 1 \Longrightarrow \log_d a = \dfrac{\log_b a}{\log_b d}$ (p. 212)

Answers to Sample Problem Set 5.7 (p. 198)

1 True. (See Definitions 5.6.1 and 5.6.2.)

2 False. This relation is given by Definition 5.6.1.

3 False. Should be $\log_9 \frac{1}{3} = -\frac{1}{2}$.

4 False. The logarithmic expression is only defined for positive bases.
   (See Definition 5.6.1.)

5 False. Domain is R_+ and range is R.

6 False. Domain is R_+ so -4 is not admissible.

7 True. (See Theorem 5.6.1(II).)

8 False. $\log_5 \frac{3}{7} = \log_5 3 - \log_5 7.$ (See Theorem 5.6.1(IV).)

CHAPTER

6

Complex Numbers — A Two-Dimensional Number System

6.1 Introduction

We have all had sufficient experience with numbers to realize how useful they are in solving many kinds of problems. Frequently the solution to a problem requires solving equations that mathematically represent the problem. The process of solving an equation involves, by various manipulations, the finding of numbers that satisfy the equation. Let us examine how increasingly more kinds of numbers are needed in order to solve certain equations. We begin with the simple equation

$$x - 8 = 2$$

and suppose the only numbers we have available are the natural (counting) numbers. For the above equation we have no problem because the solution is 10, a natural number. If, however, we take the equation

$$x + 8 = 2$$

218

we are in trouble, if all we have are natural numbers. To remedy this deficiency of numbers we expand our system to include *negative* natural numbers and *zero*; we call this the set of integers. With integers available, we can now solve the above equation—the integer -6 satisfies the equation. Now let us take the equation

$$3x = 1$$

If all we have are integers, again we are in trouble. To take care of the situation, we expand our number system further to include fractions and call this the set of rational numbers. Within this set of numbers we have a solution to the above equation; the rational $\frac{1}{3}$ is a root and once more we are saved from the impass. We next take the elementary quadratic equation

$$x^2 - 2 = 0$$

It can be easily proved that there is no rational number whose square is 2. Thus, if we were limited to just rational numbers, the above equation would remain without a solution. Therefore, we expand our system of numbers to include *irrational* numbers and call this the set of real numbers. The square root of 2, denoted $\sqrt{2}$, is an irrational number that satisfies the above equation. Again we have overcome the problem of facing a simple equation without being able to solve it. But, as you may suspect, our triumph is short-lived, for we have but to take the equally simple equation

$$x^2 + 2 = 0$$

whose solutions, if they are to exist, lie outside the set of real numbers. This can be easily seen by rewriting this equation in the form

$$x^2 = -2$$

and recalling that we have proven that the square of every real number is greater than or equal to zero. And this means, of course, that the square of no real number is ever negative—but this is precisely what is required in the above equation. Even with all the real numbers, we have again run out of numbers. Following the approach successively taken above, the natural thing to do is to again expand our number system by creating some more numbers that, hopefully, would satisfy equations such as $x^2 + 2 = 0$. This is precisely what we will do as we introduce complex numbers.

Although you may not be aware of them, there are many important problems in applied mathematics that do not have solutions expressible solely in terms of real numbers. Even when real solutions do exist, the problem of identifying them may be most difficult if one is constrained to work within the comparatively limited framework of real numbers. Many of these problems can be significantly simplified by the use of complex numbers. In fact, today complex numbers and complex functions are indispensable tools in dealing with problems in many areas of both pure and applied mathematics.

6.2 Complex Numbers

As we consider expanding our system of numbers, let us recall the very useful way real numbers can be represented geometrically as points on a line. In Chapter 2 we established, at least intuitively, a one-to-one correspondence between real numbers and points on a line, and called this collection of points the real line, as illustrated in Fig. 6.2.1. Because a line is one-dimensional, the set of real numbers is often called a one-dimensional number system. We have also worked with a two-dimensional rectangular coordinate system composed of a *plane* of points. These points are represented algebraically by ordered pairs of real numbers. (See Fig. 6.2.2.)

FIGURE 6.2.1

It is not unreasonable to conjecture, "if the members of one number system correspond in a natural way to points on a line, why could not those of another system correspond similarly to points in the plane?" The answer is, they can. In fact we shall define the set of all *ordered pairs* of real numbers and, hence, geometrically the set of all points in the plane, as the set of complex numbers. Thus, a complex number may be denoted symbolically as (a, b), where a and b are real. The more common notation, however, is $a + bi$, where by definition i, called the imaginary unit, satisfies the property $i^2 = -1$. It is this property of i that will distinguish and characterize the complex number system. We formalize our definition as follows.

DEFINITION 6.2.1   The set C of complex numbers is the set of *all ordered pairs of real numbers*, each denoted (a, b) or $a + bi$, where, in the latter notation, i is called the imaginary unit and satisfies the property $i^2 = -1$. Symbolically we have

$$C = \{(a, b) \mid a, b \in \mathbf{R}\}$$

or

$$C = \{a + bi \mid a, b \in \mathbf{R} \quad \text{and} \quad i^2 = -1\}$$

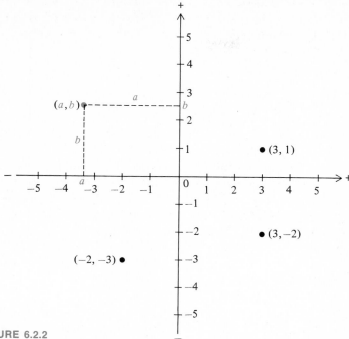

FIGURE 6.2.2

When points in the plane represent complex numbers, the plane is called the complex plane. In the complex plane of Fig. 6.2.3 we identify a few complex numbers.

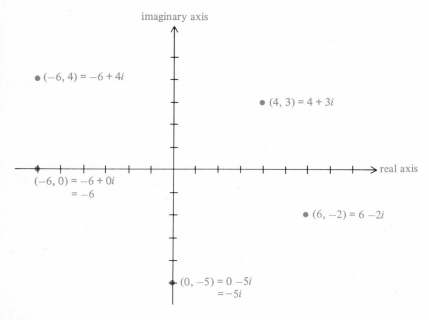

FIGURE 6.2.3 The Complex Plane

Notice that complex points lying on the horizontal axis of the complex plane are of the form $a + 0i$. It would be desirable for the set of complex numbers to contain the set of real numbers as a *subset*. To accomplish this we define the product $0i$ to be $0●$ and write

$$a + 0i = a$$

or in ordered pair notation

$$(a, 0) = a$$

The complex points on the horizontal axis thus become real points corresponding to real numbers. For this reason the horizontal axis in the complex plane is called the real axis. Notice also that points lying on the vertical axis are of the form $0 + bi$. For convenience we define

$$0 + bi = bi$$

or in ordered pair notation

$$(0, b) = bi$$

Complex numbers of this form are called pure imaginary and the vertical axis of the complex plane is called the imaginary axis. Finally, for an arbitrary complex number $a + bi$, a is called the real part and b the imaginary part. (Note that the imaginary part is not bi but simply b.)

Remark 6.2.1 You would not want to misinterpret the word "imaginary" as used above. It is a carry-over from the nineteenth century when people actually thought these numbers to be imaginary in the sense that they had no meaning or relevance in the real physical world. This, of course, has since been proven false. Complex numbers today are indispensable in applied mathematics and are just as "real" as real numbers.

We must now define an algebra for complex numbers. Note that the real numbers would be of little or no use to us if all we had were the numbers themselves and no defined operations or properties. It is because we can add, subtract, multiply, and divide them and set up equations composed of real numbers that make these numbers so useful to us. So it is with complex numbers; to just have these numbers without any relations or operations would do us little good. We need to give meaning to a complex equation and to the operations of addition, subtraction, multiplication, and division of complex numbers and to investigate the resulting properties of these; that is, we need an algebra for complex numbers. We begin with the relation equals for complex numbers.

DEFINITION 6.2.2 Two complex numbers $a + bi$ and $c + di$ are equal if their respective real and imaginary parts are equal. Symbolically we have

$$[a + bi = c + di] \Longleftrightarrow [a = c \quad \text{and} \quad b = d]$$

Remark 6.2.2 It is significant to note from the above definition that one complex equation is equivalent to two real equations. This observation will be the key to solving many complex equations.

Example 6.2.1 Examine the following equalities and nonequalities.

(a) $2 + (-3)i = \sqrt{4} + (-\frac{6}{2})i \Longleftrightarrow 2 = \sqrt{4}$ and $-3 = -\frac{6}{2}$

(b) $2 + (-3)i \neq -3 + 2i \quad \Longleftrightarrow 2 \neq -3$
(Complex numbers correspond to ordered pairs — one cannot reverse the order of the real and imaginary parts.)

(c) $2 + (-3)i \neq 2 + (-6)i \quad \Longleftrightarrow -3 \neq -6$
(Both real and imaginary parts must be equal.)

For notational convenience, we define

$$a - bi = a + (-b)i$$

so that complex numbers such as $2 + (-3)i$ may be written more simply as $2 - 3i$.

It can be shown that the relation equals for complex numbers satisfies the same properties as this relation for real numbers. These properties are identified in the following theorem.

THEOREM 6.2.1 If $a + bi$, $c + di$, and $e + fi$ are complex numbers, then the following properties hold:

I **REFLEXIVE LAW OF EQUALITY (RLE)**
$$a + bi = a + bi$$

II **SYMMETRIC LAW OF EQUALITY (SLE)**
$$a + bi = c + di \Longleftrightarrow c + di = a + bi$$

III **TRANSITIVE LAW OF EQUALITY (TLE)**
$$[a + bi = c + di \quad \text{and} \quad c + di = e + fi] \Longleftrightarrow [a + bi = e + fi]$$

IV **SUBSTITUTION LAW OF EQUALITY (SubLE)**

If $a + bi = c + di$, then $a + bi$ may be replaced by $c + di$ or, conversely, $c + di$ may be replaced by $a + bi$, in any complex mathematical statement without changing the truth or falsity of that statement.

D6.2.2

$a + bi = c + di \Longleftrightarrow a = c$ and $b = d$

E-2

$a, b \in \mathbf{R} \Longrightarrow [a = b \Longleftrightarrow b = a]$

$$a + bi = c + di \overset{\text{D6.2.2}}{\Longrightarrow} a = c \quad \text{and} \quad b = d$$

$$\overset{\text{E-2}}{\Longleftrightarrow} c = a \quad \text{and} \quad d = b$$

$$\overset{\text{D6.2.2}}{\Longleftrightarrow} c + di = a + bi \qquad \text{QED}$$

(Proofs of Parts I, III, and IV are to be done as exercises.)

We next introduce the operations of addition and multiplication.

DEFINITION 6.2.3   Let $a + bi$ and $c + di$ be two complex numbers. Then the sum of these two numbers, denoted $(a + bi) + (c + di)$, • is given by

$$(a + bi) + (c + di) = (a + c) + (b + d)i$$

The process of performing this sum is called addition.

● Notice that we are using the same symbol + for the sum of complex numbers as is used for the sum of real numbers. This is common practice where there is little chance for confusion. Some authors may use initially the symbol ⊕ for complex sums to distinguish between the two operations, one for real numbers and the other for complex.

Example 6.2.2

(a) $(2 + 5i) + (-3 + 4i) = (2 + (-3)) + (5 + 4)i$
$$= -1 + 9i$$

(b) $(-3 + 2i) + (6 - 3i) = (-3 + 6) + (2 + (-3))i$
$$= (-3 + 6) + (2 - 3)i$$
$$= 3 + (-1)i$$
$$= 3 - i$$

Of course, in practice, all of the above steps are not included. In fact, you will probably perform all of these mentally in one step and just write down the sum.

DEFINITION 6.2.4   Let $a + bi$ and $c + di$ be two complex numbers. Then the product of these two numbers, denoted $(a + bi) \cdot (c + di)$, is given by

$$(a + bi) \cdot (c + di) = (ac - bd) + (ad + bc)i$$

The process of performing this product is called multiplication.

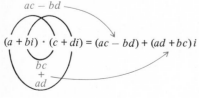

$$(a + bi) \cdot (c + di) = (ac - bd) + (ad + bc)i$$

FIGURE 6.2.4

A pattern that may assist you in remembering how to perform this product is given in Fig. 6.2.4.

Example 6.2.3

$$(2 - 3i) \cdot (1 + 4i) = (2 \cdot 1 - (-3) \cdot 4) + (2 \cdot 4 + (-3) \cdot 1)i$$
$$= (2 + 12) + (8 - 3)i$$
$$= 14 + 5i$$

Complex products can also be performed by treating the complex factors as if they were real and, remembering that $i^2 = -1$, applying the distributive, associative, and commutative laws for real numbers. The next example will illustrate this method.

Example 6.2.4   We will use the same numbers as in Example 6.2.3.

$$(2 - 3i) \cdot (1 + 4i) = 2(1 + 4i) - 3i(1 + 4i)$$
$$= 2 + 8i - 3i - 12i^2$$
$$= 2 + 5i + 12$$
$$= 14 + 5i$$

which is identical with the result obtained in Example 6.2.3. Most people find this latter method the simplest to employ.

Problem Sets   6.3

Problem Set I   ■   *Reading Comprehension*

Determine whether statements 1 through 7 are true or false and give reasons for your responses.

1   As versatile and useful as the system of real numbers is, still there are algebraic equations of the simplest kind that have no solutions among the real numbers.
2   Two complex numbers are equal if they contain exactly the same pair of real numbers.
3   All four equality axioms of real numbers hold in the complex number system as theorems.
4   The complex number $a + 0i$ is equal to the real number a.
5   Every real number is a complex number, and hence the set of real numbers forms a proper subset of the set of complex numbers.
6   The imaginary part of the complex number $a + bi$ is bi.
7   The imaginary unit i in the complex number system is the same as -1 in the real system.

Problem Set II   ■   *Skills Development*

Perform the following operations and simplify, leaving answers in the standard form $a + bi$.

1   $(-1 + 2i) + (3 + 4i)$ 2 $(2 + 3i) + (-4 + 6i)$ 3 $(-2 + 0i) + (4 - 5i)$
4   $-4 + (-3 - 5i)$ 5 $(1 - 3i) + 2i$ 6 $(-5 - 6i) + (-1 + 4i)$
7   $(-1 + 2i) \cdot (3 + 4i)$ 8 $(2 + 3i) \cdot (-4 + 6i)$ 9 $(-2 + 0i) \cdot (4 - 5i)$
10   $(-4) \cdot (-3 - 5i)$ 11 $(1 - 3i) \cdot (2i)$ 12 $(-5 - 6i) \cdot (-1 + 4i)$

Prove the following theorems:

1 Theorem 6.2.1 Part I   **2** Theorem 6.2.1 Part III   **3** Theorem 6.2.1 Part IV

6.4 Properties of Addition and Multiplication and Other Operations

We now investigate properties that the operations of addition and multiplication possess along with the relation equals. You will recall that the central properties associated with these in the real number system were the field properties, from which nearly all our basic techniques for arithmetic and algebraic manipulation were derived. It happens that all the field properties hold equally well for complex numbers. These we now state in terms of complex numbers. The proofs of most of them are to be done as exercises.

THEOREM 6.4.1
THE CLOSURE LAWS OF ADDITION AND MULTIPLICATION (Cl+ and Cl·) The sum and product of any two complex numbers is again a complex number. Symbolically, we have

(1) $(a + bi), (c + di) \in C \Longrightarrow (a + bi) + (c + di) \in C$

(2) $(a + bi), (c + di) \in C \Longrightarrow (a + bi) \cdot (c + di) \in C$

THEOREM 6.4.2
THE ASSOCIATIVE LAWS OF ADDITION AND MULTIPLICATION (A+ and A) If $(a + bi)$, $(c + di)$, and $(e + fi)$ are complex numbers, then

(1) $(a + bi) + [(c + di) + (e + fi)] = [(a + bi) + (c + di)] + (e + fi)$

(2) $(a + bi) \cdot [(c + di) \cdot (e + fi)] = [(a + bi) \cdot (c + di)] \cdot (e + fi)$

Remark 6.4.1   Note that the Associative Law of Addition and Multiplication for complex numbers, as for real numbers, indicates that when performing sums or products of three or more numbers the operations may be performed in any order—the presence of parentheses is irrelevant and may even be dropped.

THEOREM 6.4.3
THE COMMUTATIVE LAWS OF ADDITION AND MULTIPLICATION (C+ and C·)   If $a + bi$ and $c + di$ are complex numbers, then

(1) $(a + bi) + (c + di) = (c + di) + (a + bi)$

(2) $(a + bi) \cdot (c + di) = (c + di) \cdot (a + bi)$

Proof   (for C+)

$$(a + bi) + (c + di) \overset{\text{D6.2.3}}{=\!=\!=} (a + c) + (b + d)i$$

$$\overset{\text{C+}}{=\!=} (c + a) + (d + b)i$$

$$\overset{\text{D6.2.3}}{=\!=\!=} (c + di) + (a + bi)$$

$$\overset{\text{TLE}}{=\!=\!\Longrightarrow} (a + bi) + (c + di) =\!=\!= (c + di) + (a + bi) \qquad \text{QED}$$

D6.2.3   Definition of Addition

C+ $a, b \in \boldsymbol{R} \Longrightarrow a + b = b + a$

Remark 6.4.2   The Commutative Law of Addition (C+) used in step 2 of the above proof was applied to the *real* sums $a + c$ and $b + d$. This, of course, is permissible because this property for real numbers has already been established.

THEOREM 6.4.4   ADDITIVE IDENTITY LAW (Id+)   There exists a complex number called zero, denoted $0 + 0i$, or simply 0, such that the sum of any complex number $a + bi$ and $0 + 0i$ is $a + bi$. Symbolically, we have

$$(a + bi) + (0 + 0i) = (a + bi)$$

The complex number $0 + 0i$ is called the additive identity element of the complex number system.

THEOREM 6.4.5   ADDITIVE INVERSE LAW (In+)   For each complex number $a + bi$, there exists an additive inverse, $-a - bi$, also denoted $-(a + bi)$, such that

$$(a + bi) + (-a - bi) = 0 + 0i$$

THEOREM 6.4.6   DISTRIBUTIVE LAW (D)   If $a + bi$, $c + di$, and $e + fi$, are complex numbers, then

$$(a + bi) \cdot [(c + di) + (e + fi)] =$$
$$(a + bi) \cdot (c + di) + (a + bi) \cdot (e + fi)$$

Proof

$$(a + bi) \cdot [(c + di) + (e + fi)]$$

$$\underset{\text{D6.2.3}}{=\!=\!=} (a + bi) \cdot [(c + e) + (d + f)i]$$

$$\underset{\text{D6.2.4}}{=\!=\!=} [a(c + e) - b(d + f)] + [a(d + f) + b(c + e)]i$$

$$\underset{\text{D}}{=\!=} [ac + ae - bd - bf] + [ad + af + bc + be]i$$

$$\underset{\text{C+}}{\overset{\text{A+}}{=\!=}} [(ac - bd) + (ae - bf)] + [(ad + bc) + (af + be)]i$$

$$\underset{\text{D6.2.3}}{=\!=\!=} [(ac - bd) + (ad + bc)i] + [(ae - bf) + (af + be)i]$$

$$\underset{\text{D6.2.4}}{=\!=\!=} [(a + bi) \cdot (c + di)] + [(a + bi) \cdot (e + fi)]$$

$$\underset{\text{TLE}}{=\!=\!\Longrightarrow} (a + bi) \cdot [(c + di) + (e + fi)]$$

$$=\!=\!= (a + bi) \cdot (c + di) + (a + bi) \cdot (e + fi) \qquad \text{QED}$$

D6.2.3   Definition of Addition

D6.2.4
$(a + bi)(c + di) =$
    $(ac - bd) + (ad + bc)i$

Remark 6.4.3   Notice that the brackets around the products in the second-to-last step of the proof are omitted in the concluding step. In the complex number system, as in the real system, when no associative symbols, such as braces and brackets, are used to indicate otherwise, multiplication always is performed before addition.

THEOREM 6.4.7 MULTIPLICATIVE IDENTITY LAW (Id·)   There exists a complex number called one, denoted $1 + 0i$ or simply 1, such that

$$(a + bi) \cdot (1 + 0i) = a + bi$$

THEOREM 6.4.8 MULTIPLICATIVE INVERSE LAW (In·)   For each complex number $a + bi$, except $0 + 0i$, there exists a multiplicative inverse

$$\left(\frac{a}{a^2 + b^2}\right) + \left(\frac{-b}{a^2 + b^2}\right)i$$

also denoted $(a + bi)^{-1}$, such that the product of $(a + bi)$ and $(a + bi)^{-1}$ is $1 + 0i$. Symbolically this becomes

$$(a + bi) \cdot \left[\left(\frac{a}{a^2 + b^2}\right) + \left(\frac{-b}{a^2 + b^2}\right)i\right] = 1 + 0i$$

It is important to note in the above theorem that the complex zero $0 + 0i$ does not have a multiplicative inverse, just as the real zero 0 does not have such an inverse.

We now define the operations of subtraction and division for complex numbers. You will note that they will be defined exactly as they were in the real number system.

DEFINITION 6.4.1   If $a + bi$ and $c + di$ are complex numbers, then

$$(a + bi) - (c + di) = (a + bi) + [-(c + di)]$$
$$= (a + bi) + [-c - di]$$
$$= (a - c) + (b - d)i$$

The process of performing this difference is called subtraction.

Example 6.4.1 $(2 - 3i) - (4 + i) = (2 - 3i) + (-4 - i)$
$$= (2 - 4) + (-3 - 1)i$$
$$= -2 - 4i$$

DEFINITION 6.4.2   If $a + bi$ and $c + di$ are complex numbers, $c + di \neq 0 + 0i$, then

$$\frac{a + bi}{c + di} = (a + bi) \cdot (c + di)^{-1}$$

$$= (a + bi) \cdot \left[\left(\frac{c}{c^2 + d^2}\right) + \left(\frac{-d}{c^2 + d^2}\right)i\right]$$

$$= \left(\frac{ac + bd}{c^2 + d^2}\right) + \left(\frac{bc - ad}{c^2 + d^2}\right)i$$

The process of performing this quotient is called division.

Example 6.4.2 $\dfrac{4 - 3i}{-1 + 2i} = (4 - 3i) \cdot (-1 + 2i)^{-1}$

$$= (4 - 3i) \cdot \left(\frac{-1}{(-1)^2 + 2^2} + \frac{-2}{(-1)^2 + 2^2}i\right)$$

$$= \left(\frac{-4 - 6}{5} + \frac{3 - 8}{5}i\right)$$

$$= -2 - i$$

An alternative method of division is performed using the *conjugate* of a complex number. If $a + bi$ is a complex number, then the conjugate of $a + bi$, denoted $\overline{a + bi}$, is $a - bi$. Symbolically, we have

$$\overline{a + bi} = a - bi$$

The method is to multiply the numerator and the denominator of a quotient by the conjugate of the denominator. This method is illustrated in the next example.

Example 6.4.3

$$\frac{4-3i}{-1+2i} = \frac{4-3i}{-1+2i} \cdot \frac{-1-2i}{-1-2i} = \frac{(-4-6)+(3-8)i}{1+4}$$

$$= \frac{-10}{5} + \frac{-5}{5}i$$

$$= -2 - i$$

which is identical with that obtained in the previous example where division was performed by direct application of the definition. Either method, of course, is acceptable.

We have shown that the system of complex numbers not only satisfies the same equality axioms as real numbers but also satisfies the same eleven field axioms. This is most significant, because this will save us a great deal of time and effort. You may recall from Chapter 2 the many properties that were shown to be direct consequences of the equality and field axioms. It would take considerable work and patience, you would agree, to verify all those properties for complex numbers. Fortunately this will not be necessary because the proofs of those properties depended solely upon the equality and field axioms and not because we happened to be using real numbers. Thus, *any* system satisfying a given set of axioms would enjoy all the properties that follow from those axioms. Herein lies the strength and utility of an axiomatic development of a mathematical system—theorems need only be proven once. We may, therefore, accept within the complex number system all of the field properties of arithmetic and algebraic manipulation, that is, the addition and multiplication laws of equality, the term and factor cancellation laws, the laws of adding and multiplying fractions, the laws of signed numbers, and all the others. We illustrate a few of these with the following examples.

Example 6.4.4 Recall from Chapter 2 the property

$$[a, b \in \boldsymbol{R} \quad \text{and} \quad ab = 0] \Longrightarrow [a = 0 \quad \text{or} \quad b = 0]$$

This must also hold for complex numbers. We have

$$[(a+bi)(c+di) = 0] \Longrightarrow [a+bi = 0+0i \quad \text{or} \quad c+di = 0+0i]$$

Example 6.4.5 The Subtraction Law of Fractions. If $a, b, c, d \in \boldsymbol{R}$, $b, d \neq 0$, then

$$\frac{a}{b} - \frac{c}{d} = \frac{ad - bc}{bd}$$

For complex numbers we would have the following: If $a + bi$, $c + di$, $e + fi$, and $g + hi$ are complex numbers, where $c + di$ and $g + hi$ are not $0 + 0i$, then

$$\frac{a+bi}{c+di} - \frac{e+fi}{g+hi} = \frac{(a+bi)(g+hi) - (c+di)(e+fi)}{(c+di)(g+hi)}$$

In summary, with the field axioms satisfied and the operations of subtraction and division defined for complex numbers, we are able to perform complex calculations and manipulate and solve complex equations in essentially the same way as we do within the real number system. You might wonder at this point whether there is really that much difference between the two number systems. Actually there is a great deal of difference. It is just that we have only looked thus far at their similarities. In the sections that follow we will investigate some of their important differences.

Problem Sets 6.5

Problem Set I ■ *Reading Comprehension*

Determine whether statements 1 through 5 are true or false and give reasons for your answers.

1 Only one of the field axioms is not satisfied by complex numbers, that is, there do not exist additive inverses.
2 The complex number called one is $1 + 1i$, just as the complex number called zero is $0 + 0i$.
3 The multiplicative inverse of a complex number (a, b) is $(1/a, 1/b)$ if $a, b \neq 0$.
4 The set C of complex numbers is closed under the operation of division.
5 Because the complex numbers satisfy the field axioms, we are relieved from having to prove, for complex numbers, any property resulting from these axioms in the development of the real number system; they all must also hold for complex numbers.
6 Explain some of the advantages of an axiomatic development of a number system.
7 List ten field properties of real numbers and below each write its counterpart for complex numbers.

Example $a, b, c \in \mathbf{R}, c \neq 0 \Longrightarrow \dfrac{a}{c} + \dfrac{b}{c} = \dfrac{a+b}{c}$

Thus, for complex numbers we have:

$$a+bi, c+di, e+fi \in \mathbf{C}, e+fi \neq 0+0i \Longrightarrow \frac{a+bi}{e+fi} + \frac{c+di}{e+fi} = \frac{(a+bi)+(c+di)}{e+fi}$$

Problem Set II ■ *Skills Development*

Perform each of the following operations and express your answers in the form $a + bi$.

1 $(-1 + 2i) - (3 + 4i)$ 2 $(2 + 3i) - (-4 + 6i)$ 3 $(-2 + 0i) - (4 - 5i)$
4 $-4 - (-3 - 5i)$ 5 $(1 - 3i) - 2i$ 6 $(-5 - 6i) - (-1 + 4i)$
7 $(-1 + 2i) \div (3 + 4i)$ 8 $(2 + 3i) \div (-4 + 6i)$ 9 $(-2 + 0i) \div (4 - 5i)$

10 $(-4) \div (-3 - 5i)$ **11** $(1 - 3i) \div (2i)$ **12** $(-5 - 6i) \div (-1 + 4i)$

13 $(-2 + 3i) + (1 + 2i)(-3 - i)$ **14** $(3 - 2i)(2i) + 4$ **15** $(1 - 4i)(3) - 6$

16 $\dfrac{(3i) - (2 - i)}{(-2 + i)}$ **17** $\dfrac{(4 - i) + (2i)}{-2i}$ **18** $\dfrac{6 - (1 + 2i)}{4}$

19 $\dfrac{(-2 + 3i)}{(1 + 2i)} - \dfrac{(-3 - i)}{(1 - 4i)}$ **20** $\dfrac{(-2 + 3i)}{(1 + 2i)} \cdot \dfrac{(-3 - i)}{(1 - 4i)}$

21 $\dfrac{(-2 + 3i)}{(1 + 2i)} \div \dfrac{(-3 - i)}{(1 - 4i)}$ **22** $(-2 - 3i)\{(1 + 2i) - [(1 - 4i) - (4 - i)]\}$

23 $\{(1 - 2i) - (3i)[(-1 - 2i) + (3 - 2i)]\}(-i)$

24 $(1 - i)\{2i - [-(2 + 3i) - (-3 - 4i)]\}$

Solve the following complex equations for x and y, where x and y are real.

Example $(x + 2yi)^2 = xi$

Solution $(x + 2yi)^2 = xi \Longleftrightarrow x^2 + 4xyi + 4y^2i^2 = xi$

$\Longleftrightarrow (x^2 - 4y^2) + 4xyi = 0 + xi$

D6.2.2

$\Longleftrightarrow x^2 - 4y^2 = 0$ and $4xy = x$

$\Longleftrightarrow \begin{cases} x^2 = 4y^2 \quad \text{and} \quad 4y = 1 \qquad \text{if } x \neq 0 \\ \text{or} \\ x = 0 \quad \text{and} \quad y = 0 \end{cases}$

$\Longleftrightarrow \begin{cases} x = \pm 2y \quad \text{and} \quad y = \frac{1}{4} \\ \text{or} \\ x = 0 \quad \text{and} \quad y = 0 \end{cases}$

$\Longleftrightarrow \begin{cases} x = \pm \frac{1}{2} \quad \text{and} \quad y = \frac{1}{4} \\ \text{or} \\ x = 0 \quad \text{and} \quad y = 0 \end{cases}$

$\Longrightarrow S = \{0 + 0i, \frac{1}{2} + \frac{1}{4}i, -\frac{1}{2} + \frac{1}{4}i\}$

25 $-2x + yi = 4 - i$ **26** $-x + 3yi = 2$ **27** $-2i = 2x - yi$

28 $(x + 3i)^2 = 2yi$

Problem Set III ■ *Theoretical Developments*

Prove the following theorems:

1 Theorem 6.4.1   **2** Theorem 6.4.2   **3** Theorem 6.4.3

4 Theorem 6.4.4   **5** Theorem 6.4.5   **6** Theorem 6.4.7

7 Theorem 6.4.8

8 Does there exist a complex number whose additive inverse equals its multiplicative inverse? If yes, identify the number.

9 In each of the following problems find the real part of z, the imaginary part of z, and the conjugate of z.

   a $z = -1 - 3i$ **b** $z = 2i$ **c** $z = 4$

   d $z = \dfrac{2 - i}{3i}$ **e** $z = i(3 + 2i)$ **f** $z = (4 + 5i) - (-3 + 6i)$

10   Prove each of the following statements, where z represents a complex number.

    a   If $\bar{z} = z$, then z is a real number.

    b $z + \bar{z} = 0 \Longleftrightarrow$ real part of z is 0

    c $\overline{z_1 + z_2} = \bar{z_1} + \bar{z_2}$ **d** $\overline{z_1 z_2} = \bar{z_1} \bar{z_2}$

    e $\overline{z_1/z_2} = \bar{z_1}/\bar{z_2}$ **f** $\bar{\bar{z}} = z$

Summary of Definitions and Properties in Chapter 6   6.6

D6.2.1 $C = \{a + bi \mid a, b \in R \quad \text{and} \quad i^2 = -1\}$ (p. 220)

D6.2.2 $a + bi = c + di \Longleftrightarrow a = c \quad \text{and} \quad b = d$ (p. 223)

T6.2.1   (p. 223)
  I   Reflexive Law of Equality (RLE) $a + bi = a + bi$
  II   Symmetric Law of Equality (SLE) $a + bi = c + di \Longleftrightarrow c + di = a + bi$
  III   Transitive Law of Equality (TLE)
              $[a + bi = c + di \quad \text{and} \quad c + di = e + fi] \Longleftrightarrow [a + bi = e + fi]$
  IV   Substitution Law of Equality (SubLE)

D6.2.3 $(a + bi) + (c + di) = (a + c) + (b + d)i$ (p. 224)

D6.2.4 $(a + bi) \cdot (c + di) = (ac - bd) + (ad + bc)i$ (p. 224)

T6.4.1   The Closure Laws of Addition and Multiplication (Cl+ and Cl·)   (p. 226)
  (1) $(a + bi), (c + di) \in C \Longrightarrow (a + bi) + (c + di) \in C$
  (2) $(a + bi), (c + di) \in C \Longrightarrow (a + bi) \cdot (c + di) \in C$

T6.4.2   The Associative Laws of Addition and Multiplication   (p. 226)
  (1) $(a + bi) + [(c + di) + (e + fi)] = [(a + bi) + (c + di)] + (e + fi)$
  (2) $(a + bi) \cdot [(c + di) \cdot (e + fi)] = [(a + bi) \cdot (c + di)] \cdot (e + fi)$

T6.4.3   The Commutative Laws of Addition and Multiplication   (p. 227)
  (1) $(a + bi) + (c + di) = (c + di) + (a + bi)$
  (2) $(a + bi) \cdot (c + di) = (c + di) \cdot (a + bi)$

T6.4.4   Additive Identity Law (Id+)   (p. 227)
  $(a + bi) + (0 + 0i) = (a + bi)$

T6.4.5   Additive Inverse Law (In+)   (p. 227)
  $(a + bi) + (-a - bi) = 0 + 0i$

T6.4.6 Distributive Law (D)   (p. 227)
$$(a + bi) \cdot [(c + di) + (e + fi)]$$
$$= (a + bi) \cdot (c + di) + (a + bi) \cdot (e + fi)$$

T6.4.7 Multiplicative Identity Law (Id·)   (p. 228)
$$(a + bi) \cdot (1 + 0i) = a + bi$$

T6.4.8 Multiplicative Inverse Law (In·)   (p. 228)
$$(a + bi) \cdot \left[\left(\frac{a}{a^2 + b^2} \right) + \left(\frac{-b}{a^2 + b^2} \right) i \right] = 1 + 0i$$

D6.4.1 Subtraction   (p. 229)
$$(a + bi) - (c + di) = (a + bi) + [-(c + di)]$$
$$= (a + bi) + (-c - di) = (a - c) + (b - d)i$$

D6.4.2 Division   (p. 229)
$$\frac{a + bi}{c + di} = (a + bi) \cdot (c + di)^{-1} = (a + bi) \cdot \left[\left(\frac{c}{c^2 + d^2} \right) + \left(\frac{-d}{c^2 + d^2} \right) i \right]$$

Conjugate $\overline{a + bi} = a - bi$ (p. 229)

© 1974 United Feature Syndicate, Inc.

Polynomial and Rational Functions

7

CHAPTER

Introduction **7.1**

In Chapter 4 we began our study of functions by examining two of the simpler classes of functions, that is, linear, defined by $f(x) = ax + b$, and quadratic, defined by $f(x) = ax^2 + bx + c$. These, however, are only special cases of an even broader class called polynomials, which is the subject of this chapter.

Before beginning our work on polynomials, let us look at a problem whose solution requires techniques that we will develop in this chapter.

Problem 7.1.1 Suppose three weights are positioned on a frictionless surface and are joined by four springs as illustrated in Fig. 7.1.1. Suppose, further, we initially displace each from its equilibrium position a distance of 2, −1, and 1 units and then impart each with an initial velocity of 1, 2, and 0 units per second. We wish to

235

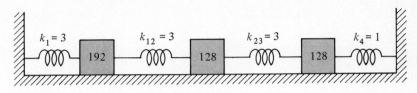

FIGURE 7.1.1

determine the subsequent motion of each of the weights within the system. The complete solution of this problem requires techniques of calculus. However, in the process, one must solve the following polynomial equation of degree 6.

$$16w^6 - 56w^4 + 49w^2 - 9 = 0$$

We have formulas for the solutions of linear and quadratic polynomial equations but nothing at this point to handle those of degree greater than two. We will return and easily solve the above equation after developing several techniques for identifying the solutions of certain higher degree polynomial equations.

7.2   Polynomial Expressions

We now formally define a polynomial.

DEFINITION 7.2.1   Let $a_0, a_1, a_2, \ldots, a_n \in C$, $a_n \neq 0$, where C is the set of complex numbers and n is a nonnegative integer. Then the algebraic expression

$$a_n x^n + a_{n-1} x^{n-1} + a_{n-2} x^{n-2} + \cdots + a_1 x + a_0 \tag{1}$$

is called a polynomial of degree n in the variable x over C. The constants $a_0, a_1, a_2, \ldots, a_n$ are called the coefficients of the polynomial. If the coefficients are all real, then the polynomial is said to be over R.

For notational convenience, polynomials will often be denoted by symbols such as $P(x)$, $D(x)$, or $Q(x)$. Thus, we may write

$$P(x) = 2x - 3$$
$$D(x) = -ix^2 - 2x + (1 - 2i)$$
$$Q(x) = 3x^5 + x^3 - 4x^2 - x - 5$$

where $P(x)$ is a polynomial of degree 1 in the variable x over R, $D(x)$ is of degree 2 in x over C, and $Q(x)$ is of degree 5 in x over R.

The choice of the variable as x was arbitrary. Any letter toward the end of the alphabet would do, for example,

$$P(z) = z^3 - 4z^2 + z - \sqrt{2}$$

is a polynomial of degree 3 in z over R. Furthermore, we will use this symbolic notation to represent values of a polynomial for specific choices of the variable. Thus, $P(3)$ represents the value of the polynomial $P(x)$ at 3 and $P(-2)$ the value at -2.

Example 7.2.1 Let $P(x) = x^3 - 2x^2 + x - 4$. Then

$$P(-1) = (-1)^3 - 2(-1)^2 + (-1) - 4 = -8,$$
$$P(2) = (2)^3 - 2(2)^2 + (2) - 4 = -2$$
$$P(i) = (i)^3 - 2(i)^2 + (i) - 4 = -i + 2 + i - 4 = -2$$

An Algebra for Polynomials   7.3

Because polynomials are composed of sums of real or complex terms, the field properties apply in algebraic manipulations comprising polynomials. The associative, commutative, and distributive laws are especially helpful in performing many calculations, as the following examples will illustrate.

Example 7.3.1 Let

$$P(x) = 2x^4 - x^3 - 3x^2 + 4$$
$$Q(x) = -3x^4 - x^3 + x^2 - 2x - 1$$

and

$$R(x) = -2x^3 + 3x^2 - 5x + 2$$

Then,

$$\begin{aligned}
P(x) + Q(x) &= (2x^4 - x^3 - 3x^2 + 4) + (-3x^4 - x^3 + x^2 - 2x - 1) \\
&= (2 - 3)x^4 + (-1 - 1)x^3 + (-3 + 1)x^2 \\
&\quad + (0 - 2)x + (4 - 1) \\
&= -x^4 - 2x^3 - 2x^2 - 2x + 3 \\
P(x) - Q(x) &= (2x^4 - x^3 - 3x^2 + 4) - (-3x^4 - x^3 + x^2 - 2x - 1) \\
&= (2 + 3)x^4 + (-1 + 1)x^3 + (-3 - 1)x^2 \\
&\quad + (0 + 2)x + (4 + 1) \\
&= 5x^4 - 4x^2 + 2x + 5
\end{aligned}$$

and

$$\begin{aligned}
P(x) + R(x) &= (2x^4 - x^3 - 3x^2 + 4) + (-2x^3 + 3x^2 - 5x + 2) \\
&= 2x^4 + (-1 - 2)x^3 + (-3 + 3)x^2 \\
&\quad + (0 - 5)x + (4 + 2) \\
&= 2x^4 - 3x^3 - 5x + 6
\end{aligned}$$

Example 7.3.2 Let

$$P(x) = 3x^2 + 2x - 1$$

and

$$Q(x) = x^3 - 2x + 3$$

Then,

$$P(x) \ Q(x) = (3x^2 + 2x - 1) \cdot (x^3 - 2x + 3)$$

$$= 3x^2(x^3) + 3x^2(-2x) + 3x^2(3) + 2x(x^3) + 2x(-2x)$$
$$+ 2x(3) + (-1)(x^3) + (-1)(-2x) + (-1)(3)$$

$$= 3x^5 - 6x^3 + 9x^2 + 2x^4 - 4x^2 + 6x - x^3 + 2x - 3$$

$$= 3x^5 + 2x^4 - 7x^3 + 5x^2 + 8x - 3$$

7.4 Problem Sets

Problem Set I ■ *Reading Comprehension*

Determine whether statements 1 through 3 are true or false and give reasons for your responses.

1 The following algebraic expression is a polynomial:
   $3x^4 - 2x^2 + 3x^{-1} + 2$
2 The following algebraic expression is a polynomial:
   $-x^3 + x^2 - 5\sqrt{x} + 3$
3 The polynomial $2x^3 - 4x^5 + 3x^2 - 3x - (1 + 2i)$ is of degree 5 over C (the set of complex numbers.)

Problem Set II ■ *Skills Development*

In each of the following problems, perform the indicated operation and simplify.

1 $(x^3 + 2x^2 - x - 3) + (x^2 - 3x + 4)$
2 $(-5x^4 - x^3 + x - 1) - (x^4 - 2x^3 + x^2 - x + 3)$
3 $(x^6 - x^4 + x - 1) - (x^5 - 2x^4 + 3x^3 + x^2 - 2x + 5)$
4 $(x^3 + x^2 - x - 1) \cdot (x^4 - x^2 + x - 3)$
5 $(x^5 - 2x^4 + x^2 + 2) \cdot (x^3 - x^2 + 2x - 4)$
6 $(x^6 - 3x^5 + x^3 + x - 1) \cdot (x^3 + x^2 - x + 3)$

In problems 7 to 9, perform the indicated operations and simplify where $P(x) = x^3 - 2x^2 + x - 1$, $Q(x) = 3x - 4x^3 + x^2 + 2$, and $R(x) = x^4 - 3x + 4x^2 - 3$.

7 $P(x) + [Q(x) - R(x)]$ 8 $[P(x) - Q(x)] + R(x)$
9 $Q(x) - [P(x) - R(x)]$

From the previous sections we found the operations of addition, subtraction, and multiplication of polynomials could be handled in a rather straightforward, routine manner. Division, on the other hand, does not come quite so easily. We will, however, develop some techniques that will simplify the process of dividing a polynomial by a polynomial.

The system of polynomials in x over C is in many ways analogous to the set of integers in the real number system. Recall that rational numbers were defined as ratios of two integers, that is,

$$Q = \left\{ x \mid x = \frac{a}{b}, \text{ where } a, b \in J \text{ and } b \neq 0 \right\}$$

Thus, the quotients $\frac{1}{2}$, $-\frac{5}{3}$, $\frac{3}{1}$, and $\frac{3}{10}$ are all rational numbers. In like manner, we define rational algebraic expressions.

DEFINITION 7.5.1   Let $P(x)$ and $D(x)$ be two polynomials in x over C; then the quotient of $P(x)$ and $D(x)$, denoted $P(x)/D(x)$, is called a rational algebraic expression in x over C.

We find a particularly interesting parallel between rational algebraic expressions and rational numbers as we analyze the algorithmic process of division. Consider the following quotients:

$$\frac{12}{4} = \frac{4 \cdot 3}{4} = 3 \qquad \frac{10}{3} = 3 + \frac{1}{3}$$

Notice, in the first, that the integer 12 in the numerator was factored into the product of two integers, one of which corresponded with that of the denominator. Hence, by the cancellation law of fractions, the quotient was reduced to the integer 3. In this case, the numerator is said to be exactly divisible by 4. In the second quotient, however, 10 is not exactly divisible by 3; nevertheless, because 10 is greater than 3, this rational may be expressed as an integer plus another rational whose numerator (called the remainder) is less than the denominator. We say, "3 divides 10 three times with a remainder of 1." Interestingly, we find similar results following for rational algebraic expressions. For example, we may write

$$\frac{x^2 + x - 6}{x - 2} = \frac{(x - 2)(x + 3)}{x - 2} = x + 3 \qquad x \neq 2$$

and

$$\frac{x^3 - 2x^2 + x - 1}{x^2 - 3x + 1} = (x + 1) + \frac{3x - 2}{x^2 - 3x + 1}$$

In the first quotient, the polynomial in the numerator factored into the product of two polynomials, one of which corresponded with that in the denominator. Hence, by the cancellation law of fractions, the quotient was reduced to the polynomial $x + 3$, for all x except $x = 2$; the polynomial $x^2 + x - 6$ is said to be exactly divisible by $x - 2$. In the second quotient, the numerator $x^3 - 2x^2 + x - 1$ was not exactly divisible by $x^2 - 3x + 1$; however, because the degree of the numerator was greater than that of the denominator, the quotient was expressible as a polynomial plus another rational expression whose numerator $3x - 2$ (called the remainder) was of degree less than that of the divisor $x^2 - 3x + 1$.

As it is sometimes convenient to express $\frac{10}{3}$ as $3 + \frac{1}{3}$, so there are times when it is desirable to express

$$\frac{x^3 - 2x^2 + x - 1}{x^2 - 3x + 1}$$

as

$$(x + 1) + \frac{3x - 2}{x^2 - 3x + 1}$$

This will be particularly true in Section 7.10 where rational functions will be defined and graphed.

It may not have been clear how the division of $x^3 - 2x^2 + x - 1$ by $x^2 - 3x + 1$, resulting in $(x + 1) + (3x - 2)/(x^2 - 3x + 1)$, was performed. Actually there are two methods in common use; the one chosen will usually depend on the form of the denominator.

If the polynomial in the denominator is a monomial (containing one term only), then we apply Theorem 2.9.13(III), which is,

$$a, b, c \in C \Longrightarrow \frac{a + b}{c} = \frac{a}{c} + \frac{b}{c}$$

Example 7.5.1

$$\frac{2x^3 - 3x^2 + x - 6}{2x} = \frac{2x^3}{2x} - \frac{3x^2}{2x} + \frac{x}{2x} - \frac{6}{2x}$$

$$= \left(x^2 - \frac{3}{2}x + \frac{1}{2}\right) - \frac{3}{x}$$

or

$$\frac{x^4 - 6x^2}{x^2} = \frac{x^4}{x^2} - \frac{6x^2}{x^2} = (x^2 - 6)$$

If the divisor of a quotient contains more than one term, then the familiar long-division algorithm ● involving successive subtractions may be applied.

● The word algorithm means "a process of computation."

Example 7.5.2

$$\require{enclose}\begin{array}{r}x + 1 \\ x^2 - 3x + 1 \enclose{longdiv}{x^3 - 2x^2 + x - 1}\\\end{array}$$

$$
\begin{array}{r}
\,x + 1 \\
x^2 - 3x + 1 \,\overline{\smash{)}\, x^3 - 2x^2 + x - 1}\\
\underline{x^3 - 3x^2 + x}\\
x^2 - 1\\
\underline{x^2 - 3x + 1}\\
3x - 2
\end{array}
$$

From this, we may write

$$\frac{x^3 - 2x^2 + x - 1}{x^2 - 3x + 1} = (x + 1) + \frac{3x - 2}{x^2 - 3x + 1}$$

just as

$$
\begin{array}{r}
3\\
3\,\overline{\smash{)}\,10}\\
\underline{9}\\
1
\end{array}
\implies \frac{10}{3} = 3 + \tfrac{1}{3}
$$

If the degrees of the polynomials in the numerator and denominator differ appreciably, then the above division can indeed become very "long" division. If, however, the polynomial in the denominator is of the form $x - c$, then this process of long division can be performed more easily and quickly through a method called *synthetic division*. We shall explain this by means of the following example of long division.

$$
\begin{array}{r}
x^3 - x^2 - 2x + 1\\
x - 2 \,\overline{\smash{)}\, x^4 - 3x^3 + + 5x - 4}\\
\underline{x^4 - 2x^3}\\
-x^3 + + 5x - 4\\
\underline{-x^3 + 2x^2}\\
-2x^2 + 5x - 4\\
\underline{-2x^2 + 4x}\\
x - 4\\
\underline{x - 2}\\
-2
\end{array}
$$

so that

$$\frac{x^4 - 3x^3 + 5x - 4}{x - 2} = (x^3 - x^2 - 2x + 1) + \frac{-2}{x - 2} \tag{1}$$

Observe in the computation that the order of the terms in the numerator and denominator were expressed in descending powers of x (highest powered term first, second to the highest second, and so forth). Notice also that the powers of x in the division process only served to identify position and that their respective coefficients were the real determiners of the outcome. Finally notice how often

x^3, x^2, and x were repeated in passing through the successive steps in the long-division process. Let us rewrite the above computation omitting the variable x and leaving only the coefficients but still lined up in their appropriate columns. We have

$$
\begin{array}{r}
1-1-2 \quad 1 \\
1-2\overline{)1-3 \quad 0 \quad 5 \quad -4} \\
\underline{1-2} \\
-1 \quad 0 \quad 5 \quad -4 \\
\underline{-1 \quad 2} \\
-2 \quad 5 \quad -4 \\
\underline{-2 \quad 4} \\
1 \quad -4 \\
\underline{1 \quad -2} \\
-2 \text{ (remainder)}
\end{array}
$$

Note that 0 was inserted as the coefficient of the missing x^2 term so that the coefficients of x^2 in the remaining portions of the computation line up properly. Observe also that the numerals in color are repetitions of those written immediately above (except for the remainder). They are also repetitions of the coefficients of the quotient $x^3 - x^2 - 2x + 1$, which is located at the top. Now if we eliminate all these repetitions, the above structure collapses into the simple form

$$
\begin{array}{rl}
(1) & \quad -2\,\lfloor\; 1 \quad -3 \quad 0 \quad 5 \quad -4 \\
(2) & \qquad\qquad -2 \quad 2 \quad 4 \quad -2 \\
(3) & \qquad \underline{1 \quad -1 \quad -2 \quad 1 \quad -2} \longleftarrow \text{(remainder)}
\end{array}
$$

(coefficients of the quotient polynomial)

where even the coefficient 1 of x in the divisor has been eliminated. Notice that the entries in line (3) were obtained by subtracting those of line (2) from the corresponding entries in line (1). We can convert that operation to addition by merely changing the sign of the generating coefficient -2 and using 2. The process may now be expressed as

$$
\begin{array}{rl}
2\,\lfloor\; & 1 \quad -3 \quad 0 \quad 5 \quad -4 \\
& \quad\;\; 2 \quad -2 \quad -4 \quad 2 \\
& \underline{1 \quad -1 \quad -2 \quad 1 \quad -2} \longleftarrow \text{(remainder)}
\end{array}
$$

(coefficients of the quotient polynomial)

The process of performing long division by means of the abbreviated form described above is called synthetic division, which, as observed, does take some of the "long" out of long division.

Example 7.5.3 Write

$$\frac{x^3 - 3x^2 + 3}{x - 1}$$

in the form $Q(x) + r/(x-1)$, where $Q(x)$ is a polynomial and r is a constant.

Solution Using synthetic division, we first write

$$\underline{1 \rvert}\quad 1 \quad -3 \quad\;\; 0 \quad\;\; 3$$

Note the sign of -1 in the divisor has been changed to 1 and the coefficient 0 of the missing first-degree term is inserted in its proper column. Continuing now with the process, we obtain

(1) $\quad\underline{1 \rvert}\quad 1 \quad -3 \quad\;\; 0 \quad\;\; 3$
(2) $\qquad\qquad\qquad\;\; 1 \quad -2 \quad -2$
(3) $\qquad\quad\;\; 1 \quad -2 \quad -2 \quad\;\; 1$

Let us review schematically the sequence of steps in the above process, numbered 1 through 7, as illustrated in Fig. 7.5.1. Steps 2, 4, and 6 are products, whereas steps 3, 5, and 7 are sums. The last entry of line (3), namely 1, is the remainder r, whereas the others form the coefficients of the quotient polynomial $Q(x)$, that is,

$$r = 1 \quad \text{and} \quad Q(x) = 1x^2 - 2x - 2 = x^2 - 2x - 2$$

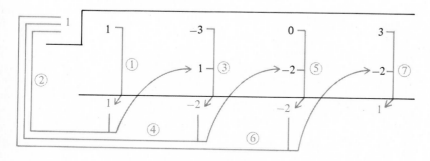

FIGURE 7.5.1

The degree of $Q(x)$ will always be one less than that of the numerator of the rational expression if the denominator is of degree 1.

Finally, we write

$$\frac{x^3 - 3x^2 + 3}{x - 1} = (x^2 - 2x - 2) + \frac{1}{x - 1}$$

Example 7.5.4 Express the quotient

$$\frac{x^5 - 2x^3 + x + 20}{x + 2}$$

in the form $Q(x) + r/(x + 2)$.

Solution By synthetic division, we have

$$
\begin{array}{r|rrrrrr}
-2 & 1 & 0 & -2 & 0 & 1 & 20 \\
   &   & -2 & 4 & -4 & 8 & -18 \\
\hline
   & 1 & -2 & 2 & -4 & 9 & 2 = r \text{ (remainder)}
\end{array}
$$

$\underbrace{\qquad\qquad}$
(coefficients of $Q(x)$)

Thus,

$$\frac{x^5 - 2x^3 + x + 20}{x + 2} = (x^4 - 2x^3 + 2x^2 - 4x + 9) + \frac{2}{x + 2}$$

Notice the sign of the coefficient 2 in the divisor was changed to -2 as the generating number in the above process.

Example 7.5.5 Express

$$\frac{3x^3 + 4x^2 + 5x - 6}{3x - 2}$$

in the form $Q(x) + r/(3x - 2)$.

Solution We first note that the divisor $3x - 2$ is not of the form $x - c$, and thus synthetic division cannot be applied to this quotient in its present form. However, the rational expression may be written as

$$\frac{1}{3} \cdot \frac{3x^3 + 4x^2 + 5x - 6}{x - \frac{2}{3}} \tag{1}$$

We may now apply synthetic division to the right-hand factor of (1).

$$
\begin{array}{r|rrrr}
\frac{2}{3} & 3 & 4 & 5 & -6 \\
            &   & 2 & 4 & 6 \\
\hline
            & 3 & 6 & 9 & 0
\end{array}
$$

Hence, $Q(x) = 3x^2 + 6x + 9$ and $r = 0$, yielding the result

$$\frac{3x^3 + 4x^2 + 5x - 6}{x - \frac{2}{3}} = (3x^2 + 6x + 9)$$

which, returning to (1), gives

$$\tfrac{1}{3} \cdot (3x^2 + 6x + 9) = x^2 + 2x + 3$$

and hence,

$$\frac{3x^3 + 4x^2 + 5x - 6}{3x - 2} = x^2 + 2x + 3$$

As illustrated in the foregoing examples, the division algorithm indicates that if $P(x)$ and $D(x)$ are two polynomials for which the degree of $P(x)$ is greater than or equal to that of $D(x)$, then there exist polynomials $Q(x)$, called the quotient, and $R(x)$, the remainder, with degree less than that of $D(x)$, such that

$$\frac{P(x)}{D(x)} = Q(x) + \frac{R(x)}{D(x)}$$

And, if $D(x) = x - c$, we have

$$\frac{P(x)}{x - c} = Q(x) + \frac{r}{x - c}$$

where r is a constant. If we multiply both sides of this latter equation by $x - c$, we obtain

$$P(x) = Q(x) \cdot (x - c) + r$$

We are now ready for the following important theorem regarding this last equation, which we state without proof.

THEOREM 7.5.1   If c is a complex number and $P(x)$ a polynomial of degree $n \geq 1$ in x over the set C of complex numbers, then there exists a unique polynomial $Q(x)$ of degree $n - 1$ and a unique complex number r such that

$$P(x) = Q(x) \cdot (x - c) + r$$

If $x \neq c$, then

$$P(x) = Q(x) \cdot (x - c) + r \iff \frac{P(x)}{x - c} = Q(x) + \frac{r}{x - c}$$

Two very important properties follow from Theorem 7.5.1, which we shall single out and state as theorems.

THEOREM 7.5.2   THE REMAINDER THEOREM

If $P(x)$ is a complex polynomial of degree $n \geq 1$ and c is any complex number, then there exists a unique polynomial $Q(x)$ such that

$$P(x) = Q(x) \cdot (x - c) + P(c)$$

where $P(c)$ is the polynomial $P(x)$ evaluated at c. Thus, we have a formula for the remainder r, that is,

$$r = P(c)$$

Proof From Theorem 7.5.1, we have

$$P(x) = Q(x)(x - c) + r$$

If $x = c$, this equation becomes

$$P(c) = Q(c) \cdot (c - c) + r = Q(c) \cdot 0 + r = r$$

$$\Longrightarrow P(c) = r$$

Hence,

$$P(x) = Q(x) \cdot (x - c) + P(c) \qquad \text{QED}$$

Example 7.5.6 Write

$$\frac{2x^4 - 3x^2 + 5x - 7}{x + 2}$$

in the form $Q(x) + r/(x + 2)$ and verify the remainder theorem.

Solution Using synthetic division we have

$$
\begin{array}{r|rrrrr}
-2 & 2 & 0 & -3 & 5 & -7 \\
   &   & -4 & 8 & -10 & 10 \\
\hline
   & 2 & -4 & 5 & -5 & 3 \quad \longleftarrow \text{(remainder)}
\end{array}
$$

$$\Longrightarrow \frac{2x^4 - 3x^2 + 5x - 7}{x + 2} = (2x^3 - 4x^2 + 5x - 5) + \frac{3}{x + 2}$$

Multiplying both sides of this equation by $x + 2$ yields

$$2x^4 - 3x^2 + 5x - 7 = (2x^3 - 4x^2 + 5x - 5)(x + 2) + 3$$

Now let $P(x) = 2x^4 - 3x^2 + 5x - 7$, then

$$P(-2) = 2(-2)^4 - 3(-2)^2 + 5(-2) - 7$$

$$= 32 - 12 - 10 - 7$$

$$= 32 - 29 = 3$$

$$\Longrightarrow P(-2) = 3$$

which verifies the Remainder Theorem for $c = -2$. That is, if $P(x)$ is divided by $x - (-2)$ (which equals $x + 2$), then the remainder 3 is equal to the value of the polynomial $P(x)$ at -2.

THEOREM 7.5.3 THE FACTOR THEOREM

If $P(x)$ is a complex polynomial of degree $n \geq 1$ and if c is a complex number for which $P(c) = 0$, then there exists a complex polynomial $Q(x)$ of degree $n - 1$ such that

$$P(x) = Q(x) \cdot (x - c)$$

The number c is called a zero or root of the polynomial.

Proof From the Remainder Theorem we have

$$P(x) = Q(x) \cdot (x - c) + P(c)$$

But here $P(c) = 0$, hence,

$$P(x) = Q(x) \cdot (x - c) \qquad \text{QED}$$

Example 7.5.7 Determine whether $x - 2$ is a factor of the expression $5x^3 + 12x^2 - 36x - 16$.

Solution We shall solve this problem using two methods. First, let us divide the polynomial $P(x) = 5x^3 + 12x^2 - 36x - 16$ by $(x-2)$, employing synthetic division, knowing that the remainder obtained by this process is equal to $P(2)$. If the remainder equals zero then we know by the Factor Theorem that $x - 2$ is a factor of $P(x)$.

$$
\begin{array}{r|rrrr}
2 & 5 & 12 & -36 & -16 \\
  &   & 10 & 44 & 16 \\
\hline
  & 5 & 22 & 8 & 0
\end{array}
\longleftarrow \text{(remainder)}
$$

(coefficients of $Q(x)$)

Thus,

$$Q(x) = 5x^2 + 22x + 8 \quad \text{and} \quad r = 0 = P(2)$$

We therefore conclude that indeed $x - 2$ is a factor of $P(x)$ and furthermore the other factor is $Q(x)$. We thus factor, obtaining

$$5x^3 + 12x^2 - 36x - 16 = (5x^2 + 22x + 8) \cdot (x - 2)$$

Next we solve the problem by a more direct approach. Simply evaluate the polynomial $P(x)$ at 2.

$$
\begin{aligned}
P(2) &= 5(2)^3 + 12(2)^2 - 36(2) - 16 \\
&= 5 \cdot 8 + 12 \cdot 4 - 36 \cdot 2 - 16 \\
&= 40 + 48 - 72 - 16 = 0
\end{aligned}
$$

$$\implies P(2) = 0$$

And thus again, by the Factor Theorem, we conclude that $x - 2$ is a factor of $P(x)$.

It may appear that the latter method is much the simpler of the two, but this is not necessarily the case. Recall that in synthetic division we use only the coefficients of the polynomial, whereas in evaluating a polynomial at c directly we require the entire polynomial, coefficients, and powers of x. In the example above, the evaluation of $P(x)$ at 2 was not difficult. However, consider the problem of determining whether $x + 5$ is a factor of

$$P(x) = 2x^6 - 49x^4 + 3x^3 - 14x^2 - 18x + 10$$

Evaluating $P(x)$ at -5 (note $x + 5 = x - (-5)$), would be a relatively long, tedious task because of the higher powers of x. On the other hand, synthetic division can accomplish the same purpose (finding $P(-5)$) by dealing only with the coefficients of the polynomial. Proceeding in this manner, we have

$$
\begin{array}{r|rrrrrrr}
-5 & 2 & 0 & -49 & 3 & -14 & -18 & 10 \\
   &   & -10 & 50 & -5 & 10 & 20 & -10 \\
\hline
   & 2 & -10 & 1 & -2 & -4 & 2 & 0 \longleftarrow \text{(remainder)}
\end{array}
$$

(coefficients of $Q(x)$)

$$\Longrightarrow r = 0 = P(-5)$$

and furthermore

$$Q(x) = 2x^5 - 10x^4 + x^3 - 2x^2 - 4x + 2$$

Obtaining $P(-5)$ by synthetic division for this example is considerably easier than by direct substitution, that is,

$$P(-5) = 2(-5)^6 - 49(-5)^4 + 3(-5)^3 - 14(-5)^2 - 18(-5) + 10$$

Moreover, the process of synthetic division also provides the other factor $Q(x)$ when $P(c) = 0$, which in many applications is equally important.

7.6 Problem Sets

Problem Set I ■ *Reading Comprehension*

Determine whether statements 1 through 8 are true or false and give reasons for your responses.

1 Any rational algebraic expression in x over C can be expressed in the form
   $$a_n x^n + a_{n-1} x^{n-1} + \cdots + a_1 x + a_0$$
   where $n \in N$ and $a_n, a_{n-1}, \ldots, a_1, a_0$ are all rational numbers.

2 Using the long-division algorithm, the quotient of two polynomials, say, $P(x)$ and $D(x)$, may be expressed in the form
   $$Q(x) + \frac{R(x)}{D(x)}$$
   where $Q(x)$ and $R(x)$ are polynomials and the degree of $R(x)$ is less than that of $D(x)$.

3 Synthetic division, as the name implies, is not real division and hence cannot be depended upon for complete accuracy.

4 The process of synthetic division may be applied to the quotient
   $$\frac{x^5 - 3x^4 + 2x - 1}{x^2 - 1}$$

5 The process of synthetic division applied to the quotient

$$\frac{x^4 - 3x^2 + 2x - 1}{x + 2}$$

is initially set up in the form

$$-2 \overline{\left|\; 1 \quad -3 \quad 2 \quad -1 \right.}$$

6 According to the Remainder Theorem, if a polynomial $P(x)$ of degree $n \geq 1$ is divided by $x + c$, then the remainder r is given by $P(c)$.

7 If $P(x)$ is a polynomial of degree $n \geq 1$ and if $P(c) = 0$, then $(x - c)$ is a polynomial factor of $P(x)$.

8 When determining whether or not $x - c$ is a factor of a polynomial $P(x)$, synthetic division is always the preferred method.

Problem Set II ■ Skills Development

Express the following quotients in the form $Q(x) + (R(x)/D(x))$, where $Q(x), R(x), D(x)$ are polynomials and the degree of $R(x)$ is less than that of $D(x)$. If $D(x)$ is linear, use synthetic division and check the remainder by direct substitution.

1 $\dfrac{x^3 - 3x^2 + 4}{x - 3}$ **2** $\dfrac{x^5 + 4x^4 - x^2 - 12x - 6}{x + 2}$

3 $\dfrac{x^6 + 4x^5 - x^4 - 10x^3 - 15x + 10}{x + 3}$

Using the Factor Theorem, determine which of statements 4 through 9 are true.

4 $(x - 2)$ is a factor of $x^3 - x^2 - 5x + 6$.
5 $(x + 2)$ is a factor of $2x^3 - 5x^2 - x + 6$.
6 $(x + 3)$ is a factor of $x^4 + 2x^3 - x^2 + 7x + 3$.
7 $(x - 3)$ is a factor of $x^5 - 2x^4 - 5x^2 - 6x + 2$.
8 $(x - 1)$ is a factor of $2x^5 - 3x^4 + x^3 + 2x - 3$.
9 $(x + 1)$ is a factor of $3x^4 + 4x^3 + 2x^2 - 1$.
10 If $P(x) = x^3 - 2x^2 + x - 1$, find $P(-1)$, $P(4)$, and $P(6)$.
11 If $Q(x) = -2x^4 - x^2 + 3x - 2610$, find $Q(1)$, $Q(-2)$, and $Q(6)$.
12 If $R(x) = 2x^4 - 3x^3 + 5x^2 + 4x - 1$, find $R(\frac{1}{2})$, $R(-2)$, and $R(3)$.
13 If $f(x) = 2x^3 - 3x^2 + 4x - 6$, find $f(2)$, $f(2i)$, and $f(i)$.
14 If $g(x) = x^4 + 2x^3 - 5x^2 - 4x + 6$, find $g(3)$, $g(\sqrt{2})$, and $g(0)$.
15 If $h(x) = x^3 - 2x^2 + x + 1$, find $h(3)$, $h(i)$, and $h(-1)$.

Problem Set III ■ Theoretical Developments

1 If $x \neq 1$ and $n \in N$, argue that

$$\frac{x^n - 1}{x - 1} = x^{n-1} + x^{n-2} + x^{n-3} + \cdots + x^2 + x + 1$$

2 Prove that $x - y$ is a factor of $x^n - y^n$ if n is a natural number.
3 Prove $x + y$ is a factor of $x^n - y^n$ if n is an even natural number.
4 For what choices of n will $x + y$ be a factor of $x^n + y^n$.
5 Find the value of k for which $(x + 2)$ is a factor of $x^3 + 3x^2 + kx - 2$.
6 Find the value of k for which $(x - 3)$ is a factor of $x^4 - 5x^3 + kx^2 + 18k + 18$.

7.7 Polynomial Functions

Let $P(x)$ be a polynomial in x over the set C of complex numbers. Set this polynomial expression equal to the variable y forming the open equation

$$y = P(x)$$

This equation, of course, generates a set of ordered pairs of numbers (real or complex) called its solution set, which we express

$$P = \{(x, y) \mid y = P(x)\}$$

This set is a relation. If, however, the polynomial expression $P(x)$ yields a unique complex number $P(c)$ for each $c \in C$, then the relation P is a function. Because a polynomial is composed of sums and products only and because these operations, in the complex number system, yield unique complex numbers, the relation P is indeed a function. We summarize these observations in the following theorem:

THEOREM 7.7.1   Let $P(x)$ be a polynomial of degree n over the set C of complex numbers. Then the relation P, defined by

$$P = \{(x, y) \mid y = P(x) \text{ over } C\}$$

is called a polynomial function of degree n.

Up to this point we have considered polynomials with complex coefficients and even evaluated polynomials at complex choices for the variable. We would call the polynomial functions defined in Theorem 7.7.1 complex valued. The graph of such functions requires, in general, four dimensions, two for the domain and two for the range. Although this can be handled using two planes, we will not in this treatment go into such graphs. We will limit our study of polynomial graphs to real-valued polynomial functions of a real variable.

Example 7.7.1   The linear function

$$L = \{(x, y) \mid y = mx + b \text{ over } R\}$$

is a polynomial function of degree 1.

Example 7.7.2   The quadratic function

$$f = \{(x, y) \mid y = ax^2 + bx + c \text{ over } R\}$$

is a polynomial function of degree 2.

We experienced little difficulty in graphing first- and second-degree polynomial functions in Chapter 4. Two points in the case of the first-degree function were sufficient, whereas finding the maximum or minimum point and one or two other points enabled us to obtain a reasonable approximation of the graph of a second-degree polynomial function. The problem, however, of graphing polynomial functions in general is not so simple. One method is to simply plot points until a discernible pattern is achieved; then join these points with a smooth curve. There are more efficient and accurate methods but most of these employ techniques that are beyond the level of this text. We will, therefore, be content with the point-plotting approach for now.

Example 7.7.3 Graph the polynomial function

$$f = \{(x, y) \mid y = x^3 - 3x^2 + x + 1 \text{ over } \boldsymbol{R}\}$$

Solution First determine several ordered pairs in f. This may be accomplished using synthetic division as follows:

$$
\begin{array}{r|rrrr}
-2 & 1 & -3 & 1 & 1 \\
   &   & -2 & 10 & -22 \\
\hline
   & 1 & -5 & 11 & -21
\end{array}
$$

$$\underset{\text{T7.5.2}}{\Longrightarrow} f(-2) = -21 \Longrightarrow (-2, -21) \in f$$

T7.5.2 The Remainder Theorem.

$$
\begin{array}{r|rrrr}
-1 & 1 & -3 & 1 & 1 \\
   &   & -1 & 4 & -5 \\
\hline
   & 1 & -4 & 5 & -4
\end{array}
$$

$$\Longrightarrow f(-1) = -4 \Longrightarrow (-1, -4) \in f$$

$$f(0) = 1 \Longrightarrow (0, 1) \in f$$

$$
\begin{array}{r|rrrr}
1 & 1 & -3 & 1 & 1 \\
  &   & 1 & -2 & -1 \\
\hline
  & 1 & -2 & -1 & 0
\end{array}
$$

$$\Longrightarrow f(1) = 0 \Longrightarrow (1, 0) \in f$$

$$
\begin{array}{r|rrrr}
2 & 1 & -3 & 1 & 1 \\
  &   & 2 & -2 & -2 \\
\hline
  & 1 & -1 & -1 & -1
\end{array}
$$

$$\Longrightarrow f(2) = -1 \Longrightarrow (2, -1) \in f$$

$$
\begin{array}{r|rrrr}
3 & 1 & -3 & 1 & 1 \\
  &   & 3 & 0 & 3 \\
\hline
  & 1 & 0 & 1 & 4
\end{array}
$$

$$\Longrightarrow f(3) = 4 \Longrightarrow (3, 4) \in f$$

$$\begin{array}{r|rrrr} 4 & 1 & -3 & 1 & 1 \\ & & 4 & 4 & 20 \\ \hline & 1 & 1 & 5 & 21 \end{array}$$

$$\Longrightarrow f(4) = 21 \Longrightarrow (4, 21) \in f$$

We now plot these points and join them by a smooth curve, as illustrated in Fig. 7.7.1.

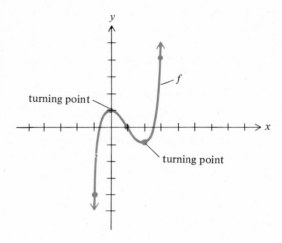

FIGURE 7.7.1

Example 7.7.4 Graph the polynomial function

$$g = \{(x, y) \mid y = x^4 + x^3 - 7x^2 - x + 2 \text{ over } \boldsymbol{R}\}$$

Solution Again we apply synthetic division successively with -4, $-3, -2, -1, 0, 1, 2, 3$, but we arrange the results in a more compact form. Each row, after the first row in the array below, represents the bottom row of the synthetic division process associated with the number at the left.

	1	1	−7	−1	2
−4	1	−3	5	−21	86
−3	1	−2	−1	2	−4
−2	1	−1	−5	9	−16
−1	1	0	−7	6	−4
0	1	1	−7	−1	2
1	1	2	−5	−6	−4
2	1	3	−1	−3	−4
3	1	4	5	14	44

$$\Longrightarrow \begin{cases} (-4, 86),\ (-3, -4), \\ (-2, -16),\ (-1, -4), \\ (0, 2),\ (1, -4), \\ (2, -4),\ (3, 44) \in f \end{cases}$$

We plot these points and join them by a smooth curve. (See Fig. 7.7.2.)

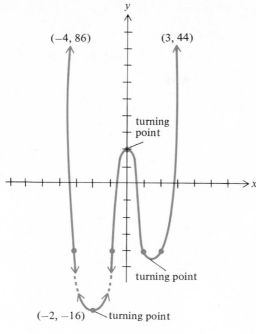

FIGURE 7.7.2

Observing Examples 7.7.3 and 7.7.4, you will note that the third-degree polynomial of 7.7.3 possessed two turning points (curve changes from decreasing to increasing or from increasing to decreasing, which are also called local minimum and maximum points, respectively), whereas the fourth-degree polynomial function of 7.7.4 had three turning points. It is a fact, attested to by the following theorem, that these functions could have possessed fewer turning points but no more than the two and the three each possessed.

THEOREM 7.7.2 If

$$P(x) = a_n x^n + a_{n-1} x^{n-1} + \cdots + a_1 x + a_0$$

is a real polynomial of degree $n \geq 1$, then the graph of the polynomial function

$$P = \{(x, y) \mid y = P(x) \text{ over } R\}$$

is a smooth curve possessing *at most* $n - 1$ turning points (local maximums and minimums).

The proof of this theorem involves techniques not treated in this text and, therefore, is omitted.

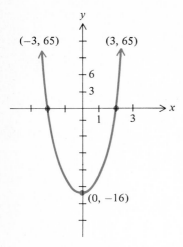

FIGURE 7.7.3

Example 7.7.5 Graph the function

$$h = \{(x, y) \mid y = x^4 - 16 \text{ over } R\}$$

Solution By synthetic division, we obtain the following points in h:

	1	0	0	0	-16
-3	1	-3	9	-27	65
-2	1	-2	4	-8	0
-1	1	-1	1	-1	-15
0	1	0	0	0	-16
1	1	1	1	1	-15
2	1	2	4	8	0
3	1	3	9	27	65

$$\implies \begin{cases} (-3, 65), (-2, 0), \\ (-1, -15), (0, -16), \\ (1, -15), (2, 0), \\ (3, 65) \in h \end{cases}$$

We plot these points and join them by a smooth curve. (See Fig. 7.7.3.) This fourth-degree polynomial function has only one turning point, a minimum at $(0, -16)$.

7.8 Real and Complex Zeros

The process of solving many theoretical and applied problems containing polynomials requires determining the solution sets of polynomial *equations*. Let

$$P(x) = a_n x^n + a_{n-1} x^{n-1} + \cdots + a_1 x + a_0$$

where $a_n, a_{n-1}, \ldots, a_1, a_0 \in C$ and $n \geq 1$, then

$$a_n x^n + a_{n-1} x^{n-1} + \cdots + a_1 x + a_0 = 0 \qquad (1)$$

is a complex polynomial equation. If the coefficients $a_n, a_{n-1}, \ldots, a_1, a_0 \in R$, then (1) is a real polynomial equation. As usual we may write the solution set of such an equation as

$$S = \{x \mid a_n x^n + a_{n-1} x^{n-1} + \cdots + a_1 x + a_0 = 0 \text{ over } C\}$$

Here $c \in S$ implies c is a real or complex number that reduces the polynomial equation to a true statement, that is, $P(c)$ is equal to zero. The number c is called a zero or root of the polynomial.

We learned earlier (the Remainder Theorem) that the remainder r, which results from dividing a polynomial $P(x)$ by $x - c$, equals $P(c)$. We also learned that such a remainder may be easily obtained by means of synthetic division. Hence, one method of discovering a zero of a polynomial is by synthetic division. If the complex number c produces a zero remainder, then c is identified as a zero of the polynomial.

However, if one were to try and discover all or any of the zeros of a general polynomial equation by arbitrarily choosing certain complex numbers and checking them one by one, the chances of ever finding any are quite remote, because there are infinitely many such numbers. Hence, we seek to determine more realistic and systematic ways of finding these zeros.

For this purpose we shall identify several properties concerning the zeros of polynomial equations, which will significantly help in determining these zeros. These properties will be presented in the form of theorems.

THEOREM 7.8.1   If $P(x)$ is a complex polynomial of degree $n \geq 1$, then there exists at least one complex number c such that $P(c) = 0$. ●
The proof of this theorem is beyond the scope of this text.

● In the late eighteenth century, this theorem was proved by the German mathematician Karl Friedrich Gauss and is known as the Fundamental Theorem of Algebra.

Now that we are assured that zeros do exist for all complex polynomial equations, you might wonder how many zeros a polynomial equation may have. This is answered in the next theorem.

THEOREM 7.8.2   Every complex polynomial equation of degree $n \geq 1$ has exactly n complex roots (counting a root of multiplicity k as k separate roots).

Proof   Let $P(x)$ be a complex polynomial of degree $n \geq 1$. Then, by Theorem 7.8.1, there exists a constant c_1 such that $P(c_1) = 0$. But this, in turn, implies by the Factor Theorem that there exists a polynomial $Q_1(x)$ of degree $n - 1$ that, along with c_1, yields the relation

$$P(x) = Q_1(x) \cdot (x - c_1) \tag{1}$$

If the degree of $Q_1(x)$ is zero, that is, $Q_1(x)$ is a constant, say, a_1, then

$$P(x) = a_1(x - c_1)$$

If, on the other hand, the degree of $Q_1(x)$ is $(n-1) \geq 1$, then, applying the Fundamental Theorem to $Q_1(x)$, there must exist a constant c_2 such that $Q_1(c_2) = 0$, which again implies by the Factor Theorem that there exists a polynomial $Q_2(x)$ of degree $n - 2$ such that

$$Q_1(x) = Q_2(x) \cdot (x - c_2) \tag{2}$$

This further implies from (1) that

$$P(x) = Q_2(x) \cdot (x - c_2) \cdot (x - c_1) \tag{3}$$

If the degree of $Q_2(x)$ is zero, then $Q_2(x)$ is a constant, say, a_2, and

$$P(x) = a_2(x - c_2) \cdot (x - c_1)$$

If, however, $(n - 2) \geq 1$, by the Fundamental Theorem, applied to $Q_2(x)$, there exists a constant c_3 such that $Q_2(c_3) = 0$, which implies that there exists a polynomial $Q_3(x)$ of degree $n - 3$ such that

$$Q_2(x) = Q_3(x) \cdot (x - c_3) \tag{4}$$

which from (3) implies

$$P(x) = Q_3(x) \cdot (x - c_3) \cdot (x - c_2) \cdot (x - c_1) \tag{5}$$

If $n - 3$ is zero, then $Q_3(x)$ is a constant, say, a_3, and

$$P(x) = a_3(x - c_3)(x - c_2)(x - c_1)$$

If we continue the above process n times, we arrive at a $Q_n(x)$ of degree 0, which we set equal to a constant a_n and obtain

$$P(x) = a_n \cdot (x - c_n) \cdot (x - c_{n-1}) \cdots (x - c_3) \cdot (x - c_2) \cdot (x - c_1) \tag{6}$$

Thus, we conclude that the polynomial $P(x)$ can be factored completely into n linear factors. Moreover, from (6) it is clear that

$$P(c_n) = P(c_{n-1}) = \cdots = P(c_3) = P(c_2) = P(c_1) = 0$$

which implies that $c_n, c_{n-1}, \ldots, c_3, c_2, c_1$ are n roots of the complex polynomial $P(x)$. Furthermore, it is clear from (6) that $P(x)$ could have no other roots distinct from $c_n, c_{n-1}, \ldots, c_3, c_2, c_1$. Thus, $P(x)$ has exactly n complex roots.   QED

There is one final property of complex zeros we wish to discuss. Recall from Chapter 6 that associated with each complex number, say, $a + bi$, is its conjugate $a - bi$, sometimes written $\overline{a + bi}$. The next theorem identifies a property of complex zeros and their conjugates.

THEOREM 7.8.3   If $P(x)$ is a real polynomial (all coefficients are real), then

$$P(c) = 0 \Longrightarrow P(\overline{c}) = 0$$

That is, complex roots occur in conjugate pairs.

Proof   If c is real ($c = a + 0i$ for some real number a) then, of course,

$$c = \overline{c} \tag{7}$$

and the above conclusion is trivial. If c is imaginary, that is, of the form $a + bi$ where $a, b \in R$ and $b \neq 0$, then $\overline{c} = a - bi$. Moreover, it was shown in Problem 10, page 233, that if c_1 and c_2 are two complex numbers, then

$$\overline{c_1 + c_2} = \overline{c_1} + \overline{c_2} \quad \text{and} \quad \overline{c_1 \cdot c_2} = \overline{c_1} \cdot \overline{c_2} \tag{8}$$

We now proceed as follows:

$$0 = P(c) \Longrightarrow \overline{0} = \overline{P(c)}$$

$$\xrightarrow{(7)} 0 = \overline{P(c)} = \overline{a_n c^n + a_{n-1} c^{n-1} + \cdots + a_1 c + a_0}$$

$$\xrightarrow{(8)} \overline{a}_n (\overline{c})^n + \overline{a}_{n-1} (\overline{c})^{n-1} + \cdots + \overline{a}_1 \overline{c} + \overline{a}_0$$

$$\xrightarrow{(7)} a_n (\overline{c})^n + a_{n-1} (\overline{c})^{n-1} + \cdots + a_1 \overline{c} + a_0$$

$$=\!=\!= P(\overline{c})$$

$$\xrightarrow{\text{TLE}} 0 = P(\overline{c}) \qquad \text{QED}$$

Example 7.8.1 Let $P(x) = 2x^3 + 3x^2 + 2x + 3$. If i is a zero of $P(x)$, find the other two zeros and verify them by synthetic division.

Solution $P(i) = 0 \xrightarrow{\text{T7.8.3}} P(-i) = 0$

T7.8.3 $P(c) = 0 \Longrightarrow P(\overline{c}) = 0$

Using synthetic division, we have

$$
\begin{array}{r|cccc}
i & 2 & 3 & 2 & 3 \\
  &   & 2i & -2+3i & -3 \\
\hline
-i & 2 & 3+2i & 3i & 0 \\
   &   & -2i & -3i & \\
\hline
3 & 2 & 3 & 0 & \\
-2 &   & -3 & & \\
\hline
   & 2 & 0 & &
\end{array}
$$

$$\xrightarrow{\text{T7.5.2}} P(i) = 0$$

$$\Longrightarrow P(-i) = 0$$

$$\Longrightarrow P\left(-\tfrac{3}{2}\right) = 0$$

T7.5.2 The Remainder Theorem

Example 7.8.2 Let $P(x) = x^3 - 6x^2 + 13x - 10$. If $2 - i$ is a zero of $P(x)$, find the other two zeros and verify them by synthetic division.

Solution $P(2 - i) = 0 \xrightarrow{\text{T7.8.3}} P(2 + i) = 0$

By synthetic division, we obtain

$$
\begin{array}{r|cccc}
2-i & 1 & -6 & 13 & -10 \\
    &   & 2-i & -9+2i & 10 \\
\hline
2+i & 1 & -4-i & 4+2i & 0 \\
    &   & 2+i & -4-2i & \\
\hline
2 & 1 & -2 & 0 & \\
  &   & 2 & & \\
\hline
  & 1 & 0 & &
\end{array}
$$

$$\xrightarrow{\text{T7.5.2}} P(2-i) = 0$$

$$\Longrightarrow P(2+i) = 0$$

$$\Longrightarrow P(2) = 0$$

The next three theorems present properties of real zeros only. The first property provides a characterization of the real zeros of a real polynomial equation regarding the number of positive and negative zeros possible.

THEOREM 7.8.4 DESCARTES' RULE OF SIGNS

If $P(x)$ is a real polynomial, then the number of positive real zeros of $P(x)$ is either equal to the number of variations in sign occurring in the coefficients of $P(x)$ or else is less than this number by an even natural number. Moreover, the number of negative real zeros of $P(x)$ is either equal to the number of variations in sign occurring in the coefficients of $P(-x)$, or else is less than this number by an even natural number.

Example 7.8.3 Let $P(x) = 3x^5 - x^3 - 2x^2 + x - 4$. Determine all possible combinations of positive and negative real zeros of $P(x)$ by Descartes' Rule of Signs.

Solution $P(x) = 3x^5 - x^3 - 2x^2 + x - 4$

which implies there are three variations in sign occurring in the coefficients of $P(x)$. Next,

$$P(-x) = 3(-x)^5 - (-x)^3 - 2(-x)^2 + (-x) - 4$$
$$= -3x^5 + x^3 - 2x^2 - x - 4$$

implying two variations in sign. We know by Theorem 7.8.2 that there are exactly five zeros of $P(x)$, including complex zeros. We must keep in mind, however, that complex numbers are neither positive nor negative—there is no ordering in the system of complex numbers. Also, complex zeros of real polynomials occur in conjugate pairs (Theorem 7.8.3)—there must be an even number of them. Descartes' Rule of Signs speaks only of real zeros. From the results of applying Descartes' Rule of Signs to $P(x)$, we construct the following table, which identifies the possible patterns for the zeros of this polynomial.

Each entry less	R_+	R_-	I	Each row
than 3 for R_+, or	3	2	0	represents a
2 for R_-, is less	1	2	2	possible com-
by an even	3	0	2	bination of
natural number.	1	0	4	zeros.

The five zeros of $P(x)$ will agree with one of the four row patterns identified above.

Let us now discuss the advantages of this information. Suppose by some means we are able to find two positive real zeros. By Descartes' Rule of Signs we know there are either three positive or one positive zero; having two, we therefore conclude there must be one more positive zero. Note also that once three positive zeros have been found, we need look no further for positive zeros—we have them all. Similarly, if two negative zeros are found, we would not look further for others.

Example 7.8.4 Let

$$P(x) = x^7 - x^6 - 2x^5 + 3x^4 + x^3 - x^2 + 4x - 1$$

Determine all possible combinations of positive, negative, and imaginary zeros of $P(x)$ by Descartes' Rule of Signs.

Solution $P(x) = x^7 - x^6 - 2x^5 + 3x^4 + x^3 - x^2 + 4x - 1$

Thus, there are at most five positive real zeros. Also

$$P(-x) = (-x)^7 - (-x)^6 - 2(-x)^5 + 3(-x)^4 + (-x)^3 - (-x)^2$$
$$+ 4(-x) - 1$$
$$= -x^7 - x^6 + 2x^5 + 3x^4 - x^3 - x^2 - 4x - 1$$

Hence, there are at most two negative zeros.

Knowing that there are exactly *seven* zeros (including those that are complex), we form Table 7.8.1. Thus, there are six possible combinations of positive, negative, and imaginary zeros of $P(x)$. The actual character of the zeros of $P(x)$ must coincide with one of these six patterns.

Another helpful characterization of the real zeros of a real polynomial is offered by the next theorem, which deals with upper and lower bounds.

Table 7.8.1

R_+	R_-	I
5	2	0
3	2	2
1	2	4
5	0	2
3	0	4
1	0	6

THEOREM 7.8.5 Let $P(x)$ be a real polynomial of degree $n \geq 1$. If we divide $P(x)$ synthetically, using a *positive* multiplier k, and if all the coefficients of the quotient $Q(x)$ and the remainder r are either nonnegative or nonpositive, then no real zero of $P(x)$ can be greater than k. Furthermore, if k is *negative* and all the coefficients of $Q(x)$ and the remainder r alternate in sign (zero taking either sign as needed), then no real zero of $P(x)$ can be less than k

Proof First take the case where $k > 0$ and all the coefficients of the quotient $Q(x)$ and r are nonnegative. By the division algorithm we have

$$P(x) = Q(x) \cdot (x - k) + r$$

Now choose m so that $m - k > 0$. Because all the coefficients of $Q(x)$ are nonnegative and $m > 0$, we have $Q(m) > 0$ (otherwise $Q(x)$ would be identically zero and $P(x)$ a constant). Next evaluate $P(x)$ at m, obtaining

$$P(m) = Q(m)(m - k) + r$$

But this expression, along with $Q(m) > 0$, $m - k > 0$, and $r \geq 0$, implies

$$P(m) > 0$$

Thus, no real number greater than k can be a zero of $P(x)$; k is an upper bound of all positive real roots.

The argument for the case $k > 0$ and all the coefficients of $Q(x)$ and r are nonpositive is exactly as above but with the inequalities reversed.

The proof of the second half concerning lower bounds is to be done as an exercise.

Example 7.8.5 Find upper and lower bounds for the real zeros of $P(x) = x^3 - 2x^2 + 3x - 4$.

Solution Using synthetic division, we have

$$
k = \begin{cases}
& \begin{array}{r|rrrr}
 & 1 & -2 & 3 & -4 \\
\hline
1 & 1 & -1 & 2 & -2 \\
2 & 1 & 0 & 3 & 2 \\
-1 & 1 & -3 & 6 & -10 \\
\end{array}
\end{cases}
$$

$\xrightarrow{\text{T7.8.5}}$ 2 upper bound

$\Longrightarrow$ -1 lower bound

Notice that the second row of coefficients alternates in sign; nevertheless, 1 is not a lower bound because it is not negative. The upper and lower bound tests of Theorem 7.8.5 only hold when $k > 0$ for an upper bound and $k < 0$ for a lower bound.

Example 7.8.6 Find upper and lower bounds for the real zeros of

$$P(x) = -2x^4 - 3x^3 + 6x^2 - 3x + 5$$

Solution Using synthetic division, we have

$$
k = \begin{cases}
& \begin{array}{r|rrrrr}
 & -2 & -3 & 6 & -3 & 5 \\
\hline
-1 & -2 & -1 & 7 & -10 & 15 \\
-2 & -2 & 1 & 4 & -11 & 27 \\
-3 & -2 & 3 & -3 & 6 & -13 \\
1 & -2 & -5 & 1 & -2 & 3 \\
2 & -2 & -7 & -8 & -19 & -33 \\
\end{array}
\end{cases}
$$

$\Longrightarrow$ -3 is a lower bound

$\Longrightarrow$ 2 is an upper bound

Up to this point all of the theorems concerning polynomial zeros have provided information about the zeros but have not given methods for finding them, except in the linear and quadratic cases where the zeros can be identified by formula. Unfortunately formulas do not exist by which polynomial equations in general can be solved. There are formulas for third and fourth degree polynomial equations, but they are rather complicated and will not be treated here. • There is, however, one important theorem that will

● It has been proved that algebraic formulas for the solutions of general polynomial equations of degree $n \geq 5$ do not exist.

enable us, under certain conditions, to find all rational roots of a polynomial equation.

THEOREM 7.8.6   THE RATIONAL ROOT THEOREM

Let $P(x)$ be a real polynomial of degree $n \geq 1$ with integral coefficients (all coefficients are integers). If p/q is a rational zero of $P(x)$, where p and q are relatively prime integers, then p is an integral factor of the constant term a_0 of $P(x)$ and q is an integral factor of the leading coefficient a_n.

Proof   If p/q is a zero of $P(x)$, then

$$a_n \frac{p^n}{q^n} + a_{n-1} \frac{p^{n-1}}{q^{n-1}} + \cdots + a_1 \frac{p}{q} + a_0 = 0 \qquad (9)$$

Multiplying both sides of (9) by q^n yields

$$a_n p^n + a_{n-1} p^{n-1} q + \cdots + a_1 p q^{n-1} + a_0 q^n = 0 \qquad (10)$$

Adding $-a_0 q^n$ to both sides of (10) results in

$$a_n p^n + a_{n-1} p^{n-1} q + \cdots + a_1 p q^{n-1} = -a_0 q^n \qquad (11)$$

Next, factor p out of the left-hand side, obtaining

$$p(a_n p^{n-1} + a_{n-1} p^{n-2} q + \cdots + a_1 q^{n-1}) = -a_0 q^n \qquad (12)$$

Because p is an integral factor of the left-hand side of (12), it must also be a factor of the right-hand side. But p and q (and hence q^n) are relatively prime (have no integral factors in common). Hence, p must be an integral factor of a_0. This proves the first part of the theorem. The fact that q is an integral factor of a_n is to be done as an exercise.

The rational root theorem enables us to find all rational zeros of polynomials with integral coefficients. You form all possible rational numbers, taking as numerators integral factors of a_0 and as denominators integral factors of a_n. If there is a rational zero of the polynomial, it must be in this collection. Although there may be several choices to check by this method, at least the number is finite.

Example 7.8.7   Find the zeros of

$$P(x) = 2x^3 - 13x^2 + 27x - 18$$

Solution   First determine all integral factors of -18 and 2. They are

$$-18 = (\pm 1)(\pm 18) = (\pm 2)(\pm 9) = (\pm 3)(\pm 6)$$
$$2 = (\pm 1)(\pm 2)$$

Now form all possible ratios of the integral factors of -18 to those of 2. We obtain

$$\{\pm 1, \pm \tfrac{1}{2}, \pm 18, \pm 9, \pm 2, \pm \tfrac{9}{2}, \pm 3, \pm \tfrac{3}{2}, \pm 6\}$$

As you begin to examine these ratios it will be to your advantage to start with those choices near zero so that if an upper or lower bound is encountered, you may eliminate several of the other possibilities.

$$
\begin{array}{r|rrrr}
 & 2 & -13 & 27 & -18 \\
\hline
-1 & 2 & -15 & 42 & -60 \Longrightarrow -1 \text{ is a lower bound} \\
\hline
1 & 2 & -11 & 16 & -2 \\
\hline
2 & 2 & -9 & 9 & 0 \Longrightarrow P(2) = 0 \\
\hline
3 & 2 & -3 & 0 & \Longrightarrow P(3) = 0 \\
\hline
\tfrac{3}{2} & 2 & 0 & \Longrightarrow P(\tfrac{3}{2}) = 0 \\
\end{array}
$$

Example 7.8.8 Find all zeros of $P(x) = x^4 + x^3 + 2x - 4$.

Solution The integral factors of -4 are

$$-4 = (\pm 1)(\mp 4) = (\pm 2)(\mp 2)$$

Because the leading coefficient is 1, the above factors represent all possible rational zeros. We proceed by synthetic division:

$$
\begin{array}{r|rrrrr}
 & 1 & 1 & 0 & 2 & -4 \\
\hline
-1 & 1 & 0 & 0 & 2 & -6 \\
\hline
-2 & 1 & -1 & 2 & -2 & 0 \Longrightarrow P(-2) = 0 \text{ and } -2 \text{ is a lower bound} \\
\hline
1 & 1 & 0 & 2 & 0 & \Longrightarrow P(1) = 0 \text{ and } 1 \text{ is an upper bound} \\
\end{array}
$$

$\underbrace{\qquad\qquad\qquad}$

(coefficients of the remaining $Q(x)$)

In the last step, the reduced polynomial $Q(x)$ becomes $x^2 + 2$, which, being quadratic, can be solved directly as follows:

$$
\begin{aligned}
x^2 + 2 = 0 &\Longleftrightarrow x^2 = -2 \\
&\Longleftrightarrow x = \pm\sqrt{-2} \\
&\Longleftrightarrow x = \pm\sqrt{2}\,i
\end{aligned}
$$

Therefore, the four zeros of $x^4 + x^3 + 2x - 4$ are $1, -2, -\sqrt{2}i$, and $\sqrt{2}i$

In Example 7.8.8 we were fortunate that all the zeros were rational, except two, and hence we were able to reduce the polynomial containing the final two zeros to one of second degree, which can always be solved by elementary methods. In cases where this is not possible, you may have to approximate the real zeros, should there be any. There are several methods to accomplish this. We will introduce but one here. It is called the method of successive approximation. This method will be demonstrated by the example that follows:

Example 7.8.9 Approximate a zero of $P(x) = x^3 + 2x^2 - 5x - 1$ to within one-tenth.

Solution Using synthetic division, we approximate successively (converge upon) a zero as follows:

	1	2	−5	−1	
1	1	3	−2	−3	$\Longrightarrow P(1) = -3$
2	1	4	3	5	$\Longrightarrow P(2) = 5$
1.6	1	3.6	0.76	0.216	$\Longrightarrow P(1.6) = 0.216$
1.5	1	3.5	0.25	−0.625	$\Longrightarrow P(1.5) = -0.625$

Because a polynomial function is continuous (no breaks), the two expressions $P(1) = -3$ and $P(2) = 5$ (one negative and the other positive) imply the graph crosses the x-axis from a negative value (below) to a positive value (above) somewhere between 1 and 2. Hence, there must exist a real zero between 1 and 2. Similarly, $P(1.5) = -0.625$ and $P(1.6) = 0.216$ imply a zero between 1.5 and 1.6. For greater accuracy you would simply continue squeezing in this manner.

We are now ready to return to our applied problem with which this chapter was introduced. Recall that we had a system of three weights, connected by springs, that was set into motion. In the process of finding functions describing the subsequent motion, it was necessary to find the six zeros of the following polynomial equation.

$$16w^6 - 56w^4 + 49w^2 - 9 = 0$$

Because all the coefficients are integers, we may apply the Rational Root Theorem. We first determined all possible integral factors of the leading and last coefficients. We obtain

$$9 = (\pm 9)(\pm 1) = (\pm 3)(\pm 3)$$

and

$$16 = (\pm 16)(\pm 1) = (\pm 8)(\pm 2) = (\pm 4)(\pm 4)$$

We next form all possible ratios of the integral factors of 9 to the integral factors of 16, obtaining the set

$$\{\pm \tfrac{9}{16}, \pm \tfrac{9}{1}, \pm \tfrac{1}{16}, \pm 1, \pm \tfrac{3}{16}, \pm 3, \pm \tfrac{9}{8}, \pm \tfrac{9}{2}, \pm \tfrac{1}{8}, \pm \tfrac{1}{2}, \pm \tfrac{3}{8}, \pm \tfrac{3}{2}, \pm \tfrac{3}{4}, \pm \tfrac{1}{4}\}$$

We know that if there are any rational zeros of the above polynomial, they must be among the elements of the above set. We will verify them by synthetic division as follows:

$$
\begin{array}{r|ccccccc}
 & 16 & 0 & -56 & 0 & 49 & 0 & -9 \\
\hline
\frac{1}{2} & 16 & 8 & -52 & -26 & 36 & 18 & 0 \\
-\frac{1}{2} & 16 & 0 & -52 & 0 & 36 & 0 & \\
1 & 16 & 16 & -36 & -36 & 0 & & \\
-1 & 16 & 0 & -36 & 0 & & & \\
\frac{3}{2} & 16 & 24 & 0 & & & & \\
-\frac{3}{2} & 16 & 0 & & & & & \\
\end{array}
$$

Hence, the solution set of the above polynomial equation is

$$\{\pm\tfrac{1}{2},\ \pm 1,\ \pm\tfrac{3}{2}\}$$

and the problem of solving this equation is completed.

7.9   Problem Sets

Problem Set I   ■   *Reading Comprehension*

Determine whether statements 1 through 8 are true or false and give reasons for your responses.

1   All polynomial relations defined by expressions of the form $y = P(x)$, where $P(x)$ is a polynomial, are functions.
2   In general, the graph of a polynomial function of degree $n \geq 1$ will have n turning points.
3   If $P(x)$ is a real polynomial of degree $n \geq 1$, then $P(x)$ possesses exactly n complex zeros, with zeros of multiplicity counted as distinct zeros.
4   Every polynomial of degree n has exactly n distinct complex roots.
5   If $P(x)$ is a polynomial and $P(c) = 0$, then $P(\bar{c}) = 0$, where $\bar{c}$ is the complex conjugate of c.
6   If a polynomial $P(x)$ has 3 variations in sign occurring in its coefficients, then $P(x)$ possesses at least 3 positive real zeros.
7   If $P(x)$ is a polynomial and $k > 0$ is an upper bound for the zeros of $P(x)$, then all real zeros of $P(x)$ are less than k.
8   If $P(x)$ is a polynomial all of whose coefficients are integers, then all rational zeros of $P(x)$ can be determined.

Problem Set II   ■   *Skills Development*

Graph the following polynomial functions:

1   $P = \{(x, y) \mid y = x^3 - 3x^2 + x - 1 \text{ over } R\}$
2   $Q = \{(x, y) \mid y = x^4 + 2x^3 - x + 2 \text{ over } R\}$
3   $D = \{(x, y) \mid y = -x^4 - x^2 + 3x - 4 \text{ over } R\}$

Determine real polynomials (polynomials with real coefficients) whose zeros are those indicated in Problems 4, 5, and 6.

Example 1 and $2i$.

Solution If $2i$ is a zero of a real polynomial, then, by Theorem 7.8.3, $-2i$ is also a zero. Furthermore, by the factor theorem, we have

$$P(x) = (x-1)(x-2i)(x+2i) = (x-1)(x^2+4) = x^3 - x^2 + 4x - 4$$
$$\Longrightarrow P(x) = x^3 - x^2 + 4x - 4$$

4 $1, -2, -2$ **5** $1-i, 3$ **6** $1-i, 2+i$

Solve the following polynomial equations given the roots indicated:

7 $x^4 - 2x^3 + 3x^2 - 2x + 2 = 0$ if $1+i$ is a zero.
8 $x^4 + 3x^3 + 4x^2 + 27x - 45 = 0$ if $-3i$ is a zero.
9 $x^6 + 3x^5 + 3x^4 + 6x^3 + 3x^2 + 3x + 1 = 0$ if i is a double root.

Determine the possible combinations of positive, negative, and imaginary zeros according to Descartes' Rule of Signs for each of the following polynomials:

10 $x^5 - 2x^4 + x^2 - x - 1$ **11** $-x^6 - 3x^5 + x^4 - 2x^3 - x^2 + x + 1$
12 $x^7 + x^6 + x^5 + x^4 - x^3 - x^2 - x - 1$

Determine upper and lower bounds for the real zeros of each of the following polynomials:

13 $x^4 - 2x^3 - x^2 + x - 2$ **14** $x^3 + 5x^2 - 6x - 4$ **15** $x^5 - 2x^4 + x^2 - x + 1$

By applying the rational root theorem and taking advantage of possible upper and lower bounds, find all zeros of the following polynomials:

16 $x^3 - x^2 - 3x + 2 = 0$ **17** $2x^4 + 5x^3 + 7x^2 + 10x + 6 = 0$
18 $2x^3 - 3x^2 - 11x + 6 = 0$ **19** $x^5 - x^4 - 7x^3 - 14x^2 - 24x = 0$
20 $3x^3 - x^2 - 2x + 24 = 0$ **21** $9x^4 - 3x^3 + 7x^2 - 3x - 2 = 0$

By successive approximations find at least one zero of each of the following equations to within one-tenth:

22 $x^3 + x^2 + 2x + 3 = 0$, between -1 and -2.
23 $x^3 - 4x + 2 = 0$, between 1 and 2.   **24** $x^3 - 2x - 7 = 0$, greater than zero.

Problem Set III ■ *Theoretical Developments*

1 Prove that if $P(x)$ is a real polynomial with integral coefficients and p/q is a rational zero of $P(x)$, then q is an integral factor of the leading coefficient a_n. (Complete the proof of Theorem 7.8.6.)
2 Prove the second half of Theorem 7.8.5 for lower bounds.
3 Prove that a real polynomial of degree 3 has either three real roots or one real and two complex conjugate roots.
4 Prove that a real polynomial of odd degree possesses at least one real root.
5 Prove that the polynomial $P(x) = 2x^4 - 15x + 2$ has exactly two real and two imaginary zeros.
6 Prove that the polynomial $x^6 - 14x^3 + x^2 + 1$ has exactly two real and four imaginary zeros.
7 Prove that $\sqrt[3]{6}$ is irrational. (HINT: Apply the rational root theorem to the equation $x^3 - 6 = 0$).
8 Prove that $\sqrt[4]{10}$ is irrational.

1 In the system shown in Fig. 7.9.1, the weights w_1 and w_2 are given an initial displacement and then the system is allowed to move from this state of rest. In the problem of finding a function describing the subsequent motion, it is necessary to solve the following polynomial equation:

$$w^4 - 5w^2 + 4 = 0$$

Determine the solution set of this equation.

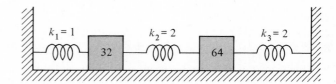

FIGURE 7.9.1

2 In the problem of finding the natural frequencies and normal modes of the torsional system shown in Fig. 7.9.2, it is necessary to solve the following polynomial equation.

$$4w^6 - 21w^4 + 21w^2 - 4 = 0$$

Determine the solution set of this equation.

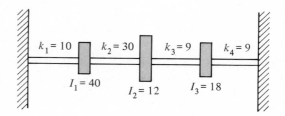

FIGURE 7.9.2

3 When solving differential equations by means of a special technique called the Laplace transform, we are frequently required to factor certain polynomials in the variable s. Factor completely the following polynomials:

a $s^3 + 6s^2 + 11s + 6$ b $s^4 + 4s^3 + 4s^2 - 4s - 5$ c $s^4 + 4s^3 + 6s^2 + 5s + 2$

7.10   Rational Functions

In Sections 7.3 and 7.5 we reviewed an algebra for polynomials including the operations of addition, subtraction, multiplication, and division. The division algorithm was discussed at some length; we noted that synthetic division is particularly useful if the divisor is of the form $x - c$. Definition 7.5.1 identified the quotient of two polynomials as a rational algebraic expression. As we have done several times earlier, we set this algebraic expression in the variable x equal to y, thereby obtaining an open equation in two variables, which defines a relation. Because the quotient of two polynomials in x admits a unique real number for each real choice of x for which the denominator is nonzero, we are assured that the relation is a function. It would be natural, of course, to call such functions rational. We summarize the above observations in the following definition:

DEFINITION 7.10.1   A function defined by an open equation of the form

$$y = \frac{P(x)}{D(x)}$$

where $P(x)$ and $D(x)$ are polynomials, is called a rational function Such a function may be denoted

$$\left\{ (x, y) \mid y = \frac{P(x)}{D(x)} \text{ over } \boldsymbol{R} \right\}$$

You may suspect the graph of a rational function to be more difficult to determine than that of a polynomial function. This, however, is not necessarily the case. In fact, to the contrary, rational graphs are frequently more easily identified. This is because of the presence of certain important guidelines called asymptotes. These give you an advantage in graphing rational functions not usually enjoyed by many other functions.

The manner by which asymptotes are identified and used will be explained in the next several examples.

Example 7.10.1   Graph the function

$$f = \left\{ (x, y) \mid y = \frac{2}{x - 3} \text{ over } \boldsymbol{R} \right\}$$

Solution   Before plotting any points, we shall examine carefully the defining equation. First, it is clear that the domain of f (the set of all $x \in \boldsymbol{R}$ for which y is defined) includes all real numbers *except* 3 — division by 0, of course, is not defined. It is significant to note, however, that although the quotient $2/(x - 3)$ is not defined at $x = 3$, it is well defined for all other choices of x, *however close to 3*. It is this latter phrase "however close to 3" that we wish to pursue. You have had enough experience with fractions to know that if the denominator of a fraction becomes increasingly small and the numerator remains constant, the quotient becomes increasingly large. Take, for example, the following:

$$\frac{2}{0.1} = 2 \cdot 10 = 20 \qquad \frac{2}{0.01} = 2 \cdot 100 = 200$$

$$\frac{2}{0.001} = 2 \cdot 1000 = 2000 \qquad \frac{2}{0.0001} = 2 \cdot 10000 = 20{,}000$$

It is important to understand, as the above examples suggest, that this quotient may be made as large as desired by taking the divisor sufficiently close to zero. That is, the quotient $2/(x-3)$ may be increased to any magnitude, both positively and negatively, by choosing x close enough to 3. Should x be chosen slightly greater than 3, the quotient would be large and positive, whereas, if x is chosen ever so slightly less than 3, then the quotient would be large in absolute value and negative. These observations are illustrated in Fig. 7.10.1. The function f approaches the line defined by $x = 3$ asymptotically from both the right — going upward — and from the left — going downward. The line defined by $x = 3$ is called a vertical asymptote. In general the graph of a rational function defined by

$$y = \frac{P(x)}{D(x)}$$

will have a vertical asymptote through each zero of the denominator $D(x)$ (assuming each such zero is not a zero of $P(x)$; i.e., assuming $P(x)$ and $D(x)$ have no common linear factors).

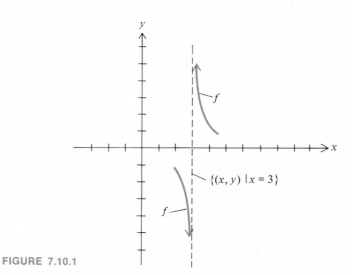

FIGURE 7.10.1

Let us continue, now, to determine the rest of the curve. Just as a fraction becomes increasingly large as the divisor alone becomes small, so it is that the fraction will become small (close to zero) as the divisor becomes large, either positively or negatively. Thus, the quotient $2/(x-3)$ becomes small, although always positive, as x takes on increasingly large, positive values; similarly, the fraction becomes small, although always negative, as x takes on large, negative values. These facts are expressed by extending the graph of f as illustrated in Fig. 7.10.2.

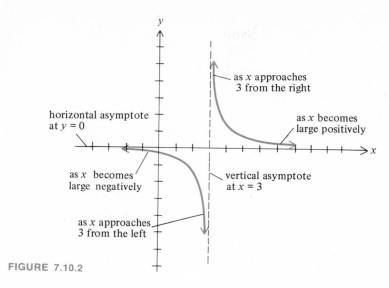

FIGURE 7.10.2

Observe that the *x*-axis is a horizontal asymptote. Note also that the above sketch has taken place without plotting a single point; hence, it is but an approximation, but the basic pattern is clear. With just a few points, a relatively accurate graph is obtained, as follows:

From $y = 2/(x-3)$, we find that

$$(1, -1), (2, -2), (4, 2), (5, 1) \in f$$

Plotting these and following the pattern of Fig. 7.10.2, we obtain the graph (Fig. 7.10.3).

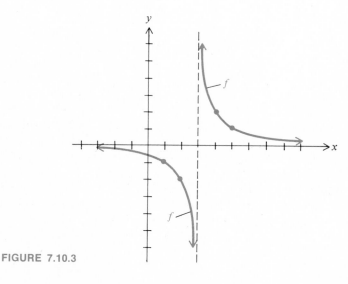

FIGURE 7.10.3

Example 7.10.2 Graph the function

$$f = \left\{ (x, y) \mid y = \frac{x^2 - 4}{x^2 + 1} \text{ over } \boldsymbol{R} \right\}$$

Solution First note that there are no vertical asymptotes—the divisor $x^2 + 1$ has no real zeros. Whenever the degree of the polynomial in the numerator is equal to or greater than that of the denominator, we may express the quotient in the alternative form

$$\boldsymbol{Q}(x) + \frac{\boldsymbol{r}(x)}{\boldsymbol{D}(x)}$$

where the degree of the polynomial remainder $r(x)$ is less than that of $\boldsymbol{D}(x)$. It is from this latter form that the nonvertical asymptotes are identified. For the function g, we obtain

$$y = \frac{x^2 - 4}{x^2 + 1} = 1 + \frac{-5}{x^2 + 1} \tag{1}$$

Let us analyze the behavior of the expression

$$\frac{-5}{x^2 + 1}$$

as x varies over $\boldsymbol{R}$. Because the numerator is negative and the denominator is positive, the quotient is always negative. That is, for every real x, we have

$$\frac{-5}{x^2 + 1} < 0$$

Because

$$y = 1 + \frac{-5}{x^2 + 1}$$

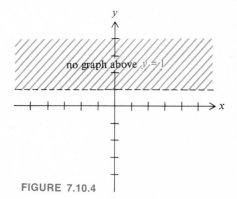

FIGURE 7.10.4

we conclude, therefore, that $y < 1$ for all x. This is illustrated in Fig. 7.10.4.

Now let us determine how much less than one y may be. Note that at $x = 0$, the divisor, $x^2 + 1$, will be minimized, and hence the quotient, $5/(x^2 + 1)$, will be maximized. Thus,

$$x = 0 \Longrightarrow \frac{-5}{x^2 + 1} = -5$$

and for every other choice of x we have

$$\frac{-5}{x^2 + 1} > -5$$

Hence, although y is always less than 1, it is also *greater than or equal to* -4 and only dips to -4 at $x = 0$. We begin our graph as illustrated in Fig. 7.10.5. Now, as x becomes increasingly *large*, both positively and negatively, the quotient, $-5/(x^2 + 1)$, approaches zero; hence *y approaches* 1. The line defined by $y = 1$ thus becomes a horizontal asymptote

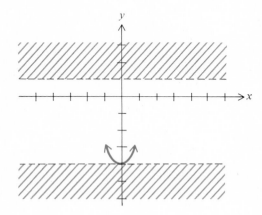

FIGURE 7.10.5

From the original form

$$y = \frac{x^2 - 4}{x^2 + 1}$$

we may determine the x-intercepts. Clearly, if $x = \pm 2$, we have $y = 0$, so that $(2, 0)$, $(-2, 0) \in g$.

Thus, with only three points and the asymptotic information determined above we graph g as illustrated in Fig. 7.10.6.

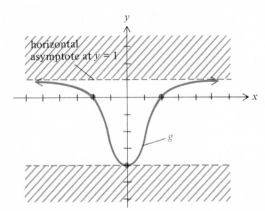

horizontal asymptote at $y = 1$

g

FIGURE 7.10.6

Example 7.10.3 Graph the function

$$h = \left\{ (x, y) \mid y = \frac{x^2 - 4}{x + 1} \text{ over } R \right\}$$

Solution Notice the rational expression defining h is almost identical to that defining g in Example 7.10.2 — the only difference being that the power of x in the divisor is reduced from 2 to 1. One may not suspect that such a small change in the rational expression would alter the graph appreciably. However, our analysis will soon dispel this illusion.

We first observe that the divisor, $x + 1$, has a zero at -1. Hence, there exists a vertical asymptote at $x = -1$. Furthermore, if $-1 < x < 0$, we have

$$y = \frac{x^2 - 4}{x + 1} < 0$$

because $x^2 - 4 < 0$ and $x + 1 > 0$. Also, if $-2 < x < -1$, we have $x^2 - 4 < 0$ and $x + 1 < 0$, so that

$$y = \frac{x^2 - 4}{x + 1} > 0$$

We thus conclude that as x approaches -1 from the *right* $(x > -1)$ y is *unbounded going negative*, and as x approaches -1 from the *left* $(x < -1)$ y is *unbounded going positive*. Therefore, the beginning of our graph would appear as in Fig. 7.10.7. To investigate the possibility of other asymptotes, we divide $x + 1$ into $x^2 - 4$:

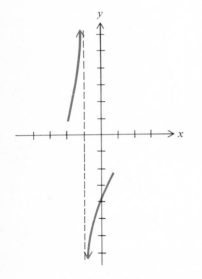

FIGURE 7.10.7

$$y = \frac{x^2 - 4}{x + 1} = (x - 1) + \frac{-3}{x + 1} \tag{2}$$

From the right-hand side of (2) it is clear that as x becomes increasingly *large*, both positively and negatively, the quotient $-3/(x + 1)$ approaches *zero*, and, hence, y approaches $(x - 1)$. The line defined by $y = x - 1$ is called an oblique asymptote. Our only question now is whether the curve approaches this asymptote from above or from below. To decide this, we argue as follows: If $x > 1$, then

$$\frac{-3}{x + 1} < 0$$

and so for large positive values of x, y will approach $(x - 1)$ from *below*; whereas if $x < 1$, then

$$\frac{-3}{x + 1} > 0$$

and thus, for large negative x, y approaches $(x - 1)$ from *above*. With this information, our graph takes the form shown in Fig. 7.10.8.

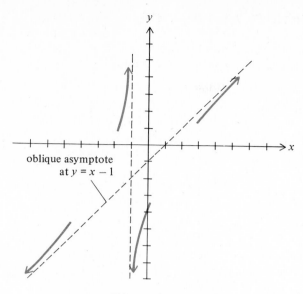

FIGURE 7.10.8

These branches can be joined by observing that $x = \pm 2$ are the *zeros* of h (*x*-intercepts). Furthermore, the *y-intercept* (obtained by setting $x = 0$ is -4. Thus,

$$(2, 0), \ (-2, 0), \ (0, -4) \in h$$

and we complete the graph. (See Fig. 7.10.9.)

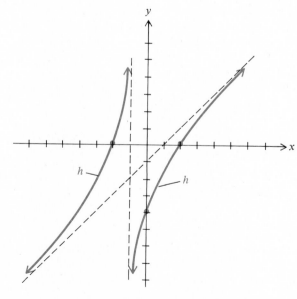

FIGURE 7.10.9

Although the defining equations for g of Example 7.10.2 and h of Example 7.10.3 were quite similar, their respective graphs possessed little resemblance to each other. This suggests the care you must take in graphing rational functions.

Example 7.10.4 Graph the rational function

$$f = \left\{ (x, y) \mid y = \frac{x^2 - 4}{x^2 - 1} \text{ over } \boldsymbol{R} \right\}$$

Solution Here again we have a rational expression very similar to those of Examples 7.10.2 and 7.10.3, yet the graph of f will be quite different from both.

We begin by observing that the denominator, $x^2 - 1$, possesses two zeros, 1 and -1, whereas the numerator $x^2 - 4$ is not zero for either of these. Thus, there exist two vertical asymptotes — one at $x = 1$ and the other at $x = -1$. For $1 < x < 2$ (right side of asymptote through $x = 1$) we have

$$y = \frac{x^2 - 4}{x^2 - 1} < 0$$

because $x^2 - 4 < 0$ and $x^2 - 1 > 0$. Therefore as x approaches 1 from the *right*, y becomes *unbounded going negative*. For $-1 < x < 1$ (between the two asymptotes), we have

$$y = \frac{x^2 - 4}{x^2 - 1} > 0$$

Thus, as x approaches either 1 from the *left* or -1 from the *right*, y becomes *unbounded going positive*. Finally, if $-2 < x < -1$ (left side of asymptote at $x = -1$), then

$$y = \frac{x^2 - 4}{x^2 - 1} < 0$$

Hence, as x approaches -1 from the *left*, y becomes *unbounded going negative*.

With this information about the vertical asymptotes, we may begin the graph of f as in Fig. 7.10.10. Our next step is to write $(x^2 - 4)/(x^2 - 1)$ in the quotient remainder form. We obtain

$$y = \frac{x^2 - 4}{x^2 - 1} = 1 + \frac{-3}{x^2 - 1}$$

which implies that $y = 1$ is a horizontal asymptote (y approaches 1 as x becomes increasingly large, both positively and negatively). The final step is to determine the intercepts. Setting $x = 0$, we get $y = 4$, and setting $y = 0$, we have $x = \pm 2$. Thus,

$$(2, 4), \ (2, 0), \ (-2, 0) \in f$$

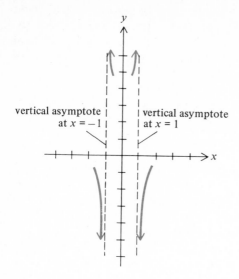

vertical asymptote at $x = -1$

vertical asymptote at $x = 1$

FIGURE 7.10.10

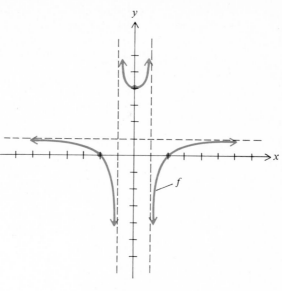

FIGURE 7.10.11

We may now finish the graph of f. (See Fig. 7.10.11.)

Example 7.10.5 Graph the rational function

$$g = \left\{ (x, y) \mid y = \frac{x^3 - 2x^2 + x - 2}{x^2 - 1} \text{ over } \boldsymbol{R} \right\}$$

Solution The divisor of the defining rational expression has two zeros, ± 1. We therefore have vertical asymptotes at $x = 1$ and $x = -1$. After factoring the numerator and denominator, we have

$$y = \frac{(x - 2)(x^2 + 1)}{(x - 1)(x + 1)}$$

This will help to determine the sign of y. For $1 < x < 2$ (x to the right of the asymptote at $x = 1$), $y < 0$; for $-1 < x < 1$, $y > 0$; and for $x < -1$ (x to the left of the asymptote at $x = -1$), $y < 0$. Hence the beginning of our graph may appear as in Fig. 7.10.12. Let us now write the defining rational expression in the usual alternative form. By long division, we obtain

$$y = \frac{x^3 - 2x^2 + x - 2}{x^2 - 1} = (x - 2) + \frac{2x - 4}{x^2 - 1}$$

From the right-hand side it is clear we have an oblique asymptote at $y = x - 2$. This asymptote has one property, however, which the others discussed thus far, do not have. Notice that the expression

$$\frac{2x - 4}{x^2 - 1}$$

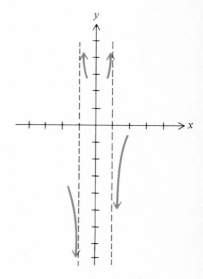

FIGURE 7.10.12

has a *zero* at $x = 2$. This fact implies that the graph of g will *cross* the asymptote defined by $y = x - 2$ before settling down to approaching it as x becomes large. Although it is the case that a vertical asymptote is never intercepted (a vertical asymptote is determined by the zeros of the divisor and division by zero is not defined), horizontal and oblique asymptotes may be crossed.

Now one final observation regarding the quotient

$$\frac{2x - 4}{x^2 - 1}$$

If $x > 2$, this expression is *positive*; thus, for large values of x, y approaches the oblique asymptote *from above*. If $x < -1$, the quotient is *negative*, and y must therefore approach the asymptote *from below* as x increases negatively.

From the original quotient in factored form

$$\frac{(x - 2)(x^2 + 1)}{(x - 1)(x + 1)}$$

we easily determine the intercepts as $(0, 2)$ and $(2, 0)$.

We are now prepared to complete the graph of g. (See Fig. 7.10.13.)

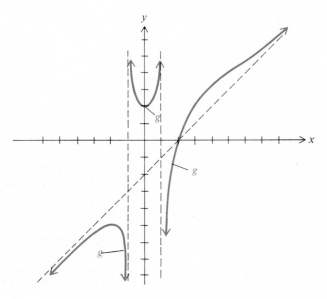

FIGURE 7.10.13

We shall now summarize the basic steps you will usually take in graphing a rational function.

(1) Find the zeros of the denominator, if there are any. Through each such zero will pass a vertical asymptote if the numerator is

nonzero for such x. Then determine whether y is positive or negative as x approaches each vertical asymptote from the right and then from the left. If $y > 0$ from the right, then y is unbounded going positive as it approaches the asymptote; whereas if $y < 0$ from the right, then it is unbounded going negative as it nears the asymptote. You argue similarly as x approaches the asymptote from the left.

(2) If the degree of the polynomial in the numerator of the defining rational expression is equal to or greater than that of the denominator, then the quotient $P(x)/D(x)$ can be expressed in the equivalent form

$$Q(x) + \frac{r(x)}{D(x)}$$

where the degree of the polynomial remainder $r(x)$ is less than that of $D(x)$. Because the quotient $r(x)/D(x)$ approaches zero for increasingly large values of x, the equation $y = Q(x)$ defines an asymptote. If the degree of $Q(x)$ is zero, $Q(x)$ is a constant and the asymptote is a horizontal line. If the degree of $Q(x)$ is 1, then the asymptote is oblique (a line whose slope is not zero but defined). One then determines whether the graph approaches these asymptotes from above or below. If $r(x)/D(x) > 0$ for large values of x, then the approach is from above; if $r(x)/D(x) < 0$, it is from below.

(3) The third and final step is to determine the x and y intercepts, if they exist. If intercepts do not exist, then at least one point of the graph on each side of each asymptote will help to obtain a reasonably accurate graph. Then sketch the graph.

Problem Sets 7.11

Problem Set I ■ *Reading Comprehension*

Determine whether statements 1 through 5 are true or false and give reasons for your responses.

1 A rational function is one whose range and domain are composed of rational numbers only.
2 The degree of a rational expression is equal to the sum of the degrees of the polynomials in the numerator and denominator.
3 Vertical asymptotes for rational functions are determined by the zeros of the numerator.
4 Just as vertical asymptotes are determined by the zeros of the denominator, horizontal asymptotes are found by finding the zeros of the polynomial in the numerator.
5 A property enjoyed by all asymptotes is that they are never crossed by their associated graphs.

Graph the rational functions defined by the following equations over **R** by first finding all asymptotes and intercepts.

1 $y = \dfrac{3}{x-1}$ 2 $y = \dfrac{2}{x+2}$ 3 $y = \dfrac{-1}{x-3}$ 4 $y = \dfrac{x+1}{x+2}$

5 $y = \dfrac{2x-1}{x+1}$ 6 $y = \dfrac{x-2}{2x-1}$ 7 $y = \dfrac{x^2}{x^2-4}$ 8 $y = \dfrac{2x^2-2}{x^2+1}$

9 $y = \dfrac{x^2-2x-8}{x^2-9}$ 10 $y = \dfrac{x^3-1}{x^2}$ 11 $y = \dfrac{x^2-x-6}{x-2}$ 12 $y = \dfrac{x^2-4}{x+1}$

13 $y = \dfrac{1}{x^2-4}$ 14 $y = \dfrac{-2}{x^2+1}$ 15 $y = \dfrac{-1}{x^2-x-6}$ 16 $y = \dfrac{3}{(x-2)^2}$

17 $y = \dfrac{-2}{(x+3)^2}$ 18 $y = \dfrac{x+1}{x^2-4x}$

1 At Wurzletown Junction an old lady put her head out of the window and shouted:

"Guard, how long will the journey be from here to Mudville?"
"All the trains take five hours, Ma'am, either way," replied the official.
"And how many trains shall I meet on the way?"

This absurd question tickled the guard, but he was ready with his reply:

"A train leaves Wurzletown for Mudville, and also one from Mudville to Wurzletown, at five minutes past every hour. Right away!"

The old lady induced one of her fellow passengers to work out the answer for her. What is the correct number of trains?

2 Here's a recent twist on an old type of logic puzzle. A logician vacationing in the South Seas finds himself on an island inhabited by the two proverbial tribes of liars and truth-tellers. Members of one tribe always tell the truth, members of the other always lie. He comes to a fork in a road and has to ask a native bystander which branch he should take to reach a village. He has no way of telling whether the native is a truth-teller or a liar. The logician thinks a moment, than asks *one* question only. From the reply he knows which road to take. What question does he ask?

7.12   Summary of Definitions and Properties in Chapter 7

© 1969 King Features Syndicate, Inc.

8

Systems of Equations — Matrices and Determinants

8.1 Introduction

We have observed many times the singular role equations play in solving problems of all sorts. With very few exceptions, we have thus far limited ourselves to a single equation in one variable. Many problems we encounter, however, contain several variables, and, furthermore, the mathematical models for these are usually composed of several equations in these variables. For example, let us consider the following accounting problem.

Problem 8.1.1 A certain company is composed of two types of departments — production and service. As the names imply, the production departments manufacture the products that are sold on the open market, whereas the service departments provide needed services within the company to the other departments. For example, the employment department would serve all departments, including itself, in assuring the highest quality people are found and hired to work for the company.

This company is in business both to serve people and to make a profit for its stockholders. For this reason, the price placed upon the products made by the company must cover the entire cost of production plus a profit. In determining the price of a product we must consider not only the *direct* costs (salaries, wages, and materials) but also the *indirect* costs (services provided by the service departments).

Suppose this company is composed of three service departments (employment, accounting, and research), and two production departments. For identification, we denote the service departments S_1, S_2, and S_3, and the production departments P_1 and P_2. We wish to determine the *total monthly costs* needed to run each of these departments. Since for the moment these costs are unknown, we denote them x_1, x_2, x_3, x_4, and x_5. The direct and indirect costs of each department are identified in Table 8.1.1. Notice the total cost of a service department is allocated on a *percentage* basis among all departments receiving its services.

Table 8.1.1

Dept.	Direct costs	Indirect costs			Total costs
		S_1	S_2	S_3	
S_1	1200	$0.20x_1$	$0.20x_2$	$0.10x_3$	x_1
S_2	800	$0.20x_1$	$0.10x_2$	$0.20x_3$	x_2
S_3	1600	$0.10x_1$	$0.40x_2$	$0.10x_3$	x_3
P_1	3000	$0.35x_1$	$0.20x_2$	$0.35x_3$	x_4
P_2	4000	$0.15x_1$	$0.10x_2$	$0.25x_3$	x_5

The total cost of each department is equal to the *sum* of the direct and indirect costs to run it. This yields, from the table, the following system of equations.

$$x_1 = 1200 + 0.20x_1 + 0.20x_2 + 0.10x_3$$
$$x_2 = 800 + 0.20x_1 + 0.10x_2 + 0.20x_3$$
$$x_3 = 1600 + 0.10x_1 + 0.40x_2 + 0.10x_3$$
$$x_4 = 3000 + 0.35x_1 + 0.20x_2 + 0.35x_3$$
$$x_5 = 4000 + 0.15x_1 + 0.10x_2 + 0.25x_3$$

This is called a linear system of five equations in five unknowns. Notice that the unknowns x_4 and x_5 are expressed in terms of the other three. Thus, *if we could solve the first three equations in these three unknowns,* we could solve the entire system.

The purpose of the next several sections will be to develop techniques by which systems of linear equations in several variables may be solved easily. For this reason, of course, we will delay solving the above system until some of these methods have been developed.

8.2   Solving Systems of Linear Equations — The Diagonalizing Method

The first technique we introduce is called the diagonalizing or Gauss reduction method. Central to the application of this method are the following two properties of equations. First, if both sides of an equation are *multiplied* by a nonzero constant, we obtain an equivalent equation. And second, two equations of a system can be *added* to obtain a third equation equally valid in the system; furthermore, this third equation may replace one or the other (but not both) of the equations used to form it. A new system obtained by either or both of these operations is *equivalent* to the original, that is, it possesses the same solutions.

We now demonstrate the method by several examples.

Example 8.2.1   Solve the following linear system of equations:

$$\begin{aligned} x - 2y + z &= 2 \\ 2x + 2y - z &= 4 \\ x + y &= 0 \end{aligned}$$

Solution   An equivalent system is obtained by keeping the first equation, while determining an alternate second one formed by adding to the present second equation -2 times the first equation. We obtain

$$\begin{array}{r} -2(x - 2y + z) = (-2)(2) \\ 2x + 2y - z = 4 \\ \hline 0 + 6y - 3z = 0 \end{array}$$

With this substitution, the system becomes

$$\begin{aligned} x - 2y + z &= 2 \\ 0 + 6y - 3z &= 0 \\ x + y &= 0 \end{aligned}$$

Next we find an alternate third equation by subtracting from the present third equation the first one. This yields

$$x - 2y + \ z = 2$$
$$0 + 6y - 3z = 0$$
$$0 + 3y - \ z = -2$$

It is important to note in the process presented above that we have reduced the coefficients of x in the second and third equations to zero.

We now take the new second and third equations and do the same but to the variable y. We replace the third equation by one obtained by adding the second equation to -2 times the third. The system now becomes

$$x - 2y + \ z = 2$$
$$0 + 6y - 3z = 0$$
$$0 \quad 0 \ - \ z = 4$$

The system is now diagonalized. In this form the system can be readily solved. Notice that the third equation has only *one* unknown. This we easily solve to obtain $z = -4$. Substituting this value of z into the second equation reduces it to an equation in *one* unknown, which is also easily solved. We have

$$6y - 3(-4) = 0 \Longleftrightarrow y = -2$$

And finally, substituting these values for y and z into the first equation reduces it to an equation in *one* unknown, which is likewise readily solved. We obtain

$$x - 2(-2) + (-4) = 2 \Longleftrightarrow x = 2$$

and the entire system is solved. That is, the choices $x = 2$, $y = -2$, and $z = -4$, satisfy all three equations of the system *simultaneously*.

Just as a solution to an equation in two variables can be expressed as an ordered pair (x, y) so a solution to an equation in three variables can be expressed as an *ordered triplet* (x, y, z). Thus, we may denote the solution of the above system by the triplet $(2, -2, -4)$, and the solution set

$$S = \{(2, -2, -4)\}$$

Example 8.2.2 Solve the following linear system.

$$x - 2y + 3z = 1$$
$$-2x + 5y - 4z = -2$$
$$x - 4y - \ z = 5$$

Solution We begin diagonalizing this system by replacing the second equation with the sum of the present second one and twice the first, and replacing the third by the difference of the first and third equations. By these two steps we obtain the following equivalent system:

$$x - 2y + 3z = 1$$
$$0 + y + 2z = 0$$
$$0 - 2y - 4z = 4$$

Next replace the third equation by the sum of the third and twice the second equation. We obtain

$$x - 2y + 3z = 1$$
$$0 + y + 2z = 0$$
$$0 + 0 + 0 = 4$$

and the diagonalizing process is complete. Notice that the third equation of the system is never true, but is, nevertheless, an *irrevocable* member of the system. Therefore, this system of equations has no solution. Such a system is said to be inconsistent.

Example 8.2.3 Solve the following system.

$$x - 2y + z = 2$$
$$-x - y + 2z = 1$$
$$x - 5y + 4z = 5$$

Solution To diagonalize this system, replace the second equation by the sum of the first and second ones, and the third by the difference of the first and third equations. Applying these steps and simplifying, we obtain

$$x - 2y + z = 2$$
$$0 - 3y + 3z = 3$$
$$0 - 3y + 3z = 3$$

Next replace the third equation by the difference of the second and third ones. This gives

$$x - 2y + z = 2$$
$$0 - 3y + 3z = 3$$
$$0 + 0 + 0 = 0$$

Notice the third equation essentially has been eliminated—it is trivially true and makes *no* contribution to the system. We write the two remaining equations in diagonalized form by taking z to the right-hand side, obtaining the system

$$x - 2y = 2 - z$$
$$0 - y = 1 - z$$

The unknown or variable z now becomes a parameter, which may be chosen arbitrarily. Any choice of z will uniquely determine y, which, along with that same choice of z, will uniquely determine x. For example, let $z = 2$, then

$$-y = 1 - (2) \Longleftrightarrow y = 1$$

Substituting these into the first equation yields

$$x - 2(1) = 2 - (2) \Longleftrightarrow x - 2 = 0 \Longleftrightarrow x = 2$$

Therefore, the ordered triplet $(2, 1, 2)$ satisfies the system. On the other hand, letting $z = -2$, we have

$$-y = 1 - (-2) \Longleftrightarrow y = -3$$

and

$$x - 2(-3) = 2 - (-2) \Longleftrightarrow x + 6 = 4$$
$$\Longleftrightarrow x = -2$$

and we see the ordered triplet $(-2, -3, -2)$ also satisfies the system. It is thus clear that there are *infinitely* many solutions to this system, for we have infinitely many choices available for z. Returning to the system, if we substitute $y = z - 1$ in the first equation, the system can be further simplified as follows:

$$x = 2(z - 1) + 2 - z$$
$$y = z - 1$$

which is equivalent to

$$x = z$$
$$y = z - 1$$

Thus, any ordered triplet of the form

$$(z, z - 1, z)$$

where z is arbitrary, will satisfy the system. Therefore, the solution set becomes

$$S = \{(z, z - 1, z) \mid z \in \mathbf{R}\}$$

We learn from the three examples above that systems of linear equations may have no solution (may be inconsistent), have only one solution, or have infinitely many solutions.

Now let us return to the applied problem with which this chapter was introduced. We were interested in finding the total cost to run each of five departments of a manufacturing firm. It was found that these could be determined if we could solve the following system of linear equations.

$$x_1 = 1200 + 0.20x_1 + 0.20x_2 + 0.10x_3$$
$$x_2 = 800 + 0.20x_1 + 0.10x_2 + 0.20x_3$$
$$x_3 = 1600 + 0.10x_1 + 0.40x_2 + 0.10x_3$$

We apply the diagonalizing method. First rewrite the system in the following form (also for convenience we interchange the first and last equations):

$$-0.10x_1 - 0.40x_2 + 0.90x_3 = 1600$$
$$-0.20x_1 + 0.90x_2 - 0.20x_3 = 800$$
$$0.80x_1 - 0.20x_2 - 0.10x_3 = 1200$$

Now we add to the second equation -2 times the first equation and to the third equation 8 times the first equation. This results in the following equivalent system:

$$-0.10x_1 - 0.40x_2 + 0.90x_3 = 1600$$
$$0 + 1.70x_2 - 2.00x_3 = -2400$$
$$0 - 3.40x_2 + 7.10x_3 = 14000$$

Next we add to the third equation 2 times the second equation, obtaining

$$-0.10x_1 - 0.40x_2 + 0.90x_3 = 1600$$
$$0 + 1.70x_2 - 2.00x_3 = -2400$$
$$0 0 + 3.10x_3 = 9200$$

Now that the system has been diagonalized, we can easily solve for the unknowns. From the third equation, we obtain $x_3 = 2967.74$. Substituting this value for x_3 into the second equation yields $x_2 = 2079.70$. And finally, substituting these two values into the first equation gives $x_1 = 2390.90$.

With these three unknowns determined, we can now find the other two. Recall that the unknowns x_4 and x_5 were expressed in terms of x_1, x_2, and x_3, by the equations

$$x_4 = 3000 + 0.35x_1 + 0.20x_2 + 0.35x_3$$
$$x_5 = 4000 + 0.15x_1 + 0.10x_2 + 0.25x_3$$

Substituting into these equations the values for x_1, x_2, and x_3, obtained above, we obtain $x_4 = 5291.47$ and $x_5 = 5308.54$ and the problem is solved. The monthly costs to run this company are summarized in Table 8.2.1.

Table 8.2.1

Dept.	Direct costs	Indirect costs			Total costs
		S_1	S_2	S_3	
S_1	1200	483.92	417.46	297.42	2390.90
S_2	800	483.92	208.73	297.42	2079.70
S_3	1600	241.86	834.91	297.42	2967.74
P_1	3000	846.86	417.46	1040.97	5291.47
P_2	4000	362.94	208.73	743.55	5308.54

Problem Set I   ■   Skills Development

Find the solution set of the following systems of linear equations. Express your answers as ordered triplets whenever solution sets are non-empty.

1
$$2x - 2y + 3z = 1$$
$$x - 3y - 2z = -9$$
$$x + y + z = 6$$

2
$$3x + y + 4z = 0$$
$$5x + y + 3z = 1$$
$$x - 3y - 4z = 5$$

3
$$2x + 2y + z = 1$$
$$x - y + 6z = 21$$
$$3x + 2y - z = -4$$

4
$$x + y + z = 0$$
$$2x - y - 4z = 15$$
$$x - 2y - z = 7$$

5
$$x + 2y - z = 0$$
$$4x - 3y + z = -5$$
$$3x - y - 2z = 7$$

6
$$x + y + z = 2$$
$$2x - z + t = 3$$
$$3x + 2y + t = 4$$
$$3y + 2z + 2t = 7$$

7
$$x - y + z = 1$$
$$2x + y + z = 2$$
$$-x - 2y - z = -1$$

8
$$x + y + z = 0$$
$$2x - y + z = 1$$
$$-x - 2y - 3z = 2$$

9
$$x - y + z = 1$$
$$2x - y = 4$$
$$-3x + 4y + z = 0$$

10
$$x + y + z = 2$$
$$x - y = 5$$
$$-x + 3y + z = -10$$

11
$$x - 2y - z = 1$$
$$2x - 3y - z = -4$$
$$x - 4y - 3z = 5$$

12
$$x + y - z - 2w = 1$$
$$-x + y + w = 1$$
$$2x + 4y - 3w = 3$$
$$-3x + y + 7z + 7w = -3$$

Problem Set II   ■   Applications

Consider the accounting problems described by the data given in the following tables and solve for the total cost of each department.

1

Dept.	Direct costs	Indirect costs		Total costs
		S_1	S_2	
S_1	1500	$0.10x_1$	$0.20x_2$	x_1
S_2	2000	$0.20x_1$	$0.30x_2$	x_2
P_1	3000	$0.40x_1$	$0.20x_2$	x_3
P_2	4000	$0.10x_1$	$0.20x_2$	x_4
P_3	6000	$0.20x_1$	$0.10x_2$	x_5

2

Dept.	Direct costs	Indirect costs				Total costs
		S_1	S_2	S_3	S_4	
S_1	1500	$0.30x_1$	$0.40x_2$	$0.10x_3$	$0.10x_4$	x_1
S_2	1100	$0.60x_1$	$0.50x_2$	$0.20x_3$	$0.20x_4$	x_2
S_3	600	0	0	$0.20x_3$	$0.10x_4$	x_3
S_4	1800	$0.10x_1$	$0.10x_2$	$0.20x_3$	$0.10x_4$	x_4
P_1	6000	0	0	$0.05x_3$	$0.20x_4$	x_5
P_2	4000	0	0	$0.20x_3$	$0.10x_4$	x_6
P_3	3000	0	0	$0.05x_3$	$0.20x_4$	x_7

The following diagrams represent the current loops of three-window electrical circuits. The electrical currents are represented by I, the resistors by R, and the voltage sources by e. We will not go into circuit theory here, except to say that by Kirchoff's voltage law, the currents I_1, I_2, and I_3 satisfy the three linear equations identified with each diagram. Find the three currents in each problem.

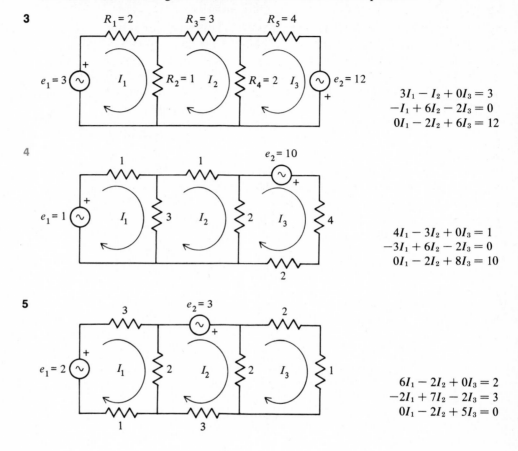

3

$$3I_1 - I_2 + 0I_3 = 3$$
$$-I_1 + 6I_2 - 2I_3 = 0$$
$$0I_1 - 2I_2 + 6I_3 = 12$$

4

$$4I_1 - 3I_2 + 0I_3 = 1$$
$$-3I_1 + 6I_2 - 2I_3 = 0$$
$$0I_1 - 2I_2 + 8I_3 = 10$$

5

$$6I_1 - 2I_2 + 0I_3 = 2$$
$$-2I_1 + 7I_2 - 2I_3 = 3$$
$$0I_1 - 2I_2 + 5I_3 = 0$$

8.4   Matrix Algebra

In the previous sections we learned to solve systems of equations by the diagonalizing method. There are other methods with advantages and disadvantages, depending upon the circumstances. Two in particular are of special importance. One makes use of matrices and the other determinants. Because both of these classes of mathematical objects are used in many mathematical analyses and applications, some to solve systems of equations and others to perform quite distinct, yet vitally important functions, we will briefly study each of these mathematical systems beginning with matrices.

Matrix theory began as a game with no apparent application but has since found widespread use not only in mathematics but also in the physical, biological, and behavioral sciences.

Many problems we face today are very complex and contain several variables influencing their outcome. One of the more significant roles of mathematics in problem solving is to provide a systematic and meaningful way to state problems symbolically; this is usually accomplished by employing a number system. The properties of that number system may then be brought to bear on the mathematical model leading frequently to a solution. This process will be particularly evident in the development and application of the matrix number system.

To introduce the elements of this system we note that in both pure and applied mathematics we encounter situations involving rectangular arrays of numbers. Consider the following example.

Example 8.4.1 A manufacturer produces two models of a certain car—four-door sedans and two-door sedans. The amount of material, labor, and advertising units required to produce and sell each of these models is described in the rectangular array below.

	Four-door sedan	Two-door sedan
Units of material	15	12
Units of labor	10	8
Units of advertisement	4	2

If the units and category positions are held constant, this array can be more simply represented as

$$\begin{bmatrix} 15 & 12 \\ 10 & 8 \\ 4 & 2 \end{bmatrix}$$

Such an array is called a matrix A matrix is a rectangular array of numbers.● Unless otherwise specified, all numbers in a matrix array will be real. Examples of matrices are:

(a) $\begin{bmatrix} 2 & -3 & 4 \\ -1 & 2 & 5 \end{bmatrix}$ **(b)** $\begin{bmatrix} 0 & \sqrt{3} \\ 8 & -4 \end{bmatrix}$ **(c)** $\begin{bmatrix} -3 & 2 & 6 \end{bmatrix}$

● Although there are more precise and sophisticated ways of defining matrices, this definition will do for our purposes.

Matrix arrays are usually enclosed by brackets or parentheses —we will use brackets, as indicated above. The members of a matrix array are called entries or elements. Notice that the elements are arranged in rows and columns. Matrix (a) above has 2 rows and 3 columns, (b), 2 rows and 2 columns, and (c), 1 row and 3 columns. The dimension of a matrix is given as the number of rows by the number of columns, in that order. Thus, the dimensions of matrices (a), (b), and (c) are, respectively, 2×3, 2×2, and 1×3.

Because of their size, matrices can be rather cumbersome to work with, especially if they are to be rewritten several times within the context of a given problem. Thus, we seek shorter, more compact ways of representing them. We will accomplish this in two ways, with two different purposes in mind. The first and simplest notation is to let a matrix be represented by a capital letter as illustrated below.

$$A = \begin{bmatrix} 2 & -1 \\ 4 & 0 \\ 3 & 2 \end{bmatrix}$$

and

$$B = \begin{bmatrix} 2 & 4 & 3 \\ -1 & 0 & 2 \end{bmatrix}$$

The *dimension* of A is 3×2 and it is sometimes denoted $A_{3 \times 2}$; B may be written as $B_{2 \times 3}$. We delay, momentarily, introducing the other abbreviated notation.

We wish to develop an algebra for matrices, just as we did for real and complex numbers. As a notational convenience, therefore, it is desirable to let unidentified constant or variable entries within a matrix be represented by letters of the alphabet. These are denoted by lowercase letters. A general 2×3 matrix may be written

$$A_{2 \times 3} = \begin{bmatrix} a_{11} & a_{12} & a_{13} \\ a_{21} & a_{22} & a_{23} \end{bmatrix}$$

The double subscript on each letter is necessary to identify the position the element occupies within the matrix array. The first or left index indicates the row occupied by the element, and the second indicates the column. Thus the element a_{21} is located in the second row and first column, and, in general, a_{ij} would be the element in the ith row and jth column.

We are now ready to present the second of the two abbreviated notational forms for a matrix. We take an arbitrary element from the array, say, a_{ij}, and let it represent the entire array. This is done by placing parentheses around it, as illustrated below.

$$A_{2 \times 3} = \begin{bmatrix} a_{11} & a_{12} & a_{13} \\ a_{21} & a_{22} & a_{23} \end{bmatrix} = (a_{ij})$$

In general, we have

$$A_{m \times n} = \begin{bmatrix} a_{11} & a_{12} & a_{13} & \cdots & a_{1n} \\ a_{21} & a_{22} & a_{23} & \cdots & a_{2n} \\ \vdots & & & & \\ a_{m1} & a_{m2} & a_{m3} & \cdots & a_{mn} \end{bmatrix} = (a_{ij})$$

where $i = 1, 2, 3, \ldots, m$ and $j = 1, 2, 3, \ldots, n$.

With matrices defined and suitable notation introduced, we are prepared to develop an algebra for matrices. We begin by defining the relation equals and the operations of addition, subtraction, and multiplication.

DEFINITION 8.4.1 Two matrices A and B are equal, denoted $A = B$, if they have the *same* dimensions and if elements in corresponding positions are *equal*. Symbolically, this may be written

$$A = B \Longleftrightarrow [a_{ij} = b_{ij} \text{ for each } i, j]$$

Example 8.4.2

(a) $\begin{bmatrix} 2 & -3 \\ 4 & 0 \end{bmatrix} = \begin{bmatrix} \sqrt{4} & -4+1 \\ \frac{8}{2} & 3-3 \end{bmatrix}$

(b) $\begin{bmatrix} 2 & -3 \\ 4 & 0 \end{bmatrix} \neq \begin{bmatrix} -3 & 2 \\ 0 & 4 \end{bmatrix}$

(c) $\begin{bmatrix} 2 & -3 \\ 4 & 0 \end{bmatrix} \neq \begin{bmatrix} 2 & -3 & 0 \\ 4 & 0 & 0 \end{bmatrix}$

In (b), we have $a_{11} = 2$ and $b_{11} = -3$, and hence $a_{11} \neq b_{11}$. If the *corresponding* entries of any position are not equal, the matrices are not equal. In (c), the dimension of the matrix on the left is 2×2, whereas the one on the right is 2×3. Hence, they are not equal. They *must* be the same size.

Now let us define the operation of addition for matrices.

DEFINITION 8.4.2 Let $A = (a_{ij})$ and $B = (b_{ij})$ be two matrices of the *same* size. Then the sum of A and B, denoted $A + B$, is defined as

$$(a_{ij}) + (b_{ij}) = (a_{ij} + b_{ij})$$

for all i and j. That is, you add corresponding elements.

Example 8.4.3

(a) $\begin{bmatrix} 1 & 2 & -3 \\ 0 & -1 & 5 \end{bmatrix} + \begin{bmatrix} -2 & 3 & 4 \\ 1 & 2 & -6 \end{bmatrix}$

$= \begin{bmatrix} 1 + (-2) & 2 + 3 & -3 + 4 \\ 0 + 1 & -1 + 2 & 5 + (-6) \end{bmatrix} = \begin{bmatrix} -1 & 5 & 1 \\ 1 & 1 & -1 \end{bmatrix}$

(b) $\begin{bmatrix} 1 & 2 & -3 \\ 0 & -1 & 5 \end{bmatrix} + \begin{bmatrix} -2 & 3 \\ 1 & 2 \end{bmatrix} \neq \begin{bmatrix} -1 & 5 & -3 \\ 1 & 1 & 5 \end{bmatrix}$

That is, one cannot assume a third column with zero entries to exist in the second matrix above so that addition may be performed. Addition is *not defined* for matrices that do not possess the same dimensions.

We wish now to investigate some of the properties of matrix addition, particularly the field properties.

THEOREM 8.4.1 Let A, B, C, be any three matrices of the same dimensions with real entries. Then

I $A + B$ is a matrix with real entries. (Matrices of the same dimensions are closed under addition.)

II $(A + B) + C = A + (B + C)$ (Associative Law)

III $A + B = B + A$ (Commutative Law)

IV There exists an *additive identity* matrix of the same dimensions, which we denote 0 and call a zero matrix, such that

$$A + 0 = A$$

Every element of a zero matrix is a *real zero*. That is

$$0 = (a_{ij})$$

where $a_{ij} = 0$ for each i, j.

V For each matrix A there exists an *additive inverse* matrix, denoted $-A$, such that

$$A + [-A] = 0$$

where 0 is the appropriate zero matrix and where

$$-A = -(a_{ij}) = (-a_{ij})$$

for each i, j.

You will recognize these five properties as being the same as the first five field axioms that characterized the system of real and complex numbers. This is very significant, for it means that all properties of addition that follow from these five field axioms (many identified and proved in Chapter 2) must hold equally well in the matrix number system. Again, as with complex numbers, this is one of the important advantages of developing axiomatic structures. When a number system is created and satisfies certain axioms of a previously developed system, all of the properties discovered from the axioms in that system must hold equally well in the newer system. Thus, under addition, we may manipulate matrices just as we do real numbers.

Because additive inverses exist for matrices, we may define the operation of subtraction, just as we have done for real and complex numbers.

DEFINITION 8.4.3 If A and B are any two matrices of the same dimensions, then the difference of A and B, in the stated order and denoted $A - B$, is the *sum* of A and $-B$. Symbolically, we have

$$A - B = A + [-B]$$

Example 8.4.4 Let $A = \begin{bmatrix} 2 & -3 \\ 4 & 6 \end{bmatrix}$ and $B = \begin{bmatrix} -1 & 3 \\ 0 & 5 \end{bmatrix}$. Then

$$\begin{bmatrix} 2 & -3 \\ 4 & 6 \end{bmatrix} - \begin{bmatrix} -1 & 3 \\ 0 & 5 \end{bmatrix} = \begin{bmatrix} 2 & -3 \\ 4 & 6 \end{bmatrix} + \left(- \begin{bmatrix} -1 & 3 \\ 0 & 5 \end{bmatrix} \right)$$

$$\underset{\text{T8.4.1(V)}}{=\!=\!=} \begin{bmatrix} 2 & -3 \\ 4 & 6 \end{bmatrix} + \begin{bmatrix} 1 & -3 \\ 0 & -5 \end{bmatrix} \qquad \text{T8.4.1(V)} \quad -(a_{ij}) = (-a_{ij})$$

$$\underset{\text{D8.4.2}}{=\!=\!=} \begin{bmatrix} 3 & -6 \\ 4 & 1 \end{bmatrix} \qquad\qquad\qquad \text{D8.4.2} \quad (a_{ij}) + (b_{ij}) = (a_{ij} + b_{ij})$$

Example 8.4.5 Apply the addition law of equality and other properties to solve the following matrix equation:

$$X + \begin{bmatrix} -1 & 2 \\ 4 & 3 \end{bmatrix} = \begin{bmatrix} 0 & 5 \\ -2 & 4 \end{bmatrix}$$

This equation is equivalent to

$$X = \begin{bmatrix} 0 & 5 \\ -2 & 4 \end{bmatrix} - \begin{bmatrix} -1 & 2 \\ 4 & 3 \end{bmatrix} = \begin{bmatrix} 0 & 5 \\ -2 & 4 \end{bmatrix} + \begin{bmatrix} 1 & -2 \\ -4 & -3 \end{bmatrix} = \begin{bmatrix} 1 & 3 \\ -6 & 1 \end{bmatrix}$$

To see how a matrix proof is frequently handled, we will prove Theorem 8.4.1, Part II.

Proof

$$[A + B] + C = [(a_{ij}) + (b_{ij})] + (c_{ij})$$

D8.4.2 $\quad (a_{ij}) + (b_{ij}) = (a_{ij} + b_{ij})$

A+ $\quad (a + b) + c = a + (b + c)$

$$\underset{\text{D8.4.2}}{=} (a_{ij} + b_{ij}) + (c_{ij})$$

$$\underset{\text{D8.4.2}}{=} ([a_{ij} + b_{ij}] + c_{ij})$$

$$\underset{\text{A+}}{=} (a_{ij} + [b_{ij} + c_{ij}])$$

$$\underset{\text{D8.4.2}}{=} (a_{ij}) + (b_{ij} + c_{ij})$$

$$\underset{\text{D8.4.2}}{=} (a_{ij}) + [(b_{ij}) + (c_{ij})]$$

$$= A + [B + C]$$

$$\implies [A + B] + C = A + [B + C] \qquad \text{QED}$$

(The proofs of Parts I, III, IV, V are to be done as exercises.)

Let us now turn our attention to multiplication. We shall define two kinds of products because we are dealing with two kinds of numbers — matrix and real.

To motivate our first definition for multiplication, we return to our production matrix of automobiles vs. material, labor, and advertising, with which we introduced this section. Recall, we have the matrix

$$A = \begin{array}{c} \\ \\ \end{array} \begin{array}{cc} \text{four-door} & \text{two-door} \\ \text{sedan} & \text{sedan} \end{array}$$

$$A = \begin{bmatrix} 15 & 12 \\ 10 & 8 \\ 4 & 2 \end{bmatrix} \begin{array}{l} \text{units of material} \\ \text{units of labor} \\ \text{units of advertisement} \end{array}$$

This matrix represents the number of units of material, labor, and advertisement required to produce one four-door sedan and one two-door sedan. It is clear that a matrix representing the production of five automobiles of each category would be the sum of five of these matrices; that is $[A + A + A + A + A]$ or more simply $5 \cdot A$. But this means

$$5 \cdot \begin{bmatrix} 15 & 12 \\ 10 & 8 \\ 4 & 2 \end{bmatrix} = \overbrace{\begin{bmatrix} 15 & 12 \\ 10 & 8 \\ 4 & 2 \end{bmatrix} + \cdots + \begin{bmatrix} 15 & 12 \\ 10 & 8 \\ 4 & 2 \end{bmatrix}}^{\text{five terms}}$$

D8.4.2 $\quad (a_{ij}) + (b_{ij}) = (a_{ij} + b_{ij})$

$$\underset{\text{D8.4.2}}{=} \begin{bmatrix} 5 \cdot 15 & 5 \cdot 12 \\ 5 \cdot 10 & 5 \cdot 8 \\ 5 \cdot 4 & 5 \cdot 2 \end{bmatrix} = \begin{bmatrix} 75 & 60 \\ 50 & 40 \\ 20 & 10 \end{bmatrix}$$

We shall take this lead and define the product of a matrix and a real number (now to be called a *scalar*) as follows:

DEFINITION 8.4.4 If c is a *scalar* and A is a matrix, then the product of c and A, denoted cA and called a scalar product, is a matrix each of whose entries is the product of c and the corresponding entry of A. That is,

$$cA = c(a_{ij}) = (ca_{ij})$$

for each i, j.

Example 8.4.6 Let $c = 3$ and $A = \begin{bmatrix} -1 & 2 & 4 \\ 3 & 0 & -5 \end{bmatrix}$. Then

$$3 \cdot \begin{bmatrix} -1 & 2 & 4 \\ 3 & 0 & -5 \end{bmatrix} \overset{\text{D8.4.4}}{=\!=\!=} \begin{bmatrix} 3(-1) & 3 \cdot 2 & 3 \cdot 4 \\ 3 \cdot 3 & 3 \cdot 0 & 3 \cdot (-5) \end{bmatrix}$$

$$=\!= \begin{bmatrix} -3 & 6 & 12 \\ 9 & 0 & -15 \end{bmatrix}$$

The product operation of a scalar with a matrix satisfies the field axioms and other properties dealing with multiplication, as indicated by the following theorem.

THEOREM 8.4.2 Let A and B be two matrices of the same dimensions and a and b two scalars. Then

I aA is a matrix   (Closure Law)

II $(ab)A = a(bA)$ (Associative Law)

III   **(a)** $(a + b)A = aA + bA$ (Distributive Law)
          (b) $a(A + B) = aA + aB$ (Distributive Law)

IV   **(a)** $1A = A$ (Multiplicative Identity)
          (b) $(-1)A = -A$ (Additive Inverse Property)

V   **(a)** $0A = 0$ (Additive Identity Property under Multiplication)
          (b) $a0 = 0$ (Additive Identity Property under Multiplication)

Now let us turn to the task of defining the product of two matrices. This, too, we shall motivate by reasoning with our production matrix.

$$
\begin{array}{cc}
\text{four-door} & \text{two-door} \\
\text{sedan} & \text{sedan}
\end{array}
$$

$$
A = \begin{bmatrix} 15 & 12 \\ 10 & 8 \\ 4 & 2 \end{bmatrix} \begin{array}{l} \text{units of material} \\ \text{units of labor} \\ \text{units of advertisement} \end{array}
$$

Suppose the cost of each unit of material is $120.00, each unit of labor $160.00, and each unit of advertisement $20.00. This information we represent by the row matrix (sometimes referred to as a vector):

$$U = [120 \quad 160 \quad 20]$$

We wish to determine the cost of each model. First, it is clear that the cost of the two models may be represented by

$$
\begin{array}{ll}
120 \cdot 15 + 160 \cdot 10 + 20 \cdot 4 & \text{(four-door)} \\
120 \cdot 12 + 160 \cdot 8 \ + 20 \cdot 2 & \text{(two-door)}
\end{array}
$$

It would be notationally convenient if this result could be represented as the product of the cost matrix U and the production matrix A, that is, $U \cdot A$. This we accomplish by *defining it so*. Thus

$$
[120 \quad 160 \quad 20] \cdot \begin{bmatrix} 15 & 12 \\ 10 & 8 \\ 4 & 2 \end{bmatrix} = \begin{bmatrix} 120 \cdot 15 + 160 \cdot 10 + 20 \cdot 4 \\ 120 \cdot 12 + 160 \cdot 8 + 20 \cdot 2 \end{bmatrix}
$$

$$
= \begin{bmatrix} 3480 \\ 2760 \end{bmatrix}
$$

Thus the cost to produce and sell a four-door sedan is $3480.00, whereas the cost for a two-door model is $2760.00.

Notice that the dimensions of these matrices are *not* the same, that is, U is 1×3 and A is 3×2. However, the number of columns of U and the number of rows of A are *equal*. This will be significant in our definition. We define a matrix product as follows:

DEFINITION 8.4.5 Let A be an $m \times p$ matrix and B a $p \times n$ matrix. Then the product of A and B in the stated order, denoted $A \cdot B$ or just AB, is an $m \times n$ matrix C where

$$c_{ij} = a_{i1}b_{1j} + a_{i2}b_{2j} + a_{i3}b_{3j} + \cdots + a_{ip}b_{pj}$$

Matrices for which multiplication is defined, that is, the number of columns of the left-hand factor A *equals* the number of rows in the right-hand factor B, are said to be conformable.

Note that this restriction on matrix multiplication does not limit matrix products to those of the same dimensions. That is, if A is 2×3 and B is 3×4, then the product $A_{2\times3} \cdot B_{3\times4}$ exists as a matrix whose dimension is 2×4.

Let us now carefully examine the manner in which a product is performed. Notice the element in the i,jth place of the product, denoted c_{ij}, is

all entries in the ith row of A

$$c_{ij} = a_{i1}b_{1j} + a_{i2}b_{2j} + \cdots + a_{ip}b_{pj}$$

all entries in the jth column of B

The left-hand factors in the expression above are the elements, in order from left to right, of the ith row of A, whereas the right-hand factors are the elements, from top to bottom, of the jth column of B. Thus, to find the element in the i,jth place of a product AB, one simply takes the ith row of A, the jth column of B, and computes the sum of the products of their respective elements. This is, of course, why there must be as many rows in A as there are columns in B for matrix multiplication to be meaningful, as defined above.

The next two schematic diagrams will help you to visualize the process of performing matrix multiplication.

$$\begin{bmatrix} 2 & -1 & 1 \\ -3 & 4 & 0 \end{bmatrix} \cdot \begin{bmatrix} 1 & 4 \\ -2 & 1 \\ 3 & 2 \end{bmatrix} = \begin{bmatrix} 7 & 9 \\ -11 & -8 \end{bmatrix}$$

$$(-3) \cdot 1 + 4 \cdot (-2) + 0 \cdot 3 = -11$$

In general

$$\begin{bmatrix} a_{11} & a_{12} & \cdots & a_{1n} \\ & & & \\ a_{i1} & a_{i2} & \cdots & a_{in} \\ & & & \\ a_{m1} & a_{m2} & \cdots & a_{mn} \end{bmatrix} \cdot \begin{bmatrix} b_{11} & \cdots & b_{1j} & \cdots & b_{1k} \\ b_{21} & \cdots & b_{2j} & \cdots & b_{2k} \\ & & & & \\ b_{n1} & \cdots & b_{nj} & \cdots & b_{nk} \end{bmatrix} = \begin{bmatrix} c_{11} & \cdots & c_{1j} & \cdots & c_{1k} \\ & & & & \\ c_{i1} & \cdots & c_{ij} & \cdots & c_{ik} \\ & & & & \\ c_{m1} & \cdots & c_{mj} & \cdots & c_{mk} \end{bmatrix}$$

$$a_{i1}b_{1j} + a_{i2}b_{2j} + \cdots + a_{in}b_{nj} = c_{ij}$$

Example 8.4.7   Let $A = \begin{bmatrix} 1 & 2 \\ -3 & 0 \\ 5 & -2 \end{bmatrix}$ and $B = \begin{bmatrix} -1 & 2 \\ 5 & 1 \end{bmatrix}$. Then

$$\begin{bmatrix} 1 & 2 \\ -3 & 0 \\ 5 & -2 \end{bmatrix} \cdot \begin{bmatrix} -1 & 2 \\ 5 & 1 \end{bmatrix} = \begin{bmatrix} 1 \cdot (-1) + 2 \cdot 5 & 1 \cdot 2 + 2 \cdot 1 \\ -3 \cdot (-1) + 0 \cdot 5 & -3 \cdot 2 + 0 \cdot 1 \\ 5 \cdot (-1) + (-2) \cdot 5 & 5 \cdot 2 + (-2) \cdot 1 \end{bmatrix}$$

$$= \begin{bmatrix} 9 & 4 \\ 3 & -6 \\ -15 & 8 \end{bmatrix}$$

Example 8.4.8

$$\begin{bmatrix} 1 & -2 \\ 0 & -1 \end{bmatrix} \cdot \begin{bmatrix} 1 & -3 \\ -1 & 0 \end{bmatrix} = \begin{bmatrix} (1)(1) + (-2)(-1) & (1)(-3) + (-2)(0) \\ (0)(1) + (-1)(-1) & (0)(-3) + (-1)(0) \end{bmatrix}$$

$$= \begin{bmatrix} 3 & -3 \\ 1 & 0 \end{bmatrix}$$

However unusual this definition for multiplication at first may seem, it is one of the main reasons matrices are so useful in both theoretical and applied mathematics. This will become clearer as we apply matrix techniques to systems of linear equations and to other applied problems later in this chapter. We observed earlier that the five field axioms dealing with the operation of addition hold for matrices. With matrix multiplication now defined, we wish to determine which of the remaining six field axioms regarding multiplication hold for matrices. Three of the six will be taken care of by the following theorem.

THEOREM 8.4.3   If $A_{m \times p}$, $B_{p \times q}$, $C_{q \times n}$, $D_{q \times n}$ and $E_{n \times r}$ are matrices with real entries and dimensions as indicated, then:

I $A \cdot B$ exists as an $m \times q$ matrix (*Closure*● for conformable matrices)

II $(A \cdot B) \cdot C = A \cdot (B \cdot C)$ (*Associative Law*)

III   **(a)** $B[C + D] = BC + BD$ (*Left-Hand Distributive Law*)
(b) $[C + D]E = CE + DE$ (*Right-Hand Distributive Law*)

● We are using *closure* here rather loosely since the dimensions of the product $A \cdot B$ would not necessarily correspond with those of A or B and hence would not be in the same dimension class as either.

It is important to note in the above expressions that the dimensions of the matrices were critical for both operations of addition and multiplication. When these operations are meaningful (same dimensions for addition and conformable for multiplication), we have the closure, associative, and distributive laws holding for matrices.

Now let us consider the possibility of *multiplicative identity matrices*. Recall that the multiplicative identity element in the real number system, denoted 1, possessed the property

$$1 \cdot a = a \cdot 1 = a$$

If such an identity matrix or matrices were to exist in the matrix number system, we would expect that the product of a matrix A and an appropriately dimensioned identity matrix I, would equal A; that is

$$A \cdot I = A$$

or

$$I \cdot A = A$$

That such identity matrices do exist will now be shown. However, before presenting them, we introduce the following terminology. If A is an $m \times m$ (square) matrix, then the elements of A with positional subscripts the same (same row and column) are said to be on the main diagonal. For example, suppose

$$A = \begin{bmatrix} 1 & -2 \\ 3 & 4 \end{bmatrix}$$

Then $a_{11} = 1$ and $a_{22} = 4$ are on the main diagonal. In general, we have

THEOREM 8.4.4 If A is an $m \times n$ matrix, then there exist $m \times m$ and $n \times n$ multiplicative identity matrices, denoted $I_{m \times m}$ and $I_{n \times n}$, respectively, such that

$$I_{m \times m} A_{m \times n} = A_{m \times n}$$

and

$$A_{m \times n} I_{n \times n} = A_{m \times n}$$

where $I_{p \times p}$, $p \in N$, is a square matrix each of whose elements on the main diagonal is 1 and all others are 0.

Notice that when A is not square, the left-hand and right-hand multiplicative identity matrices have different dimensions. This is required, of course, for the products to be conformable.

Example 8.4.9   Let $A = \begin{bmatrix} 1 & 2 & 3 \\ 2 & 0 & 4 \end{bmatrix}$. Then we have

$$I_{2\times2} = \begin{bmatrix} 1 & 0 \\ 0 & 1 \end{bmatrix}$$

and

$$I_{3\times3} = \begin{bmatrix} 1 & 0 & 0 \\ 0 & 1 & 0 \\ 0 & 0 & 1 \end{bmatrix}$$

Thus, multiplying on the left by $I_{2\times2}$ leaves A unchanged:

$$\begin{bmatrix} 1 & 0 \\ 0 & 1 \end{bmatrix} \cdot \begin{bmatrix} 1 & 2 & 3 \\ 2 & 0 & 4 \end{bmatrix} = \begin{bmatrix} 1\cdot1+0\cdot2 & 1\cdot2+0\cdot0 & 1\cdot3+0\cdot4 \\ 0\cdot1+1\cdot2 & 0\cdot2+1\cdot0 & 0\cdot3+1\cdot4 \end{bmatrix}$$

$$= \begin{bmatrix} 1 & 2 & 3 \\ 2 & 0 & 4 \end{bmatrix}$$

whereas multiplying with $I_{3\times3}$ on the right likewise leaves A unchanged:

$$\begin{bmatrix} 1 & 2 & 3 \\ 2 & 0 & 4 \end{bmatrix} \cdot \begin{bmatrix} 1 & 0 & 0 \\ 0 & 1 & 0 \\ 0 & 0 & 1 \end{bmatrix}$$

$$= \begin{bmatrix} 1\cdot1+2\cdot0+3\cdot0 & 1\cdot0+2\cdot1+3\cdot0 & 1\cdot0+2\cdot0+3\cdot1 \\ 2\cdot1+0\cdot0+4\cdot0 & 2\cdot0+0\cdot1+4\cdot0 & 2\cdot0+0\cdot0+4\cdot1 \end{bmatrix}$$

$$= \begin{bmatrix} 1 & 2 & 3 \\ 2 & 0 & 4 \end{bmatrix}$$

You may have observed in the properties listed above that the commutative law of multiplication is noticeably absent. This should not surprise you, considering the way multiplication was defined. For one thing, the product $A_{m\times p} \cdot B_{p\times q}$ is not even conformable in the reverse order, if $m \neq q$. And furthermore, even if $m = q$, it would be rather rare that the element in the i,jth place of $A \cdot B$ (obtained using the ith row of A and jth column of B) would be equal to the element in the corresponding i,jth place of the commuted product

$B \cdot A$ (obtained using the ith row of B and the jth column of A), especially for every i and j pair. No wonder the commutative law of multiplication does not hold in general for matrices.

You may have noticed also that the multiplicative inverse of a matrix, if such exists, has not been introduced. This, too, is not an oversight. It is not, however, because these matrices do not exist under certain circumstances but rather because a convenient formula by which multiplicative inverses may be determined contains a mathematical expression called a determinant. Determinants will be treated in Section 8.6, and we shall postpone until after this section the presentation of multiplicative inverses.

Problem Sets 8.5

Problem Set I ■ Reading Comprehension

Determine whether statements 1 through 9 are true or false and give reasons for your responses.

1 An $m \times n$ matrix is one with m rows and n columns.
2 The element a_{23} of a matrix A is located in the third row and second column.
3 Two matrices $A_{m \times n}$ and $B_{p \times q}$ are equal if $m = p$ and $n = q$.
4 Of the field properties, the commutative law of addition does not hold in general for matrices.
5 A zero matrix is one whose elements are all zero.
6 The product of a matrix and a scalar is performed by multiplying every element of any one row or any one column of the matrix by the scalar.
7 The matrix product $A \cdot B$ is only defined if the number of rows in A corresponds with the number of columns in B.
8 The element in the i, jth place of a matrix product $A \cdot B$ is the product of the element a_{ij} and the element b_{ij}.
9 A multiplicative identity matrix has for each of its elements the real multiplicative identity 1.

Problem Set II ■ Skills Development

Write each of the following sums as a single matrix:

1 $\begin{bmatrix} 1 & -2 \\ 3 & 4 \end{bmatrix} + \begin{bmatrix} 2 & -3 \\ -4 & -1 \end{bmatrix}$ 2 $\begin{bmatrix} 0 & -4 \\ 3 & 5 \end{bmatrix} + \begin{bmatrix} -2 & 3 \\ 1 & -2 \end{bmatrix}$

3 $\begin{bmatrix} -1 & -3 & 4 \\ 2 & 0 & 3 \end{bmatrix} + \begin{bmatrix} 1 & 2 & -3 \\ -4 & -2 & 1 \end{bmatrix}$ 4 $\begin{bmatrix} 1 \\ -2 \\ 3 \end{bmatrix} + \begin{bmatrix} 2 \\ 3 \\ -5 \end{bmatrix}$

5 $\begin{bmatrix} -1 & 2 & 3 \\ 4 & -2 & 0 \end{bmatrix} + \begin{bmatrix} 2 & 3 & 4 \\ -5 & 6 & 8 \end{bmatrix}$ **6** $[3 \quad 4 \quad -2] + [2 \quad -1 \quad 4]$

Write each of the following differences as a single matrix:

7 $\begin{bmatrix} 1 & -2 \\ 3 & 4 \end{bmatrix} - \begin{bmatrix} 2 & -3 \\ -4 & -1 \end{bmatrix}$ **8** $\begin{bmatrix} 0 & -4 \\ 3 & 5 \end{bmatrix} - \begin{bmatrix} -2 & 3 \\ 1 & -2 \end{bmatrix}$

9 $\begin{bmatrix} -1 & -3 & 4 \\ 2 & 0 & 3 \end{bmatrix} - \begin{bmatrix} 1 & 2 & -3 \\ -4 & -2 & 1 \end{bmatrix}$ **10** $\begin{bmatrix} 1 \\ -2 \\ 3 \end{bmatrix} - \begin{bmatrix} 2 \\ 3 \\ -5 \end{bmatrix}$

11 $\begin{bmatrix} -1 & 2 & 3 \\ 4 & -2 & 1 \end{bmatrix} - \begin{bmatrix} 2 & 4 & -5 \\ -3 & 1 & 4 \end{bmatrix}$

12 $[3 \quad 4 \quad -2] - [2 \quad -1 \quad 4]$

Write the following products as a single matrix:

13 $\begin{bmatrix} 1 & -1 \\ 2 & 3 \end{bmatrix} \cdot \begin{bmatrix} 2 & -3 \\ 4 & -1 \end{bmatrix}$ **14** $\begin{bmatrix} 0 & -1 \\ 2 & 3 \end{bmatrix} \cdot \begin{bmatrix} -3 \\ 4 \end{bmatrix}$ **15** $\begin{bmatrix} -1 \\ -2 \end{bmatrix} \cdot [2 \quad 3 \quad 4]$

16 $[-1 \quad 2 \quad 3] \cdot \begin{bmatrix} 2 \\ -4 \\ 5 \end{bmatrix}$ **17** $\begin{bmatrix} 2 & -1 \\ 3 & -4 \\ -1 & 3 \end{bmatrix} \cdot \begin{bmatrix} 1 & 0 \\ 0 & 1 \end{bmatrix}$

18 $\begin{bmatrix} 1 & 2 & -3 \\ -2 & -1 & 4 \end{bmatrix} \cdot \begin{bmatrix} 1 & 0 & 0 \\ 0 & 1 & 0 \\ 0 & 0 & 1 \end{bmatrix}$ **19** $\begin{bmatrix} 1 & 2 & -3 \\ -2 & -1 & 4 \end{bmatrix} \cdot \begin{bmatrix} 0 & 0 & 1 \\ 0 & 1 & 0 \\ 1 & 0 & 0 \end{bmatrix}$

20 $\begin{bmatrix} 1 & -1 & 0 \\ 0 & 1 & 1 \\ 1 & 1 & 1 \end{bmatrix} \cdot \begin{bmatrix} -1 & -1 & 1 \\ 2 & 1 & 1 \\ -1 & 1 & -1 \end{bmatrix}$ **21** $\begin{bmatrix} 1 & 1 & 0 \\ 1 & 0 & 1 \\ 0 & 1 & 1 \end{bmatrix} \cdot \begin{bmatrix} 1 & 1 & 0 \\ 1 & 0 & 1 \\ 0 & 1 & 1 \end{bmatrix}$ **22** $2 \begin{bmatrix} -1 & 3 \\ 4 & -2 \end{bmatrix}$

23 $-3 \begin{bmatrix} 2 & -1 & 3 \\ -4 & -2 & 1 \end{bmatrix}$ **24** $4 \begin{bmatrix} 1 & 2 \\ -1 & -3 \\ 4 & -5 \end{bmatrix}$

25 $\begin{bmatrix} 1 & -2 \\ -1 & 1 \end{bmatrix} \cdot \left(\begin{bmatrix} 2 & 3 \\ -4 & 1 \end{bmatrix} + \begin{bmatrix} -1 & -5 \\ 3 & 1 \end{bmatrix} \right)$ **26** $\begin{bmatrix} -1 & 2 \\ 3 & 0 \end{bmatrix} \cdot \left(\begin{bmatrix} 2 & 3 \\ 4 & -5 \end{bmatrix} - \begin{bmatrix} 3 & -4 \\ 1 & 0 \end{bmatrix} \right)$

27 $\begin{bmatrix} 1 & 1 \\ -1 & 1 \end{bmatrix} \cdot \left(\begin{bmatrix} 2 & 3 \\ 4 & -5 \end{bmatrix} - \begin{bmatrix} 1 & 0 \\ 3 & -2 \end{bmatrix} \right)$

Solve the following matrix equations:

28 $\begin{bmatrix} -1 & 2 \\ 4 & -5 \end{bmatrix} - X = \begin{bmatrix} 3 & 4 \\ 2 & -6 \end{bmatrix}$ **29** $\begin{bmatrix} 1 & -1 \\ 1 & 0 \end{bmatrix} - X = \begin{bmatrix} 2 & -1 \\ -2 & 1 \end{bmatrix}$

Solve for x and y:

Example $\begin{bmatrix} x-1 \\ y+2 \end{bmatrix} = \begin{bmatrix} 3 \\ -4 \end{bmatrix}$

Solution

$\begin{bmatrix} x-1 \\ y+2 \end{bmatrix} = \begin{bmatrix} 3 \\ -4 \end{bmatrix} \xLeftrightarrow{\text{D8.4.1}} x-1=3 \quad \text{and} \quad y+2=-4$

$\Longleftrightarrow x=4 \quad \text{and} \quad y=-6$

D8.4.1
$(a_{ij}) = (b_{ij}) \Longleftrightarrow a_{ij} = b_{ij}$

30 $\begin{bmatrix} 2-x \\ y+3 \end{bmatrix} = \begin{bmatrix} 3 \\ -2 \end{bmatrix}$

31 $\begin{bmatrix} x \\ y-2 \end{bmatrix} = \begin{bmatrix} 3 \\ -4 \end{bmatrix}$

32 $\begin{bmatrix} x-2y \\ -x+y \end{bmatrix} = \begin{bmatrix} 2 \\ 3 \end{bmatrix}$

33 $\begin{bmatrix} x+y \\ x-3y \end{bmatrix} = \begin{bmatrix} 1 \\ 0 \end{bmatrix}$

34 $\begin{bmatrix} x & y \\ x^2 & y^2 \end{bmatrix} = \begin{bmatrix} -1 & 1 \\ 1 & 1 \end{bmatrix}$

35 $\begin{bmatrix} x^2 & y \\ y^2 & x \end{bmatrix} = \begin{bmatrix} -1 & 2 \\ 4 & 1 \end{bmatrix}$

Problem Set III ■ *Theoretical Developments*

1 Prove the relation of *equality* for matrices is an equivalence relation. That is, if A, B, C are $m \times n$ matrices then:

    I $A = A$ (Reflexive Property)
    II $A = B \Longrightarrow B = A$ (Symmetric Property)
    III $[A = B \quad \text{and} \quad B = C] \Longrightarrow A = C$ (Transitive Property)

2 Prove $0_{2\times2}$ is the only additive identity matrix of dimension 2×2.
3 Prove $-A_{2\times3}$ is the only additive inverse of the matrix $A_{2\times3}$.

Prove the following theorems:

4 Theorem 8.4.1
    a Part I   **b** Part III   **c** Part IV   **d** Part V
5 Theorem 8.4.2
    a Part I   **b** Part II   **c** Part III(b)   **d** Part IV(a)
    e Part IV(b)   **f** Part V(a)   **g** Part V(b)
6 Theorem 8.4.3
    a Part I   **b** Part II   **c** Part III(b)
7 Theorem 8.4.4
8 If $A = \begin{bmatrix} 1 & -1 \\ 2 & 3 \end{bmatrix}$ and $B = \begin{bmatrix} 2 & -1 \\ 1 & 4 \end{bmatrix}$, show
    a $(A-B)^2 \neq A^2 - 2AB + B^2$, and explain why.
    b $(A-B)(A+b) \neq A^2 - B^2$, and explain why.
9 Show that for the matrices

$$A = \begin{bmatrix} a & b \\ -b & a \end{bmatrix} \quad \text{and} \quad B = \begin{bmatrix} c & d \\ -d & c \end{bmatrix}$$

the commutative law of multiplication holds. Does this contradict the statement that the commutative law does not hold in general for matrices?

10 If $A = \begin{bmatrix} -2 & 3 \\ 4 & -1 \end{bmatrix}$, $B = \begin{bmatrix} -1 & 4 \\ 0 & 3 \end{bmatrix}$, and $C = \begin{bmatrix} 2 & 4 \\ -6 & 1 \end{bmatrix}$, test the statements:

    a $(AB)C = A(BC)$ **b** $A \cdot [B + C] = A \cdot B + A \cdot C$ **c** $A \cdot B \neq B \cdot A$

11 If possible, find a matrix A so that

    a $A \cdot \begin{bmatrix} 2 & -3 \\ 0 & 1 \end{bmatrix} = \begin{bmatrix} -2 & 5 \\ 8 & -7 \end{bmatrix}$ **b** $A \cdot \begin{bmatrix} 2 & -3 \\ 4 & -6 \end{bmatrix} = \begin{bmatrix} 1 & 0 \\ 0 & 1 \end{bmatrix}$

    c $A \cdot \begin{bmatrix} 1 & -2 \\ 3 & 4 \end{bmatrix} = \begin{bmatrix} 1 & 0 \\ 0 & 1 \end{bmatrix}$

    (HINT: Let $A = \begin{bmatrix} a & b \\ c & d \end{bmatrix}$. Then perform the indicated operations and use the definition of equality.)

12 Solve for x in the following:

    a $[x \quad 2] \cdot \begin{bmatrix} 2 & -1 \\ 3 & 4 \end{bmatrix} \cdot \begin{bmatrix} x \\ -4 \end{bmatrix} = -19$ **b** $[-4 \quad x \quad 0] \cdot \begin{bmatrix} 2 & 0 \\ 3 & -5 \\ x & 3 \end{bmatrix} \cdot \begin{bmatrix} 2 \\ 4 \end{bmatrix} = -2$

    c $\begin{bmatrix} -2 & 3 \\ 4 & -1 \end{bmatrix} \cdot \begin{bmatrix} 1 & x & 5 \\ 2 & 4 & x \end{bmatrix} \cdot \begin{bmatrix} -3 \\ 1 \\ 0 \end{bmatrix} = \begin{bmatrix} 2 \\ -14 \end{bmatrix}$

Problem Set IV ■ Applications

1 John went to a grocery store and purchased 5 loaves of bread, 8 dozen eggs, and 6 cans of beans. Jim purchased 3 loaves of bread, 4 dozen eggs, and 2 cans of beans. If bread is 32 cents a loaf, eggs 55 cents a dozen, and beans 25 cents a can, by matrix methods find the total spent by the two boys.

2 A furniture manufacturing firm makes three different styles or models of sofas that, for notational convenience, we call S_1, S_2, S_3. All three models require various quantities of wood, cloth, steel, and hours of labor to produce. These amounts are identified in appropriate units for each style by the matrix

	wood	cloth	steel	hours
S_1	7	8	2	20
S_2	9	9	2	25
S_3	12	11	3	32

The firm anticipates yearly sales of the models S_1, S_2, and S_3 to be, respectively, 2500, 1800, and 1200 units. Furthermore, the respective costs per material units are $1.50 for wood, $2.25 for cloth, $0.50 for steel, and $5.00 an hour for labor.

    a Write the anticipated sales of the three models as a 1×3 matrix. By matrix multiplication, find a 1×4 matrix giving the total amount of materials and labor required for the yearly production of these three styles of sofas.

    b Write the per unit cost of materials as a 4×1 matrix and, by matrix multiplication, find a 3×1 matrix that indicates the total production costs for each model.

    c Determine, by an appropriate matrix multiplication, a 1×1 matrix giving the total production cost of all three models combined.

In Section 8.4 it was noted that a formula for the inverse of a matrix, if one exists, would be delayed until after determinants were introduced. The reason for this will be clear as we investigate the question of an inverse for a 2×2 matrix. From our experiences with real and complex numbers, we know that if a 2×2 matrix A has an inverse, say, A^{-1}, then

$$A \cdot A^{-1} = A^{-1} \cdot A = I$$

where I is the 2×2 multiplicative identity matrix. Let us examine these products more carefully. Let

$$A^{-1} = \begin{bmatrix} b_{11} & b_{12} \\ b_{21} & b_{22} \end{bmatrix}$$

where $b_{11}, b_{12}, b_{21}, b_{22}$ are yet to be determined. We have

$$A \cdot A^{-1} = \begin{bmatrix} a_{11} & a_{12} \\ a_{21} & a_{22} \end{bmatrix} \cdot \begin{bmatrix} b_{11} & b_{12} \\ b_{21} & b_{22} \end{bmatrix}$$

$$= \begin{bmatrix} a_{11}b_{11} + a_{12}b_{21} & a_{11}b_{12} + a_{12}b_{22} \\ a_{21}b_{11} + a_{22}b_{21} & a_{21}b_{12} + a_{22}b_{22} \end{bmatrix}$$

But, supposedly

$$A \cdot A^{-1} = \begin{bmatrix} 1 & 0 \\ 0 & 1 \end{bmatrix}$$

hence, by the definition of equality, we have the two systems

$$\begin{cases} a_{11}b_{11} + a_{12}b_{21} = 1 \\ a_{21}b_{11} + a_{22}b_{21} = 0 \end{cases} \text{ and } \begin{cases} a_{11}b_{12} + a_{12}b_{22} = 0 \\ a_{21}b_{12} + a_{22}b_{22} = 1 \end{cases}$$

Solving these two sets of equations in two unknowns yields

$$\begin{cases} b_{11} = \dfrac{a_{22}}{a_{11}a_{22} - a_{12}a_{21}} \\[2ex] b_{21} = \dfrac{-a_{21}}{a_{11}a_{22} - a_{12}a_{21}} \\[2ex] b_{12} = \dfrac{-a_{12}}{a_{11}a_{22} - a_{12}a_{21}} \\[2ex] b_{22} = \dfrac{a_{11}}{a_{11}a_{22} - a_{12}a_{21}} \end{cases} \tag{1}$$

It is significant to note that each denominator above is the same; namely,

$$a_{11}a_{22} - a_{12}a_{21} \tag{2}$$

Thus the inverse of a 2×2 matrix A exists if

$$a_{11}a_{22} - a_{12}a_{21} \neq 0$$

That is, the expression (2) being nonzero determines the existence of A^{-1} for a given 2×2 matrix A. Now the expression $a_{11}a_{22} - a_{12}a_{21}$ contains each and every element of the matrix A; hence, such an expression exists and is unique for each 2×2 matrix. This expression is now given a special name by the following definition.

DEFINITION 8.6.1   If A is a 2×2 matrix, then the determinant of A, denoted $|A|$, is a real number equal to $a_{11}a_{22} - a_{12}a_{21}$. Symbolically we have

$$|A| = \begin{vmatrix} a_{11} & a_{12} \\ a_{21} & a_{22} \end{vmatrix} = a_{11}a_{22} - a_{12}a_{21}$$

With this definition we may say that a 2×2 matrix A has an inverse if the determinant of A is *nonzero*. It can be shown this is a necessary condition as well.

Example 8.6.1   Evaluate the following determinants:

(a) $\begin{vmatrix} -1 & 2 \\ -3 & 4 \end{vmatrix} = -1 \cdot 4 - 2(-3) = 2$

(b) $\begin{vmatrix} 2 & 3 \\ 4 & 6 \end{vmatrix} = 2 \cdot 6 - 3 \cdot 4 = 0$

From the results of (a) and (b) it follows that the matrix

$$\begin{bmatrix} -1 & 2 \\ -3 & 4 \end{bmatrix}$$

has an inverse, whereas

$$\begin{bmatrix} 2 & 3 \\ 4 & 6 \end{bmatrix}$$

does not. Furthermore, the set of equations (1) could be used to determine the inverse of

$$\begin{bmatrix} -1 & 2 \\ -3 & 4 \end{bmatrix}$$

However, we shall not pursue that at this time. Our present concern, rather, will be to define higher-order determinants and investigate some of their properties.

The manner in which we shall define the determinant of an arbitrary square matrix of any order may at first seem unmotivated. It will, however, reduce to the definition already given for a 2×2 matrix, and, furthermore, this definition will possess the property that a nonzero determinant will be a necessary and sufficient condition for the existence of a unique inverse. Our definition of determinant will be simplified by the following preliminary definition:

DEFINITION 8.6.2 If A is a 3×3 matrix, then the minor of the element a_{ij} of A, denoted M_{ij}, is the determinant of the 2×2 array remaining after removing the ith row and jth column of A. The cofactor of a_{ij}, denoted A_{ij}, is given by the formula

$$A_{ij} = (-1)^{i+j} \cdot M_{ij}$$

Remark 8.6.1 From the formula for a cofactor it is clear that if the sum of the positional subscripts of an element is even, then the cofactor equals the minor. If, on the other hand, the sum of the subscripts is odd, the cofactor is equal to the *negative* of the minor.

Example 8.6.2 Given the matrix

$$A = \begin{bmatrix} 1 & -2 & 5 \\ 3 & 0 & -1 \\ 5 & 2 & 0 \end{bmatrix}$$

(a) Find the cofactor of a_{13}.

Solution Because a_{13} is the element located in the first row and third column (here it is 5), and because $1 + 3 = 4$, an *even* number, the cofactor is *equal* to the minor. The minor is the determinant of the array remaining upon removing the first row and third column. That is

$$\begin{vmatrix} 1 & -2 & 5 \\ 3 & 0 & -1 \\ 5 & 2 & 0 \end{vmatrix}$$

D8.6.1
$$\begin{vmatrix} a & b \\ c & d \end{vmatrix} = ad - bc$$

Thus,

$$M_{13} = \begin{vmatrix} 3 & 0 \\ 5 & 2 \end{vmatrix} \xlongequal{\text{D8.6.1}} 3 \cdot 2 - 0 \cdot 5 = 6$$

and hence the cofactor is

$$A_{13} = 6$$

(b) Find the cofactor of a_{21}.

Solution Because $2 + 1 = 3$, an *odd* number, the cofactor will equal the *negative* of the minor. Removing the second row and first column of A and taking the determinant, yields

$$M_{21} = \begin{vmatrix} -2 & 5 \\ 2 & 0 \end{vmatrix} \xlongequal{\text{D8.6.1}} (-2) \cdot 0 - 5 \cdot 2 = -10$$

In this case we take the negative of the minor to obtain the cofactor. Thus

$$A_{21} = 10$$

We now define the determinant of a 3×3 matrix.

DEFINITION 8.6.3 If A is a 3×3 matrix, then the determinant of A, denoted $|A|$, is obtained by taking the elements of any one row or any one column, forming the *product* of each of these with its respective cofactor, and then *summing* the result. Symbolically, this may be stated

$$|A| = a_{i1}A_{i1} + a_{i2}A_{i2} + a_{i3}A_{i3}$$

where $i = 1, 2,$ or 3, or

$$|A| = a_{1j}A_{1j} + a_{2j}A_{2j} + a_{3j}A_{3j}$$

where $j = 1, 2,$ or 3.

Remark 8.6.2 According to Definition 8.6.3 there are six different ways to expand and compute a 3×3 determinant. One would naturally ask, "Does each of the six ways yield the same result?" Otherwise the definition would be inconsistent. Surprisingly, the answer is yes. This will be proved for a 3×3 determinant in the next problem set (Section 8.7).

Example 8.6.3 Evaluate the determinant

$$\begin{vmatrix} 1 & -1 & 2 \\ 3 & 1 & 0 \\ 1 & 0 & -1 \end{vmatrix}$$

Solution   We expand using the elements of the first row.

$$\begin{vmatrix} 1 & -1 & 2 \\ 3 & 1 & 0 \\ 1 & 0 & -1 \end{vmatrix} \xupover{D8.6.3} 1 \cdot A_{11} + (-1) \cdot A_{12} + 2 \cdot A_{13}$$

$$\xupover{D8.6.2} 1 \cdot (-1)^{1+1} \begin{vmatrix} 1 & 0 \\ 0 & -1 \end{vmatrix}$$

$$+ (-1)(-1)^{1+2} \begin{vmatrix} 3 & 0 \\ 1 & -1 \end{vmatrix} + 2(-1)^{1+3} \begin{vmatrix} 3 & 1 \\ 1 & 0 \end{vmatrix}$$

$$\xupover{D8.6.1} (-1)^2(-1) + (-1)^4(-3) + 2(-1)^4(-1)$$

$$= -1 - 3 - 2 = -6$$

D8.6.3
$|A| = a_{11}A_{11} + a_{12}A_{12} + a_{13}A_{13}$

D8.6.2 $A_{ij} = (-1)^{i+j} \cdot M_{ij}$

D8.6.1 $\begin{vmatrix} a & b \\ c & d \end{vmatrix} = ad - bc$

We now expand the same determinant but using the second column, just to verify that the same result is obtained.

$$\begin{vmatrix} 1 & -1 & 2 \\ 3 & 1 & 0 \\ 1 & 0 & -1 \end{vmatrix} \xupover{D8.6.3} (-1)A_{12} + 1 \cdot A_{22} + 0 \cdot A_{32}$$

$$\xupover{D8.6.2} (-1)(-1)^{1+2} \begin{vmatrix} 3 & 0 \\ 1 & -1 \end{vmatrix}$$

$$+ 1 \cdot (-1)^{2+2} \begin{vmatrix} 1 & 2 \\ 1 & -1 \end{vmatrix} + 0 \cdot (-1)^{3+2} \begin{vmatrix} 1 & 2 \\ 3 & 0 \end{vmatrix}$$

$$\xupover{D8.6.1} (-1)^4(-3) + (-1)^4(-3) + 0$$

$$= -3 - 3 = -6$$

D8.6.3
$|A| = a_{12}A_{12} + a_{22}A_{22} + a_{32}A_{32}$

D8.6.2 $A_{ij} = (-1)^{i+j} \cdot M_{ij}$

D8.6.1 $\begin{vmatrix} a & b \\ c & d \end{vmatrix} = ad - bc$

With 3×3 determinants defined, we may define the cofactor of an element in a 4×4 matrix just as we did for a 3×3 matrix in Definition 8.6.2. The only difference is that the minors now are 3×3 determinants.

Example 8.6.4   Let

$$A = \begin{bmatrix} 1 & -1 & 2 & 3 \\ 0 & -2 & 1 & 0 \\ 3 & 4 & 2 & 1 \\ 1 & 0 & 0 & 2 \end{bmatrix}$$

Find the cofactor of a_{32}.

Solution   Because $3 + 2 = 5$, an *odd* number, the cofactor will equal the *negative* of the minor. The minor of a_{32} is obtained by taking the determinant of the array remaining upon deleting the third row and second column of A as illustrated below.

$$A = \begin{bmatrix} 1 & -1 & 2 & 3 \\ 0 & -2 & 1 & 0 \\ 3 & 4 & 2 & 1 \\ 1 & 0 & 0 & 2 \end{bmatrix}$$

Therefore, the minor is

$$M_{32} = \begin{vmatrix} 1 & 2 & 3 \\ 0 & 1 & 0 \\ 1 & 0 & 2 \end{vmatrix} = 1 \cdot A_{11} + 0 \cdot A_{21} + 1 \cdot A_{31}$$

$$= 1 \cdot 2 + 0 \cdot (-4) + 1 \cdot (-3) = -1$$

The determinant of a 4×4 matrix may be expanded by cofactors similar to the way 3×3 determinants were expanded.

Example 8.6.5   Evaluate the determinant

$$\begin{vmatrix} 1 & -1 & 2 & 0 \\ -2 & 0 & 1 & 3 \\ 4 & 1 & 0 & 0 \\ 2 & -1 & 1 & 0 \end{vmatrix}$$

Solution   Because we may choose any row or column by which to expand, we naturally choose the one with the most zero entries to minimize our calculations. We choose to expand, therefore, by the fourth column.

$$\begin{vmatrix} 1 & -1 & 2 & 0 \\ -2 & 0 & 1 & 3 \\ 4 & 1 & 0 & 0 \\ 2 & -1 & 1 & 0 \end{vmatrix} = 0 \cdot A_{14} + 3 \cdot A_{24} + 0 \cdot A_{34} + 0 \cdot A_{44}$$

$$= 3(-1)^{2+4} \begin{vmatrix} 1 & -1 & 2 \\ 4 & 1 & 0 \\ 2 & -1 & 1 \end{vmatrix}$$

$$= 3 \cdot (-1)^6 \left[4(-1)^{2+1} \begin{vmatrix} -1 & 2 \\ -1 & 1 \end{vmatrix} \right.$$

$$\left. + 1 \cdot (-1)^{2+2} \begin{vmatrix} 1 & 2 \\ 2 & 1 \end{vmatrix} + 0 \cdot A_{23} \right]$$

$$= 3[-4[-1 - (-2)] + (1 - 4)]$$

$$= 3(-4 - 3) = -21$$

From this example, we can appreciate that the work involved in evaluating higher-order determinants, with few if any zero entries, can be long and tedious. Fortunately there are methods of eliminating much of this work, which we shall discover as we investigate some of the properties of determinants later in this section.

With the recurring pattern of defining 3×3 determinants using 2×2 minor determinants, and then 4×4 determinants in terms of 3×3 minors, we are prepared to define the determinant of a general $n \times n$ matrix.

DEFINITION 8.6.4 If A is an $n \times n$ matrix, then the determinant of A, denoted $|A|$, is given by the expression

$$|A| = a_{i1}A_{i1} + a_{i2}A_{i2} + \cdots + a_{in}A_{in}$$

where $i = 1, 2, 3, \ldots,$ or n.
Also $A_{ij} = (-1)^{i+j} \cdot M_{ij}$ where M_{ij} is the minor of the element in the i,jth place. Furthermore,

$$|A| = a_{1j}A_{1j} + a_{2j}A_{2j} + \cdots + a_{nj}A_{nj}$$

where $j = 1, 2, 3, \ldots,$ or n.

The proof that this definition is consistent, expanding by any row or column yields the same result, will not be given in this text.

We shall now investigate some of the more obvious and useful properties of determinants.

THEOREM 8.6.1 If A is an $n \times n$ matrix with all entries in one row or one column equal to zero, then the determinant of A equals zero, that is,

$$|A| = 0$$

Proof Suppose all entries in, say, the ith row are zero. Because we may expand the determinant of A by the elements of any row or any column, we choose, naturally, the ith row. We obtain

$$|A| = 0 \cdot A_{i1} + 0 \cdot A_{i2} + \cdots + 0 \cdot A_{in} = 0 \qquad \text{QED}$$

THEOREM 8.6.2 If an $n \times n$ matrix B is obtained from A by interchanging any two rows or any two columns of A, then

$$|B| = -|A|$$

Proof To be done as an exercise.

Example 8.6.6 Evaluate the determinant

$$\begin{vmatrix} 1 & 2 & -1 \\ 3 & 0 & 2 \\ 1 & -1 & 0 \end{vmatrix}$$

and then interchange the second and third columns and reevaluate, thus verifying Theorem 8.6.2 for this case.

Solution

D8.6.3
$$|A| = a_{13}A_{13} + a_{23}A_{23} + a_{33}A_{33}$$

$$\begin{vmatrix} 1 & 2 & -1 \\ 3 & 0 & 2 \\ 1 & -1 & 0 \end{vmatrix} \overset{\text{D8.6.3}}{=\!=\!=\!=} (-1)(-1)^{1+3} \begin{vmatrix} 3 & 0 \\ 1 & -1 \end{vmatrix}$$

$$+ 2(-1)^{2+3} \begin{vmatrix} 1 & 2 \\ 1 & -1 \end{vmatrix} + 0 \cdot A_{33}$$

D8.6.2
$$\begin{vmatrix} a & b \\ c & d \end{vmatrix} = ad - bc$$

$$\overset{\text{D8.6.2}}{=\!=\!=\!=} -1(-3-0) - 2(-1-2) = 9$$

Interchanging the second and third columns and evaluating, we obtain

$$\begin{vmatrix} 1 & -1 & 2 \\ 3 & 2 & 0 \\ 1 & 0 & -1 \end{vmatrix} \overset{\text{D8.6.3}}{=\!=\!=\!=} 1 \cdot (-1)^{3+1} \begin{vmatrix} -1 & 2 \\ 2 & 0 \end{vmatrix} + 0 \cdot A_{32}$$

$$+ (-1)(-1)^{3+3} \begin{vmatrix} 1 & -1 \\ 3 & 2 \end{vmatrix}$$

$$\overset{\text{D8.6.2}}{=\!=\!=\!=} (0-4) - (2-[-3]) = -4 - 5 = -9$$

Thus

$$\begin{vmatrix} 1 & 2 & -1 \\ 3 & 0 & 2 \\ 1 & -1 & 0 \end{vmatrix} = - \begin{vmatrix} 1 & -1 & 2 \\ 3 & 2 & 0 \\ 1 & 0 & -1 \end{vmatrix}$$

verifying Theorem 8.6.2.

THEOREM 8.6.3 If A is an $n \times n$ matrix having two rows or two columns with corresponding entries, then

$$|A| = 0$$

Proof Suppose the entries in the ith and jth columns of A are equal. Let B be the matrix obtained by interchanging the ith and jth columns of A. Then by Theorem 8.6.2,

$$|B| = -|A| \tag{3}$$

but, because the entries in the ith and jth columns of A are identical, we also have

$$|B| = |A| \qquad\qquad (4)$$

From equations (3) and (4) we conclude

$$|A| = -|A| \Longleftrightarrow 2|A| = 0$$

$$\Longleftrightarrow |A| = 0 \qquad \text{QED}$$

Example 8.6.7 Evaluate the determinant

$$\begin{vmatrix} -1 & 2 & 0 & 3 \\ 4 & 6 & \sqrt{3} & 2 \\ 8 & 7 & -4 & 9 \\ -1 & 2 & 0 & 3 \end{vmatrix}$$

Solution

$$\begin{vmatrix} -1 & 2 & 0 & 3 \\ 4 & 6 & \sqrt{3} & 2 \\ 8 & 7 & -4 & 9 \\ -1 & 2 & 0 & 3 \end{vmatrix} \underset{\text{T8.6.3}}{=\!=\!=\!=} 0$$

T8.6.3 $|A| = 0$

THEOREM 8.6.4 If A is an $n \times n$ matrix and B is obtained from A by multiplying *every* entry in any *one* row or any *one* column of A by a real constant k, then

$$|B| = k|A|$$

Proof   To be done as an exercise.

Example 8.6.8 Evaluate first the product

$$3 \cdot \begin{vmatrix} 1 & -1 & 2 \\ 0 & 3 & -2 \\ 1 & 0 & -1 \end{vmatrix}$$

and then

$$\begin{vmatrix} 1 & -1 & 2 \\ 3 \cdot 0 & 3 \cdot 3 & 3(-2) \\ 1 & 0 & -1 \end{vmatrix}$$

to verify Theorem 8.6.4 for this case.

Solution   Expanding the first product, we obtain

$$3 \cdot \begin{vmatrix} 1 & -1 & 2 \\ 0 & 3 & -2 \\ 1 & 0 & -1 \end{vmatrix} = 3 \left[1 \cdot (-1)^{1+1} \begin{vmatrix} 3 & -2 \\ 0 & -1 \end{vmatrix} + 0 \cdot A_{21} \right.$$

$$\left. + 1 \cdot (-1)^{3+1} \begin{vmatrix} -1 & 2 \\ 3 & -2 \end{vmatrix} \right]$$

$$= 3[(-3 - 0) + 0 + (2 - 6)] = 3(-7) = -21$$

Expanding the second determinant yields

$$\begin{vmatrix} 1 & -1 & 2 \\ 3 \cdot 0 & 3 \cdot 3 & 3(-2) \\ 1 & 0 & -1 \end{vmatrix} = \begin{vmatrix} 1 & -1 & 2 \\ 0 & 9 & -6 \\ 1 & 0 & -1 \end{vmatrix}$$

$$= 1 \cdot (-1)^{3+1} \begin{vmatrix} -1 & 2 \\ 9 & -6 \end{vmatrix} + 0 \cdot A_{32}$$

$$+ (-1)(-1)^{3+3} \begin{vmatrix} 1 & -1 \\ 0 & 9 \end{vmatrix}$$

$$= (6 - 18) + 0 + (-1)(9 - 0)$$
$$= -12 - 9 = -21$$

which verifies Theorem 8.6.4.

THEOREM 8.6.5   Let A be an $n \times n$ matrix. Suppose the elements in the ith row, $i = 1, 2, \ldots, n$, are expressed as the sums

$$q_{i1} = b_{i1} + c_{i1}$$
$$a_{i2} = b_{i2} + c_{i2}$$
$$\vdots$$
$$a_{in} = b_{in} + c_{in}$$

Then

$$\begin{vmatrix} a_{11} & a_{12} & & a_{1n} \\ a_{21} & a_{22} & \cdots & a_{2n} \\ \vdots & & & \\ b_{i1} + c_{i1} & b_{i2} + c_{i2} & \cdots & b_{in} + c_{in} \\ \vdots & & & \\ a_{n1} & a_{n2} & \cdots & a_{nn} \end{vmatrix}$$

$$= \begin{vmatrix} a_{11} & a_{12} & \cdots & a_{1n} \\ a_{21} & a_{22} & \cdots & a_{2n} \\ & \vdots & & \\ b_{i1} & b_{i2} & \cdots & b_{in} \\ & \vdots & & \\ a_{n1} & a_{n2} & \cdots & a_{nn} \end{vmatrix} + \begin{vmatrix} a_{11} & a_{12} & \cdots & a_{1n} \\ a_{21} & a_{22} & \cdots & a_{2n} \\ & \vdots & & \\ c_{i1} & c_{i2} & \cdots & c_{in} \\ & \vdots & & \\ a_{n1} & a_{n2} & \cdots & a_{nn} \end{vmatrix}$$

The same is true for columns.

Proof   To be done as an exercise.

Example 8.6.9   Evaluate first the determinant

$$\begin{vmatrix} 1 & -2 & 3 \\ 4 & -3 & 1 \\ 2 & 1 & 0 \end{vmatrix} = \begin{vmatrix} 1 & -2 & 3 \\ 2+2 & -2-1 & 4-3 \\ 2 & 1 & 0 \end{vmatrix}$$

and then the sum

$$\begin{vmatrix} 1 & -2 & 3 \\ 2 & -2 & 4 \\ 2 & 1 & 0 \end{vmatrix} + \begin{vmatrix} 1 & -2 & 3 \\ 2 & -1 & -3 \\ 2 & 1 & 0 \end{vmatrix}$$

to verify Theorem 8.6.5 for this case.

Solution   Expanding the first by the third row yields

$$\begin{vmatrix} 1 & -2 & 3 \\ 4 & -3 & 1 \\ 2 & 1 & 0 \end{vmatrix} = 2(-1)^{3+1} \begin{vmatrix} -2 & 3 \\ -3 & 1 \end{vmatrix} + 1 \cdot (-1)^{3+2} \begin{vmatrix} 1 & 3 \\ 4 & 1 \end{vmatrix} + 0 \cdot A_{33}$$

$$= 2(-2 - [-9]) - (1 - 12) + 0 = 14 + 11 = 25$$

We simplify the determinant sum as follows:

$$\begin{vmatrix} 1 & -2 & 3 \\ 2 & -2 & 4 \\ 2 & 1 & 0 \end{vmatrix} + \begin{vmatrix} 1 & -2 & 3 \\ 2 & -1 & -3 \\ 2 & 1 & 0 \end{vmatrix} = 2(-1)^{3+1} \begin{vmatrix} -2 & 3 \\ -2 & 4 \end{vmatrix}$$

$$+ 1(-1)^{3+2} \begin{vmatrix} 1 & 3 \\ 2 & 4 \end{vmatrix} + 0 \cdot A_{33}$$

$$+ 3(-1)^{1+3} \begin{vmatrix} 2 & -1 \\ 2 & 1 \end{vmatrix} - 3(-1)^{2+3} \begin{vmatrix} 1 & -2 \\ 2 & 1 \end{vmatrix} + 0 \cdot A_{33}$$

$$= 2(-8 - [-6]) - 1(4 - 6) + 3(2 - [-2]) + 3(1 - [-4])$$

$$= -4 + 2 + 12 + 15 = 25$$

verifying Theorem 8.6.5.

THEOREM 8.6.6 Let A be an $n \times n$ matrix. If a matrix B is obtained from A by adding to the elements of some row of A a constant multiple of the corresponding elements of any row of A, then $|B| = |A|$. The determinant is unchanged. Symbolically, this may be represented as

$$
\begin{vmatrix} a_{11} & a_{12} & \cdots & a_{1n} \\ a_{21} & a_{22} & \cdots & a_{2n} \\ \vdots \\ a_{n1} & a_{n2} & \cdots & a_{nn} \end{vmatrix}
=
\begin{vmatrix} a_{11} & a_{12} & \cdots & a_{1n} \\ a_{21} & a_{22} & \cdots & a_{2n} \\ \vdots \\ a_{j1} + ka_{i1} & a_{j2} + ka_{i2} & \cdots & a_{jn} + ka_{in} \\ \vdots \\ a_{n1} & a_{n2} & \cdots & a_{nn} \end{vmatrix}
$$

where k times the elements of the ith row have been added, respectively, to the corresponding elements in the jth row. The same property holds true for columns.

Proof   Let $|B|$ represent the right-hand determinant of the equation above. Then

$$
|B| = \begin{vmatrix} a_{11} & a_{12} & \cdots & a_{1n} \\ a_{21} & a_{22} & \cdots & a_{2n} \\ \vdots \\ a_{j1} + ka_{i1} & a_{j2} + ka_{i2} & \cdots & a_{jn} + ka_{in} \\ \vdots \\ a_{n1} & a_{n2} & \cdots & a_{nn} \end{vmatrix}
$$

D8.6.3

$$= (a_{j1} + ka_{i1})A_{j1} + (a_{j2} + ka_{i2})A_{j2} \\ + \cdots + (a_{jn} + ka_{in})A_{jn}$$

$\dfrac{\ }{\text{A+}}$ D

$$= (a_{j1}A_{j1} + a_{j2}A_{j2} + \cdots + a_{jn}A_{jn}) \\ + k(a_{i1}A_{j1} + a_{i2}A_{j2} + \cdots + a_{in}A_{jn})$$

D8.6.3

$$= |A| + k \begin{vmatrix} a_{11} & a_{12} & \cdots & a_{1n} \\ \vdots \\ a_{i1} & a_{i2} & \cdots & a_{in} & \longleftarrow i\text{th row} \\ \vdots \\ a_{i1} & a_{i2} & \cdots & a_{in} & \longleftarrow j\text{th row} \\ \vdots \\ a_{n1} & a_{n2} & \cdots & a_{nn} \end{vmatrix}$$

are identical

T8.6.3

$$= |A| + k \cdot 0 = |A| \qquad \text{QED}$$

D8.6.3
$|A| = a_{j1}A_{j1} + a_{j2}A_{j2} + \cdots + a_{jn}A_{jn}$

D $(a + b)c = ac + bc$
A+ $(a + b) + c = a + (b + c)$

T8.6.3
If A has two rows that are identical then $|A| = 0$.

This latter theorem is the one that will greatly simplify the process of evaluating determinants of higher order. By repeated application of Theorem 8.6.6, you can reduce the determinant of an $n \times n$ matrix to a 2×2 determinant, which, to be sure, is quite a simplification. We shall demonstrate how this is accomplished in the next example.

Example 8.6.10 Evaluate the determinant:

$$\begin{vmatrix} 1 & -1 & 2 & 3 \\ 0 & -2 & 1 & 1 \\ 1 & -1 & 0 & 2 \\ 3 & 2 & -1 & 0 \end{vmatrix}$$

Solution Adding to the elements of the third row -1 times the corresponding elements of the first row and to the fourth row -3 times the elements of the first row yields

$$\begin{vmatrix} 1 & -1 & 2 & 3 \\ 0 & -2 & 1 & 1 \\ 1 & -1 & 0 & 2 \\ 3 & 2 & -1 & 0 \end{vmatrix} = \begin{vmatrix} 1 & -1 & 2 & 3 \\ 0 & -2 & 1 & 1 \\ 1-1 & -1-(-1) & 0-2 & 2-3 \\ 3-3\cdot 1 & 2-3\cdot(-1) & -1-3(2) & 0-3(3) \end{vmatrix}$$

$$= \begin{vmatrix} 1 & -1 & 2 & 3 \\ 0 & -2 & 1 & 1 \\ 0 & 0 & -2 & -1 \\ 0 & 5 & -7 & -9 \end{vmatrix} \xrightarrow{\textbf{D8.6.3}} \begin{vmatrix} -2 & 1 & 1 \\ 0 & -2 & -1 \\ 5 & -7 & -9 \end{vmatrix}$$

D8.6.3
$|A| = a_{11}A_{11} + a_{21}A_{21} + a_{31}A_{31} + a_{41}A_{41}$

Now add to the elements of the second column -2 times the corresponding elements of the third column. We obtain

$$\begin{vmatrix} -2 & 1-2(1) & 1 \\ 0 & -2-2(-1) & -1 \\ 5 & -7-2(-9) & -9 \end{vmatrix} = \begin{vmatrix} -2 & -1 & 1 \\ 0 & 0 & -1 \\ 5 & 11 & -9 \end{vmatrix}$$

$$\xrightarrow{\textbf{D8.6.3}} -1(-1)^{2+3} \begin{vmatrix} -2 & -1 \\ 5 & 11 \end{vmatrix} = (-22 - [-5]) = -17$$

It is convenient to have within the determinant array an entry which is 1. By adding the proper *integral* multiples of the row in which 1 lies to the other rows, all other entries in this column may be reduced to zero. Similarly, by column addition, all other entries in the row in which 1 occurs can be reduced to zero. And, interestingly, all of these changes have no effect upon the ultimate value of the expanded determinant. One note of caution: When all the entries in some row or column have been reduced to zero except the single entry 1 and if the position of 1 is odd (sum of the positional subscripts is odd), then the negative of the minor of that element will be the answer. If, however, -1 should occupy this odd position, then the minor alone would do, because the product of two negatives is a positive. Because the minor of an element is of dimension one less than that of the original determinant, by repeated application of this principle, all determinants can be reduced to a single 2×2 determinant. We again demonstrate this process by the following example.

Example 8.6.11 Evaluate the determinant

$$
\begin{vmatrix}
-2 & 1 & 2 & -1 & 0 \\
1 & 3 & -1 & 0 & 1 \\
2 & -2 & 1 & -1 & 1 \\
-3 & 0 & 1 & 0 & 2 \\
2 & 0 & -3 & -2 & 1
\end{vmatrix}
$$

Solution Let us use the third column to reduce all entries in the fourth row to zero except the 1 in the third column and fourth row.

$$
\begin{vmatrix}
-2 & 1 & 2 & -1 & 0 \\
1 & 3 & -1 & 0 & 1 \\
2 & -2 & 1 & -1 & 1 \\
-3 & 0 & 1 & 0 & 2 \\
2 & 0 & -3 & -2 & 1
\end{vmatrix}
\underset{\text{T8.6.6}}{=\!=\!=}
\begin{vmatrix}
-2+3(2) & 1 & 2 & -1 & 0-2(2) \\
1+3(-1) & 3 & -1 & 0 & 1-2(-1) \\
2+3(1) & -2 & 1 & -1 & 1-2(1) \\
-3+3(1) & 0 & 1 & 0 & 2-2(1) \\
2+3(-3) & 0 & -3 & -2 & 1-2(-3)
\end{vmatrix}
$$

● Note that 1 is in an odd position and hence we multiply the minor by -1.

$$
=
\begin{vmatrix}
4 & 1 & 2 & -1 & -4 \\
-2 & 3 & -1 & 0 & 3 \\
5 & -2 & 1 & -1 & -1 \\
0 & 0 & 1 & 0 & 0 \\
-7 & 0 & -3 & -2 & 7
\end{vmatrix}
\underset{\text{D8.6.3}}{=\!=\!=}
(-1) \bullet
\begin{vmatrix}
4 & 1 & -1 & -4 \\
-2 & 3 & 0 & 3 \\
5 & -2 & -1 & -1 \\
-7 & 0 & -2 & 7
\end{vmatrix}
$$

$$
\underset{\text{T8.6.6}}{=\!=\!=}
-
\begin{vmatrix}
4 & 1 & -1 & -4 \\
-2 & 3 & 0 & 3 \\
5-(4) & -2-(1) & -1-(-1) & -1-(-4) \\
-7-2(4) & 0-2(1) & -2-2(-1) & 7-2(-4)
\end{vmatrix}
$$

$$
=
-
\begin{vmatrix}
4 & 1 & -1 & -4 \\
-2 & 3 & 0 & 3 \\
1 & -3 & 0 & 3 \\
-15 & -2 & 0 & 15
\end{vmatrix}
\underset{\text{D8.6.3}}{=\!=\!=}
-(-1)
\begin{vmatrix}
-2 & 3 & 3 \\
1 & -3 & 3 \\
-15 & -2 & 15
\end{vmatrix}
$$

$$
\underset{\text{T8.6.6}}{=\!=\!=}
\begin{vmatrix}
-2 & 3+3(-2) & 3-3(-2) \\
1 & -3+3(1) & 3-3(1) \\
-15 & -2+3(-15) & 15-3(-15)
\end{vmatrix}
=
\begin{vmatrix}
-2 & -3 & 9 \\
1 & 0 & 0 \\
-15 & -47 & 60
\end{vmatrix}
$$

$$
\underset{\text{D8.6.3}}{=\!=\!=}
-
\begin{vmatrix}
-3 & 9 \\
-47 & 60
\end{vmatrix}
\underset{\text{T8.6.4}}{=\!=\!=}
(-1)3 \cdot
\begin{vmatrix}
-1 & 3 \\
-47 & 60
\end{vmatrix}
$$

$$\underline{\underline{\textbf{T8.6.4}}} \quad -3\,(3)\cdot\begin{vmatrix} -1 & 1 \\ -47 & 20 \end{vmatrix} = -9(-20-[-47])$$

$$= -9(27) = -243$$

Here we succeeded in reducing a 5×5 determinant to a single 2×2.

Problem Set I ■ *Reading Comprehension*

Determine whether the following statements are true or false and give reasons for your responses.

1 A determinant is a special kind of matrix.
2 The minor of a matrix element is a matrix obtained by removing the row and column the element occupies.
3 The cofactor of a matrix element either equals the minor of that element or the negative of the minor.
4 The value of a determinant does not depend on the row or column used to form the expansion.
5 The value of a determinant is unchanged if two rows and two columns are interchanged.
6 If all the entries in one row of a determinant are exactly twice the entries in another row, then the value of the determinant is zero.
7 The value of a determinant is unchanged even though to the elements of one row are added a constant multiple of the respective elements of another row.

Problem Set II ■ *Skills Development*

1 Let

$$A = \begin{bmatrix} 1 & -1 & 2 \\ 3 & 2 & 0 \\ 1 & 0 & -2 \end{bmatrix}$$

Find and evaluate the minor M_{ij} and cofactor A_{ij} of each of the following entries:

   a a_{11} **b** a_{21} **c** a_{23} **d** a_{31} **e** a_{32} **f** a_{33}

2 Let

$$A = \begin{bmatrix} 2 & -1 & 0 & 3 \\ -2 & 1 & 1 & 0 \\ 0 & 3 & -1 & 2 \\ 1 & -1 & 1 & -1 \end{bmatrix}$$

Find and evaluate the minor M_{ij} and cofactor A_{ij} of each of the following entries:

   a a_{21} **b** a_{22} **c** a_{32} **d** a_{43} **e** a_{23} **f** a_{24}

Without expanding indicate the reason for the following equalities:

3 $\begin{vmatrix} 1 & -2 & 8 \\ 10 & 4 & 6 \\ 0 & 0 & 0 \end{vmatrix} = 0$
     4 $\begin{vmatrix} 4 & 0 & 7 \\ 8 & 0 & 3 \\ -5 & 0 & 2 \end{vmatrix} = 0$
     5 $\begin{vmatrix} 1 & -2 & 8 \\ 3 & 6 & -4 \\ 1 & -2 & 8 \end{vmatrix} = 0$

6 $\begin{vmatrix} -4 & 2 & -8 \\ 3 & 3 & 6 \\ -2 & 6 & -4 \end{vmatrix} = 0$
     7 $\begin{vmatrix} 1 & 8 & -4 \\ -2 & 7 & 8 \\ 3 & 3 & -12 \end{vmatrix} = 0$
     8 $\begin{vmatrix} 1 & 4 & 3 \\ -3 & -12 & -9 \\ 6 & 7 & 12 \end{vmatrix} = 0$

9 $\begin{vmatrix} 2 & 6 & -10 \\ 3 & 4 & 7 \\ 1 & -1 & 1 \end{vmatrix} = 2 \cdot \begin{vmatrix} 1 & 3 & -5 \\ 3 & 4 & 7 \\ 1 & -1 & 1 \end{vmatrix}$
     10 $\begin{vmatrix} 1 & 3 & 6 \\ 2 & 12 & 7 \\ -4 & -15 & 3 \end{vmatrix} = 3 \cdot \begin{vmatrix} 1 & 1 & 6 \\ 2 & 4 & 7 \\ -4 & -5 & 3 \end{vmatrix}$

Evaluate the following determinants taking advantage, when appropriate, of Theorem 8.6.6 and others:

11 $\begin{vmatrix} 2 & -5 \\ 4 & 8 \end{vmatrix}$
     12 $\begin{vmatrix} 3 & 7 \\ -1 & 2 \end{vmatrix}$
     13 $\begin{vmatrix} -2 & 4 \\ 7 & -3 \end{vmatrix}$
     14 $\begin{vmatrix} 1 & 1 & -1 \\ 4 & 3 & -2 \\ 0 & 2 & -3 \end{vmatrix}$

15 $\begin{vmatrix} 2 & 1 & 3 \\ 4 & -5 & 6 \\ -7 & 8 & -9 \end{vmatrix}$
     16 $\begin{vmatrix} 2 & 4 & -6 \\ 3 & 1 & -2 \\ -4 & 2 & 0 \end{vmatrix}$
     17 $\begin{vmatrix} 1 & 1 & -1 & 1 \\ -1 & -1 & 1 & 1 \\ -1 & -1 & -1 & 1 \\ 1 & 1 & 1 & 1 \end{vmatrix}$

18 $\begin{vmatrix} 1 & -2 & 3 & 2 \\ 4 & 0 & 1 & -3 \\ 0 & 5 & 1 & 2 \\ 2 & 1 & 0 & 1 \end{vmatrix}$
     19 $\begin{vmatrix} 3 & 2 & 1 & 0 \\ -2 & 3 & 0 & 1 \\ 0 & -3 & 1 & 2 \\ 1 & 0 & 3 & 2 \end{vmatrix}$

Solve for x:

20 $\begin{vmatrix} x & 2 \\ 3 & 4 \end{vmatrix} = 2$
     21 $\begin{vmatrix} 3 & x \\ -2 & 4 \end{vmatrix} = 4$
     22 $\begin{vmatrix} x & 2 & 3 \\ 0 & 4 & 8 \\ 0 & -2 & 7 \end{vmatrix} = 0$

23 $\begin{vmatrix} x & x & 1 \\ -1 & 2 & 3 \\ 1 & 0 & -2 \end{vmatrix} = 1$
     24 $\begin{vmatrix} 1 & -2 & x \\ 1 & 0 & -1 \\ 2 & -3 & x \end{vmatrix} = -3$

Problem Set III   ■   *Theoretical Developments*

1 If one were to expand a 3×3 determinant by Definition 8.6.3, how many 2×2 minors would be evaluated? How many 2×2's for a 5×5 determinant? Conjecture how many 2×2's for an $n \times n$ determinant.

2 Let

$$|A| = \begin{vmatrix} a_{11} & a_{12} & a_{13} \\ a_{21} & a_{22} & a_{23} \\ a_{31} & a_{32} & a_{33} \end{vmatrix}$$

Show

a $a_{11}A_{11} + a_{12}A_{12} + a_{13}A_{13} = a_{11}A_{11} + a_{21}A_{21} + a_{31}A_{31}$

b $a_{11}A_{11} + a_{12}A_{12} + a_{13}A_{13} = a_{12}A_{12} + a_{22}A_{22} + a_{32}A_{32}$

c $\quad a_{11}A_{11} + a_{12}A_{12} + a_{13}A_{13} = a_{31}A_{31} + a_{32}A_{32} + a_{33}A_{33}$

and hence conclude that it doesn't matter which row is used to expand a 3×3 determinant.

3 If A is an $n \times n$ matrix, prove
$$|kA| = k^n|A|$$
where k is a real constant.

4 If A and B are 2×2 matrices, prove
$$|A \cdot B| = |A| \cdot |B|$$

5 Show that the determinant equation

$$\begin{vmatrix} x & y & 1 \\ x_1 & y_1 & 1 \\ x_2 & y_2 & 1 \end{vmatrix} = 0$$

represents the equation of a line passing through two points (x_1, y_1) and (x_2, y_2).

6 Using the results of Problem 5, find the equation of the line passing through the points $(-1, 2)$ and $(3, -2)$.

7 Show $\quad \begin{vmatrix} 1 & a & a^2 \\ 1 & b & b^2 \\ 1 & c & c^2 \end{vmatrix} = (a - b)(b - c)(c - a)$

Prove the following theorems:

8 Theorem 8.6.2.

9 Theorem 8.6.4.

10 Theorem 8.6.5.

The Inverse of a Square Matrix $\qquad$ 8.8

We now return to the problem of determining an inverse, when it exists, for a square matrix. Recall that in the real and complex number systems, every number except zero possessed a multiplicative inverse. The property of a multiplicative inverse a^{-1} of an element a is that

$$a \cdot a^{-1} = a^{-1} \cdot a = 1$$

where 1 is the multiplicative identity element of the system. It is reasonable to suspect that inverses might exist for at least some matrices. This fact was demonstrated for 2×2 matrices in Section 10.6.

If A is an $n \times n$ matrix for which a multiplicative inverse exists, we shall denote this inverse in the customary manner as A^{-1} and require that it satisfy the relation

$$A \cdot A^{-1} = A^{-1} \cdot A = I$$

where I is the $n \times n$ multiplicative identity matrix.

Let us review and take our lead from the 2×2 inverse developed in Section 8.6. Recall that if A is a 2×2 matrix whose determinant is nonzero, then

$$A^{-1} = \begin{bmatrix} \dfrac{a_{22}}{a_{11}a_{22} - a_{12}a_{21}} & \dfrac{-a_{12}}{a_{11}a_{22} - a_{12}a_{21}} \\[3mm] \dfrac{-a_{21}}{a_{11}a_{22} - a_{12}a_{21}} & \dfrac{a_{11}}{a_{11}a_{22} - a_{12}a_{21}} \end{bmatrix}$$

$$= \frac{1}{|A|} \begin{bmatrix} a_{22} & -a_{12} \\ -a_{21} & a_{11} \end{bmatrix}$$

Although it may not have occurred to you yet, each entry in this latter matrix is the cofactor of an entry in the original. In fact, we have

$$\begin{bmatrix} a_{22} & -a_{12} \\ -a_{21} & a_{11} \end{bmatrix} = \begin{bmatrix} A_{11} & A_{21} \\ A_{12} & A_{22} \end{bmatrix}$$

● It must be explained here that the determinant of a 1×1 "array," as the minors would be for a 2×2 matrix, is just the element remaining when the appropriate row and column have been removed. We do not use the notation $|a_{11}|$ for a 1×1 determinant because this would then be confused with absolute value.

where A_{11}, A_{12}, A_{21}, and A_{22} ● are, respectively, the cofactors of a_{11}, a_{12}, a_{21}, and a_{22}. Thus, we may write

$$A^{-1} = \frac{1}{|A|} \begin{bmatrix} A_{11} & A_{21} \\ A_{12} & A_{22} \end{bmatrix}$$

This may be even further simplified by the following definition.

DEFINITION 8.8.1 If an $n \times m$ matrix B is obtained from an $m \times n$ matrix A by interchanging the rows and columns of A, that is, the first row becomes first column, the second row becomes second column, and so forth, then B is called the transpose of A and is denoted

$$B = A^t$$

Example 8.8.1 Determine the transpose of

$$\begin{vmatrix} 1 & 2 & -3 & -1 \\ 2 & 0 & 1 & 3 \\ -1 & 1 & 0 & 2 \end{vmatrix}$$

Solution

$$A^t = \begin{vmatrix} 1 & 2 & -3 & -1 \\ 2 & 0 & 1 & 3 \\ -1 & 1 & 0 & 2 \end{vmatrix}^t = \begin{vmatrix} 1 & 2 & -1 \\ 2 & 0 & 1 \\ -3 & 1 & 0 \\ -1 & 3 & 2 \end{vmatrix}$$

With this definition, the formula for a 2×2 inverse may be written

$$A^{-1} = \frac{1}{|A|} \begin{bmatrix} A_{11} & A_{12} \\ A_{21} & A_{22} \end{bmatrix}^t$$

$$= \frac{1}{|A|} (A_{ij})^t$$

where the matrix (A_{ij}) is obtained from (a_{ij}) by simply replacing a_{ij} by its cofactor. This latter form for A^{-1} is relatively easy to remember. Replace each entry of A by its respective cofactor, take the transpose, and multiply by $1/|A|$.

Example 8.8.2 If $A = \begin{vmatrix} 1 & 2 \\ -1 & 3 \end{vmatrix}$, find A^{-1} and show $A \cdot A^{-1} = I$.

Solution Applying the formula developed above we have

$$A^{-1} = \frac{1}{|A|} \begin{bmatrix} A_{11} & A_{12} \\ A_{21} & A_{22} \end{bmatrix}^t$$

where $A_{11} = 3$, $A_{12} = 1$, $A_{21} = -2$, $A_{22} = 1$, and $|A| = 5$. Hence,

$$A^{-1} = \frac{1}{5} \begin{bmatrix} 3 & 1 \\ -2 & 1 \end{bmatrix}^t = \frac{1}{5} \begin{bmatrix} 3 & -2 \\ 1 & 1 \end{bmatrix}$$

To check, we perform the product

$$A \cdot A^{-1} = \begin{bmatrix} 1 & 2 \\ -1 & 3 \end{bmatrix} \cdot \begin{bmatrix} 3 & -2 \\ 1 & 1 \end{bmatrix} \cdot \frac{1}{5}$$

$$= \frac{1}{5} \cdot \begin{bmatrix} 1 \cdot 3 + 2 \cdot 1 & 1(-2) + 2 \cdot 1 \\ -1(3) + 3 \cdot 1 & -1(-2) + 3 \cdot 1 \end{bmatrix}$$

$$= \frac{1}{5} \cdot \begin{bmatrix} 5 & 0 \\ 0 & 5 \end{bmatrix} = \begin{bmatrix} 1 & 0 \\ 0 & 1 \end{bmatrix}$$

It is not uncommon in mathematics for patterns or formulas that hold for limited systems to hold also for more general ones. It is not unreasonable to conjecture, therefore, that if for a 2×2 matrix, where $|A| \neq 0$, we find the inverse to be

$$A^{-1} = \frac{1}{|A|} \begin{bmatrix} A_{11} & A_{12} \\ A_{21} & A_{22} \end{bmatrix}^t$$

then possibly the inverse for an $n \times n$ matrix, for which $|A| \neq 0$, may be

$$A^{-1} = \frac{1}{|A|} \begin{bmatrix} A_{11} & A_{12} & \cdots & A_{1n} \\ A_{21} & A_{22} & \cdots & A_{2n} \\ \vdots & & & \\ A_{n1} & A_{n2} & \cdots & A_{nn} \end{bmatrix}^{t}$$

The following theorem verifies this as a fact.

THEOREM 8.8.1 If A is an $n \times n$ matrix for which $|A| \neq 0$, then

$$A^{-1} = \frac{1}{|A|} \begin{bmatrix} A_{11} & A_{12} & \cdots & A_{1n} \\ A_{21} & A_{22} & \cdots & A_{2n} \\ \vdots & & & \\ A_{n1} & A_{n2} & \cdots & A_{nn} \end{bmatrix}^{t}$$

where A_{ij} is the cofactor of the element a_{ij} of A for each i and j. Furthermore, if $|A| = 0$, A does not have an inverse. A matrix that possesses an inverse is said to be nonsingular, otherwise it is singular.

The proof of this theorem will not be given here.

Example 8.8.3 Find the inverse of

$$A = \begin{bmatrix} 1 & -1 & 2 \\ 3 & -1 & 0 \\ -2 & 1 & 1 \end{bmatrix}$$

if it exists and check by showing $A^{-1} \cdot A = I$.

Solution According to Theorem 8.8.1, if $|A| \neq 0$, then

$$A^{-1} = \frac{1}{|A|} \begin{bmatrix} A_{11} & A_{12} & A_{13} \\ A_{21} & A_{22} & A_{23} \\ A_{31} & A_{32} & A_{33} \end{bmatrix}^{t}$$

In this case

$$A_{11} = \begin{vmatrix} -1 & 0 \\ 1 & 1 \end{vmatrix} = -1 \qquad A_{12} = -\begin{vmatrix} 3 & 0 \\ -2 & 1 \end{vmatrix} = -3$$

$$A_{13} = \begin{vmatrix} 3 & -1 \\ -2 & 1 \end{vmatrix} = 1 \qquad A_{21} = -\begin{vmatrix} -1 & 2 \\ 1 & 1 \end{vmatrix} = 3$$

$$A_{22} = \begin{vmatrix} 1 & 2 \\ -2 & 1 \end{vmatrix} = 5 \qquad A_{23} = - \begin{vmatrix} 1 & -1 \\ -2 & 1 \end{vmatrix} = 1$$

$$A_{31} = \begin{vmatrix} -1 & 2 \\ -1 & 0 \end{vmatrix} = 2 \qquad A_{32} = - \begin{vmatrix} 1 & 2 \\ 3 & 0 \end{vmatrix} = 6$$

$$A_{33} = \begin{vmatrix} 1 & -1 \\ 3 & -1 \end{vmatrix} = 2$$

and

$$|A| = \begin{vmatrix} 1 & -1 & 2 \\ 3 & -1 & 0 \\ -2 & 1 & 1 \end{vmatrix} = \begin{vmatrix} 1-2(-2) & -1-2(1) & 2-2(1) \\ 3 & -1 & 0 \\ -2 & 1 & 1 \end{vmatrix}$$

$$= \begin{vmatrix} 5 & -3 & 0 \\ 3 & -1 & 0 \\ -2 & 1 & 1 \end{vmatrix} = \begin{vmatrix} 5 & -3 \\ 3 & -1 \end{vmatrix} = -5 - (-9) = 4$$

Thus,

$$A^{-1} = \frac{1}{4} \begin{bmatrix} -1 & -3 & 1 \\ 3 & 5 & 1 \\ 2 & 6 & 2 \end{bmatrix}^t = \frac{1}{4} \begin{bmatrix} -1 & 3 & 2 \\ -3 & 5 & 6 \\ 1 & 1 & 2 \end{bmatrix}$$

Checking, we have

$$A^{-1} \cdot A = \frac{1}{4} \cdot \begin{bmatrix} -1 & 3 & 2 \\ -3 & 5 & 6 \\ 1 & 1 & 2 \end{bmatrix} \cdot \begin{bmatrix} 1 & -1 & 2 \\ 3 & -1 & 0 \\ -2 & 1 & 1 \end{bmatrix}$$

$$= \frac{1}{4} \begin{bmatrix} -1 \cdot 1 + 3 \cdot 3 + 2(-2) & -1(-1) + 3(-1) + 2 \cdot 1 & -1(2) + 3 \cdot 0 + 2 \cdot 1 \\ -3 \cdot 1 + 5(3) + 6(-2) & -3(-1) + 5(-1) + 6 \cdot 1 & -3(2) + 5 \cdot 0 + 6 \cdot 1 \\ 1 \cdot 1 + 1(3) + 2(-2) & 1(-1) + 1 \cdot (-1) + 2 \cdot 1 & 1 \cdot 2 + 1 \cdot 0 + 2 \cdot 1 \end{bmatrix}$$

$$= \frac{1}{4} \cdot \begin{bmatrix} 4 & 0 & 0 \\ 0 & 4 & 0 \\ 0 & 0 & 4 \end{bmatrix} = \begin{bmatrix} 1 & 0 & 0 \\ 0 & 1 & 0 \\ 0 & 0 & 1 \end{bmatrix}$$

We conclude this section with an application in cryptography. Cryptography is the art of writing or deciphering secret codes. The following example will illustrate how matrices are used in coding and decoding messages.

Suppose we wish to send the message

CEASE FIRE

First establish a *one-to-one* correspondence between the letters of the alphabet and a set of 26 numbers. For simplicity here, we take the first 26 natural numbers in order as follows:

With the above prescribed correspondence, the command CEASE FIRE becomes

3 5 1 19 5 6 9 18 5

which in itself is sort of a coding, but one that could be easily deciphered. Note that the space separating the two words is not identified in the code. We now group these numbers in sets of three (any size would do). If the last set has fewer than three members, just fill in with Z's.

3 5 1 19 5 6 9 18 5

and then form a matrix comprised of these sets as columns. We obtain

$$\begin{bmatrix} 3 & 19 & 9 \\ 5 & 5 & 18 \\ 1 & 6 & 5 \end{bmatrix} \text{ which decoded is } \begin{bmatrix} C & S & I \\ E & E & R \\ A & F & E \end{bmatrix}$$

Finally, we multiply the above coded matrix on the left side by any 3×3 nonsingular matrix (called the *mixer*), which really garbles up the message. This is performed as follows:

$$\begin{bmatrix} 1 & 2 & -1 \\ -1 & -2 & 3 \\ 2 & 1 & -4 \end{bmatrix} \cdot \begin{bmatrix} 3 & 19 & 9 \\ 5 & 5 & 18 \\ 1 & 6 & 5 \end{bmatrix} = \begin{bmatrix} 12 & 23 & 40 \\ -10 & -11 & -30 \\ 7 & 19 & 16 \end{bmatrix}$$

The Mixer   The Message   The Coded Message

The coded matrix is sent to front-line headquarters, where it is decoded by multiplying this matrix on the left by the *inverse mixer* matrix. Computing the inverse of the mixer and multiplying, we obtain

$$\frac{1}{6} \cdot \begin{bmatrix} 5 & 7 & 4 \\ 2 & -2 & -2 \\ 3 & 3 & 0 \end{bmatrix} \cdot \begin{bmatrix} 12 & 23 & 40 \\ -10 & -11 & -30 \\ 7 & 19 & 16 \end{bmatrix} = \begin{bmatrix} 3 & 19 & 9 \\ 5 & 5 & 18 \\ 1 & 6 & 5 \end{bmatrix}$$

which, of course, is the original message matrix. Having the more simple coded relationship between numbers and letters of the alphabet, they then obtain the lettered matrix

$$\begin{bmatrix} C & S & I \\ E & E & R \\ A & F & E \end{bmatrix}$$

from which the message CEASE FIRE is determined.

Let us now review the steps you take using this method to code a message. First, prescribe a *one-to-one* correspondence between the 26 letters of the alphabet and 26 numbers. Next *group* in order the letters of the message in two's, three's, or more (if the last such grouping is short a letter or two, fill in with Z's). Then form a matrix whose columns are composed of these groups — first column is the first group, second column is the second group, and so forth. Let this *message* matrix be denoted A. Finally, pick a nonsingular *mixer* matrix M and multiply A by M on the left, obtaining the *coded* matrix B. The party receiving the coded matrix B must, of course, have the *inverse mixer* M^{-1}. By multiplying B by M^{-1} they obtain A, as follows:

$$M^{-1}B = M^{-1}(MA) = (M^{-1}M)A = IA = A$$

The message matrix A is then translated into a letter matrix, assuming the receiving party also knows the number letter correspondence prescribed above. From this latter matrix the message is identified, knowing that the letters of each column from top to bottom arranged horizontally form the message.

<h2>Solving Systems of Linear Equations by Matrix and Determinant Methods   8.9</h2>

Another important application of matrices is in analyzing and solving systems of equations. Take for example, the following system of three equations in three unknowns:

$$\begin{aligned} x - 2y + z &= 2 \\ 2x + 2y - z &= 4 \\ x + y &= 0 \end{aligned} \tag{1}$$

We wish to find an ordered triplet (x_1, y_1, z_1) that satisfies all three equations simultaneously. Because of the special way matrix multiplication was defined, the above system can be written as a single matrix equation. This is accomplished as follows:

Form a 3×3 matrix A composed of the coefficients of the unknowns x, y, and z in the order they appear in the system array. (The order in which x, y, and z occur in each equation must be the same — otherwise, you must so arrange them.) For the above system we obtain

$$A = \begin{bmatrix} 1 & -2 & 1 \\ 2 & 2 & -1 \\ 1 & 1 & 0 \end{bmatrix}$$

Next, form the matrix equation

$$\begin{bmatrix} 1 & -2 & 1 \\ 2 & 2 & -1 \\ 1 & 1 & 0 \end{bmatrix} \cdot \begin{bmatrix} x \\ y \\ z \end{bmatrix} = \begin{bmatrix} 2 \\ 4 \\ 0 \end{bmatrix} \tag{2}$$

where the 3×1 matrix on the right is obtained using the constants on the right of (1), in their respective order. The matrix equation (2) is equivalent to the system (1). To see this, perform the indicated multiplication as follows:

$$\begin{bmatrix} 1 & -2 & 1 \\ 2 & 2 & -1 \\ 1 & 1 & 0 \end{bmatrix} \begin{bmatrix} x \\ y \\ z \end{bmatrix} = \begin{bmatrix} 2 \\ 4 \\ 0 \end{bmatrix} \iff \begin{bmatrix} 1 \cdot x - 2y + 1 \cdot z \\ 2x + 2y - 1 \cdot z \\ 1 \cdot x + 1 \cdot y + 0 \cdot z \end{bmatrix} = \begin{bmatrix} 2 \\ 4 \\ 0 \end{bmatrix}$$

But by definition of matrix equality, this implies

$$\begin{aligned} x - 2y + z &= 2 \\ 2x + 2y - z &= 4 \\ x + y &= 0 \end{aligned}$$

which is precisely the system (1). Thus, the matrix equation (2) is indeed equivalent to the system (1). For notational convenience, we write (2) as

$$AX = B \text{ where } X = \begin{bmatrix} x \\ y \\ z \end{bmatrix}, B = \begin{bmatrix} 2 \\ 4 \\ 0 \end{bmatrix} \tag{3}$$

and A is defined as above. Because matrices satisfy all the field axioms except the commutative law of multiplication, equation (3) may be solved using the general manipulative properties of a field to which you are already accustomed, only remembering never to arbitrarily commute matrices. We proceed as follows:

If A is nonsingular, then

$$AX = B \iff A^{-1}(AX) = A^{-1}B \iff (A^{-1}A)X = A^{-1}B$$

$$\iff IX = A^{-1}B \iff X = A^{-1}B$$

The unknown matrix X is now determined as the product $A^{-1}B$. Because

$$A = \begin{bmatrix} 1 & -2 & 1 \\ 2 & 2 & -1 \\ 1 & 1 & 0 \end{bmatrix}$$

we have

$$A^{-1} = \frac{1}{|A|} \cdot \begin{bmatrix} A_{11} & A_{12} & A_{13} \\ A_{21} & A_{22} & A_{23} \\ A_{31} & A_{32} & A_{33} \end{bmatrix} = \frac{1}{3} \cdot \begin{bmatrix} 1 & -1 & 0 \\ 1 & -1 & -3 \\ 0 & 3 & 6 \end{bmatrix} = \frac{1}{3} \cdot \begin{bmatrix} 1 & 1 & 0 \\ -1 & -1 & 3 \\ 0 & -3 & 6 \end{bmatrix}$$

So that

$$X = \frac{1}{3} \cdot \begin{bmatrix} 1 & 1 & 0 \\ -1 & -1 & 3 \\ 0 & -3 & 6 \end{bmatrix} \cdot \begin{bmatrix} 2 \\ 4 \\ 0 \end{bmatrix} = \frac{1}{3} \begin{bmatrix} 1 \cdot 2 + 1 \cdot 4 + 0 \cdot 0 \\ -1 \cdot 2 - 1 \cdot 4 + 3 \cdot 0 \\ 0 \cdot 2 - 3 \cdot 4 + 6 \cdot 0 \end{bmatrix}$$

$$= \frac{1}{3} \cdot \begin{bmatrix} 6 \\ -6 \\ -12 \end{bmatrix} = \begin{bmatrix} 2 \\ -2 \\ -4 \end{bmatrix}$$

$$\implies \begin{bmatrix} x \\ y \\ z \end{bmatrix} = \begin{bmatrix} 2 \\ -2 \\ -4 \end{bmatrix}$$

Therefore, $x = 2$, $y = -2$, and $z = -4$. The matrix method for solving a system is limited, however, to systems for which the coefficient matrix A is nonsingular. You will recall we did not experience such a limitation when applying the diagonalization method of Section 8.2.

We summarize the above observations as follows:

THEOREM 8.9.1 The following linear system composed of n equations in n unknowns

$$\begin{aligned} a_{11}x_1 + a_{12}x_2 + \cdots + a_{1n}x_n &= b_1 \\ a_{21}x_1 + a_{22}x_2 + \cdots + a_{2n}x_n &= b_2 \\ &\vdots \\ a_{n1}x_1 + a_{n2}x_2 + \cdots + a_{nn}x_n &= b_n \end{aligned} \tag{4}$$

is equivalent to the matrix equation

$$AX = B \tag{5}$$

where

$$A = \begin{bmatrix} a_{11} & a_{12} & \cdots & a_{1n} \\ a_{21} & a_{22} & \cdots & a_{2n} \\ \vdots & & & \\ a_{n1} & a_{n2} & \cdots & a_{nn} \end{bmatrix}$$

$$X = \begin{bmatrix} x_1 \\ x_2 \\ \vdots \\ x_n \end{bmatrix} \quad \text{and} \quad B = \begin{bmatrix} b_1 \\ b_2 \\ \vdots \\ b_n \end{bmatrix}$$

If A is nonsingular, the solution of (5), and hence (4), is given by

$$X = A^{-1}B \tag{6}$$

The solution product $A^{-1}B$ can also be shown to possess a form composed of determinants only. The steps leading to this significant form are as follows:

$$A^{-1}B = \frac{1}{|A|} \cdot \begin{bmatrix} A_{11} & A_{21} & A_{31} & \cdots & A_{n1} \\ A_{12} & A_{22} & A_{32} & \cdots & A_{n2} \\ \vdots & & & & \\ A_{1n} & A_{2n} & A_{3n} & \cdots & A_{nn} \end{bmatrix} \cdot \begin{bmatrix} b_1 \\ b_2 \\ \vdots \\ b_n \end{bmatrix}$$

$$= \frac{1}{|A|} \cdot \begin{bmatrix} b_1 A_{11} + b_2 A_{21} + \cdots + b_n A_{n1} \\ b_1 A_{12} + b_2 A_{22} + \cdots + b_n A_{n2} \\ \vdots \\ b_1 A_{1n} + b_2 A_{2n} + \cdots + b_n A_{nn} \end{bmatrix}$$

But

$$b_1 A_{1j} + b_2 A_{2j} + \cdots + b_n A_{nj} = \begin{vmatrix} a_{11} & a_{12} & \cdots & b_1 & \cdots & a_{1n} \\ a_{21} & a_{22} & \cdots & b_2 & \cdots & a_{2n} \\ \vdots & & & & & \\ a_{n1} & a_{n2} & \cdots & b_n & \cdots & a_{nn} \end{vmatrix}$$

*j*th column ⤵

Hence, the element in the *j*th row of the $n \times 1$ matrix $A^{-1}B$, which by (6) is equal to the *j*th unknown x_j, is

$$x_j = \frac{\begin{vmatrix} a_{11} & a_{12} & \cdots & b_1 & \cdots & a_{1n} \\ a_{21} & a_{22} & \cdots & b_2 & \cdots & a_{2n} \\ \vdots & & & & & \\ a_{n1} & a_{n2} & \cdots & b_n & \cdots & a_{nn} \end{vmatrix}}{|A|} \qquad j = 1, 2, \ldots, n$$

This important result, called Cramer's rule, provides a method for solving systems of n linear equations in n unknowns using determinants exclusively, if $|A| \neq 0$. This rule may be expressed briefly as follows:

CRAMER'S RULE If the determinant of the coefficient matrix A of the system (4) is not zero, then the unknown x_j is equal to the quotient of two determinants. The denominator is the determinant of the coefficient matrix A, and the determinant in the numerator corresponds with that of the denominator except for the *j*th column, which is replaced by the column matrix B.

Example 8.9.1 By Cramer's rule, solve the linear system

$$
\begin{aligned}
x - 2y + z &= 2 \\
2x + 2y - z &= 4 \\
x + y &= 0
\end{aligned}
$$

This system is the same as that solved by matrix techniques earlier in this section.

Solution

$$
x = \frac{\begin{vmatrix} 2 & -2 & 1 \\ 4 & 2 & -1 \\ 0 & 1 & 0 \end{vmatrix}}{\begin{vmatrix} 1 & -2 & 1 \\ 2 & 2 & -1 \\ 1 & 1 & 0 \end{vmatrix}} = \frac{\begin{vmatrix} 2 & -2 & 1 \\ 4 & 2 & -1 \\ 0 & 1 & 0 \end{vmatrix}}{\begin{vmatrix} 1 & -2 & 1 \\ 3 & 0 & 0 \\ 1 & 1 & 0 \end{vmatrix}} = \frac{-\begin{vmatrix} 2 & 1 \\ 4 & -1 \end{vmatrix}}{\begin{vmatrix} 3 & 0 \\ 1 & 1 \end{vmatrix}} = \frac{6}{3} = 2
$$

$$
y = \frac{\begin{vmatrix} 1 & 2 & 1 \\ 2 & 4 & -1 \\ 1 & 0 & 0 \end{vmatrix}}{3} = \frac{\begin{vmatrix} 2 & 1 \\ 4 & -1 \end{vmatrix}}{3} = \frac{-6}{3} = -2
$$

$$
z = \frac{\begin{vmatrix} 1 & -2 & 2 \\ 2 & 2 & 4 \\ 1 & 1 & 0 \end{vmatrix}}{3} = \frac{\begin{vmatrix} 1 & -3 & 2 \\ 2 & 0 & 4 \\ 1 & 0 & 0 \end{vmatrix}}{3} = \frac{3\begin{vmatrix} 2 & 4 \\ 1 & 0 \end{vmatrix}}{3} = \frac{-12}{3} = -4
$$

The results, of course, are the same as we obtained using matrices. Generally Cramer's rule is simpler to apply than matrix techniques.

Example 8.9.2 Solve the system

$$
\begin{aligned}
-x + 2y &= 3 \\
3x - 5y &= -5
\end{aligned}
$$

Solution   By Cramer's rule,

$$
x = \frac{\begin{vmatrix} 3 & 2 \\ -5 & -5 \end{vmatrix}}{\begin{vmatrix} -1 & 2 \\ 3 & -5 \end{vmatrix}} = \frac{-15 - (-10)}{5 - 6} = \frac{-5}{-1} = 5
$$

$$
y = \frac{\begin{vmatrix} -1 & 3 \\ 3 & -5 \end{vmatrix}}{-1} = \frac{5 - 9}{-1} = \frac{-4}{-1} = 4
$$

To check these answers, we substitute 5 and 4 for x and y, respectively, in the system to obtain

$$-(5) + 2(4) = 3$$
$$3(5) - 5(4) = -5$$

which reduces to the true statements

$$3 = 3$$
$$-5 = -5$$

Thus both equations are satisfied simultaneously by the solution pair $(5, 4)$.

The two methods for solving a linear system discussed above require the determinant of the coefficient matrix A be nonzero. If $|A| \neq 0$, then there is a *unique* solution given by either of the methods.

8.10 Problem Sets

Problem Set I ■ *Reading Comprehension*

Determine whether statements 1 through 8 are true or false and give reasons for your responses.

1 Inverses are not defined for nonsquare matrices.
2 A necessary and sufficient condition for the existence of a matrix inverse is that the determinant be nonzero.
3 The transpose of a matrix is a matrix with two rows transposed.
4 The transpose of a 3×4 matrix is a 3×4 matrix whose rows and columns have been interchanged.
5 The determinant of the 1×1 matrix $[-2]$ is 2.
6 The formula for the inverse of a matrix whose determinant is nonzero is

$$A^{-1} = \frac{1}{|A|} (A_{ij})$$

7 Both the matrix method and Cramer's rule for solving linear systems of equations is limited to square systems—the number of equations and unknowns equal.
8 If the determinant of a coefficient matrix from a system of linear equations exists and is not equal to zero, then the system is assured one solution and possibly infinitely many solutions.

Problem Set II ■ *Skills Development*

Show that

1 $(A^{-1})^{-1} = A$ if $A = \begin{bmatrix} -1 & 2 \\ 1 & 4 \end{bmatrix}$

2 $(AB)^{-1} = B^{-1} \cdot A^{-1}$ if $A = \begin{bmatrix} -1 & 2 \\ 1 & 4 \end{bmatrix}$ and $B = \begin{bmatrix} 3 & 2 \\ 4 & 3 \end{bmatrix}$

3 $|A^{-1}| = \dfrac{1}{|A|}$ if $A = \begin{bmatrix} 2 & -1 \\ 4 & 3 \end{bmatrix}$

Problem Set III ■ Theoretical Developments

1 If A is an $m \times p$ matrix and B a $p \times n$ matrix, prove $(AB)^t = B^t A^t$.
2 If A is nonsingular, prove $(A^{-1})^t = (A^t)^{-1}$.
3 If A is nonsingular, prove $(A^{-1})^2 = (A^2)^{-1}$.
4 If A is nonsingular, prove $(A^{-1})^{-1} = A$.
5 Prove that $|B^{-1}AB| = |A|$ if A and B are nonsingular.
6 Given the system

$$a_{11}x_1 + a_{12}x_2 + a_{13}x_3 = 0$$
$$a_{21}x_1 + a_{22}x_2 + a_{23}x_3 = 0$$
$$a_{31}x_1 + a_{32}x_2 + a_{33}x_3 = 0$$

If $|A| \neq 0$, prove that the only solution is the trivial solution, namely, $x = 0$, $y = 0$, and $z = 0$.
7 Prove matrix inverses, when they exist, are unique.
8 If $|A| \neq 0$ and $|B| \neq 0$, prove $(AB)^{-1} = B^{-1}A^{-1}$.

Problem Set IV ■ Applications

1 An oil company produces gasoline, oil, and natural gas. The process of producing these three products requires the use of some of these products themselves. Suppose to produce 1 unit of gasoline, the company uses 0 units of gasoline, 2 units of oil, and 1 unit of natural gas. To produce 1 unit of oil the company uses 0 units of gasoline, $\frac{1}{10}$ unit of oil, and $\frac{1}{5}$ unit of natural gas. And finally, to produce 1 unit of natural gas, the company uses $\frac{1}{8}$ unit of gasoline, $\frac{1}{5}$ unit of oil, and $\frac{1}{10}$ unit of natural gas. We arrange these data in the following matrix:

$$U = \begin{matrix} & \begin{matrix} \text{Gas} & \text{Oil} & \text{Nat.Gas} \end{matrix} & \\ & \begin{bmatrix} 0 & 0 & \frac{1}{8} \\ 2 & \frac{1}{10} & \frac{1}{5} \\ 1 & \frac{1}{5} & \frac{1}{10} \end{bmatrix} & \begin{matrix} \text{Gas} \\ \text{Oil} \\ \text{Nat.Gas} \end{matrix} \end{matrix} \Big\} \text{Used}$$

One Unit of

Now let x, y, and z represent the number of units, respectively, of gasoline, oil, and natural gas produced daily by the company, and represent these by the production matrix

$$P = \begin{bmatrix} x \\ y \\ z \end{bmatrix}$$

It should be clear that the matrix product UP provides the amount of gasoline, oil, and natural gas used to produce x, y, and z units of these three commodities. Suppose the company receives an order for 1000 units of gasoline, 300 units of oil, and 500 units of natural gas. Let the matrix A represent the amount ordered. That is

$$A = \begin{bmatrix} 1000 \\ 300 \\ 500 \end{bmatrix}$$

Calculate the number of units of each of the three products the company will have to produce to fill this order. (HINT: Some of the products will be used in the production process and hence must be accounted for.) An equation for this and the amount ordered would be

$$P - UP = A$$

This can be written as

$$(I - U)P = A$$

where I is the 3×3 multiplicative identity matrix.

2 Code the following messages using the correspondence

and the *mixer* matrix

$$\begin{bmatrix} 1 & 2 & -1 \\ 3 & 0 & 1 \\ 2 & -1 & 1 \end{bmatrix}$$

a Attack   b Come home   c Give up

3 Decode the following coded matrices that were determined using the information of Problem 2 above.

a $\begin{bmatrix} 14 & 23 \\ 44 & 35 \\ 22 & 11 \end{bmatrix}$ b $\begin{bmatrix} 10 & 17 \\ 92 & 71 \\ 64 & 42 \end{bmatrix}$ c $\begin{bmatrix} 8 & 27 & 46 \\ 74 & 55 & 86 \\ 51 & 32 & 40 \end{bmatrix}$

Problem Set V ■ *Just For Fun*

1 a Smith, Jones, and Robinson are the engineer, brakeman, and fireman on a train, but not necessarily in that order. Riding the train are three passengers with the same three surnames, to be identified in the following premises by a "Mr." before their names.

   b Mr. Robinson lives in Los Angeles.

   c The brakeman lives in Omaha.

   d Mr. Jones long ago forgot all the algebra he learned in high school.

   e The passenger whose name is the same as the brakeman's lives in Chicago.

   f The brakeman and one of the passengers, a distinguished mathematical physicist, attend the same church.

   g Smith beat the fireman at billiards. Who is the engineer? (HINT: Use the following two matrices.)

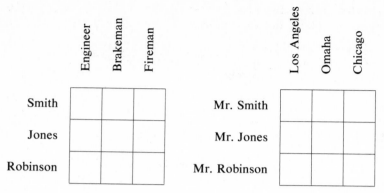

2 Professor Merle White of the mathematics department, Professor Leslie Black of philosophy, and Jan Brown, a young stenographer who worked in the university's office of admissions, were lunching together.

"Isn't it remarkable," observed the lady, "that our last names are Black, Brown, and White and that one of us has black hair, one brown hair, and one white?"

"It is indeed," replied the person with black hair, "and have you noticed that not one of us has hair that matches his or her name?"

"By golly, you're right!" exclaimed Professor White.

If the lady's hair isn't brown, what is the color of Professor Black's hair?

Summary of Definitions and Properties in Chapter 8 8.11

Gauss Reduction (Diagonalizing) Method (p. 282)

Matrix (p. 288)

D8.4.1 $A = B \Longleftrightarrow [a_{ij} = b_{ij}$ for each $i, j]$ (p. 291)

D8.4.2 $(a_{ij}) + (b_{ij}) = (a_{ij} + b_{ij})$ (p. 291)

T8.4.1 (p. 292)
I Matrices of the same dimensions are *closed* under addition.
II $(A + B) + C = A + (B + C)$ (Associative Law)
III $A + B = B + A$ (Commutative Law)
IV $A + 0 = A$ (Additive Identity Matrix)
V $A + [-A] = 0$ (Additive Inverse Matrix)

D8.4.3 $A - B = A + [-B]$ (p. 293)

D8.4.4 $cA = c(a_{ij}) = (ca_{ij})$ (p. 295)

T8.4.2 (p. 295)
I aA is a matrix (Closure Law)
II $(ab)A = a(bA)$ (Associative Law)
III a $(a + b)A = aA + bA$ (Distributive Law)
    b $a(A + B) = aA + aB$ (Distributive Law)

IV　　**a**　$1A = A$　(Multiplicative Identity)　　　　　　　　　(p. 295)
　　　　b　$(-1)A = -A$　(Additive Inverse Property)

V　　**a**　$0A = 0$　(Additive Identity Property under Multiplication)
　　　　b　$a0 = 0$　(Additive Identity Property under Multiplication)

D8.4.5　$AB = C$, where
　　　　$c_{ij} = a_{i1}b_{1j} + a_{i2}b_{2j} + a_{i3}b_{3j} + \cdots + a_{ip}b_{pj}$　　　　　(p. 296)

T8.4.3　　　　　　　　　　　　　　　　　　　　　　　　　(p. 298)
I　　Matrices that are conformable are *closed* under multiplication.
II　　$(A \cdot B) \cdot C = A \cdot (B \cdot C)$　(Associative Law)
III　**a**　$B[C + D] = BC + BD$　(Left-Hand Distributive Law)
　　　　b　$[C + D]E = CE + DE$　(Right-Hand Distributive Law)

T8.4.4　$I_{m \times m}A_{m \times n} = A_{m \times n}$　and　$A_{m \times n}I_{n \times n} = A_{m \times n}$　(p. 299)

D8.6.1　$\begin{vmatrix} a_{11} & a_{12} \\ a_{21} & a_{22} \end{vmatrix} = a_{11}a_{22} - a_{12}a_{21}$　　　　　　　(p. 306)

D8.6.2　$A_{ij} = (-1)^{i+j} \cdot M_{ij}$　　　　　　　　　　　　(p. 307)

D8.6.3　$|A| = a_{i1}A_{i1} + a_{i2}A_{i2} + a_{i3}A_{i3}$　　　　　　　(p. 308)

D8.6.4　$|A| = a_{i1}A_{i1} + a_{i2}A_{i2} + \cdots + a_{in}A_{in}$　　　　　(p. 311)

T8.6.1　If all entries in one column or one row are zero, then $|A| = 0$.　(p. 311)

T8.6.2　If two rows or two columns of a determinant are interchanged,　(p. 311)
　　　　the sign of the determinant is reversed.

T8.6.3　　　　　　　　　　　　　　　　　　　　　　　　　(p. 312)

T8.6.4　　　　　　　　　　　　　　　　　　　　　　　　　(p. 313)

T8.6.5　　　　　　　　　　　　　　　　　　　　　　　　　(p. 314)

T8.6.6　　　　　　　　　　　　　　　　　　　　　　　　　(p. 316)

The Inverse of a Square Matrix　　　　　　　　　　　　　　(p. 321)

The Transpose of a Matrix　　　　　　　　　　　　　　　　(p. 322)

T8.8.1　$A^{-1} = \dfrac{1}{|A|} (A_{ij})^t$　　　　　　　　　　　　(p. 324)

Cryptography　　　　　　　　　　　　　　　　　　　　　(p. 325)

T8.9.1　　　　　　　　　　　　　　　　　　　　　　　　　(p. 329)

Cramer's Rule　　　　　　　　　　　　　　　　　　　　　(p. 330)

Mathematical Induction — Sequences and Series

9 CHAPTER

Introduction 9.1

We introduce the material in this chapter with the following problem.

Problem 9.1.1   A man borrows $10,000.00 at a rate of 10 percent per annum simple interest. He agrees to pay $250.00 plus interest each month until the indebtedness is paid off. How much total interest will he pay for this loan?

Solution   We first must understand the meaning of simple interest If a *principal* of P dollars is borrowed for a period of n years at a *simple* interest rate r (expressed as a decimal) per annum, then the *interest* I for that period is

$$I = P \cdot n \cdot r$$

Because he is paying $250.00 per month against the initial principal of $10,000.00, it will take 40 pay periods (months) to pay off the loan $(40 \cdot 250 = 10{,}000)$. According to the above formula, the interest to be paid with the *first* payment is

$$I_1 = (10{,}000)\left(\tfrac{1}{12}\right)(0.1) = 40\left(\tfrac{25}{12}\right)$$

For the second month the principal is \$9,750.00. Hence, the interest is

$$I_2 = (9{,}750)\left(\tfrac{1}{12}\right)(0.1) = 39\left(\tfrac{25}{12}\right)$$

Continuing in this manner, we obtain

$$I_3 = (9{,}500)\left(\tfrac{1}{12}\right)(0.1) = 38\left(\tfrac{25}{12}\right)$$
$$\vdots$$
$$I_{39} = (500)\left(\tfrac{1}{12}\right)(0.1) = 2\left(\tfrac{25}{12}\right)$$
$$I_{40} = (250)\left(\tfrac{1}{12}\right)(0.1) = \tfrac{25}{12}$$

The total interest is, of course, the sum of these, that is

$$I = I_1 + I_2 + I_3 + \cdots + I_{39} + I_{40}$$

This sum may be expressed as follows (beginning with the last term):

$$I = \tfrac{25}{12} + 2\left(\tfrac{25}{12}\right) + 3\left(\tfrac{25}{12}\right) + \cdots + 39\left(\tfrac{25}{12}\right) + 40\left(\tfrac{25}{12}\right)$$
$$= \tfrac{25}{12}(1 + 2 + 3 + \cdots + 39 + 40)$$

Further simplification of this answer requires our determining the sum of the first 40 natural numbers. Although this can certainly be done by performing the sum one term at a time until all 40 have been added together, there is a much simpler and more satisfying way to compute this sum. We shall delay, therefore, the completion of this problem until these simpler summing techniques have been developed in the next section.

9.2 Mathematical Induction

As suggested by Problem 9.1.1, it would be convenient to have a formula for the sum

$$1 + 2 + 3 + \cdots + n$$

where n represents the last term in this series of counting numbers (in Problem 9.1.1, $n = 40$). Let us examine these sums for $n = 1, 2, 3, 4, 5$. We have

$$n = 1 \Longrightarrow 1$$
$$n = 2 \Longrightarrow 1 + 2 = 3$$

$$n = 3 \Longrightarrow 1 + 2 + 3 = 6$$
$$n = 4 \Longrightarrow 1 + 2 + 3 + 4 = 10$$
$$n = 5 \Longrightarrow 1 + 2 + 3 + 4 + 5 = 15$$

It may not yet be apparent to you from the sums $1, 3, 6, 10, 15$ that there is a recurring pattern that they are exhibiting. Observe that these sums may be written

$$n = 1 \Longrightarrow 1 = \frac{1 \cdot 2}{2}$$
$$n = 2 \Longrightarrow 1 + 2 = 3 = \frac{2 \cdot 3}{2}$$
$$n = 3 \Longrightarrow 1 + 2 + 3 = 6 = \frac{3 \cdot 4}{2}$$
$$n = 4 \Longrightarrow 1 + 2 + 3 + 4 = 10 = \frac{4 \cdot 5}{2}$$
$$n = 5 \Longrightarrow 1 + 2 + 3 + 4 + 5 = 15 = \frac{5 \cdot 6}{2}$$

From the suggested pattern, one would suspect that

$$n = 6 \Longrightarrow 1 + 2 + 3 + 4 + 5 + 6 = \frac{6 \cdot 7}{2} = 21$$

and, interestingly, 21 is the correct answer. In fact, what appears to be true in general is the following formula

$$1 + 2 + 3 + \cdots + n = \frac{n(n + 1)}{2}$$

where the counting number n (representing the last term) is arbitrary. For example, if $n = 40$ (as in Problem 9.1.1), we have

$$1 + 2 + 3 + \cdots + 39 + 40 = \frac{40 \cdot 41}{2} = 820$$

We are able, by this formula, to arrive at the answer without extensive calculations. However, we are not certain yet whether or not the formula holds for $n = 40$. We have only verified it for the integers 1 through 6. And just because a pattern holds true for a few counting numbers does not necessarily imply that it will hold true indefinitely. Take for example the sequence of numbers

$$2, \tfrac{5}{4}, 1, \tfrac{7}{8}, \tfrac{4}{5}, \cdots$$

With some careful thought and perhaps a little luck you might recognize that this could also be written

$$\tfrac{4}{2}, \tfrac{5}{4}, \tfrac{6}{6}, \tfrac{7}{8}, \tfrac{8}{10}, \cdots$$

from which you would suspect the next few members in sequence would logically be

$$\frac{9}{12}, \frac{10}{14}, \frac{11}{16}, \cdots$$

It would appear a general formula for each term would be

$$\frac{n+3}{2n}$$

where for $n = 1$ we get the first term $\frac{4}{2}$, for $n = 2$ the second $\frac{5}{4}$, and so on. Note that a formula that would render the *same* sequence of terms as the above for $n = 1, 2, 3, 4, 5$ is

$$\frac{n+3}{2n} + (n-1)(n-2)(n-3)(n-4)(n-5)$$

For $n = 6$, however, this formula gives

$$\frac{9}{12} + 5 \cdot 4 \cdot 3 \cdot 2 \cdot 1 = \frac{9}{12} + 120$$

a term quite distinct from the suggested one of $\frac{9}{12}$. Although this is an over-simplified example from which it was obvious by the nature of the latter generating formula that there would be a shift in the pattern from $n = 6$ on, it is not always this clear.

You have probably experienced questions on aptitude tests regarding a sequence of three or four numbers wherein you were to discover a pattern or formula from which you could predict the succeeding entries in the sequence. Technically, questions such as this on aptitude tests are not answerable. One never knows for sure what the next entry will be when just a finite number (usually three or four) of the members of a sequence are given. True, the answer such tests seek for a correct response is the most logical sequence pattern, not necessarily what must hold. In mathematics, however, it is important that we determine ways to verify whether or not a suggested formula pattern *must* hold true for *all* positive integers n.

If we were to *conjecture* that

$$1 + 2 + 3 + \cdots + n = \frac{n(n+1)}{2}$$

for *all* positive integers n, it could not be verified by simply checking the formula out for each such integer, there being infinitely many. Even if there were only a billion or two counting numbers to check, no one would be very interested in the project. Fortunately, there is a relatively simple way to prove the validity of formulas like the one above. The technique is called mathematical induction. We state this important principle in the theorem that follows.

THEOREM 9.2.1 PRINCIPLE OF MATHEMATICAL INDUCTION

Let $\quad P(n) = Q(n)$ $\hfill$ (1)

be an *open* sentence in the variable n. If the following two conditions hold

(a) $P(1) = Q(1)$ $\quad$ (1 is a solution)

and

(b) $[P(k) = Q(k)] \Longrightarrow [P(k+1) = Q(k+1)]$

(meaning *if* the number k is a solution of (1) then the number $k+1$ is also a solution), then the open sentence (1) is true for all natural numbers—it is an *identity* over the set of natural numbers.

Proof The proof we shall give depends upon a special property of the natural numbers—a property so simple that most people would regard it as insignificant, yet many mathematical arguments are based upon the property. It is called the well-ordering principle of the set of natural numbers. This principle states that every non-empty subset of natural numbers contains an element which is less then every other member of the set. For example the set

$$\{10, 6, 21, 42, 173\}$$

of natural numbers contains 6, which is less than all other members of this set. The element 1, of course, possesses this property for the entire set of natural numbers. With this intuitively obvious property, we shall now prove that the two conditions (a) and (b) above are sufficient to assure the truth of (1) for all natural numbers.

Let S be the solution set of (1). We know S is not empty, for $1 \in S$ by hypothesis (a). *Suppose S does not* include all natural numbers. Let M denote the set of all natural numbers for which (1) *does not* hold. This is not empty by assumption. By the well-ordering principle, M possesses a *smallest* element. Since this element of M is a natural number greater than 1 (because $1 \in S$), it can be written in the form $k+1$, where k is a natural number ($3 = 2 + 1$, $15 = 14 + 1$, and so forth). Because $k+1$ is the *smallest* element in M, k (which of course is less than $k+1$) must be an element in S. But our second hypothesis (b) assures us that

$$[P(k) = Q(k)] \Longrightarrow [P(k+1) = Q(k+1)]$$

that is, if $k \in S$ then $(k+1) \in S$, which contradicts the *assumption* that $(k+1) \in M$. This contradiction resulted from our assuming the solution set S of (1) did not include all natural numbers. We must therefore conclude that that assumption is false. Hence, S does include *all* positive integers. $\qquad$ QED

It is important to understand from this theorem that if you can show conditions (a) and (b) hold for an open sentence, then the open sentence is true for all natural numbers.

Let us now apply the principle of mathematical induction to the problem with which we began.

We wish to show

$$1 + 2 + 3 + \cdots + (n - 1) + n = \frac{n(n + 1)}{2}$$

for all natural numbers. Here

$$P(n) = 1 + 2 + 3 + \cdots + (n - 1) + n$$

and

$$Q(n) = \frac{n(n + 1)}{2}$$

We first check condition (a). If $n = 1$, we have

$$P(1) = 1$$

and

$$Q(1) = \frac{1 \cdot (1 + 1)}{2} = 1$$

which implies $P(1) = Q(1)$, and hence condition (a) holds.

To check condition (b) requires that we prove the *conditional* proposition

$$[P(k) = Q(k)] \Longrightarrow [P(k + 1) = Q(k + 1)]$$

We must show that if

$$P(k) = Q(k)$$

that is

$$1 + 2 + 3 + \cdots + (k - 1) + (k) = \frac{k(k + 1)}{2} \qquad (2)$$

then $P(k + 1) = Q(k + 1)$. That is,

$$1 + 2 + 3 + \cdots + k + (k + 1) = \frac{(k + 1)(k + 2)}{2} \qquad (3)$$

This may be accomplished by noting that the only difference between the left-hand sides of (2) and (3) is the extra term (3) possesses, namely, $(k + 1)$. Thus, if we add $(k + 1)$ to both sides of (2), we will at least have obtained the left-hand side of (3). If adding $(k + 1)$ to the right side of (2) also reduces it to the right side of (3), then we will have proved proposition (b). This latter step usually requires some algebraic manipulation. We now proceed with the suggested steps.

$$P(k) = 1 + 2 + 3 + \cdots + (k-1) + k = \frac{k(k+1)}{2} = Q(k)$$

$$\Downarrow \text{ ALE}$$

ALE $\quad a = b \Longrightarrow a + c = b + c$

$$P(k+1) = 1 + 2 + 3 + \cdots + k + (k+1) = \frac{k(k+1)}{2} + (k+1)$$

$$= \frac{(k+1)(k+2)}{2}$$

$$= Q(k+1)$$

$$\xrightarrow{\text{TLE}} P(k+1) = Q(k+1)$$

Hence, we have shown $[P(k) = Q(k)] \Longrightarrow [P(k+1) = Q(k+1)]$. Therefore, by the principle of mathematical induction, we know

$$1 + 2 + 3 + \cdots + n = \frac{n(n+1)}{2} \tag{1}$$

for all $n \in \mathbf{N}$.

Thus, with two relatively simple steps, we are able to prove that (1) holds true for all natural numbers; and we only checked it out specifically for one natural number, namely, 1.

The principle of mathematical induction is sometimes compared with a common game played with dominoes. If these thin rectangular blocks are lined up on end such that whenever one falls it pushes over the next domino in line (which is analogous to $[P(k) = Q(k)] \Longrightarrow [P(k+1) = Q(k+1)]$), then if the first one is tipped over in the direction of the second (which is in turn analogous to $P(1) = Q(1)$), all dominoes will eventually fall. I like to think of the first step in mathematical induction, that is, showing $P(1) = Q(1)$, as the starter, and the second step, that is, $[P(k) = Q(k)] \Longrightarrow [P(k+1) = Q(k+1)]$, as the perpetuator. (See Fig. 9.2.1.)

FIGURE 9.2.1

Example 9.2.1 Prove by mathematical induction that

$$\frac{1}{1 \cdot 2} + \frac{1}{2 \cdot 3} + \frac{1}{3 \cdot 4} + \cdots + \frac{1}{(n-1)n} + \frac{1}{n(n+1)} = \frac{n}{n+1}$$

for each $n \in \mathbf{N}$.

Solution   Here

$$P(n) = \frac{1}{1 \cdot 2} + \frac{1}{2 \cdot 3} + \cdots + \frac{1}{(n-1)n} + \frac{1}{n(n+1)}$$

and

$$Q(n) = \frac{n}{n+1}$$

For $n = 1$, we have

$$\left. \begin{array}{l} P(1) \ = \dfrac{1}{1 \cdot 2} = \dfrac{1}{2} \\[2mm] \text{and} \\[2mm] Q(1) = \dfrac{1}{1+1} = \dfrac{1}{2} \end{array} \right\} \Longrightarrow P(1) = Q(1)$$

completing step (a). Next we must verify the *perpetuator*. For $n = k$ we have

$$P(k) = \frac{1}{1 \cdot 2} + \frac{1}{2 \cdot 3} + \cdots + \frac{1}{(k-1)k} + \frac{1}{k(k+1)} = \frac{k}{(k+1)}$$

$$= Q(k)$$

and for $n = k + 1$ we have

$$P(k+1) = \frac{1}{1 \cdot 2} + \frac{1}{2 \cdot 3} + \cdots + \frac{1}{k(k+1)} + \frac{1}{(k+1)(k+2)}$$

$$= \frac{k+1}{k+2} = Q(k+1)$$

We always write this latter form (for $k + 1$) to get a *clue* as to what we must do to the expression $P(k) = Q(k)$ to convert it into $P(k+1) = Q(k+1)$. This clue will nearly always come from the *left-hand sides* of the two. Note, in this case, they only differ by the term

$$\frac{1}{(k+1)(k+2)}$$

Hence, we add this to both sides of $P(k) = Q(k)$ as follows:

$$P(k) = \underbrace{\frac{1}{1 \cdot 2} + \frac{1}{2 \cdot 3} + \cdots + \frac{1}{(k-1)k} + \frac{1}{k(k+1)}}_{} = \frac{k}{k+1} = Q(k)$$

$$\big\Downarrow \text{ ALE}$$

$$P(k+1) = \overbrace{\frac{1}{1 \cdot 2} + \frac{1}{2 \cdot 3} + \cdots \frac{1}{k(k+1)} + \frac{1}{(k+1)(k+2)}}^{} = \frac{k}{k+1} + \frac{1}{(k+1)(k+2)}$$

$$= \frac{k(k+2) + 1}{(k+1)(k+2)}$$

ALE $a = b \Longrightarrow a + c = b + c$

$$= \frac{k^2 + 2k + 1}{(k+1)(k+2)}$$

$$= \frac{(k+1)^2}{(k+1)(k+2)}$$

$$= \frac{k+1}{k+2}$$

$$= Q(k+1)$$

$\xRightarrow{\text{TLE}} P(k+1) = Q(k+1)$ TLE $a = b = c \Longrightarrow a = c$

Hence, with conditions (a) and (b) both satisfied, we have shown by the principle of mathematical induction that

$$\frac{1}{1 \cdot 2} + \frac{1}{2 \cdot 3} + \cdots + \frac{1}{n(n+1)} = \frac{n}{n+1}$$

for each $n \in N$.

It is important to realize that *both* steps in a proof by mathematical induction are essential. The next two examples will bear this out.

Example 9.2.2   Prove

$$1 + 5 + 9 + \cdots + (4n - 3) = n(2n - 1) + (n - 1)$$

for each $n \in N$.

Solution   Here we have

$$P(n) = 1 + 5 + 9 + \cdots + (4n - 3)$$

and

$$Q(n) = n(2n - 1) + (n - 1)$$

If $n = 1$, we obtain

$$P(1) = 1$$

and

$$Q(1) = 1(2 \cdot 1 - 1) + (1 - 1) = 1$$

$\left.\rule{0pt}{24pt}\right\} \Longrightarrow P(1) = Q(1)$

and condition (a) checks out. However, if we were to attempt to prove condition (b), that is

$$[P(k) = Q(k)] \Longrightarrow [P(k+1) = Q(k+1)]$$

we would find it impossible. In fact, this condition would be contradicted by just taking $n = 2$, the successor to 1. We obtain

$$P(2) = 1 + 5 = 6$$

and

$$Q(2) = 2(2 \cdot 2 - 1) + (2 - 1) = 6 + 1 = 7$$

$\left.\rule{0pt}{24pt}\right\} \Longrightarrow P(2) \neq Q(2)$

Thus

$$[P(1) = Q(1)] \xRightarrow{\;/\;} [P(2) = Q(2)]$$

Thus, we must remember that the *first* step alone in mathematical induction is *not sufficient*.

Example 9.2.3 Prove

$$1 + 5 + 9 + \cdots + (4n - 7) + (4n - 3) = n(2n - 1) + 1 \quad (4)$$

for each $n \in N$.

Solution Let

$$P(n) = 1 + 5 + 9 + \cdots + (4n - 7) + (4n - 3)$$

and

$$Q(n) = n(2n - 1) + 1$$

Then the equation $P(k) = Q(k)$ is

$$1 + 5 + 9 + \cdots + (4k - 7) + (4k - 3) = k(2k - 1) + 1$$

whereas $P(k + 1) = Q(k + 1)$ is

$$1 + 5 + 9 + \cdots + (4k - 3) + [4(k + 1) - 3]$$
$$= (k + 1)[2(k + 1) - 1] + 1$$

As with earlier examples, the only difference between the left-hand sides of $P(k)$ and $P(k + 1)$ is the extra term $4(k + 1) - 3$ in $P(k+1)$. Hence, if we add this term to both sides of $P(k) = Q(k)$, we obtain the expression $P(k + 1)$ on the left. Specifically, we have

$$P(k) = \underbrace{1 + 5 + \cdots + (4k - 3)}_{} = k(2k - 1) + 1 = Q(k)$$

$$\Downarrow \text{ ALE}$$

$$\begin{aligned}
P(k + 1) = 1 + 5 + \cdots + (4k - 3) + [4(k + 1) - 3] &= k(2k - 1) + 1 + [4(k + 1) - 3] \\
&= [2k^2 - k + 4k + 4 - 3] + 1 \\
&= [2k^2 + 3k + 1] + 1 \\
&= (k + 1)(2k + 1) + 1 \\
&= (k + 1)[2(k + 1) - 1] + 1 \\
&= Q(k + 1)
\end{aligned}$$

ALE $\quad a = b \Longrightarrow a + c = b + c$

Hence, we have shown that

$$[P(k) = Q(k)] \Longrightarrow [P(k + 1) = Q(k + 1)]$$

Thus, if (4) is *ever* true for *some* positive integer k, it must hold for *all* succeeding integers. The problem here is that it is never true for *some* positive integer. Notice that when $n = 1$ we have

$$\left. \begin{aligned} P(1) &= 1 \\ \text{and} \quad \\ Q(1) &= 1(1 \cdot 2 - 1) + 1 = 2 \end{aligned} \right\} \Longrightarrow P(1) \neq Q(1)$$

The perpetuating process of condition (b) just never gets started. Step (a) of mathematical induction is essential to get the process started, whereas step (b) is essential to keep it going. One without the other is not sufficient.

The principle of mathematical induction works equally well for inequalities, as the next example will illustrate. Since this example introduces a mathematical symbol (! , called factorial), which may be new to you, let us examine its meaning first.

It is not infrequent in mathematical formulas to encounter *products* of natural numbers 1 through n. Take for example the products

$$1 \cdot 2 \cdot 3 \cdot 4$$
$$1 \cdot 2 \cdot 3 \cdot 4 \cdot 5 \cdot 6$$

and

$$1 \cdot 2 \cdot 3 \cdot 4 \cdot 5 \cdot 6 \cdot 7 \cdot 8 \cdot 9 \cdot 10 \cdot 11 \cdot 12$$

where n (the last factor) equals 4, 6, and 12, respectively. A more compact symbolic representation for these is

$$4!$$
$$6!$$

and

$$12!$$

read "four factorial," "six factorial," and "twelve factorial." In general, "n factorial" is defined as

$$n! = 1 \cdot 2 \cdot 3 \cdots (n-1) \cdot n$$

where n is a positive integer.

We are now ready for the next example.

Example 9.2.4 Prove

$$2^{n+3} < (n+3)! \tag{5}$$

for each $n \in N$.

Solution Let

$$P(n) = 2^{n+3}$$

and

$$Q(n) = (n+3)!$$

Then

$$P(1) = 2^{1+3} = 2^4 = 16$$

and

$$Q(1) = (1+3)! = 4! = 1 \cdot 2 \cdot 3 \cdot 4 = 24$$

$$\left. \right\} \Longrightarrow P(1) < Q(1)$$

verifying condition (a) that the *inequality* holds for $n = 1$. For $n = k$, (5) becomes

$$2^{k+3} < (k+3)! \tag{6}$$

whereas for $n = k + 1$, it becomes

$$2^{k+4} < (k+4)! \tag{7}$$

Because the only difference between the left-hand sides of (6) and (7) is a factor of 2, we need only multiply both sides of (6) by 2 to obtain at least the left side of (7). Proceeding with this we obtain

$$P(k) = \underbrace{2^{k+3} < (k+3)!}_{} = Q(k)$$

T2.11.2(III)
$a < b, c > 0 \Longrightarrow ac < bc$

$$\Big\Downarrow \text{T2.11.2(III)}$$

$$P(k+1) = 2^{k+4} = \overbrace{2^{k+3} \cdot 2 < (k+3)! \cdot 2}^{} < (k+3)! \cdot (k+4)$$
$$= (k+4)!$$
$$= Q(k+1)$$

$$\Longrightarrow P(k+1) < Q(k+1)$$

Hence,

$$[P(k) < Q(k)] \Longrightarrow [P(k+1) < Q(k+1)]$$

and we conclude by the principle of mathematical induction that

$$2^{n+3} < (n+3)!$$

for each $n \in N$.

9.3   Problem Sets

Problem Set I   ■   *Reading Comprehension*

Determine whether statements 1 through 4 are true or false and give reasons for your responses.

1   If it can be shown for the open equation $P(n) = Q(n)$ that
$$[P(k) = Q(k)] \Longrightarrow [P(k+1) = Q(k+1)]$$
then we may conclude that $P(n) = Q(n)$ is identically true for all n.

2   If the equation $P(n) = Q(n)$ is true for $n = 1, 2, 3, 4, 5, 6, 7, 8, 9$, it is probably true for all natural numbers.

3   Proof of a formula $P(n) = Q(n)$ by mathematical induction could be accomplished with the following two steps:
   (a) $P(1) = Q(1)$
   (b) $P(k+1) \neq Q(k+1) \Longrightarrow P(k) \neq Q(k)$

4   The expression $n!$ (read "n factorial") means to factor n completely.

By the principle of mathematical induction, prove that formulas 1 through 15 below are valid for all natural numbers.

1 $1 + 3 + 5 + \cdots + (2n - 1) = n^2$ **2** $2 + 4 + 6 + \cdots + 2n = n(n + 1)$

3 $2 + 6 + 10 + \cdots + (4n - 2) = 2n^2$

4 $1^2 + 2^2 + 3^2 + \cdots + n^2 = \dfrac{n(n + 1)(2n + 1)}{6}$

5 $5 + 10 + 15 + \cdots + 5n = \dfrac{5n(n + 1)}{2}$

6 $1 \cdot 2 + 2 \cdot 3 + 3 \cdot 4 + \cdots + n(n + 1) = \dfrac{n(n + 1)(n + 2)}{3}$

7 $1^3 + 2^3 + 3^3 + \cdots + n^3 = \dfrac{n^2(n + 1)^2}{4}$ **8** $3 + 3^2 + 3^3 + \cdots + 3^n = \dfrac{3^{n+1} - 3}{2}$

9 $5 + 8 + 11 + \cdots + (3n + 2) = \dfrac{n(3n + 7)}{2}$

10 $a + ar + ar^2 + \cdots + ar^{n-1} = \dfrac{a(1 - r^n)}{1 - r}, \; r \neq 1$

11 $a + (a + d) + (a + 2d) + \cdots + (a + [n - 1]d) = \dfrac{n(2a + [n - 1]d)}{2}$

12 $2^n > n$ **13** $3^{n+7} < (n + 7)!$ **14** $\dfrac{1}{2} + \dfrac{1}{2^2} + \dfrac{1}{2^3} + \cdots + \dfrac{1}{2^n} = 1 - \dfrac{1}{2^n}$

15 $1 + 2 + 3 + \cdots + n < \frac{1}{8}(2n + 1)^2$

16 Prove by mathematical induction that $(a - b)$ is a factor of $a^n - b^n$ for all $n \in N$. (HINT: $a^{k+1} - b^{k+1} = (a^{k+1} - ab^k) + (ab^k - b^{k+1})$)

17 Prove

$$0 < a < b \Longrightarrow \left(\frac{a}{b}\right)^{n+1} < \left(\frac{a}{b}\right)^n$$

for all $n \in N$.

18 Prove

$$0 < b < a \Longrightarrow \left(\frac{a}{b}\right)^{n+1} > \left(\frac{a}{b}\right)^n$$

for all $n \in N$.

19 Prove that 2 is a factor of $n(n + 1)$ for all $n \in N$.

20 Prove that 3 is a factor of $n^3 - n + 3$ for all $n \in N$.

21 Prove that $1 + n^2 < (1 + n)^2$ for all $n \in N$.

Sequences and Series   **9.4**

Until now we have used the word *sequence* rather loosely, though not inappropriately, because people have a reasonable understanding of the word through common usage. Our work in this section, however, will require a more precise definition. We begin by defining a sequence function.

DEFINITION 9.4.1 A function whose domain is the set N of natural numbers is called a sequence function.

Example 9.4.1 Let the function defined by $f(n) = 2n - 1$ be a sequence function. For $n = 1, 2, 3$, and 4, we obtain

$$f(1) = 1$$
$$f(2) = 3$$
$$f(3) = 5$$
$$f(4) = 7$$

The range of f is clearly the set of odd natural numbers. The notation more commonly used for a sequence function $f(n)$ is f_n. Thus we write

$$f_1 = 1$$
$$f_2 = 3$$
$$f_3 = 5$$
$$f_4 = 7$$

DEFINITION 9.4.2 Let f_n be a sequence function. If the elements comprising the range of f_n are arranged in the *order*

$$f_1, f_2, f_3, \cdots, f_n, \cdots$$

then such an arrangement is called a sequence. In this arrangement f_n is called the nth term of the sequence.

If the domain of f is restricted to the set $\{1, 2, 3, \cdots, n\}$, where n is fixed, then the sequence

$$f_1, f_2, f_3, \cdots, f_n$$

is called finite.

Example 9.4.2 Identify the sequences defined by the following sequence functions:

(a) $f_n = \dfrac{n}{n + 1}$

Solution $f_1 = \frac{1}{2}, f_2 = \frac{2}{3}, f_3 = \frac{3}{4}$, and so forth. Hence the sequence generated by this function is

$$\frac{1}{2}, \frac{2}{3}, \frac{3}{4}, \frac{4}{5}, \frac{5}{6}, \cdots, \frac{n}{n + 1}, \cdots$$

(b) $g_n = 1 - \dfrac{n}{n + 1}$

Solution From part (a) it is clear that the sequence would be

$$\frac{1}{2}, \frac{1}{3}, \frac{1}{4}, \frac{1}{5}, \frac{1}{6}, \cdots, \frac{1}{n}, \cdots$$

(c) $h_n = \dfrac{1}{n(n+1)}$

Solution Evaluating h_n at $n = 1, 2, 3, \cdots$, renders the sequence

$$\frac{1}{1 \cdot 2}, \frac{1}{2 \cdot 3}, \frac{1}{3 \cdot 4}, \frac{1}{4 \cdot 5}, \cdots, \frac{1}{n(n+1)}, \cdots$$

As you have observed, a sequence generally has infinitely many members; hence, we usually list only the first few members, followed by the three "and so forth" dots. This is especially true when from the first few terms the *logical* sequence pattern is established. Nevertheless, we usually insert, when it is known, the generating formula as the *n*th term in the sequence, leaving no doubt as to how successive terms are formed. Recall from Section 9.2, that when the formula is not given, we can never know for sure what terms will follow a given sequence of three or four.

The previous sections pointed out that it is frequently important to determine a formula, if possible, for the *sum* of the terms of a finite sequence. Recall, for example, the sums

$$1 + 2 + 3 + \cdots + n = \frac{n(n+1)}{2}$$

and

$$\frac{1}{1 \cdot 2} + \frac{1}{2 \cdot 3} + \frac{1}{3 \cdot 4} + \cdots + \frac{1}{n(n+1)} = \frac{n}{n+1}$$

The sum of the elements of a finite sequence is called a finite *series*. More formally, we have

---

DEFINITION 9.4.3 Let f_n be a function defining the sequence

$$f_1, f_2, f_3, \cdots, f_n$$

Then the sum

$$f_1 + f_2 + f_3 + \cdots + f_n$$

is called a finite series.

A series is often symbolized by the sigma or summation notation

$$\sum_{k=1}^{n} f_k$$

where k is called the index of summation and the set of integers $\{1, 2, 3, \cdots , n\}$ is called the domain of summation. The notation is only possible when the generating function f_n is known. To expand a series when expressed in summation notation, you evaluate the function f_n at each integer in the domain of summation and then sum the resulting terms.

Example 9.4.3 Express the following series in expanded form:

(a) $\displaystyle\sum_{k=1}^{5} \left(\frac{k+2}{2k}\right)$

Solution

$$\sum_{k=1}^{5} \left(\frac{k+2}{2k}\right) = \frac{3}{2} + \frac{4}{4} + \frac{5}{6} + \frac{6}{8} + \frac{7}{10}$$

(b) $\displaystyle\sum_{k=1}^{4} (-1)^k \left[\frac{k}{(1+k)^2}\right]$

Solution

$$\sum_{k=1}^{4} (-1)^k \left[\frac{k}{(1+k)^2}\right] = -\frac{1}{4} + \frac{2}{9} - \frac{3}{16} + \frac{4}{25}$$

(c) $\displaystyle\sum_{k=1}^{6} \left(\frac{1}{2^k}\right)$

Solution

$$\sum_{k=1}^{6} \left(\frac{1}{2^k}\right) = \frac{1}{2} + \frac{1}{4} + \frac{1}{8} + \frac{1}{16} + \frac{1}{32} + \frac{1}{64}$$

If there are many terms in a series the process of computing the sum can be tedious and difficult. For this reason we usually attempt to discover a formula that represents the series sum. This is generally accomplished by writing the sums for $n = 1, 2, 3,$ and 4, or more if necessary, as follows:

$$S_1 = \sum_{k=1}^{1} f_k = f_1$$

$$S_2 = \sum_{k=1}^{2} f_k = f_1 + f_2$$

$$S_3 = \sum_{k=1}^{3} f_k = f_1 + f_2 + f_3$$

$$S_4 = \sum_{k=1}^{4} f_k = f_1 + f_2 + f_3 + f_4$$

and attempting to discover a pattern in the sequence S_1, S_2, S_3, S_4 from which

$$S_n = \sum_{k=1}^{n} f_k$$

can be predicted. We then verify the formula pattern by mathematical induction, as was done several times in Section 9.2. The sequence $S_1, S_2, S_3, \cdots, S_n, \cdots$ is called a sequence of partial sums.

Example 9.4.4 Find the sum of the series

$$\sum_{k=1}^{50} (4k - 2)$$

Solution First form a sequence of partial sums as follows:

$S_1 = 2$
$S_2 = 2 + 6 = 8$
$S_3 = 2 + 6 + 10 = 18$
$S_4 = 2 + 6 + 10 + 14 = 32$

Then attempt to discover a pattern in the sequence of sums. Here, with a bit of care, you would find the following holds:

$S_1 = 2 \cdot 1$
$S_2 = 2 \cdot 4 = 2 \cdot 2^2$
$S_3 = 2 \cdot 9 = 2 \cdot 3^2$
$S_4 = 2 \cdot 16 = 2 \cdot 4^2$

and you begin to suspect that

$$S_n = 2 \cdot n^2$$

This can be verified by mathematical induction, but will not be done here. We now have the handy formula

$$S_n = \sum_{k=1}^{n} (4k - 2) = 2 \cdot n^2$$

We may now easily solve the problem. Taking $n = 50$, we obtain

$$\sum_{k=1}^{50} (4k - 2) = 2(50)^2$$
$$= 2 \cdot (2500)$$
$$= 5000$$

a result requiring considerably less effort than finding the 50 terms of the sequence and then one by one adding them up. This example illustrates one of the significant roles of mathematics — to simplify problems that otherwise may be extremely difficult and tedious.

We will now single out two very important series that frequently occur in a variety of applications. We shall motivate the introduction of each by an applied problem.

Problem 9.4.1 A new manufacturing firm produces 10 units of a certain commodity during the first week of production. Each succeeding week the firm is able to produce 5 more than the preceeding week. If they continue in this manner for a year, what will be their total output for the year?

Solution We first express the desired sum in series form.

$$10 + (10 + 5) + (10 + 2 \cdot 5) + (10 + 3 \cdot 5) + \cdots + (10 + 51 \cdot 5)$$
$$= \sum_{n=1}^{52} [10 + (n-1)5]$$

You will note that there is a common difference between each pair of consecutive terms in this series, namely, 5. Also the first term, 10, is repeated in each succeeding term of the series. We now leave this problem, temporarily, to determine a formula by which the above sum may be easily computed.

The general form of a sequence whose terms possess properties such as those of the series in Problem 9.4.1 is identified in the following definition.

DEFINITION 9.4.4 A sequence defined by

$$f_n = a + (n-1)d$$

● Also called an arithmetic progression

is called an arithmetic sequence,● where a is the first term of the sequence and d is called the common difference.

THEOREM 9.4.1 The sum of the first n terms of an arithmetic sequence is given by

$$\sum_{k=1}^{n} [a + (k-1)d] = \frac{n}{2}[2a + (n-1)d]$$

or

$$= \frac{n}{2}[a + f_n]$$

where $f_n = a + (n-1)d$.

● This could be proved by mathematical induction. (See problem 11, p. 349.) However, we shall present a proof historically attributed to the great German mathematician, Karl Friedrich Gauss, who at the age of 10 discovered the above formula to avoid having to perform certain long, tedious arithmetic sums. This argument also leads naturally to the form of the solution.

Proof ● The key to this proof is in writing the series, which we denote S_n, in two different forms, namely,

$$S_n = a + [a + d] + [a + 2d] + \cdots + [a + (n-1)d]$$

and

$$S_n = f_n + [f_n - d] + [f_n - 2d] + \cdots + [f_n - (n-1)d]$$

This latter form is similar to the first but in reverse order; that is, you start with the last term f_n and *subtract* the *common difference* d from each succeeding term, rather than taking the first term a and *adding* d to each succeeding term. Of course, the result must be the same. We now add the two series term by term in the indicated order and simplify, obtaining the desired result as follows:

$$S_n + S_n = [a + f_n] + [a + f_n] + \cdots + [a + f_n] = n[a + f_n]$$
$$\Longleftrightarrow 2S_n = n[a + f_n]$$
$$\Longleftrightarrow S_n = \frac{n}{2}[a + f_n]$$

By substituting $a + (n-1)d$ for f_n, we obtain the alternate form

$$S_n = \frac{n}{2}[2a + (n-1)d] \qquad \text{QED}$$

Let us return now to the problem of computing the sum

$$\sum_{k=1}^{52} [10 + (k-1)5]$$

which represents the total output per year of a certain manufacturing firm. Note that this sum is an arithmetic series, where $a = 10$, $d = 5$, and $n = 52$. Hence, from Theorem 9.4.1, we have

$$\sum_{k=1}^{52} [10 + (k-1)5] = \frac{52}{2}[2 \cdot 10 + (52-1)5]$$
$$= 26[20 + 255]$$
$$= 7150$$

Thus, the number of units produced at the end of the first year is 7150.

Now let us look at another problem.

Problem 9.4.2 A young, industrious boy made a promise to his rather clever father that he would deposit one penny in the bank at the beginning of a certain week and then double the size of his deposit each week thereafter for a year. How much would he have deposited at the end of this year if he were to succeed in fulfilling his promise?

Solution The sum of the sequence of deposits may be written

$$0.01 + (0.01)2 + (0.01)2^2 + (0.01)2^3 + \cdots + (0.01)2^{52}$$

or, using summation notation,

$$\sum_{k=1}^{53} (0.01)2^{k-1}$$

Observe in this series a *common ratio* of 2 between each pair of consecutive terms. We have

$$\frac{(0.01)2}{(0.01)} = 2 \qquad \frac{(0.01)2^2}{(0.01)2} = 2 \qquad \frac{(0.01)2^3}{(0.01)2^2} = 2$$

and in general,

$$\frac{(0.01)2^k}{(0.01)2^{k-1}} = 2 \qquad \text{for } k = 1, 2, \cdots, 53$$

Let us now leave this problem, temporarily, to determine a formula by which sums of this form may be more easily computed.

The general form of a sequence whose terms possess properties such as those in the series above is identified in the following definition.

DEFINITION 9.4.5   A sequence defined by

$$f_n = ar^{n-1}$$

● Also called a geometric progression

is called a geometric sequence,● where a is the *first* term of the sequence and r is the common ratio.

THEOREM 9.4.2   The sum of the first n terms of a geometric sequence is given by

$$S_n = \sum_{k=1}^{n} ar^{n-1} = \frac{a(1 - r^n)}{1 - r} \qquad \text{for } r \neq 1$$

Proof   This also could be proved by mathematical induction, but we prefer to take a different approach, which is more illuminating insofar as the resulting formula is concerned. We take first the initial sum

$$S_n = a + ar + ar^2 + \cdots + ar^{n-1}$$

and multiply both sides of this by the common ratio r, obtaining

$$rS_n = ar + ar^2 + \cdots + ar^{n-1} + ar^n$$

Subtracting the second equation from the first yields

$$S_n - rS_n = a - ar^n$$

which is equivalent to

$$S_n = \frac{a(1 - r^n)}{1 - r} \qquad \text{if } r \neq 1 \qquad \text{QED}$$

The case for $r = 1$ is trivial; that is,

$$S_n = a + a + a + \cdots + a = na$$

Returning now to our problem of doubling an investment of pennies each week for a year, we have the series

$$\sum_{k=1}^{53} (0.01)2^{k-1}$$

This you will note is a geometric series where $a = 0.01$, $r = 2$, and $n = 53$. Hence, by Theorem 9.4.2 we have

$$\sum_{k=1}^{53} (0.01)2^{k-1} = \frac{0.01(1 - 2^{53})}{1 - 2}$$

$$= 0.01(2^{53} - 1)$$

$$\overset{\bullet}{=} 0.01(8.974 \times 10^{15} - 1)$$

● Using logarithms, we find $2^{53} = 8.974 \times 10^{15}$.

$$= 8.974 \times 10^{13} - 0.01$$

which is about \$90 million million. Although pennies may not seem much to begin with, investing them in this manner results in an astronomical sum.

Problem Sets 9.5

Problem Set I ■ *Reading Comprehension*

Determine whether statements 1 through 8 are true or false and give reasons for your responses.

1 A sequence function is a function whose domain is the set N of natural numbers.
2 The range of a sequence function is called a sequence.
3 The expression f_n represents the nth term of a sequence.
4 The sigma symbol Σ is used to shorten a sequence notationally as follows:

$$\sum_{k=1}^{n} f_k = f_1, f_2, f_3, \cdots, f_n$$

5 The terms *sequence* and *series* are mathematically synonymous.
6 A sequence of partial sums is a sequence of numbers expressed as a sum; that is

$$\sum_{k=1}^{n} f_k = f_1 + f_2 + \cdots + f_n$$

is a sequence of partial sums.
7 The difference between any two terms of an arithmetic sequence is a constant.
8 The ratio of any two consecutive terms of a given geometric sequence is constant.

1 Find the first five terms of the sequences defined by

    a $f_n = 2(n-3)$ **b** $g_n = \dfrac{(-1)^n}{n+1}$

    c $h_n = \dfrac{n^2+1}{n}$ **d** $f_n = \dfrac{4}{n^2-1}$

    e $g_n = \dfrac{n(n-1)}{2}$ **f** $h_n = \dfrac{(-1)^n(n-1)}{n^2+1}$

2 Determine the suggested formula for the nth term of each of the following sequences:
    a $-1, 1, 3, 5, \cdots$ **b** $6, 11, 16, 21, \cdots$ **c** $-1, -\frac{1}{3}, 0, \frac{1}{5}, \cdots$
    d $-1, 2, 8, 17, \cdots$ **e** $0, \frac{1}{4}, \frac{2}{9}, \frac{3}{16}, \cdots$ **f** $0, 1, 3, 6, 10, 15, \cdots$

3 Find the common difference, the last term, and the sum of the following finite sequences:
    a $2, 5, 8, 11, \cdots$ to fourteen terms
    b $3, 1, -1, -3, \cdots$ to twelve terms
    c $-5, -1, 3, 7, \cdots$ to twenty terms
    d $\frac{1}{3}, \frac{5}{3}, \frac{9}{3}, \frac{13}{3}, \cdots$ to fifteen terms
    e $2, 1\frac{1}{4}, \frac{1}{2}, -\frac{1}{4}, \cdots$ to ten terms
    f $3, 3\frac{2}{3}, 4\frac{1}{3}, 5, \cdots$ to eighteen terms

4 Find the common ratio and the sum of the following sequences. (Use logarithms where appropriate in performing calculations.)
    a $3, 6, 12, 24, \cdots$ to ten terms
    b $0.1, 0.03, 0.009, 0.0027, \cdots$ to eight terms
    c $2, 1, \frac{1}{2}, \frac{1}{4}, \cdots$ to twelve terms
    d $20, 0.2, 0.002, 0.00002, \cdots$ to twenty terms
    e $9, 3, 1, \frac{1}{3}, \cdots$ to fifteen terms
    f $2, 0.8, 0.32, 0.0128, \cdots$ to fourteen terms

5 Find the sum of the first forty positive integers divisible by
    a 3   **b** 5   **c** 6

6 How many odd numbers lie between 12 and 120? (HINT: Use the formula for the last term of an arithmetic sequence.)

7 How many numbers divisible by 4 lie between 10 and 250?

8 How many numbers divisible by 7 lie between 25 and 263?

9 Find the sum of the first fifteen positive integral multiples of
    a 3   **b** 5   **c** 8

10 Find the thirtieth term in the arithmetic sequences
    a $1, 6, 11, 16, \cdots$ **b** $-3, 1, 5, 9, \cdots$ **c** $2, 2\frac{2}{3}, 3\frac{1}{3}, 4, \cdots$

11 Find the tenth term in the geometric sequences:
    a $1, 3, 9, 27, \cdots$ **b** $-2, 0.2, -0.02, 0.002, \cdots$ **c** $\frac{1}{3}, \frac{1}{6}, \frac{1}{12}, \frac{1}{24}, \cdots$

12 Find the following sums:

Example $\displaystyle\sum_{k=1}^{15} (2k-3)$

Solution   Write out the first few terms of the series and the last term.

$$\sum_{k=1}^{15} (2k-3) = -1 + 1 + 3 + 5 + \cdots + 27$$

Note that this is an arithmetic series with $a = -1$, $d = 2$, and $n = 15$. Therefore,

$$\sum_{k=1}^{15} (2k - 3) = \tfrac{15}{2}(-1 + 27)$$
$$= 15 \cdot 13$$
$$= 195$$

a $\displaystyle\sum_{k=1}^{20} (3k - 2)$ b $\displaystyle\sum_{k=1}^{14} (2k + 4)$ c $\displaystyle\sum_{k=1}^{16} (\tfrac{1}{2}k + 1)$

d $\displaystyle\sum_{k=1}^{22} (\tfrac{1}{3}k - 2)$ e $\displaystyle\sum_{k=1}^{12} (-2k + 1)$ f $\displaystyle\sum_{k=1}^{50} (-\tfrac{1}{2}k + 10)$

Problem Set III ■ Theoretical Developments

Prove the statements of problems 1 through 5.

1 $\displaystyle\sum_{k=1}^{n} f_k + \sum_{k=1}^{n} g_k = \sum_{k=1}^{n} [f_k + g_k]$ 2 $\displaystyle\sum_{k=1}^{n} af_k = a \sum_{k=1}^{n} f_k$ 3 $\displaystyle\sum_{k=1}^{n} f_k = \sum_{p=1}^{n} f_p$

4 $\displaystyle\sum_{k=1}^{n} f_k = \sum_{k=0}^{n-1} f_{k+1}$ (The index letter used is irrelevant.)

5 Find the value of

$$\sum_{k=1}^{n} (\tfrac{1}{2})^k$$

for $n = 1, 2, 3, 4$, and 5. What do you suspect the above sum approximates with increasing accuracy as a larger and larger n is chosen?

Problem Set IV ■ Applications

1 The first year a man is employed he saves $500.00. In each succeeding year he saves $200.00 more than the previous year. How much will he have saved after 15 years of employment?

2 A super-ball rebounds four-fifths of the distance it falls. Through what distance will the ball have traveled as it strikes the ground for the fifteenth time, if it is initially dropped from a height of 30 feet?

3 A certain radioactive substance has a *half-life* of 3000 years; that is, half of the substance disintegrates every 3000 years. Suppose that initially there were 5 lb of the substance present. How much would still exist after 30,000 years? How long would it take for the substance to be reduced to 0.0005 lb?

4 A person borrows $1000.00 and agrees to pay it back by paying $50.00 per month plus 10 percent interest on the unpaid balance for the month. How much will be the total amount paid in interest when the loan is paid off?

5 A certain culture of bacteria doubles in number every 2 hours. How many will be in the culture after 2 days?

6 People are immigrating to a certain community with an initial population of 10,000 at the rate of 500 per month. What will be the population of the community after one year if these immigrants represent the only manner in which the community's population is changing?

9.6 The Binomial Theorem

In mathematics we frequently encounter products of the form

$$(a + b)^n \tag{1}$$

where $a, b \in \mathbf{R}$ and $n \in \mathbf{N}$. To expand this product by direct multiplication, for small n, poses no serious problem. We have, for example,

$$(a + b)^1 = a + b$$
$$(a + b)^2 = a^2 + 2ab + b^2$$
$$(a + b)^3 = a^3 + 3a^2b + 3ab^2 + b^3$$
$$(a + b)^4 = a^4 + 4a^3b + 6a^2b^2 + 4ab^3 + b^4$$
$$(a + b)^5 = a^5 + 5a^4b + 10a^3b^2 + 10a^2b^3 + 5ab^4 + b^5$$

The *coefficients* in the above successive expansions form a remarkable triangular pattern known as Pascal's triangle. If we begin with $(a + b)^0 = 1$ and include the leading and terminal coefficients of 1 in each expansion, the triangular array would appear as follows:

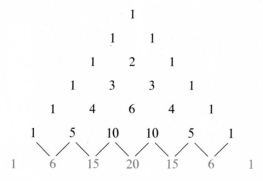

The interesting property of this triangle is that each interior element in the array can be obtained as the sum of the two elements just to the right and left in the row immediately above, as indicated in red in the seventh row. For example, without having to expand $(a + b)^6$ by direct multiplication, we see from Pascal's triangle that

$$(a + b)^6 = a^6 + 6a^5b + 15a^4b^2 + 20a^3b^3 + 15a^2b^4 + 6ab^5 + b^6$$

By continuing in this manner, we could determine the coefficients in the expansions of $(a + b)^7$, $(a + b)^8$, $(a + b)^9$, and so on. The only disadvantage of this method is that all the coefficients in the expansion preceding the one desired are necessary to generate the next set of coefficients. For n equal to, say, 50 or 100, the process of building the triangle to the fiftieth or hundredth row would be long and tedious.

The binomial theorem will provide a formula for the expansion of $(a + b)^n$ for any positive integer n independent of previous expansions.

The formula we shall use for the binomial expansion will make extensive use of the mathematical expression "n factorial," denoted $n!$, introduced earlier. To sharpen your understanding of this mathematical term, let us review the following examples:

By definition

$$n! = 1 \cdot 2 \cdot 3 \ldots (n-2)(n-1)(n)$$

Thus

$$5! = 1 \cdot 2 \cdot 3 \cdot 4 \cdot 5 = 120$$
$$3! = 1 \cdot 2 \cdot 3 = 6$$

and

$$6! = 1 \cdot 2 \cdot 3 \cdot 4 \cdot 5 \cdot 6 = 5! \cdot 6$$

Also

$$\frac{6!}{4!} = \frac{4! \cdot 5 \cdot 6}{4!} = 5 \cdot 6 = 30$$

and

$$\frac{8!}{3! \cdot 5!} = \frac{5! \cdot 6 \cdot 7 \cdot 8}{3! \cdot 5!} = \frac{6 \cdot 7 \cdot 8}{1 \cdot 2 \cdot 3} = 7 \cdot 8 = 56$$

We now state the binomial theorem:

THEOREM 9.6.1   If n is a positive integer, then

$$(a+b)^n = a^n + \frac{n!}{(n-1)!1!} a^{n-1}b + \frac{n!}{(n-2)!2!} a^{n-2}b^2 +$$

$$\cdots + \frac{n!}{(n-r)!r!} a^{n-r}b^r + \cdots + \frac{n!}{1!(n-1)!} ab^{n-1} + b^n$$

Proof   To be done as an exercise.

Observe that the $(r+1)$st term is given by the formula

$$\frac{n!}{(n-r)!r!} a^{n-r}b^r \tag{1}$$

where $r = 0, 1, 2, 3, \cdots, n - 1$. For example, suppose we wish to determine the eighth term in this expansion where $n = 12$. You must remember to choose $r = 7$, not 8 for the eighth term. Substituting $r = 7$ in formula (1) yields the desired term, as follows:

$$\frac{12!}{(12-7)!7!} a^{12-7}b^7 = \frac{12!}{5! \cdot 7!} a^5b^7 = 792a^5b^7$$

Prior to this section in this chapter our work has been to determine simple, compact formulas to represent certain finite sums. Now, by contrast, we are doing just the reverse with the binomial theorem. That is, we are taking the simple expression $(a + b)^n$ and are expanding it into a finite series. Surprisingly, this reverse process for certain problems is itself a means of simplification, as we shall see in the examples and exercises that follow.

Example 9.6.1 Expand $(x - 2xy^2)^5$ by the binomial theorem.

Solution Note

$$a = x$$
$$b = -2xy^2$$
$$n = 5$$

Therefore,

$$(x - 2xy^2)^5 \stackrel{\text{T9.6.1}}{=\!=\!=} x^5 + \frac{5!}{4!1!} x^4[-2xy^2] + \frac{5!}{3!2!} x^3[-2xy^2]^2 +$$

$$\frac{5!}{2!3!} x^2[-2xy^2]^3 + \frac{5!}{1!4!} x[-2xy^2]^4 + [-2xy^2]^5$$

$$=\!=\!= x^5 - 10x^5y^2 + 40x^5y^4 - 80x^5y^6 + 80x^5y^8 - 32x^5y^{10}$$

Example 9.6.2 Find the fourteenth term in the binomial expansion of $(a - b)^{16}$.

Solution For the fourteenth term, $r = 13$. Thus,

$$\frac{16!}{(16 - 13)!13!} a^{16-13}(-b)^{13} = \frac{13! 14 \cdot 15 \cdot 16}{13! 1 \cdot 2 \cdot 3} a^3(-b)^{13}$$

$$= -7 \cdot 5 \cdot 16 \, a^3 b^{13} = -560 a^3 b^{13}$$

Example 9.6.3 In the binomial expansion of

$$\left(x - \frac{2}{x^2}\right)^8$$

find the term that when simplified contains x^{-4}.

Solution Because initially we do not know the term desired (we do not know r), we write out the general $(r + 1)$st term for this expansion with r yet to be determined. We have

$$\frac{8!}{(8 - r)!r!} x^{8-r}\left[-\frac{2}{x^2}\right]^r \tag{2}$$

To determine r, we need examine only that portion of (2) containing x. We have

$$x^{8-r} \cdot x^{-2r} = x^{8-3r}$$

Because this must reduce to x^{-4}, it follows that

$$8 - 3r = -4 \Longleftrightarrow 3r = 12 \Longleftrightarrow r = 4$$

Setting $r = 4$ in (2) yields

$$\frac{8!}{4!4!} \, x^4 \left[-\frac{2}{x^2} \right]^4 = \frac{4!5 \cdot 6 \cdot 7 \cdot 8}{4!1 \cdot 2 \cdot 3 \cdot 4} \cdot x^4 (-2)^4 \, x^{-8} = 1120 x^{-4}$$

And because $r = 4$, this would be the fifth term in the expansion.

Example 9.6.4 Simplify the expression $(0.99)^{10}$ to within $\frac{1}{1000}$ of the exact value.

Solution Write 0.99 as the sum $(1 - 0.01)$ and apply the binomial theorem. We have

$$(0.99)^{10} = (1 - 0.01)^{10} = 1^{10} + \frac{10!}{9!1!} \, 1^9 (-0.01)$$

$$+ \frac{10!}{8!2!} \, 1^8 (-0.01)^2 + \frac{10!}{7!3!} \, 1^7 (-0.01)^3 + \cdots$$

$$= 1 - 0.10 + 0.0045 - 0.00007 + \cdots$$

$$\cong 0.90443$$

It is significant in the above expansion to note that most of the value of $(0.99)^{10}$ is contained within the first few terms of its binomial expansion. This is due to the presence of the factor (-0.01), which when raised to increasingly large positive integral powers becomes very close to zero. In fact, all terms after the fourth term contribute less than 0.00001, so that all may be neglected in most applications.

The above procedure is effective when a number can be expressed as a sum of two others, one of which is easily raised to integral powers, such as 1 or multiples of 10, whereas the other is less than 1, so that successively higher powers of this latter factor render it increasingly negligible, as was experienced in the previous example.

Problem Sets 9.7

Problem Set I ■ *Reading Comprehension*

Determine whether statements 1 through 4 are true or false and give reasons for your responses.

1 The binomial theorem provides an expansion for the expression
    $(a + b)^n$
  where $a, b, n \in \mathbf{R}$.

2 All coefficients in the binomial expansion may be obtained by Pascal's triangle.

3 The general rth term in the binomial expansion of $(a + b)^n$ is given by

$$\frac{n!}{(n - r)!r!} a^{n-r}b^r$$

4 A convenient way of simplifying 76^{12} would be to express this as $(70 + 6)^{12}$ and applying the binomial theorem to expand.

Problem Set II ■ Skills Development

Clear the following expressions of factorials:

1 $\dfrac{12!}{8!4!}$ **2** $\dfrac{16!}{13!3!}$ **3** $\dfrac{8!}{4!4!}$ **4** $\dfrac{n!}{(n - 2)!2!}$ **5** $\dfrac{n!}{(n - r)!r!}$

6 $\dfrac{(n - 1)!}{(n - 3)!}$ **7** $\dfrac{(2n + 3)!}{(2n + 1)!}$ **8** $\dfrac{n(n + 1)!}{(n + 2)!}$ **9** $\dfrac{(2n)!(n - 1)!}{(2n - 2)!(n + 1)!}$

Write at least the first four terms in the binomial expansion of each of the following expressions:

10 $(a - 3b)^6$ **11** $(x + 2y^2)^8$ **12** $(-x + xy)^{10}$ **13** $(a - 3)^9$

14 $\left(x - \dfrac{y}{2}\right)^{12}$ **15** $\left(\dfrac{x}{y} - 2y\right)^7$

Write and simplify the specified term in the binomial expansion of problems 16 through 21.

16 The seventh term of $\left(a - \dfrac{b}{2}\right)^{11}$.

17 The ninth term of $(x - xy^2)^{15}$.

18 The twelfth term of $\left(\dfrac{x}{y} - 2y\right)^{16}$.

19 The term containing x^2 in $\left(\dfrac{1}{x} - x^2y\right)^{10}$.

20 The term containing x^{18} in $\left(\dfrac{x^2}{y} - xy\right)^{12}$.

21 The term containing a^{15} in $\left(a^2 - \dfrac{2b}{a^3}\right)^{15}$.

Simplify expressions 22 through 24 by the binomial theorem, but add only a sufficient number of terms to obtain accuracy to within $\frac{1}{1000}$.

22 $(1.01)^{10}$ **23** $(0.98)^8$ **24** $(10.1)^9$

25 Given that the binomial theorem holds as an "infinite sum" for $(1 + x)^n$, where n is rational and $|x| < 1$, and further where the binomial coefficients are written as

$$\frac{n(n - 1) \cdots (n - r + 1)}{r!}$$

rather than

$$\frac{n!}{(n - r)!r!} \bullet$$

● This is necessary because we have not defined "n factorial" for any other numbers than positive integers.

find to within $1/100$ the decimal value of

a $\sqrt{0.99}$ **b** $\sqrt{1.02}$

1 If an amount of money A is invested at 5 percent interest compounded annually, then the amount present P at the end of n years is given by the formula
$$P = A(1 + 0.05)^n$$
Find the amount present to the nearest cent after
a 4 years if $1000.00 is originally invested.
b 10 years if $5000.00 is originally invested.

2 If A dollars is paid at the beginning of each interest period, with the amount on deposit drawing interest at a rate of i per period, then the amount on deposit just after the nth payment is made is given by
$$S = \frac{A(1 + i)^n - 1}{i}$$
a Find the amount accumulated after 30 annual deposits of $100.00 at an interest rate of 5 percent, compounded annually.
b Find the amount accumulated after 40 quarterly deposits of $200.00 at an interest rate of 4 percent, compounded quarterly.

Problem Set IV ■ *Just For Fun*

1 Imagine that you have three boxes, one containing two black marbles, one containing two white marbles, and the third containing one black marble and one white marble. The boxes were labeled for their contents—BB, WW and BW—but someone has switched the labels so that every box is now incorrectly labeled. You are allowed to take one marble at a time out of any box, without looking inside, and by this process of sampling you are to determine the contents of all three boxes. What is the smallest number of drawings needed to do this?

Summary of Definitions and Properties in Chapter 9   9.8

T9.2.1 Principle of Mathematical Induction. If $P(n) = Q(n)$ is an open (p. 341) sentence, $P(1) = Q(1)$, and $[P(k) = Q(k) \Longrightarrow P(k + 1) = Q(k + 1)]$, then $P(n) = Q(n)$ for each $n \in N$.

D9.4.1 A function whose domain is the set N of natural numbers is called (p. 350) a sequence function.

D9.4.2 The range of a sequence function in the order (p. 350)
$$f_1, f_2, f_3, \cdots, f_n, \cdots$$
is called a sequence.

D9.4.3 If f_n is a sequence function, then the sum (p. 351)
$$f_1 + f_2 + f_3 + \cdots + f_n$$
abbreviated
$$\sum_{k=1}^{n} f_k$$
is called a finite series.

If (p. 353)

$$S_n = \sum_{k=1}^{n} f_k$$

then $S_1, S_2, S_3, \cdots, S_n, \cdots$ is called a sequence of partial sums.

D9.4.4 A sequence defined by (p. 354)
$$f_n = a + (n-1)d$$
is called an arithmetic sequence.

T9.4.1 (p. 354)
$$\sum_{k=1}^{n} [a + (k-1)d] = \frac{n}{2}[2a + (n-1)d]$$

$$= \frac{n}{2}[a + f_n]$$

D9.4.5 A sequence defined by (p. 356)
$$f_n = ar^{n-1}$$
is called a geometric sequence.

T9.4.2 (p. 356)
$$\sum_{k=1}^{n} ar^{n-1} = \frac{a(1-r^n)}{1-r}, \; r \neq 1$$

T9.6.1 The Binomial Theorem (p. 361)

$$(a+b)^n = a^n + \frac{n!}{(n-1)!1!}a^{n-1}b + \frac{n!}{(n-2)!2!}a^{n-2}b^2 + \cdots + \frac{n!}{(n-r)!r!}a^{n-r}b^r + \cdots + b^n$$

where $n \in N$.

By permission of John Hart and Field Enterprises, Inc.

Coordinate Geometry — The Conics

10 CHAPTER

Introduction 10.1

Although the title of this chapter is Coordinate Geometry, we have actually been treating this subject since Chapter 4. The difference here will be a more in-depth study treating a very important class of geometric figures called *the conics*. (See Fig. 10.1.1 for examples of these.) We shall also take a different approach. Earlier we defined relations first algebraically and then examined their geometric properties. Here our approach will be just the reverse—we shall define a relation first geometrically and then discover its algebraic properties.

We introduce our study of the conics with the following problem.

Problem 10.1.1 A radar antenna focuses the outgoing radio waves into a *beam* and also collects the returning echoes. We wish to

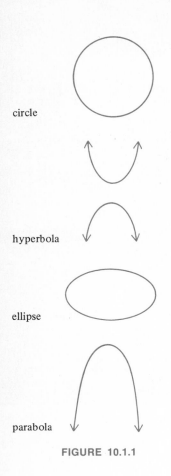

circle

hyperbola

ellipse

parabola

FIGURE 10.1.1

design a dish-shaped antenna so that it focuses a beam which is *parallel* to the axis of symmetry, as illustrated in Fig. 10.1.2.

Solution It is clear that we must determine the *curvature* of an antenna that will reflect the transmitted waves all in the same (parallel) direction. (The mirror of a searchlight or spotlight possesses these same characteristics.)

Upon familiarizing ourselves with the properties of conics, particularly those of *parabolas,* this problem will be easily solved. Therefore, we shall return later in this chapter to complete the solution.

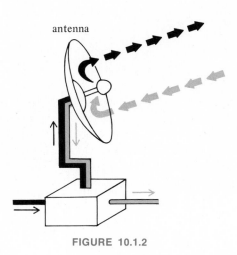

antenna

FIGURE 10.1.2

10.2   The Parabola

We begin with a definition.

DEFINITION 10.2.1   A parabola is the set of all points in a plane *equidistant* from a *fixed point* and a *fixed line*. The fixed point is called the focus and the fixed line the directrix.

Given a *focus* and a *directrix,* you can easily construct the graph of the resulting parabolas, as illustrated in Fig. 10.2.1. Properties of relations can often be made more evident by discovering and analyzing their algebraic representations in the form of equations. Our first task, therefore, will be to take the geometric definition of a parabola and obtain from it an algebraic equation

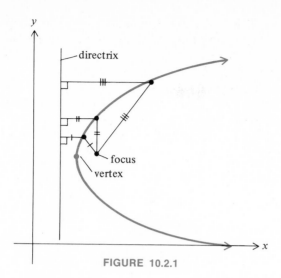

FIGURE 10.2.1

in two variables whose solution set will be the parabola. Because a parabola may be oriented in several different directions, we choose first one with *vertical* directrix and focus to the *right* of this directrix and place these within a rectangular coordinate system. For convenience, we denote the vertex of the parabola (h, k) and the distance between the focus and the directrix $2p$. With these designations, the coordinates of the focus become $(h + p, k)$ and the equation of the directrix becomes $x = h - p$. (See Fig. 10.2.2.) We now apply Definition 10.2.1 using the distance formula between two points.● We know the distances between an arbitrary point (x, y) on the parabola and the directrix and between (x, y) and the focus must be *equal*. With this information, we proceed as follows:

● $d = \sqrt{(x_2 - x_1)^2 + (y_2 - y_1)^2}$

$$d_1{}^2 = [x - (h - p)]^2$$
$$d_2{}^2 = [x - (h + p)]^2 + [y - k]^2$$

Setting $d_2{}^2 = d_1{}^2$, we obtain

$$[x - (h + p)]^2 + [y - k]^2 = [x - (h - p)]^2$$
$$\Longleftrightarrow [y - k]^2 = [x^2 - 2(h - p)x + (h - p)^2]$$
$$- [x^2 - 2(h + p)x + (h + p)^2]$$
$$= 4p(x - h)$$
$$\Longrightarrow (y - k)^2 = 4p(x - h)$$

Consequently, we have the following result:

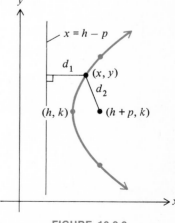

FIGURE 10.2.2

The equation of a cup-right parabola in *standard form* with vertex (h, k), focus $(h + p, k)$, and directrix $x = h - p$ is

$$(y - k)^2 = 4p(x - h) \tag{1}$$

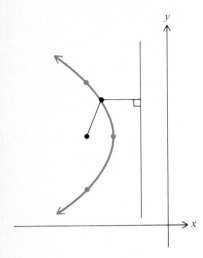

FIGURE 10.2.3

Similarly, if the focus is located to the *left* of the directrix (see Fig. 10.2.3), we obtain the following result (the proof is to be done as an exercise):

The equation of a cup-left parabola in *standard form* with vertex (h, k), focus $(h - p, k)$, and directrix $x = h + p$ is

$$(y - k)^2 = -4p(x - h) \tag{2}$$

Furthermore, if the focus is *above* a *horizontal* directrix (see Fig. 10.2.4), we have the following result:

The equation of a cup-up parabola in *standard form* with vertex (h, k), focus $(h, k + p)$, and directrix $y = h - p$ is

$$(x - h)^2 = 4p(y - k) \tag{3}$$

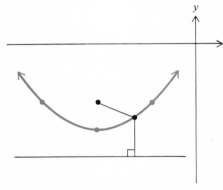

FIGURE 10.2.4

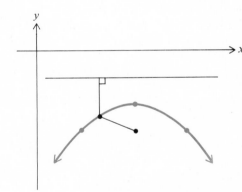

FIGURE 10.2.5

And finally, if the focus is *below* a *horizontal* directrix (see Fig. 10.2.5) we have:

The equation of a cup-down parabola in *standard form* with vertex (h, k), focus $(h, k - p)$, and directrix $y = h + p$ is

$$(x - h)^2 = -4p(y - k) \tag{4}$$

The line segment extending from one branch of a parabola to the other and passing through the focus perpendicular to the axis of symmetry is called the focal chord ● (See Fig. 10.2.6.) To find

● Also called *latus rectum*.

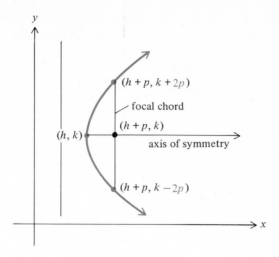

FIGURE 10.2.6

the length of the focal chord, we take standard form (1) (directrix vertical and focus to the right) and substitute the x-coordinates of the endpoints of the focal chord. Proceeding thus, we obtain

$$(y - k)^2 = 4p(h + p - h)$$

$$= 4p^2$$

$$\Longleftrightarrow (y - k)^2 = 4p^2 \Longleftrightarrow y - k = \pm 2p$$

$$\Longleftrightarrow \qquad y = k \pm 2p$$

Thus, the coordinates of the endpoints of the focal chord are $(h + p, k + 2p)$ and $(h + p, k - 2p)$. Hence, the length of the focal chord is $4p$. The same would follow for standard forms (2), (3), and (4). This information is especially helpful when graphing a parabola.

Example 10.2.1 Determine the vertex, focus, and directrix of the parabola defined by $(y - 3)^2 = 8(x + 4)$ and graph.

Solution First express the equation in *standard form*. We obtain

$$(y - 3)^2 = 4 \cdot 2 (x - [-4])$$

It is clearly a cup-right parabola of form (1) with $p = 2$. Furthermore, the vertex is $(-4, 3)$, the focus $(-2, 3)$, the directrix $x = -6$, and the endpoints of the focal chord are $(-2, 7)$ and $(-2, -1)$. With this information we sketch the parabola as illustrated in Fig. 10.2.7.

Example 10.2.2 Determine the vertex, focus, and directrix of the parabola defined by $x^2 - 14x + 16y + 33 = 0$ and graph.

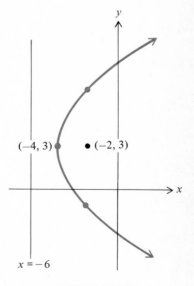

FIGURE 10.2.7

Solution We first convert the above equation into standard form by completing the square and rearranging, as follows:

$$x^2 - 14x + 16y + 33 = 0 \Longleftrightarrow x^2 - 14x + 7^2 = -16y - 33 + 7^2$$

$$\Longleftrightarrow (x-7)^2 = -16(y-1)$$

$$\Longleftrightarrow (x-7)^2 = -4 \cdot 4(y-1)$$

This standard form corresponds with equation (4), where $p = 4$. Hence, it is a *cup-down* parabola with vertex $(7, 1)$, focus $(7, -3)$, directrix $y = 5$, and endpoints of focal chord $(-1, -3)$ and $(15, -3)$. The graph is illustrated in Fig. 10.2.8.

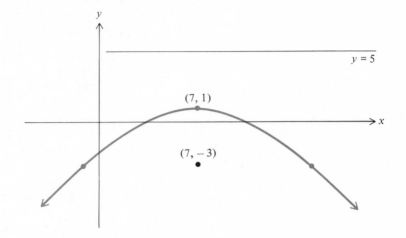

FIGURE 10.2.8

Example 10.2.3 Find an equation defining the parabola with vertex $(1, -3)$ and focus $(1, 2)$.

Solution Because the focus is *above* the vertex, the parabola is *cup-up* and hence the standard form is equation (3). The distance p from the vertex to the focus is clearly 5. With p determined and the vertex given, we have

$$(x-1)^2 = 4 \cdot 5(y - [-3])$$
$$= 20(y+3)$$
$$\Longrightarrow (x-1)^2 = 20(y+3)$$

The most important property of a parabola is probably best illustrated by a mirror having the shape of a *paraboloid of revolution,* a surface formed by revolving a parabola about its axis. If a light source is placed at the focus, all of the light rays (incident rays) emanating from this point will be reflected in lines parallel to the axis of the parabola. (See Fig. 10.2.9.)

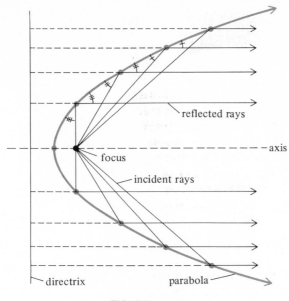

reflected rays

axis

focus

incident rays

directrix

parabola

FIGURE 10.2.9

We see this principle used in the design of automobile head-lights where the light bulb is placed at the focus of a parabolic mirror to form a beam of light parallel to the axis. Parabolic mirrors are also used in reflecting telescopes that receive parallel light rays from a distant light source, such as a star, that are reflected to form a sharply defined image at the focus.

The solution to Problem 10.1.1 concerning the design of a radar antenna that reflects transmitted microwaves into a straight beam is now clear. The antenna must be shaped in the form of a *paraboloid of revolution* with the transmitted waves originating at the *focus*. Also, because of this shape, the returning echo waves are concentrated at the focus from which they are sent to a receiver.

Problem Sets 10.3

Problem Set I ■ *Reading Comprehension*

Determine whether statements 1 through 4 are true or false and give reasons for your responses.

1 A parabola is the set of all points in a plane equidistant from a fixed point and a fixed line.
2 The focal chord is the line segment joining the focus and the directrix and is of length $2p$.

3   The equation in standard form of a cup-left parabola is
$$(x - h)^2 = -4p(y - k)$$
4   The vertex of a parabola is always midway between the focus and the directrix.

Problem Set II ■ Skills Development

Find an equation defining each parabola described in problems 1 through 15.

1   Vertex $(3, 6)$ and focus $(3, 8)$.   2   Vertex $(-2, 4)$ and focus $(-2, 0)$.
3   Vertex $(-1, -3)$ and focus $(4, -3)$.   4   Vertex $(4, -1)$ and focus $(6, -1)$.
5   Vertex $(2, -5)$ and directrix $x = -2$.   6   Vertex $(3, 2)$ and directrix $y = 6$.
7   Vertex $(-4, 1)$ and directrix $y = -4$.   8   Vertex $(-2, -5)$ and directrix $x = -8$.
9   Focus $(2, 4)$ and directrix $y = 5$.   10   Focus $(2, 4)$ and directrix $x = 3$.
11   Focus $(-1, -3)$ and directrix $x = 1$.
12   Focus $(5, -2)$ and directrix $y = -4$.
13   Vertex $(3, 2)$ and ends of focal chord $(5, 6)$ and $(5, -2)$.
14   Vertex $(-1, 4)$ and ends of focal chord $(-7, 1)$ and $(5, 1)$.
15   Vertex $(3, -3)$ and ends of focal chord $(-3, 0)$ and $(9, 0)$.

From each equation in problems 16 through 24, determine the vertex, focus, and directrix of the defined parabola and sketch the graph.

16   $y^2 - 6y + 8x - 23 = 0$ 17 $y^2 - 4y + 16x - 92 = 0$
18   $x^2 - 25x + 8y + 1 = 0$ 19 $y^2 + 4y + 32x - 92 = 0$
20   $y^2 - 8y + 16x + 64 = 0$ 21 $x^2 - 8x - 16y - 32 = 0$
22   $x^2 + 4x + 4y - 12 = 0$ 23 $y^2 + 6y + 4x + 5 = 0$
24   $x^2 - 16y + 16 = 0$

Problem Set III ■ Theoretical Developments

1   Use Definition 10.2.1 to prove the following standard forms for parabolas of differing orientations.
    a   Cup-left $(y - k)^2 = -4p(x - h)$
    b   Cup-up $(x - h)^2 = 4p(y - k)$
    c   Cup-down $(x - h)^2 = -4p(y - k)$
2   Find an equation of the parabola with axis parallel to the y-axis, vertex $(-2, 3)$, and passing through the point $(2, 1)$.
3   Prove that a parabola is symmetric about the line passing through the vertex and perpendicular to the directrix.
4   What class of parabolas are functions? Are they one-to-one?

Problem Set IV ■ Applications

1   A disk-shaped parabolic reflector has a diameter of 16 inches and a maximum depth of 4 inches. Find the depth of the focus.
2   Find the focus of a disk-shaped antenna that is 12 feet in diameter with a maximum depth of 1 foot.

We begin with a problem.

Problem 10.4.1 Artificial satellites circle the earth in *elliptic* orbits with the center of the earth at one of the focal points. Suppose a satellite is in orbit above the equator with a perigee (nearest point of the path above the earth) of 2,037 miles and an apogee (furthest point on the path from the surface of the earth) of 20,037 miles. (See Fig. 10.4.1.) Find an equation describing the entire elliptic path of the satellite. (The radius of the earth at the equator is 3,963 miles.)

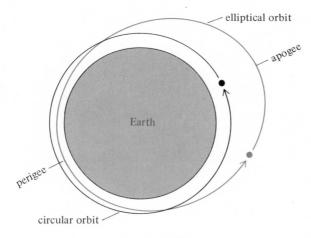

FIGURE 10.4.1

Solution (Following our study of ellipses, this will be a very simple problem to solve.)

The definition of an ellipse is as follows:

DEFINITION 10.4.1 An ellipse is the set of all points in a plane, each of which possesses the property that the *sum* of its distances from two *fixed* points, called foci, is a *constant*.

Probably the simplest way to sketch an ellipse is by picking two points representing the foci, pinning the ends of a string at each of the two points (making sure the string is *longer* than the distance between the foci) and then, after pulling the string taut with a pencil, move the pencil around in a circular motion allowing

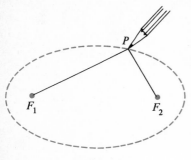

FIGURE 10.4.2

the string to slip along the point of the pencil as it moves, as illustrated in Fig. 10.4.2.

Unfortunately, the above procedure for sketching an ellipse sheds little light on the important properties of this conic. Several of these properties will become evident as we discover an equation for an ellipse. This we shall do as follows.

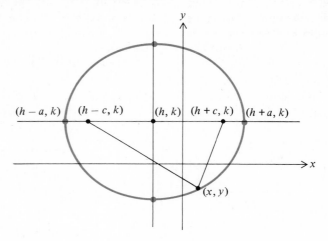

FIGURE 10.4.3

First, place the ellipse within a rectangular coordinate system with center (h, k), foci $(h - c, k)$ and $(h + c, k)$, and endpoints of *major* (longest) axis $(h - a, k)$ and $(h + a, k)$, as illustrated in Fig. 10.4.3. Note that we are choosing c to be the distance from center to focus and that $0 < c < a$. Next, we pick an arbitrary point (x, y) on the ellipse and apply Definition 10.4.1 using this point and the two foci. We know the *sum* of the distances between (x, y) and the foci is a *constant,* which we are denoting $2a$. Using the formula for the distance between two points, this property may be expressed

$$\sqrt{[(x-h)+c]^2+(y-k)^2}+\sqrt{[(x-h)-c]^2+(y-k)^2}=2a$$

Moving the second radical to the right side, squaring both sides, and simplifying, we obtain

$$\sqrt{[(x-h)-c]^2+(y-k)^2}=a-\frac{c}{a}(x-h)$$

Again, squaring both sides, this time to clear the final radical, we get

$$(x-h)^2-2c(x-h)+c^2+(y-k)^2=a^2-2c(x-h)+\frac{c^2}{a^2}(x-h)^2$$

By simplifying and rearranging, this equation becomes

$$\frac{(x-h)^2}{a^2}+\frac{(y-k)^2}{a^2-c^2}=1$$

Because $a > c$, the difference $a^2 - c^2$ is positive. For notational convenience, we let $b^2 = a^2 - c^2$. Our equation becomes, finally,

$$\frac{(x-h)^2}{a^2} + \frac{(y-k)^2}{b^2} = 1 \tag{5}$$

Observe when $x = h$, we have $y = k - b$ and $y = k + b$. Hence, the endpoints of the *minor* (shortest) axis are $(h, k - b)$ and $(h, k + b)$. Knowing the center and the endpoints of the major and minor axes of an ellipse allows you to sketch this conic rather quickly.

We summarize these results as follows:

The equation of an ellipse in *standard form* with center (h, k), horizontal foci $(h - c, k)$ and $(h + c, k)$, and *major* axis endpoints $(h - a, k)$ and $(h + a, k)$, is

$$\frac{(x-h)^2}{a^2} + \frac{(y-k)^2}{b^2} = 1$$

where $a > c$ and $b^2 = a^2 - c^2$. Also the *minor* axis endpoints are $(h, k - b)$ and $(h, k + b)$.

The line segment joining the center and an endpoint of the major axis of an ellipse is called a semimajor axis. The length of the semimajor axis is a. Similarly, the length of a semiminor axis joining the center and a minor axis endpoint is b. (See Fig. 10.4.4.)

The procedure for finding an equation of an ellipse with foci lined up vertically (Fig. 10.4.5) is similar to that used above for

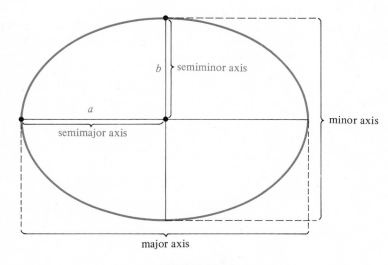

FIGURE 10.4.4

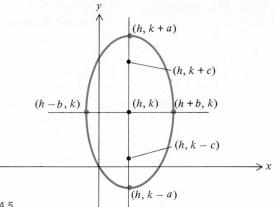

FIGURE 10.4.5

horizontally placed foci. We will leave the details as an exercise. The results, however, are as follows:

The equation of an ellipse in *standard form* with center (h, k), vertical foci $(h, k - c)$ and $(h, k + c)$, and *major* axis endpoints $(h, k - a)$ and $(h, k + a)$ is

$$\frac{(x - h)^2}{b^2} + \frac{(y - k)^2}{a^2} = 1$$

where $a > c$ and $b^2 = a^2 - c^2$. Also the *minor* axis endpoints are $(h - b, k)$ and $(h + b, k)$.

Remark 10.4.1 The direction of the major axis (vertical or horizontal) can be easily identified by observing whether a, which is always greater than or equal to b, is under the squared expression containing y or the one containing x. If under the y expression, it is *vertical*, while if under the x expression, it is *horizontal*.

Example 10.4.1 Find the center, vertices, and foci and graph the ellipse defined by the equation $4x^2 - 16x + 25y^2 + 200y + 316 = 0$.

Solution First we must convert the equation into standard form by completing the square on both the variables x and y and rearranging terms. We proceed as follows:

$$4x^2 - 16x + 25y^2 + 200y + 316 = 0$$

$$\Longleftrightarrow \quad 4(x^2 - 4x) + 25(y^2 + 8y) = -316$$

$$\Longleftrightarrow 4(x^2 - 4x + 4) + 25(y^2 + 8y + 16) = 16 + 400 - 316$$

$$\Longleftrightarrow \quad 4(x - 2)^2 + 25(y + 4)^2 = 100$$

$$\Longleftrightarrow \quad \frac{(x - 2)^2}{5^2} + \frac{(y - [-4])^2}{2^2} = 1$$

From this form it is clear that the center is $(2, -4)$, $a = 5$, and $b = 2$. Because $5 > 2$, we know the major axis is horizontal, hence,

$$2^2 = 5^2 - c^2 \iff c^2 = 21 \iff c = \sqrt{21}$$

With the center and a, b, c identified, the vertices are $(-3, -4)$, $(7, -4)$, $(2, -6)$, $(2, -2)$, and the foci are $(2 - \sqrt{21}, -4)$ and $(2 + \sqrt{21}, -4)$. The graph is illustrated in Fig. 10.4.6.

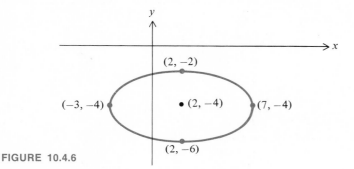

FIGURE 10.4.6

If the semimajor and semiminor axes of an ellipse are *equal,* the ellipse becomes a circle. That is, a circle is a special case of an ellipse where the two foci become a single focus located at the center. Setting $b = a$ in the standard form for an ellipse, we obtain the standard form for a circle as follows:

$$\frac{(x - h)^2}{a^2} + \frac{(y - k)^2}{a^2} = 1$$

This is equivalent to

$$(x - h)^2 + (y - k)^2 = a^2$$

Consequently, we have the following result:

The equation of a circle in *standard form* with center (h, k) and radius r is

$$(x - h)^2 + (y - k)^2 = r^2$$

Example 10.4.2 Find the center and radius of the circle defined by $x^2 + 4x + y^2 - 6y - 12 = 0$ and graph.

Solution We first convert the equation to standard form by completing the square on both variables x and y.

$$x^2 + 4x + y^2 - 6y - 12 = 0$$
$$\iff (x^2 + 4x + 4) + (y^2 - 6y + 9) = 12 + 4 + 9$$
$$\iff (x - [-2])^2 + (y - 3)^2 = 5^2$$

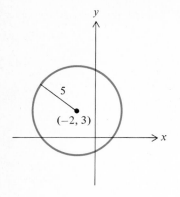

FIGURE 10.4.7

With the equation in standard form it is clear that the center is $(-2, 3)$ and the radius is 5. The graph of the circle is illustrated in Fig. 10.4.7.

Now let us return to Problem 10.4.1 with which we introduced this section. Recall we had a satellite in an elliptic orbit around the earth (of radius 3,963 miles) with a perigee of 2,037 miles and an apogee of 20,037 miles. (See Fig. 10.4.8.) Therefore, the major axis is given by

$$\underset{\text{Perigee}}{2a =} \quad \underset{\downarrow}{2037} \quad + \quad \underset{\substack{\text{Diameter} \\ \text{of Earth} \\ \downarrow}}{7926} \quad + \quad \underset{\substack{\text{Apogee} \\ \downarrow}}{20037}$$

$$= \quad 30,000$$
$$\Longrightarrow a = 15,000 \text{ miles}$$

Because $a - c = \text{perigee} + \text{radius of earth}$, we have

$$15,000 - c = 2,037 + 3,963$$
$$\Longrightarrow c = 9,000$$

Finally, using the formula $b^2 = a^2 - c^2$, we have

$$b^2 = (15,000)^2 - (9,000)^2 = 144,000,000$$
$$= (12,000)^2$$
$$\Longrightarrow b = 12,000 \text{ miles}$$

With a and b determined, the equation identifying every point on the elliptic orbit is

$$\frac{x^2}{(15,000)^2} + \frac{y^2}{(12,000)^2} = 1$$

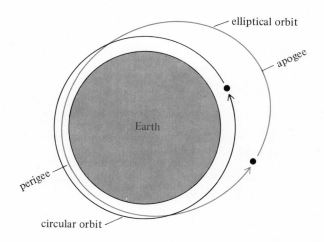

FIGURE 10.4.8

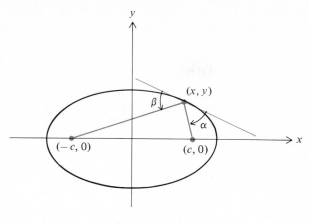

FIGURE 10.4.9

Notice here we have chosen the center of the ellipse to be the center of the coordinate system, that is, $(0, 0)$. The problem is solved.

Ellipses have important reflective properties just as parabolas do. It can be proved that a wave source (radio, light, sound, etc.) emanating from one of the foci will be reflected by an elliptic surface directly to the other focus; that is, angles α and β of Fig. 10.4.9 are equal. This principle is utilized in the design of ellipsoidal "whispering galleries" where the ceiling is constructed in the shape of an *ellipsoid of revolution*. In such a gallery a person whispering softly near one focus can be clearly heard at the other focus, whereas those located in between the two foci may not hear anything.

Problem Sets 10.5

Problem Set I ■ *Reading Comprehension*

Determine whether statements 1 through 4 are true or false and give reasons for your responses.

1 An ellipse is the set of all points in a plane equidistant from two fixed points called foci.
2 For the ellipse defined by
$$\frac{(x - h)^2}{a^2} + \frac{(y - k)^2}{b^2} = 1$$
the length of the major axis is $2a$ and the minor axis $2b$.
3 In the equation of Problem 2, if $a > b$ then a formula for c from which the positions of the foci are identified is $c^2 = b^2 - a^2$.
4 A circle is an ellipse.

Find the equation of each ellipse described in problems 1 through 10. Then determine the vertices and foci of each and sketch the graph.

1 Center $(2, 4)$, $a = 3$, $b = 2$, and major axis horizontal.
2 Center $(-3, -4)$, $a = 6$, $b = 4$, and major axis vertical.
3 Center $(-3, 1)$, $a = 5$, $b = 2$, and major axis vertical.
4 Center $(4, -2)$, $a = 10$, $c = 6$, and major axis horizontal.
5 Center $(3, 5)$, $b = 2$, $c = 1$, and major axis vertical.
6 Center $(-5, -2)$, $a = 4$, $c = 2$, and major axis vertical.
7 Vertices $(-4, -1)$ and $(6, -1)$ and one focus at $(1 - \sqrt{21}, -1)$.
8 Foci at $(-1, -3)$ and $(5, -3)$ and one vertex at $(2, 1)$.
9 Foci at $(4, 1)$ and $(4, 5)$ and one vertex at $(4 - 4\sqrt{2}, 3)$.
10 Vertices $(-4, 1)$ and $(1, -3)$ and one focus at $(-1, -3)$.

Find an equation describing each of the point sets of problems 11 and 12.

11 The set of all points satisfying the property that the sum of the distances from each to the points $(1, 3)$ and $(7, 3)$ is 10.
12 The set of all points satisfying the property that the sum of the distances from each to the points $(-2, 4)$ and $(-2, -4)$ is 12.

In problems 13 through 18, convert the equations to standard form and determine the center, vertices, and foci, and then sketch the graph.

13 $9x^2 - 18x + 4y^2 + 16y - 11 = 0$ 14 $25x^2 + 100x + 9y^2 + 72y + 19 = 0$
15 $x^2 + 6x + 16y^2 - 32y + 9 = 0$ 16 $4x^2 - 32x + 25y^2 - 300y + 864 = 0$
17 $16x^2 + 9y^2 - 54y - 63 = 0$ 18 $4x^2 + 24x + 36y^2 - 72y - 72 = 0$

1 After a satellite has been placed in orbit around the earth, it is determined that the perigee is 4,037 miles and the apogee is 10,037 miles. Find an equation describing every position of the satellite as it travels around the earth. (Assume an orbit near the equator where the radius of the earth is 3,963 miles.)
2 The moon travels around the earth in an elliptic orbit. It is known that the perigee is 221,463 miles and the apogee is 252,710 miles. Find an equation describing the entire elliptic path. (The radius of the moon is 1,080 miles.)

10.6   The Hyperbola

The final conc we shall study is the hyperbola, defined as follows:

DEFINITION 10.6.1   A hyperbola is the set of all points in a plane, each of which possesses the property that the *difference* of its distances from two fixed points, called foci, is a *constant*.

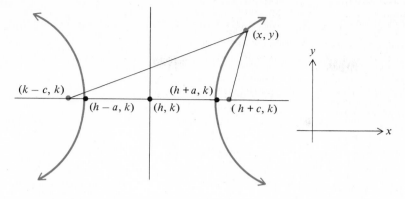

FIGURE 10.6.1

Our first task is to determine an equation for a hyperbola. This we do, as with the previous conics, by orienting the configuration within a rectangular coordinate system. We choose the center (h, k) the foci $(h - c, k)$ and $(h + c, k)$, and the two vertices $(h - a, k)$ and $(h + a, k)$, as illustrated in Fig. 10.6.1. Next, we pick an arbitrary point (x, y) on the hyperbola and apply Definition 10.6.1 to this point and the two foci. We know the *difference* of the distances between (x, y) and the foci is a constant, which we are denoting $\pm 2a$. Using the formula for the distance between two points, this property may be expressed

$$\sqrt{[(x - h) + c]^2 + (y - k)^2} - \sqrt{[(x - h) - c]^2 + (y - k)^2} = \pm 2a$$

Moving the second radical to the right side, squaring both sides, and simplifying, we obtain

$$\pm \sqrt{[(x - h) - c]^2 + (y - k)^2} = \frac{c}{a}(x - h) - a$$

Again squaring both sides we get

$$[(x - h) - c]^2 + (y - k)^2 = \frac{c^2}{a^2}(x - h)^2 - 2c(x - h) + a^2$$

By simplifying and rearranging, this equation becomes

$$\frac{(x - h)^2}{a^2} - \frac{(y - k)^2}{c^2 - a^2} = 1$$

Because $c > a$, the difference $c^2 - a^2$ is positive. For notational convenience, we let $b^2 = c^2 - a^2$. Our equation now takes the form

$$\frac{(x - h)^2}{a^2} - \frac{(y - k)^2}{b^2} = 1$$

We summarize as follows:

The equation of a hyperbola in *standard form* with center (h, k), horizontally aligned foci $(h - c, k)$ and $(h + c, k)$, and vertices $(h - a, k)$ and $(h + a, k)$ is

$$\frac{(x - h)^2}{a^2} - \frac{(y - k)^2}{b^2} = 1$$

where $c > a$ and $b^2 = c^2 - a^2$.

The procedure for finding an equation of a hyperbola with foci lined up vertically is similar to that used above for horizontally placed foci. We will leave the details as an exercise. The results, however, are as follows:

The equation of a hyperbola in *standard form* with center (h, k), vertically aligned foci $(h, k - c)$ and $(h, k + c)$, and vertices $(h, k - a)$ and $(h, k + a)$ is

$$\frac{(y - k)^2}{a^2} - \frac{(x - h)^2}{b^2} = 1$$

where $c > a$ and $b^2 = c^2 - a^2$.

A *rule of thumb* by which you may determine whether the foci of a hyperbola are oriented vertically or horizontally after converting the equation to standard form is "a *minus* preceding the term containing x indicates *vertical* foci, whereas a *minus* preceding the term containing y indicates *horizontal* foci."

Surprisingly, the hyperbola is the easiest of the conics to graph. This is because of the presence of *asymptotes,* which along with the vertices enable us to very simply yet accurately graph a hyperbola. The manner by which the asymptotes are determined and the hyperbola sketched will be explained, with the proof left as an exercise.

From the equation of a horizontally oriented hyperbola in standard form, identify a, b and the center (h, k). Then form a *rectangle* of width $2a$ and height $2b$ with center (h, k). Next draw both diagonals through the corners extending out in both directions indefinitely. (See Fig. 10.6.2.) These two extended diagonals are the asymptotes. The hyperbola is then readily sketched as illustrated in Fig. 10.6.2. Similarly, a vertically oriented hyperbola is sketched as illustrated in Fig. 10.6.3.

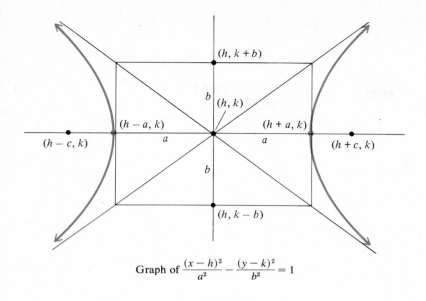

Graph of $\dfrac{(x-h)^2}{a^2} - \dfrac{(y-k)^2}{b^2} = 1$

FIGURE 10.6.2

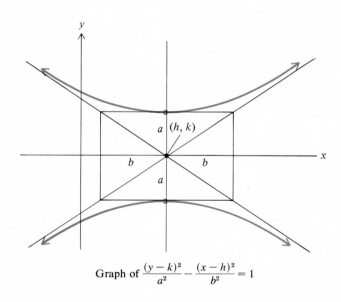

Graph of $\dfrac{(y-k)^2}{a^2} - \dfrac{(x-h)^2}{b^2} = 1$

FIGURE 10.6.3

Example 10.6.1 Find the center, vertices, foci, and asymptotes of the hyperbola defined by the equation

$$-25x^2 + 150x + 9y^2 - 72y - 306 = 0$$

and graph.

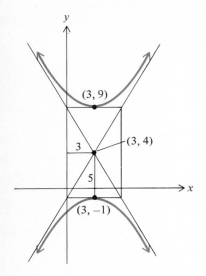

(3, 9)

(3, 4)

3

5

(3, −1)

FIGURE 10.6.4

Solution We first convert this equation into standard form by completing the square on both variables x and y.

$$-25x^2 + 150x + 9y^2 - 72y - 306 = 0$$

$$\Longleftrightarrow -25(x^2 - 6x + 9) + 9(y^2 - 8y + 16) = 306 - 225 + 144$$

$$\Longleftrightarrow -25(x - 3)^2 + 9(y - 4)^2 = 225$$

$$\Longleftrightarrow -\frac{(x - 3)^2}{3^2} + \frac{(y - 4)^2}{5^2} = 1$$

It is now clear that the center is $(3, 4)$ and $b = 3$, $a = 5$. Because the *minus* symbol precedes the term containing x, we know the foci are oriented vertically. Thus, applying the formula $b^2 = c^2 - a^2$, we obtain

$$3^2 = c^2 - 5^2 \Longleftrightarrow c^2 = 34 \Longrightarrow c = \sqrt{34}$$

From the above information the vertices are $(3, -1)$ and $(3, 9)$, and the foci are $(3, 4 - \sqrt{34})$ and $(3, 4 + \sqrt{34})$. Knowing the center and a and b, we construct the rectangle from which the asymptotes are defined. Remembering that the foci are vertically oriented, we sketch the hyperbola as illustrated in Fig. 10.6.4.

Our development of the conics in this and the preceding sections has been by techniques of *analytic geometry*. That is, we used *both* geometric and algebraic principles to discover properties of the conics and also to graph them. There is, however, a purely geometric way to construct the conics. This is accomplished by the intersecting surfaces of a *cone* with two nappes and a *plane*, as illustrated in Fig. 10.6.5.

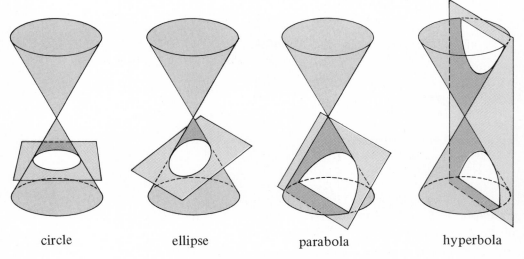

circle   ellipse   parabola   hyperbola

FIGURE 10.6.5

Problem Set I ■ Reading Comprehension

Determine whether statements 1 through 5 are true or false and give reasons for your responses.

1 The only difference in the definitions of an ellipse and a hyperbola is that for a hyperbola the *difference* of the distances to the foci is a constant rather than the *sum*.

2 For a hyperbola, the relationship between the three constants a, b, c, is always $b^2 = c^2 - a^2$.

3 The simplest way to determine whether the foci of a hyperbola are vertical or horizontal is the rule "from the equation in standard form, a minus preceding the term containing x indicates horizontal foci, whereas a minus preceding the term containing y indicates vertical foci."

4 To determine the asymptotes of a given hyperbola, we take a rectangle of width $2a$ and height $2b$ and draw the diagonals through the vertices.

5 From the equation of a hyperbola in standard form, if $a > b$ then the branches cup right and left, whereas if $a < b$ they cup up and down.

Problem Set II ■ Skills Development

In problems 1 through 6 find an equation of the hyperbola satisfying the given conditions.

1 Foci $(5, 4)$ and $(-5, 4)$ and vertices $(2, 4)$ and $(-2, 4)$.
2 Foci $(5, 4)$ and $(5, -6)$ and vertices $(5, 3)$ and $(5, -5)$.
3 Foci $(-11, 4)$ and $(1, 4)$ and vertices $(-8, 4)$ and $(-2, 4)$.
4 Center $(1, 3)$, focus $(-4, 3)$, and vertex $(4, 3)$.
5 Center $(-4, -1)$, focus $(-4, -1 - 2\sqrt{5})$, and vertex $(-4, 3)$.
6 $a = 5$, $b = 12$, center $(-2, 3)$, and foci vertical.

In problems 7 through 12, convert the equations into standard form by completing the square on both variables. Identify the center, vertices, foci, and asymptotes, and then sketch the graph.

7 $9x^2 - 36x - 16y^2 + 128y - 364 = 0$
8 $144y^2 + 864y - 25x^2 + 50x - 2329 = 0$
9 $225y^2 - 900y - 64x^2 - 13{,}500 = 0$
10 $4y^2 - 32y - x^2 + 6x = 9$
11 $9x^2 - 36x - 16y^2 - 32y - 124 = 0$
12 $36y^2 - 216y - 13x^2 + 156x = 612$

Find an equation that determines the points satisfying the conditions in problems 13 through 15.

13 The difference of the distances of each point from $(-1, -3)$ and $(5, -3)$ is 4.
14 The difference of the distances of each point from $(5, 8)$ and $(5, 0)$ is 6.
15 The difference of the distances of each point from $(-11, 4)$ and $(1, 4)$ is 8.

1 Show that the asymptotes of the hyperbola

$$\frac{(x-h)^2}{a^2} - \frac{(y-k)^2}{b^2} = 1 \tag{1}$$

are defined by the lines

$$y - k = \pm\frac{b}{a}(x - h)$$

(HINT: Solve (1) for $y - k$, factor $(x - h)$ out of the radical, and examine as x becomes increasingly large.)

2 Prove that every line parallel to an asymptote of a hyperbola intersects the hyperbola in exactly one point.

3 Prove that the length of a focal chord of the hyperbola (1) of Problem 1 above is $2b^2/a$.

4 Describe the set of points satisfying the inequality

$$\frac{x^2}{a^2} - \frac{y^2}{b^2} < 1$$

5 Sketch the graphs of the following:

a $\left|\dfrac{x^2}{2^2} - \dfrac{y^2}{3^2}\right| = 1$ b $\dfrac{x\,|\,x\,|}{2^2} - \dfrac{y\,|\,y\,|}{3^2} = 1$

6 Write the equation of a hyperbola in determinant form.

1 During World War II the *Loran* system of navigation was developed. In this system two *pairs* of transmitting stations were located at fixed points. A ship at sea would receive signals from each pair of stations from which it was possible to determine the *difference* in the distances from the ship to each pair. These stations would be interpreted as two pairs of foci each determining, with the above information, a hyperbola. The ship's location would be the *intersection* of two branches of these hyperbola.

Suppose in a coordinate system where each unit represents 25 miles we have two pairs of *Loran* shore stations located at points $(5 - \sqrt{10}, 5)$ and $(5 + \sqrt{10}, 5)$, and $(5 - \sqrt{5}, -10)$ and $(5 + \sqrt{5}, -10)$. Further, suppose a ship determines that the difference in its distances from the first pair is 150 miles and from the second pair is 100. Determine the position of the ship within this coordinate system. (Assume y-axis is shore line.)

10.8   Translation of Axes

In the preceding sections of this chapter, we treated quadratic equations in two variables of the form

$$ax^2 + bx + cy^2 + dy + e = 0 \tag{1}$$

where a and c were not both zero and $a, b, c, d, e \in \mathbf{R}$. If one of the two constants, a or c, is zero, then the resulting conic is a parabola. If both are nonzero, then in general we either have an ellipse or a hyperbola. To convert (1) into standard form we complete the square, obtaining one of the following forms:

$$(x - h)^2 = \pm 4p(y - k)$$
$$(y - k)^2 = \pm 4p(x - h)$$
$$\pm \frac{(x - h)^2}{a^2} \pm \frac{(y - k)^2}{b^2} = 1$$

In every case we have the two expressions

$$x - h \quad \text{and} \quad y - k$$

present. If we set

$$x' = x - h \quad \text{and} \quad y' = y - k$$

then the above equations become

$$(x')^2 = \pm 4py'$$
$$(y')^2 = \pm 4px'$$
$$\pm \frac{(x')^2}{a^2} \pm \frac{(y')^2}{b^2} = 1$$

That is, in an $x'y'$ coordinate system the vertex of the parabolas would be $(0', 0')$ and the center of the ellipses and hyperbolas would also be $(0', 0')$. To visualize the relationship between these two coordinate systems, under the conditions $x' = x - h$ and $y' = y - k$, we superimpose one upon the other as illustrated in Fig. 10.8.1. We see that the *origin* in the $x'y'$ system is the point (h, k) of the xy system. Another way of interpreting this is to imagine that the origin of the xy system has been translated to the point (h, k) to form the origin of the $x'y'$ system. For this reason the substitution

$$x' = x - h \quad \text{and} \quad y' = y - k$$

is called a translation of axes.

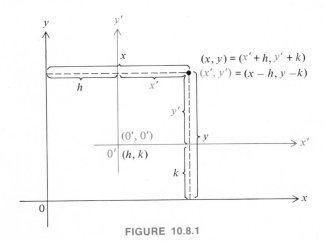

FIGURE 10.8.1

Example 10.8.1 Determine a translation of axes that will bring the center of the following conic to the origin of the transformed system..

$$9x^2 + 54x + 5y^2 - 40y - 19 = 0$$

Solution We convert this to standard form by completing the square and rearranging as follows:

$$9x^2 + 54x + 5y^2 - 40y - 19 = 0$$
$$\Longleftrightarrow 9(x^2 + 6x + 9) + 5(y^2 - 8y + 16) = 19 + 81 + 80$$
$$\Longleftrightarrow \frac{(x+3)^2}{20} + \frac{(y-4)^2}{36} = 1$$

It is now clear that if

$$x' = x + 3 \quad \text{and} \quad y' = y - 4$$

then the center of the ellipse will be *translated* from $(-3, 4)$ in the xy system to the origin of the $x'\,y'$ system, as illustrated in Fig. 10.8.2.

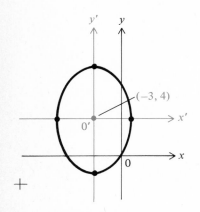

FIGURE 10.8.2

Remark 10.8.1 It is important to remember that when the two coordinate systems are superimposed one upon the other the conic itself is *not* moved or translated but rather the *xy axes* are translated to become the $x'y'$ axes with origin $(0', 0')$ located at (h, k).

As noted when we began this section, we have thus far treated quadratic equations in two variables of the form

$$ax^2 + bx + cy^2 + dy + e = 0 \qquad (1)$$

This form, however, is not completely general. We have purposely been omitting the quadratic term containing the *product* of the two variables, that is, the term containing xy. The effect of this term upon the geometric orientation of the conics is one of *rotation*. Because, however, the transformation producing a rotation of axes that would eliminate the term containing the product xy requires the use of trigonometric functions, we will not treat this case in this text.

10.9 Problem Sets

Problem Set I ■ *Reading Comprehension*

Determine whether statements 1 through 2 are true or false and give reasons for your responses.

1 The following equations are used to translate axes.
$$x = x' + h \quad \text{and} \quad y = y' + k$$
2 By the translation formulas
$$x' = x - 2 \quad \text{and} \quad y' = y - 3$$
the vertex of the parabola defined by $(x-2)^2 = 4(y-3)$ is translated from the point $(2, 3)$ to the origin.

Problem Set II ■ Skills Development

Transform the following equations into standard form by a translation of axes, then sketch the graph.

1 $x^2 - 4y^2 + 2x + 8y - 7 = 0$ **2** $x^2 + y^2 - 2x - 4y - 4 = 0$
3 $16x^2 + 4y^2 - 32x - 4y + 13 = 0$ **4** $-4x^2 + 9y^2 - 16x - 18y - 43 = 0$
5 $4x^2 - y^2 + 24x + 8y + 4 = 0$ **6** $x^2 - 8x - 8y - 8 = 0$

Problem Set III ■ Just For Fun

1 An old riddle runs as follows. An explorer walks one mile due south, turns and walks one mile due east, turns again and walks one mile due north. He finds himself back where he started. He shoots a bear. What color is the bear? The time-honored answer is white, because the explorer must have started at the North Pole. But not long ago someone made the discovery that the North Pole is not the only starting point that satisfies the given conditions! Can you think of any other spot on the globe from which one could walk a mile south, a mile east, a mile north and find himself back at his original location?

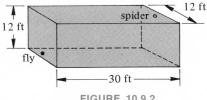

FIGURE 10.9.1

2 One of the oldest of topological puzzles, familiar to many a pupil, consists of drawing a continuous line across the closed network shown in Fig. 10.9.1 so that the line crosses each of the 16 segments of the network only once. The curved line shown here does not solve the puzzle because it leaves one segment uncrossed. No trick solutions are allowed, such as passing the line through a vertex or along one of the segments, or folding the paper.

    It is not difficult to prove that the puzzle cannot be solved on a plane surface. Two questions: Can it be solved on the surface of a sphere? On the surface of a torus (doughnut)?

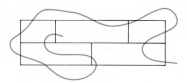

FIGURE 10.9.2

3 A rectangular room has the dimensions shown in Fig. 10.9.2. A spider is at the middle of an end wall, one foot from the ceiling. A fly is at the middle of the opposite end wall, one foot above the floor, and too paralyzed with fear to move. What is the shortest distance the spider must crawl in order to reach the fly?

10.10   Summary of Definitions and Properties in Chapter 10

Appendixes

Greek Alphabet

A	α	Alpha	N	ν	Nu
B	β	Beta	Ξ	ξ	Xi
Γ	γ	Gamma	O	o	Omicron
Δ	δ	Delta	Π	π	Pi
E	ϵ	Epsilon	P	ρ	Rho
Z	ζ	Zeta	Σ	σ	Sigma
H	η	Eta	T	τ	Tau
Θ	θ	Theta	Y	υ	Upsilon
I	ι	Iota	Φ	ϕ	Phi
K	κ	Kappa	X	χ	Chi
Λ	λ	Lambda	Ψ	ψ	Psi
M	μ	Mu	Ω	ω	Omega

Factoring Quadratic Polynomials

A polynomial of the form $ax^2 + bx + c$, $a \neq 0$, is of the *second degree* and called quadratic. It is frequently desirable to express such polynomials as the product of two linear factors. The process of achieving this is called factoring. A few techniques and formulas that may assist you in factoring certain quadratic polynomials are described below.

We devise first a semi-general form and then a completely general one:

$$\text{I.} \quad (x + a)(x + b) = x^2 + (a + b)x + a \cdot b$$

Example Factor $x^2 + 3x - 4$

Solution Let $a = 4$ and $b = -1$, then $a \cdot b = -4$ and $a + b = 3$. Hence, $x^2 + 3x - 4 = (x + 4)(x - 1)$.

Example Factor $x^2 - 7xy + 10y^2$

Solution Let $a = -2y$ and $b = -5y$, then $a \cdot b = 10y^2$ and $a + b = -7y$. Hence, $x^2 - 7xy + 10y^2 = (x - 2y)(x - 5y)$.

II. $(ax + b)(cx + d) = (ac)x^2 + (ad + bc)x + bd$

Example $6x^2 + x - 15$

Solution   Let $a = 2$, $b = -3$, $c = 3$, and $d = 5$, then $ac = 6$, $ad + bc = 1$, and $bd = -15$. Hence, $6x^2 + x - 15 = (2x - 3)(3x + 5)$.

Remark   The proper selection of a, b, c, d in the example above is usually made after determining all possible integral factor pairs for 6 and -15, and then choosing the two pairs that fit the required conditions. Proceeding in this manner with the example above, we obtain

$$\begin{matrix} & a & c \\ 6 = 6 \cdot 1 = & 2 & \cdot 3 \end{matrix}$$

$$\left. \right\} ad + bc = 1$$

$$\begin{matrix} -15 = (-15)(1) = (15)(-1) = (-3)(5) = (3)(-5) \\ \qquad\qquad\qquad\qquad\qquad b \qquad d \end{matrix}$$

Example $8x^2 - 34x + 33$

Solution   Factoring 8 and 33, we obtain

$$\begin{matrix} & a & c \\ 8 = 8 \cdot 1 = & 2 & \cdot 4 \end{matrix}$$

$$\left. \right\} ad + bc = -34$$

$$\begin{matrix} 33 = 33 \cdot 1 = (-33)(-1) = (-11)(-3) = 11 \cdot 3 \\ \qquad\qquad\qquad\qquad\qquad d \qquad b \end{matrix}$$

Hence, $8x^2 - 34x + 33 = (2x - 3)(4x - 11)$.
$$\qquad\qquad\qquad\qquad\qquad a \quad\ b \quad\ c \quad\ d$$

III.   Formulas

A.   Perfect Squares

$$x^2 + 2ax + a^2 = (x + a)^2$$
$$x^2 - 2ax + a^2 = (x - a)^2$$

Example $x^2 - 6x + 9 = x^2 - 2(3)x + 3^2 = (x - 3)^2$

B.   Difference of Two Squares

$$x^2 - a^2 = (x - a)(x + a)$$

Example $16x^2 - 49y^2 = (4x)^2 - (7y)^2 = (4x - 7y)(4x + 7y)$

Table I   Common Logarithms (base 10)

log x	0	1	2	3	4	5	6	7	8	9
log 1.0	.0000	.0043	.0086	.0128	.0170	.0212	.0253	.0294	.0334	.0374
1.1	.0414	.0453	.0492	.0531	.0569	.0607	.0645	.0682	.0719	.0755
1.2	.0792	.0828	.0864	.0899	.0934	.0969	.1004	.1038	.1072	.1106
1.3	.1139	.1173	.1206	.1239	.1271	.1303	.1335	.1367	.1399	.1430
1.4	.1461	.1492	.1523	.1553	.1584	.1614	.1644	.1673	.1703	.1732
log 1.5	.1761	.1790	.1818	.1847	.1875	.1903	.1931	.1959	.1987	.2014
1.6	.2041	.2068	.2095	.2122	.2148	.2175	.2201	.2227	.2253	.2279
1.7	.2304	.2330	.2355	.2380	.2405	.2430	.2455	.2480	.2504	.2529
1.8	.2553	.2577	.2601	.2625	.2648	.2672	.2695	.2718	.2742	.2765
1.9	.2788	.2810	.2833	.2856	.2878	.2900	.2923	.2945	.2967	.2989
log 2.0	.3010	.3032	.3054	.3075	.3096	.3118	.3139	.3160	.3181	.3201
2.1	.3222	.3243	.3263	.3284	.3304	.3324	.3345	.3365	.3385	.3404
2.2	.3424	.3444	.3464	.3483	.3502	.3522	.3541	.3560	.3579	.3598
2.3	.3617	.3636	.3655	.3674	.3692	.3711	.3729	.3747	.3766	.3784
2.4	.3802	.3820	.3838	.3856	.3874	.3892	.3909	.3927	.3945	.3962
log 2.5	.3979	.3997	.4014	.4031	.4048	.4065	.4082	.4099	.4116	.4133
2.6	.4150	.4166	.4183	.4200	.4216	.4232	.4249	.4265	.4281	.4298
2.7	.4314	.4330	.4346	.4362	.4378	.4393	.4409	.4425	.4440	.4456
2.8	.4472	.4487	.4502	.4518	.4533	.4548	.4564	.4579	.4594	.4609
2.9	.4624	.4639	.4654	.4669	.4683	.4698	.4713	.4728	.4742	.4757
log 3.0	.4771	.4786	.4800	.4814	.4829	.4843	.4857	.4871	.4886	.4900
3.1	.4914	.4928	.4942	.4955	.4969	.4983	.4997	.5011	.5024	.5038
3.2	.5051	.5065	.5079	.5092	.5105	.5119	.5132	.5145	.5159	.5172
3.3	.5185	.5198	.5211	.5224	.5237	.5250	.5263	.5276	.5289	.5302
3.4	.5315	.5328	.5340	.5353	.5366	.5378	.5391	.5403	.5416	.5428
log 3.5	.5441	.5453	.5465	.5478	.5490	.5502	.5514	.5527	.5539	.5551
3.6	.5563	.5575	.5587	.5599	.5611	.5623	.5635	.5647	.5658	.5670
3.7	.5682	.5694	.5705	.5717	.5729	.5740	.5752	.5763	.5775	.5786
3.8	.5798	.5809	.5821	.5832	.5843	.5855	.5866	.5877	.5888	.5899
3.9	.5911	.5922	.5933	.5944	.5955	.5966	.5977	.5988	.5999	.6010
log 4.0	.6021	.6031	.6042	.6053	.6064	.6075	.6085	.6096	.6107	.6117
4.1	.6128	.6138	.6149	.6160	.6170	.6180	.6191	.6201	.6212	.6222
4.2	.6232	.6243	.6253	.6263	.6274	.6284	.6294	.6304	.6314	.6325
4.3	.6335	.6345	.6355	.6365	.6375	.6385	.6395	.6405	.6415	.6425
4.4	.6435	.6444	.6454	.6464	.6474	.6484	.6493	.6503	.6513	.6522
log 4.5	.6532	.6542	.6551	.6561	.6571	.6580	.6590	.6599	.6609	.6618
4.6	.6628	.6637	.6646	.6656	.6665	.6675	.6684	.6693	.6702	.6712
4.7	.6721	.6730	.6739	.6749	.6758	.6767	.6776	.6785	.6794	.6803
4.8	.6812	.6821	.6830	.6839	.6848	.6857	.6866	.6875	.6884	.6893
4.9	.6902	.6911	.6920	.6928	.6937	.6946	.6955	.6964	.6972	.6981
log 5.0	.6990	.6998	.7007	.7016	.7024	.7033	.7042	.7050	.7059	.7067
5.1	.7076	.7084	.7093	.7101	.7110	.7118	.7126	.7135	.7143	.7152
5.2	.7160	.7168	.7177	.7185	.7193	.7202	.7210	.7218	.7226	.7235
5.3	.7243	.7251	.7259	.7267	.7275	.7284	.7292	.7300	.7308	.7316
5.4	.7324	.7332	.7340	.7348	.7356	.7364	.7372	.7380	.7388	.7396
log x	0	1	2	3	4	5	6	7	8	9

Table I Common Logarithms (*continued*)

log *x*	0	1	2	3	4	5	6	7	8	9
log 5.5	.7404	.7412	.7419	.7427	.7435	.7443	.7451	.7459	.7466	.7474
5.6	.7482	.7490	.7497	.7505	.7513	.7520	.7528	.7536	.7543	.7551
5.7	.7559	.7566	.7574	.7582	.7589	.7597	.7604	.7612	.7619	.7627
5.8	.7634	.7642	.7649	.7657	.7664	.7672	.7679	.7686	.7694	.7701
5.9	.7709	.7716	.7723	.7731	.7738	.7745	.7752	.7760	.7767	.7774
log 6.0	.7782	.7789	.7796	.7803	.7810	.7818	.7825	.7832	.7839	.7846
6.1	.7853	.7860	.7868	.7875	.7882	.7889	.7896	.7903	.7910	.7917
6.2	.7924	.7931	.7938	.7945	.7952	.7959	.7966	.7973	.7980	.7987
6.3	.7993	.8000	.8007	.8014	.8021	.8028	.8035	.8041	.8048	.8055
6.4	.8062	.8069	.8075	.8082	.8089	.8096	.8102	.8109	.8116	.8122
log 6.5	.8129	.8136	.8142	.8149	.8156	.8162	.8169	.8176	.8182	.8189
6.6	.8195	.8202	.8209	.8215	.8222	.8228	.8235	.8241	.8248	.8254
6.7	.8261	.8267	.8274	.8280	.8287	.8293	.8299	.8306	.8312	.8319
6.8	.8325	.8331	.8338	.8344	.8351	.8357	.8363	.8370	.8376	.8382
6.9	.8388	.8395	.8401	.8407	.8414	.8420	.8426	.8432	.8439	.8445
log 7.0	.8451	.8457	.8463	.8470	.8476	.8482	.8488	.8494	.8500	.8506
7.1	.8513	.8519	.8525	.8531	.8537	.8543	.8549	.8555	.8561	.8567
7.2	.8573	.8579	.8585	.8591	.8597	.8603	.8609	.8615	.8621	.8627
7.3	.8633	.8639	.8645	.8651	.8657	.8663	.8669	.8675	.8681	.8686
7.4	.8692	.8698	.8704	.8710	.8716	.8722	.8727	.8733	.8739	.8745
log 7.5	.8751	.8756	.8762	.8768	.8774	.8779	.8785	.8791	.8797	.8802
7.6	.8808	.8814	.8820	.8825	.8831	.8837	.8842	.8848	.8854	.8859
7.7	.8865	.8871	.8876	.8882	.8887	.8893	.8899	.8904	.8910	.8915
7.8	.8921	.8927	.8932	.8938	.8943	.8949	.8954	.8960	.8965	.8971
7.9	.8976	.8982	.8987	.8993	.8998	.9004	.9009	.9015	.9020	.9025
log 8.0	.9031	.9036	.9042	.9047	.9053	.9058	.9063	.9069	.9074	.9079
8.1	.9085	.9090	.9096	.9101	.9106	.9112	.9117	.9122	.9128	.9133
8.2	.9138	.9143	.9149	.9154	.9159	.9165	.9170	.9175	.9180	.9186
8.3	.9191	.9196	.9201	.9206	.9212	.9217	.9222	.9227	.9232	.9238
8.4	.9243	.9248	.9253	.9258	.9263	.9269	.9274	.9279	.9284	.9289
log 8.5	.9294	.9299	.9304	.9309	.9315	.9320	.9325	.9330	.9335	.9340
8.6	.9345	.9350	.9355	.9360	.9365	.9370	.9375	.9380	.9385	.9390
8.7	.9395	.9400	.9405	.9410	.9415	.9420	.9425	.9430	.9435	.9440
8.8	.9445	.9450	.9455	.9460	.9465	.9469	.9474	.9479	.9484	.9489
8.9	.9494	.9499	.9504	.9509	.9513	.9518	.9523	.9528	.9533	.9538
log 9.0	.9542	.9547	.9552	.9557	.9562	.9566	.9571	.9576	.9581	.9586
9.1	.9590	.9595	.9600	.9605	.9609	.9614	.9619	.9624	.9628	.9633
9.2	.9638	.9643	.9647	.9652	.9657	.9661	.9666	.9671	.9675	.9680
9.3	.9685	.9689	.9694	.9699	.9703	.9708	.9713	.9717	.9722	.9727
9.4	.9731	.9736	.9741	.9745	.9750	.9754	.9759	.9763	.9768	.9773
log 9.5	.9777	.9782	.9786	.9791	.9795	.9800	.9805	.9809	.9814	.9818
9.6	.9823	.9827	.9832	.9836	.9841	.9845	.9850	.9854	.9859	.9863
9.7	.9868	.9872	.9877	.9881	.9886	.9890	.9894	.9899	.9903	.9908
9.8	.9912	.9917	.9921	.9926	.9930	.9934	.9939	.9943	.9948	.9952
9.9	.9956	.9961	.9965	.9969	.9974	.9978	.9983	.9987	.9991	.9996
log *x*	0	1	2	3	4	5	6	7	8	9

Table II   Exponential Functions e^x and e^{-x}

x	e^x	e^{-x}	x	e^x	e^{-x}
0.00	1.0000	1.0000	1.5	4.4817	0.2231
0.01	1.0101	0.9901	1.6	4.9530	0.2019
0.02	1.0202	0.9802	1.7	5.4739	0.1827
0.03	1.0305	0.9702	1.8	6.0496	0.1653
0.04	1.0408	0.9608	1.9	6.6859	0.1496
0.05	1.0513	0.9512	2.0	7.3891	0.1353
0.06	1.0618	0.9418	2.1	8.1662	0.1225
0.07	1.0725	0.9324	2.2	9.0250	0.1108
0.08	1.0833	0.9331	2.3	9.9742	0.1003
0.09	1.0942	0.9139	2.4	11.023	0.0907
0.10	1.1052	0.9048	2.5	12.182	0.0821
0.11	1.1163	0.8958	2.6	13.464	0.0743
0.12	1.1275	0.8869	2.7	14.880	0.0672
0.13	1.1388	0.8781	2.8	16.445	0.0608
0.14	1.1503	0.8694	2.9	18.174	0.0550
0.15	1.1618	0.8607	3.0	20.086	0.0498
0.16	1.1735	0.8521	3.1	22.198	0.0450
0.17	1.1853	0.8437	3.2	24.533	0.0408
0.18	1.1972	0.8353	3.3	27.113	0.0369
0.19	1.2092	0.8270	3.4	29.964	0.0334
0.20	1.2214	0.8187	3.5	33.115	0.0302
0.21	1.2337	0.8106	3.6	36.598	0.0273
0.22	1.2461	0.8025	3.7	40.447	0.0247
0.23	1.2586	0.7945	3.8	44.701	0.0224
0.24	1.2712	0.7866	3.9	49.402	0.0202
0.25	1.2840	0.7788	4.0	54.598	0.0183
0.30	1.3499	0.7408	4.1	60.340	0.0166
0.35	1.4191	0.7047	4.2	66.686	0.0150
0.40	1.4918	0.6703	4.3	73.700	0.0136
0.45	1.5683	0.6376	4.4	81.451	0.0123
0.50	1.6487	0.6065	4.5	90.017	0.0111
0.55	1.7333	0.5769	4.6	99.484	0.0101
0.60	1.8221	0.5488	4.7	109.95	0.0091
0.65	1.9155	0.5220	4.8	121.51	0.0082
0.70	2.0138	0.4966	4.9	134.29	0.0074
0.75	2.1170	0.4724	5.0	148.41	0.0067
0.80	2.2255	0.4493	5.5	244.69	0.0041
0.85	2.3396	0.4274	6.0	403.43	0.0025
0.90	2.4596	0.4066	6.5	665.14	0.0015
0.95	2.5857	0.3867	7.0	1096.6	0.0009
1.0	2.7183	0.3679	7.5	1808.0	0.0006
1.1	3.0042	0.3329	8.0	2981.0	0.0003
1.2	3.3201	0.3012	8.5	4914.8	0.0002
1.3	3.6693	0.2725	9.0	8103.1	0.0001
1.4	4.0552	0.2466	10.0	22026	0.00005

Table III   Natural Logarithms of Numbers (base *e*)

n	log*e* *n*	*n*	log*e* *n*	*n*	log*e* *n*
		4.5	1.5041	9.0	2.1972
0.1	−2.3026	4.6	1.5261	9.1	2.2083
0.2	−1.6094	4.7	1.5476	9.2	2.2192
0.3	−1.2040	4.8	1.5486	9.3	2.2300
0.4	−0.9163	4.9	1.5892	9.4	2.2407
0.5	−0.6931	5.0	1.6094	9.5	2.2513
0.6	−0.5108	5.1	1.6292	9.6	2.2618
0.7	−0.3567	5.2	1.6487	9.7	2.2721
0.8	−0.2231	5.3	1.6677	9.8	2.2824
0.9	−0.1054	5.4	1.6864	9.9	2.2925
1.0	0.0000	5.5	1.7047	10	2.3026
1.1	0.0953	5.6	1.7228	11	2.3979
1.2	0.1823	5.7	1.7405	12	2.4849
1.3	0.2624	5.8	1.7579	13	2.5649
1.4	0.3365	5.9	1.7750	14	2.6391
1.5	0.4055	6.0	1.7918	15	2.7081
1.6	0.4700	6.1	1.8083	16	2.7726
1.7	0.5306	6.2	1.8245	17	2.8332
1.8	0.5878	6.3	1.8405	18	2.8904
1.9	0.6419	6.4	1.8563	19	2.9444
2.0	0.6931	6.5	1.8718	20	2.9957
2.1	0.7419	6.6	1.8871	25	3.2189
2.2	0.7885	6.7	1.9021	30	3.4012
2.3	0.8329	6.8	1.9169	35	3.5553
2.4	0.8755	6.9	1.9315	40	3.6889
2.5	0.9163	7.0	1.9459	45	3.8067
2.6	0.9555	7.1	1.9601	50	3.9120
2.7	0.9933	7.2	1.9741	55	4.0073
2.8	1.0296	7.3	1.9879	60	4.0943
2.9	1.0647	7.4	2.0015	65	4.1744
3.0	1.0986	7.5	2.0149	70	4.2485
3.1	1.1314	7.6	2.0281	75	4.3175
3.2	1.1632	7.7	2.0412	80	4.3820
3.3	1.1939	7.8	2.0541	85	4.4427
3.4	1.2238	7.9	2.0669	90	4.4998
3.5	1.2528	8.0	2.0794	100	4.6052
3.6	1.2809	8.1	2.0919	110	4.7005
3.7	1.3083	8.2	2.1041	120	4.7875
3.8	1.3350	8.3	2.1163	130	4.8676
3.9	1.3610	8.4	2.1282	140	4.9416
4.0	1.3863	8.5	2.1401	150	5.0106
4.1	1.4110	8.6	2.1518	160	5.0752
4.2	1.4351	8.7	2.1633	170	5.1358
4.3	1.4586	8.8	2.1748	180	5.1930
4.4	1.4816	8.9	2.1861	190	5.2470

Table IV   Powers and Roots (from 1 to 100)

n	n^2	$\sqrt{n}$	n^3	$\sqrt[3]{n}$	n	n^2	$\sqrt{n}$	n^3	$\sqrt[3]{n}$
1	1	1.000	1	1.000	51	2,601	7.141	132,651	3.708
2	4	1.414	8	1.260	52	2,704	7.211	140,608	3.733
3	9	1.732	27	1.442	53	2,809	7.280	148,877	3.756
4	16	2.000	64	1.587	54	2,916	7.348	157,464	3.780
5	25	2.236	125	1.710	55	3,025	7.416	166,375	3.803
6	36	2.449	216	1.817	56	3,136	7.483	175,616	3.826
7	49	2.646	343	1.913	57	3,249	7.550	185,193	3.849
8	64	2.828	512	2.000	58	3,364	7.616	195,112	3.871
9	81	3.000	729	2.080	59	3,481	7.681	205,379	3.893
10	100	3.162	1,000	2.154	60	3,600	7.746	216,000	3.915
11	121	3.317	1,331	2.224	61	3,721	7.810	226,981	3.936
12	144	3.464	1,728	2.289	62	3,844	7.874	238,328	3.958
13	169	3.606	2,197	2.351	63	3,969	7.937	250,047	3.979
14	196	3.742	2,744	2.410	64	4,096	8.000	262,144	4.000
15	225	3.873	3,375	2.466	65	4,225	8.062	274,625	4.021
16	256	4.000	4,096	2.520	66	4,356	8.124	287,496	4.041
17	289	4.123	4,913	2.571	67	4,489	8.185	300,763	4.062
18	324	4.243	5,832	2.621	68	4,624	8.246	314,432	4.082
19	361	4.359	6,859	2.668	69	4,761	8.307	328,509	4.102
20	400	4.472	8,000	2.714	70	4,900	8.367	343,000	4.121
21	441	4.583	9,261	2.759	71	5,041	8.426	357,911	4.141
22	484	4.690	10,648	2.802	72	5,184	8.485	373,248	4.160
23	529	4.796	12,167	2.844	73	5,329	8.544	389,017	4.179
24	576	4.899	13,824	2.884	74	5,476	8.602	405,224	4.198
25	625	5.000	15,625	2.924	75	5,625	8.660	421,875	4.217
26	676	5.099	17,576	2.962	76	5,776	8.718	438,976	4.236
27	729	5.196	19,683	3.000	77	5,929	8.775	456,533	4.254
28	784	5.292	21,952	3.037	78	6,084	8.832	474,552	4.273
29	841	5.385	24,389	3.072	79	6,241	8.888	493,039	4.291
30	900	5.477	27,000	3.107	80	6,400	8.944	512,000	4.309
31	961	5.568	29,791	3.141	81	6,561	9.000	531,441	4.327
32	1,024	5.657	32,768	3.175	82	6,724	9.055	551,368	4.344
33	1,089	5.745	35,937	3.208	83	6,889	9.110	571,787	4.362
34	1,156	5.831	39,304	3.240	84	7,056	9.165	592,704	4.380
35	1,225	5.916	42,875	3.271	85	7,225	9.220	614,125	4.397
36	1,296	6.000	46,656	3.302	86	7,396	9.274	636,056	4.414
37	1,369	6.083	50,653	3.332	87	7,569	9.327	658,503	4.431
38	1,444	6.164	54,872	3.362	88	7,744	9.381	681,472	4.448
39	1,521	6.245	59,319	3.391	89	7,921	9.434	704,969	4.465
40	1,600	6.325	64,000	3.420	90	8,100	9.487	729,000	4.481
41	1,681	6.403	68,921	3.448	91	8,281	9.539	753,571	4.498
42	1,764	6.481	74,088	3.476	92	8,464	9.592	778,688	4.514
43	1,849	6.557	79,507	3.503	93	8,649	9.644	804,357	4.531
44	1,936	6.633	85,184	3.530	94	8,836	9.695	830,584	4.547
45	2,025	6.708	91,125	3.557	95	9,025	9.747	857,375	4.563
46	2,116	6.782	97,336	3.583	96	9,216	9.798	884,736	4.579
47	2,209	6.856	103,823	3.609	97	9,409	9.849	912,673	4.595
48	2,304	6.928	110,592	3.634	98	9,604	9.899	941,192	4.610
49	2,401	7.000	117,649	3.659	99	9,801	9.950	970,299	4.626
50	2,500	7.071	125,000	3.684	100	10,000	10.000	1,000,000	4.642

Answers to
Selected Problems

CHAPTER 0, Section 0.4 (p. 9)

Problem Set I ■ *Reading Comprehension*

1 False. The major developments in mathematics in years past are still valid and very useful today.   **2** False. The field of mathematics today is growing and expanding at an unprecedented rate.   **3** True. Much of the current progress in the biological and social sciences is due to the increased use of mathematics in these disciplines.   **4** False. This area of mathematics is insignificant relative to the rest.   **5** Pure mathematics is not directly concerned with nor motivated by physical reality and utility, whereas applied mathematics is.   **6** Two examples are non-Euclidean geometry and complex numbers.   **7** First, it is very concise and abbreviated in form, and second, it can easily be manipulated into a variety of rearranged forms, each perhaps bringing to light new information about a problem.   **8** The rules and grammar are quite simple. (See also answer to Problem 7 above.)   **9** One does not multiply men and cars—for ambiguities often result. A safe rule of thumb is to never mix units.

Problem Set II ■ *Challenges in Reason and Logic*

1(a) One possible solution is as shown.   **(b)** Not possible.

2 Yes.   **4** 1124 games will be played since 1124 teams must lose one game each.

CHAPTER 1, Section 1.5 (p. 24)

Problem Set I ■ *Reading Comprehension*

1 True. By definition a proposition is not both true and false.   **2** False. The statement is not a *conditional* proposition, but a simple proposition.   **3** False. The statement is an implication.
4 False. "A and B are equivalent" is written $A \Longleftrightarrow B$.   **5** True. $A \Longrightarrow B$ and $B \Longrightarrow A$ are converse statements.   **6** True. A does not imply B.   **7** False. The contrapositive of $[A \Longrightarrow B]$ is [not $B \Longrightarrow$ not A].   **8** False. Mathematically "or" means one or the other or both.
9 False. Not(A or B) $\Longleftrightarrow$ (not A and not B).   **10** False. A and not A always have opposite truth values. If A is true, not A is false; while if A is false, then not A is true.   **11** True. If a counterexample exists, then the statement is not true.   **12(a)** Proposition. The statement is true in general. **(b)** Proposition. The statement is true. **(c)** Not a proposition. No truth value.
(d) Not a proposition. Not a complete sentence. **(e)** Proposition. The statement is false.
(f) Proposition. True if we are using Euclidean geometry. **(g)** Proposition. The statement is false.
(h) Not a proposition. No truth value. **(i)** An open proposition. Its truth value can not be determined until we know when and where it is given. **(j)** Proposition. The statement is true.
13 Not(not A) $\Longleftrightarrow A$. One example is "We succeed if, and only if, we do not fail."   **14** Many examples are possible. The form of each will be $A \Longrightarrow B$ and $B \Longrightarrow C$, so $A \Longrightarrow C$.   **15** Let

each proposition be denoted by a capital letter and replace implications and equivalences by appropriate symbols.   **16** The main advantages are brevity, clarity, and manipulability.
17 The example should include some statement about all members of a class or set; a counter-example is determined by exhibiting a member of the class or set for which the statement is false.

Problem Set II ■ Skills Development

1(a) Let A represent the statement "Jack went to the show," and B the statement, "Harry went to the show." The diagrammed form is A and B. Classifications: conjunction and compound. **(d)** Let A represent "Susan will pass," and B, "She hands in an acceptable paper." The diagrammed form is $B \Longrightarrow A$. Classifications: conditional and compound. **(g)** Let A, B, and C represent "Carol has maintained a B average," "She has taken all the required courses," and "She will be admitted to graduate school," respectively. The diagrammed form is $(A$ and $B) \Longleftrightarrow C$. Classifications: equivalence and compound overall, with $(A$ and $B)$ a conjunction. **(j)** Let A and B represent "He has the correct address," and "He will find her house," respectively. The diagrammed form is $(A \Longrightarrow B)$ and $(B \not\Longrightarrow A)$. Classifications: conjunction and compound overall, with $(A \Longrightarrow B)$ conditional and $(B \not\Longrightarrow A)$ negative conditional. **2(a)** True. Because of the inclusive use of or. **(b)** True. **(d)** False. Because neither statement is true. **(e)** False. **3(a)** If Mr. Politician wins the majority of the votes from Mississippi, Kansas, and Oregon, then he will become President of the United States. **(d)** If Mr. Politician becomes President, then he will become the Commander in Chief of all our armed forces and he will not be a member of the judicial branch of government.
4 No. Let A, B, and C represent, respectively, "All the generators of the city power plant are running simultaneously," "The street lights are on," and "The load is over 10,000 kw." We are given $A \Longrightarrow B$ and $C \Longrightarrow A$. We are asked if $B \Longrightarrow C$. This does not necessarily follow. One valid conclusion is $C \Longrightarrow B$. That is, if the load is over 10,000 kw, then the street lights are on.

Problem Set III ■ Challenges in Reason and Logic

1(a) Valid. **(b)** Invalid. **(d)** Invalid. **2(a)** Valid. Let A represent "The company will produce more cars," B represent "The company will hire more men," and C "The company will make trucks." The problem is to show $[\text{not}(A$ and not $B)$ and $(\text{not } C$ or $A)$ and $A] \Longrightarrow B$. First, we know by **L-4** that $\text{not}(A$ and not $B) \Longleftrightarrow \text{not } A$ or B. But by the meaning of "and" and "or" we also have A and $(\text{not } A$ or $B) \Longrightarrow B$. And hence, we know B does follow from the given information—that is, "The company will hire more men." **(b)** Valid.

Problem Set IV ■ Theoretical Developments

1 $[A$ and $(A \Longrightarrow B)] \xrightarrow[\text{Def. of} \Longrightarrow]{\textbf{D1.4.3}} [A$ and $B]$.   **4** $[A \Longrightarrow (B$ and not $B)] \xleftrightarrow{\textbf{L-2}}$

$[\text{not}(B$ and not $B) \Longrightarrow \text{not } A] \xleftrightarrow{\textbf{L-4}} [(\text{not } B$ or $B) \Longrightarrow \text{not } A] \xrightarrow[\text{Def. of} \Longrightarrow]{\textbf{D1.4.4}} \text{not } A$.

7 $[\text{not}(\text{not } A$ or $B)] \xleftrightarrow{\textbf{L-4}} [A$ and not $B]$.

CHAPTER 1, Section 1.8 (p. 35)

Problem Set I ■ *Reading Comprehension*

1 False. Set theory is an innovation of the past century.   **2** True. This avoids circular definitions.   **3** False. This is Definition 1.7.1.   **4** False. Set C has only three distinct members
and they correspond with those of A, hence, $A = C$.   **5** True. Roster form simply lists all the
members of a set.   **6** False. The correct notation is $a \in A$.   **7** False. $[A \subseteq B] \Longleftrightarrow$
$[A \subset B \text{ or } A = B]$.   **8** True.   **9** A set is well defined if its members are identifiable.
10 Set builder notation is almost always used for sets containing a large number of elements.
11 Venn diagrams are useful for graphical representations of set relationships.   **12** A universal
set could be empty but this would rarely be significant and rarely occur.

Problem Set II ■ *Skills Development*

1 $\{2, 4, 6, 8, 10\}$.   **2** $\{\text{Truman, Eisenhower, Kennedy, Johnson, Nixon, Ford}, \ldots\}$.
4 $\{4, 8, 12\}$.   **5** This set will be different for each student.   **7** $\{x \mid x \text{ is an integer and }$
$1 < x < 100 \text{ and } x \text{ is a multiple of } 3\}$.   **8** $\{x \mid x \text{ is a perfect square and } x \text{ is a positive integer}\}$
or $\{x \mid x = n^2 \text{ for some integer } n\}$.   **10** $\{x \mid x \text{ is a friend over six feet tall who has blue eyes}$
and sings well$\}$.   **11** $\{x \mid x = p/q, \text{ where } p \text{ and } q \text{ are positive integers}\}$.   **13** $B \subset C$.
14 $[A \subseteq B \text{ and } B \nsubseteq A] \Longrightarrow [A \subset B]$.   **16** $a \in A \text{ and } a \notin D$.   **17** $[(a \in B \Longrightarrow a \in C)$
and $C \subset D] \Longrightarrow [\exists\, a \in D \ni a \notin B]$.   **19** 4.   **20** The set of positive integers between 2
and 7.   **22** A is a subset of D does not imply that there exists some member d of set D that is
not a member of set A.   **23** A is a subset of B and B is a subset of A if and only if sets A and B
are equal.   **24(a)** Correct. **(b)** Incorrect. **(d)** Correct.   **25** All four sets are equal.
26 Sets (1), (5), and (6) are equal.

CHAPTER 1, Section 1.10 (p. 41)

Problem Set I ■ *Reading Comprehension*

1 True. The repetitions of letters e and g are usually omitted.   **2** False. Two non-empty sets
are disjoint if their intersection is empty.   **3** False. $A - B$ is the set of all elements of A which
are not elements of B. Notice that $A - B \neq B - A$, in general.   **4** False. The intersection of two
sets is the set of all elements common to both sets.

Problem Set II ■ *Skills Development*

1(a) $\{4\}$. **(b)** $\{1, 2, 3, 4, 5\}$. **(d)** $\{1, 2, 4, 5\}$. **(e)** $\varnothing$. $U' = \varnothing$. **(g)** $\{1, 4\} = A$. **(h)** $\{1, 3, 4\}$.
$(A \cup B) \cap (A \cup C) = A \cup (B \cap C)$. **2(a)** U. **(b)** $\varnothing$. **(d)** A. **(e)** $\varnothing$. **(g)** $\varnothing$. **(h)** U.
(j) $\varnothing$. **(k)** A.   **4(a)** $[B - (A \cup C)] \cup [A - (B \cup C)]$. Other answers are also possible.
(b) $[C - (B \cup A')] \cup [A' \cap B \cap C]$. Other answers are also possible. **(d)** $B - (A \cup C)$.
Other answers are also possible.   **5** $A \cap [C' \cap (B \cup D)]$. Other answers are also possible.

1 $A = \varnothing$. Proof: $A \cup \varnothing = \varnothing \overset{\textbf{D1.7.1}}{\longleftrightarrow} [a \in (A \cup \varnothing) \Longleftrightarrow a \in \varnothing] \overset{\textbf{D1.9.1}}{\longleftrightarrow} [(a \in A$ or

$a \in \varnothing) \Longleftrightarrow a \in \varnothing] \overset{\textbf{Def. of } \varnothing}{\longrightarrow} [a \in A \Longleftrightarrow a \in \varnothing] \overset{\textbf{D1.7.1}}{\longleftrightarrow} [A = \varnothing]$.   **2** $A = U$.   **4** $A \subseteq B$.

Proof: $A \cap B = A \overset{\textbf{D1.7.1}}{\longleftrightarrow} [a \in (A \cap B) \Longleftrightarrow a \in A] \overset{\textbf{D1.9.2}}{\longleftrightarrow} [(a \in A$ and $a \in B) \Longleftrightarrow$

$a \in A] \overset{\textbf{D1.4.1}}{\longleftrightarrow} [a \in A \Longrightarrow a \in A$ and $a \in B] \overset{\textbf{Def. of } and}{\longleftrightarrow} [a \in A \Longrightarrow a \in B] \overset{\textbf{D1.7.2}}{\longrightarrow} A \subseteq B$.

5 $A = B = \varnothing$.   **7** $A = \varnothing$. Proof: $A' \cap U = U \overset{\textbf{D1.7.1}}{\longleftrightarrow} [a \in (A' \cap U) \Longleftrightarrow a \in U] \overset{\textbf{D1.9.2}}{\longleftrightarrow}$

$[(a \in A'$ and $a \in U) \Longleftrightarrow a \in U] \overset{\textbf{D1.7.1}}{\longleftrightarrow} A' = U \overset{\textbf{D1.9.5}}{\longleftrightarrow} A = \varnothing$.   **8** $A = U$.

10(a) $\{x \in B \overset{\textbf{Def. of } or}{\longrightarrow} [x \in A$ or $x \in B] \overset{\textbf{D1.9.1}}{\longleftrightarrow} x \in (A \cup B)\} \overset{\textbf{L-3}}{\underset{\textbf{D1.7.2}}{\longrightarrow}} [B \subseteq (A \cup B)]$.

(d) $[x \in (A - B) \overset{\textbf{D1.9.4}}{\longleftrightarrow} (x \in A$ and $x \notin B)] \overset{\textbf{L-3}}{\underset{\textbf{D1.7.2}}{\longrightarrow}} [(A - B) \subseteq A]$.   **11(a)** $A \subseteq B \Longrightarrow$

$(A \cap B) = A$. Proof: (1) By Problem 10(b), $(A \cap B) \subseteq A$. (2) $A \subseteq B \overset{\textbf{D1.7.2}}{\longleftrightarrow} [x \in A \Longrightarrow$

$x \in B]$.   (3) $[x \in A \overset{\textbf{(2)}}{\longrightarrow} (x \in A$ and $x \in B) \overset{\textbf{D1.9.2}}{\longrightarrow} x \in (A \cap B)] \overset{\textbf{D1.7.2}}{\underset{\textbf{L-3}}{\longrightarrow}}$

$A \subseteq (A \cap B) \overset{\textbf{(1)}}{\underset{\textbf{D1.7.1}}{\longrightarrow}} A \cap B = A$.   **11(d)** $[A \cup B' = A] \Longrightarrow [A' \cap B = A']$. Proof: (1) We first

prove the following: $(A \cup B)' = A' \cap B'$. Proof: $x \in (A \cup B)' \overset{\textbf{D1.9.5}}{\longleftrightarrow} x \notin A \cup B \overset{\textbf{D1.9.1}}{\longleftrightarrow}$

$x \notin A$ and $x \notin B \overset{\textbf{D1.9.5}}{\longleftrightarrow} x \in A'$ and $x \in B' \overset{\textbf{D1.9.2}}{\longleftrightarrow} x \in (A' \cap B') \overset{\textbf{D1.7.1}}{\longrightarrow} (A \cup B)' =$

$A' \cap B'$ QED.   (2) $[A \cup B' = A] \Longrightarrow [(A \cup B')' = A'] \overset{\textbf{(1)}}{\longleftrightarrow} [A' \cap (B')'] = A' \overset{\textbf{D1.9.5}}{\longleftrightarrow}$

$A' \cap B = A'$ QED.

CHAPTER 2, Section 2.5 (p. 51)

Problem Set I ■ *Reading Comprehension*

1 False. A right triangle with two sides of length 1 unit has a hypotenuse that cannot be measured by a rational fraction.   **2** False. The Pythagorean theorem states that the sum of the squares of the lengths of the two shorter sides of a right triangle is equal to the square of the length of the hypotenuse.   **3** True.   **4** False. $1/3 = 0.\overline{3}$.   **5** False. They are now known as irrational numbers.   **6** False. Any number that can be expressed as a ratio of two integers is rational.
7 True.   **8** False. Technically, numerals are symbols used to denote numbers. It is common practice, however, to refer to modern Arabic numerals as numbers.   **9** True.   **10** True.
11 False. 3.1416 is only an approximation of π.   **12** False. They do exist with fixed magnitudes, although it is true they cannot be expressed in finite form by means of Arabic numerals.
13 True. The natural numbers are the positive integers.

1 False. $n \in N \Longrightarrow n > 0$.   **2** True.   **4** True.   **5** True.   **7** False. $Q \cap H = \emptyset$.

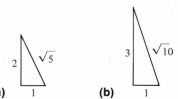

8 False.   **10** True.   **11** True.   **13(a)**   **(b)**
14(a) $3/8 = 0.375\overline{0}$.   **(b)** $2/3 = 0.66\overline{6}$.

CHAPTER 2, Section 2.8 (p. 57)

Problem Set I ■ *Reading Comprehension*

1 True.   **2** False. $a < b \Longleftrightarrow a$ is left of b.   **3** False. We also consider half-open intervals such as open-closed or closed-open intervals. (See page 54 for detailed discussion.)   **4** False. $a \in R \Longrightarrow |a| = |-a|$ and $|a| = a$, if $a \geq 0$, or $-a$, if $a < 0$.

Problem Set II ■ *Skills Development*

1(a) <.   **(b)** <.   **(d)** >.   **(e)** <.   **2(a)** $\{x \mid -1 \leq x \leq 3\}$.   **(b)** $\{x \mid -6 \leq x < -2\}$.
(d) $\{x \mid x > 1\}$.   **3(a)** $|3| = 3$.   **(b)** 2.   **(d)** $4 - x$, if $x \leq 4$; or $x - 4$, if $x > 4$.   **(e)** 2/3.
4(a) $a > b$.   **(b)** $a < b$.   **(d)** $a < x < b$.   **(e)** $|x| \leq 3$.   **(g)** $-2 < x < 5$.   **(h)** $0 \leq x < 9$.

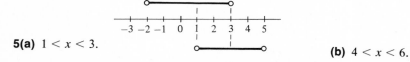

5(a) $1 < x < 3$.

(b) $4 < x < 6$.

Problem Set III ■ *Theoretical Developments*

1(a) If $a < b$ then a is to the left of b on the real line. If the same number c is added to both a and b, then $a + c$ must also be to the left of $b + c$. Therefore, $a + c < b + c$.   **(d)** If a is to the left of or equal to b and b is likewise to the left of or equal to a, then the only alternative is that $a = b$.
2 When the intersection is non-empty.

CHAPTER 2, Section 2.10 (p. 79)

Problem Set I ■ *Reading Comprehension*

1 True.   **2** False.   **3** False.   **4** False.   **5** False. Most of the creation of mathematics depends on inductive reasoning.   **6** True.   **7** A field is a mathematical system that satisfies the field axioms.   **8** Subtraction, negative number, and additive inverse.

Problem Set II ■ Skills Development

1 $\dfrac{1}{4} - \dfrac{1}{6} \overset{\textbf{T2.9.13}}{\underset{\textbf{(V)}}{=\!=\!=\!=}} \dfrac{6-4}{24} = \dfrac{1}{12}.$ **4** $\dfrac{1}{4} \cdot \dfrac{2}{5} \overset{\textbf{T2.9.13}}{\underset{\textbf{(II)}}{=\!=\!=\!=}} \dfrac{2}{20} \overset{\textbf{T2.9.10}}{=\!=\!=\!=} \dfrac{1}{10}.$ **7** $2[-3 + 4(-2-5)]$

$\overset{\textbf{T2.9.8(II)}}{=\!=\!=\!=} 2[-3+4(-7)] \overset{\textbf{T2.9.8}}{\underset{\textbf{(III)}}{=\!=\!=\!=}} 2[-3-28] \overset{\textbf{T2.9.8}}{\underset{\textbf{(II)}}{=\!=\!=\!=}} 2[-31] \overset{\textbf{T2.9.8}}{\underset{\textbf{(III)}}{=\!=\!=\!=}} -62.$

10 $(x-3)(x-1) \overset{\textbf{D}}{=\!=\!=} x(x-1) - 3(x-1) \overset{\textbf{D}}{=\!=\!=} (x^2 - x) + (-3x + 3) \overset{\textbf{A+}}{\underset{\textbf{T2.9.8(II)}}{=\!=\!=\!=}} x^2 - 4x + 3.$

13 $(2x+1)(3x-2) \overset{\textbf{D}}{=\!=\!=} 2x(3x-2) + 1 \cdot (3x-2) \overset{\textbf{D}}{=\!=\!=} (6x^2 - 4x) + (3x-2) \overset{\textbf{A+}}{\underset{\textbf{DSub}}{=\!=\!=\!=}} 6x^2 - x - 2.$

14 RLE.   **16** A+.   **17** Id+.   **19** D.   **20** In ·.   **22** FCLE.   **23** TCLE.

25 T2.9.5.   **26** T2.9.9.   **28** $2x+4 = 8 \overset{\textbf{TCLE}}{=\!=\!=\!\Longrightarrow} 2x = 4 \overset{\textbf{FCLE}}{=\!=\!=\!\Longrightarrow} x = 2.$ **29** $x = 9/4.$

31 $4(x+2) = 0 \overset{\textbf{T2.9.7}}{=\!=\!=\!\Longrightarrow} x + 2 = 0 \overset{\textbf{T2.9.5}}{=\!=\!=\!\Longrightarrow} x = -2.$ **32** $x = -3.$ **34** $\dfrac{x+2}{4} - \dfrac{6}{x-2} = 0$

$\overset{\textbf{T2.9.13}}{\underset{\textbf{(V)}}{\Longleftarrow\!=\!=\!\Longrightarrow}} \dfrac{(x+2)(x-2) - 24}{4(x-2)} = 0 \overset{\textbf{D}}{\Longleftrightarrow} \dfrac{x^2 - 28}{4(x-2)} = 0 \overset{\textbf{MLE}}{\underset{\textbf{Id·}}{=\!=\!\Longrightarrow}} x^2 - 28 = 0 \text{ and } x \neq 2 \overset{\textbf{D}}{\Longleftrightarrow}$

$(x - \sqrt{28})(x + \sqrt{28}) = 0 \overset{\textbf{T2.9.7}}{\Longleftarrow\!=\!\Longrightarrow} x - \sqrt{28} = 0 \text{ or } x + \sqrt{28} = 0 \overset{\textbf{T2.9.5}}{\Longleftarrow\!=\!\Longrightarrow} x = \sqrt{28} \text{ or } x = -\sqrt{28}.$

Problem Set III ■ Theoretical Developments

1 Not closed; e.g., $4 - 5 \notin N.$ **2** Closed.   **4** Closed.   **5** Not closed; e.g., $1 + 3 =$

$4 \notin \{\text{odd } n\}.$ **7** No. $8/4 \neq 4/8.$ **8** Yes. $a/1 = a.$ **10** $[a = b \text{ and } b = c] \overset{\textbf{E-4}}{\underset{a\,=\,b}{\Longleftarrow\!=\!\Longrightarrow}}$

$a = c.$ **13** T2.9.4. *Proof:* $ac = bc \overset{\textbf{MLE}}{=\!=\!\Longrightarrow} (ac)c^{-1} = (bc)c^{-1} \overset{\textbf{A·}}{\Longrightarrow} a(cc^{-1}) = b(cc^{-1}) \overset{\textbf{In·}}{\Longrightarrow}$

$a \cdot 1 = b \cdot 1 \overset{\textbf{Id·}}{\Longrightarrow} a = b$ QED.   **16(a)** T2.9.8(II). *Proof:* $0 \overset{\textbf{In+}}{\underset{\textbf{Id+}}{=\!=\!=}} [a + (-a)] + [b + (-b)] \overset{\textbf{A+}}{\underset{\textbf{C+}}{=\!=\!=}}$

$[(-a) + (-b)] + (a+b) \overset{\textbf{TLE}}{=\!=\!=\!\Longrightarrow} [(-a) + (-b)] + (a+b) = 0 \overset{\textbf{T2.9.5}}{=\!=\!=\!\Longrightarrow} (-a) + (-b) =$

$-(a+b)$ QED.   **19(a)** T2.9.12(I). *Proof:* $\dfrac{a}{a} \overset{\textbf{DDiv}}{=\!=\!=} aa^{-1} \overset{\textbf{In·}}{=\!=\!=} 1$ QED.   **(d)** T2.9.12(V).

Proof: (1) $\dfrac{-a}{b} \overset{\textbf{DDiv}}{=\!=\!=} (-a)b^{-1} \overset{\textbf{T2.9.8}}{\underset{\textbf{(III)}}{=\!=\!=}} -(ab^{-1}) \overset{\textbf{DDiv}}{=\!=\!=} -\dfrac{a}{b}.$ (2) $\dfrac{-a}{b} = \dfrac{a}{-b} \overset{\textbf{T2.9.9}}{\Longleftarrow\!=\!\Longrightarrow}$

$(-a)(-b) = ab \overset{\textbf{T2.9.8}}{\underset{\textbf{(IV)}}{\Longleftarrow\!=\!\Longrightarrow}} ab = ab.$ By (1), (2), and TLE, $-\dfrac{a}{b} = \dfrac{-a}{b} = \dfrac{a}{-b}$ QED.   **20(a)** T2.9.13(II).

Proof: $\dfrac{a}{b} \cdot \dfrac{c}{d} \overset{\textbf{DDiv}}{=\!=\!=} (ab^{-1})(cd^{-1}) \overset{\textbf{A·}}{\underset{\textbf{C·}}{=\!=\!=}} (ac)(b^{-1}d^{-1}) \overset{\textbf{T2.9.13(I)}}{\underset{\textbf{T2.9.12(III)}}{=\!=\!=\!=}} (ac)(bd)^{-1} \overset{\textbf{DDiv}}{=\!=\!=} \dfrac{ac}{bd}$ QED.

(d) T2.9.13(VI). *Proof:* $\dfrac{1}{a/b} \overset{\textbf{DDiv}}{\underset{\textbf{T2.9.12(III)}}{=\!=\!=\!=}} \dfrac{1}{a(1/b)} \overset{\textbf{T2.9.13(I)}}{=\!=\!=\!=} \dfrac{1}{a} \cdot \dfrac{1}{1/b} \overset{\textbf{T2.9.12(IV)}}{=\!=\!=\!=} \dfrac{1}{a} \cdot b$

$\overset{\textbf{T2.9.12(III)}}{\underset{\textbf{DDiv}}{=\!=\!=\!=}} \dfrac{b}{a}$ QED.   **22** *Proof:* Suppose 1 and $1'$ are two identity elements. Then by Id·,

$a \cdot 1 = a$ and $a \cdot 1' = a$ for each $a \in R \overset{\textbf{TLE}}{=\!=\!=\!\Longrightarrow} a \cdot 1 = a \cdot 1' \overset{\textbf{FCLE}}{\underset{a\,\neq\,0}{\Longleftarrow\!=\!\Longrightarrow}} 1 = 1'$ QED.   **25** *Proof:*

$\dfrac{a}{b} = c \overset{\textbf{T2.9.12(II)}}{\Longleftarrow\!=\!=\!\Longrightarrow} \dfrac{a}{b} = \dfrac{c}{1} \overset{\textbf{T2.9.9}}{\Longleftarrow\!=\!=\!\Longrightarrow} a \cdot 1 = bc \overset{\textbf{Id·}}{\Longleftrightarrow} a = bc$ QED.   **28** $a(a - b - c) =$

$b(a - b - c) \Longleftrightarrow a = b,$ since $a - b - c = 0.$

CHAPTER 2, Section 2.12 (p. 86)

Problem Set I ■ Skills Development

1(a) $a \leq b$. **(b)** $a \neq b$ or $b > a$. **(d)** $a \not< b$ or $b \geq a$. **(e)** $a \not\equiv b$ or $b < a$. **2(a)** $-2 < 1$.
(b) $-5 < -2$. **(d)** $\sqrt{4} = 2$. **(e)** $2 > -4$. **3(a)** $a > b$. **(b)** $c < x$. **(d)** $|x| > 4$.
(e) $|y - 3| \leq 5$. **4(a)** $-3 < x + 3 < 4 \Longleftrightarrow -6 < x < 1$. **(b)** $3 < x < 6$. **5(a)** $|x - 3| < 7$
$\Longleftrightarrow -7 < x - 3 < 7 \Longleftrightarrow -4 < x < 10$. **(b)** $x > 7/2$ or $x < -1/2$. **(d)** $|-2 - 3x| \geq -5$ is true
for all real x because $|a| \geq 0$ for every $a \in R$. **(e)** $-7/2 < x < 1/2$. **(g)** $-|x - 3| > 4 \Longleftrightarrow$
$|x - 3| < -4$, which is false for all real x; i.e., solution set is empty. **(h)** $x > 5$ or $x < -2$.

Problem Set II ■ Theoretical Developments

1(a) D2.11.1. **(b)** T2.11.2(I). **(d)** T2.11.2(II). **(e)** T2.11.2(IV). **2(a)** $[a > 0 \overset{\textbf{D2.11.1}}{\Longleftrightarrow}$

$(a - 0) \in \boldsymbol{R}_+ \overset{\textbf{Id+}}{\Longleftrightarrow} a \in \boldsymbol{R}_+] \overset{\textbf{L-4}}{\Longrightarrow} [a > 0 \Longleftrightarrow a \in \boldsymbol{R}_+]$ QED. **(d)** T2.11.1(II). *Proof:*

$[a < 0$ and $b < 0] \overset{\textbf{L2.11.1}}{\Longleftrightarrow} a \in \boldsymbol{R}_-$ and $b \in \boldsymbol{R}_- \overset{\textbf{N}_2}{\Longleftrightarrow} -a \in \boldsymbol{R}_+$ and $-b \in \boldsymbol{R}_+ \overset{\textbf{O-2}}{\Longrightarrow}$

$(-a)(-b) \in \boldsymbol{R}_+ \overset{\textbf{T2.9.8(V)}}{\Longrightarrow} ab \in \boldsymbol{R}_+ \overset{\textbf{L2.11.1}}{\Longleftrightarrow} ab > 0$. T2.11.1(III). *Proof:* (1) $a = 0 \overset{\textbf{T2.9.6}}{\Longrightarrow}$

$a^2 = 0$. (2) $a > 0 \overset{\textbf{O-2}}{\Longrightarrow} a^2 > 0$. (3) $a < 0 \overset{\textbf{T2.11.1(II)}}{\Longrightarrow} a^2 > 0$. Thus, (1), (2), and (3) imply

$a^2 \geq 0$ for every $a \in \boldsymbol{R}$ QED. **(g)** T2.11.2(IV). *Proof:* $a < b$ and $c < 0 \overset{\textbf{D2.11.1}}{\underset{\textbf{L2.11.1}}{\Longleftrightarrow}} (b - a)$,

$(0 - c) \in \boldsymbol{R}_+ \overset{\textbf{O-2}}{\Longrightarrow} (b - a)(-c) \in \boldsymbol{R}_+ \overset{\textbf{D}}{\underset{\textbf{T2.9.8(III,IV)}}{\Longleftrightarrow}} (ac - bc) \in \boldsymbol{R}_+ \overset{\textbf{D2.11.1}}{\Longleftrightarrow} ac > bc$ QED.

(j) $a < 0$ and $b < 0 \overset{\textbf{L2.11.1}}{\Longleftrightarrow} a \in \boldsymbol{R}_-$ and $b \in \boldsymbol{R}_- \overset{\textbf{N}_2}{\Longleftrightarrow} -a \in \boldsymbol{R}_+$ and $-b \in \boldsymbol{R}_+ \overset{\textbf{O-2}}{\Longrightarrow}$

$[(-a) + (-b)] \in \boldsymbol{R}_+ \overset{\textbf{T2.9.8(II)}}{\Longleftrightarrow} -(a + b) \in \boldsymbol{R}_+ \overset{\textbf{N}_2}{\Longleftrightarrow} (a + b) \in \boldsymbol{R}_- \overset{\textbf{L2.11.1}}{\Longrightarrow} a + b < 0$ QED.

(m) $1 < a \overset{\textbf{T2.11.2(III)}}{\Longrightarrow} a < a^2$ QED.

CHAPTER 3, Section 3.3 (p. 104)

Problem Set I ■ Reading Comprehension

1 False. A variable may represent any of the members of the replacement set. **2** True.
3 True. This is given as part of Definition 3.2.2. **4** False. The solution set may be empty.
5 False. False roots introduced by not maintaining equivalence are not roots.

Problem Set II ■ Skills Development

1 $x - 3 = 4 \Longleftrightarrow x = 7$. **2** $x = -6$. **4** $x/2 - 4 = 2x/3 \Longleftrightarrow 3x - 24 = 4x \Longleftrightarrow x = -24$.
5 $x = -9/7$. **7** $2x - 3 < 3x + 4 \Longleftrightarrow -7 < x$. **8** $1 < x < 2$. **10** $x^2 - 7x + 12 > 0 \Longleftrightarrow$

$(x-3)(x-4) > 0 \Longleftrightarrow [x-3 > 0 \text{ and } x-4 > 0] \text{ or } [x-3 < 0 \text{ and } x-4 < 0] \Longleftrightarrow$
$[x > 3 \text{ and } x > 4] \text{ or } [x < 3 \text{ and } x < 4] \Longleftrightarrow x > 4 \text{ or } x < 3.$ **11** $x > 3.$

Problem Set III ■ Theoretical Developments

1   T3.2.1(II). *Proof:* $a = b \text{ and } c \neq 0 \xLeftrightarrow[\text{FCLE}]{\text{MLE}} ac = bc \text{ and } c \neq 0$ QED.

Problem Set IV ■ Applications

1   Let x represent the number of minutes required to fill the tank using both pumps. Then
$1/45 + 1/30 = 1/x \Longleftrightarrow 2/90 + 3/90 = 1/x \Longleftrightarrow 1/18 = 1/x \Longleftrightarrow x = 18.$ **2** $T = 285/8 \text{ or } 35\frac{5}{8}$ days.
4   Let x represent the amount invested at 8 percent and $20{,}000 - x$, the amount invested at $5\frac{1}{2}$
percent. Then $0.08x + 0.055(20{,}000 - x) = 1400 \Longleftrightarrow 0.08x + 1100 - 0.055x = 1400 \Longleftrightarrow 0.025x = 300 \Longleftrightarrow x = 12{,}000.$ Therefore, \$8,000 is invested at 5.5 percent and \$12,000 at 8 percent.
5 $x = 140/27 \text{ or } 5\frac{5}{27}$ quarts.

CHAPTER 3, Section 3.9 (p. 120)

Problem Set I ■ Reading Comprehension

1   False. Only a linear open sentence in two variables describes a line.   **2**   False. $|-a| = a$
only if $a \geq 0.$ **3**   False. $\sqrt{4} = 2$; i.e., radical means principal square root only.   **4**   False.
The square root of a negative number does not exist in the real number system.   **5**   False.
$a^2 = b^2 \Longleftrightarrow a = \pm b.$

Problem Set II ■ Skills Development

1 $3x + 4 = x - 5 \Longleftrightarrow 2x = -9 \Longleftrightarrow x = -9/2.$ **2** $x = 3.$ **4** $1/2 + 2/3 = 1/x \Longleftrightarrow 7/6 = 1/x$
$\Longleftrightarrow x = 6/7.$ **5** $x = 7.$ **7** $x^2 - x - 6 = 0 \Longleftrightarrow (x-3)(x+2) = 0 \Longleftrightarrow x = 3 \text{ or } x = -2.$
8 $x = -4 \text{ or } x = 2.$ **10** $12x^2 + 11x - 15 = 0 \Longleftrightarrow (3x+5)(4x-3) = 0 \Longleftrightarrow x = -5/3 \text{ or }$
$x = 3/4.$ **11** $x = \pm 2\sqrt{2} \text{ or } x = \pm 2.$ **13** $x^2 - 3x - 10 = 0 \Longleftrightarrow (x-5)(x+2) = 0 \Longleftrightarrow$
$x = 5 \text{ or } x = -2.$ **14** $x = 3/2.$ **16** $\sqrt{3}x^2 - 2x - \sqrt{12} = 0 \Longleftrightarrow x = [2 \pm \sqrt{3 + 4 \cdot 6}]/2\sqrt{3}$
$\Longleftrightarrow x = \sqrt{3}/3 \pm 3/2.$ **17**   No real solutions.   **19** $x/(x-1) = 2/(x-4) \Longleftrightarrow x(x-4) = 2(x-1) \text{ and } x \neq 1, 4 \Longleftrightarrow x^2 - 6x + 2 = 0 \text{ and } x \neq 1, 4 \Longleftrightarrow x = [6 \pm \sqrt{36 - 8}]/2 = 3 \pm \sqrt{7}.$
20 $x = [5 \pm \sqrt{17}]/4.$ **22** $|2x - 4| = 6 \Longleftrightarrow [2x - 4 = 6 \text{ if } 2x - 4 \geq 0] \text{ or } [-2x + 4 = 6 \text{ if }$
$2x - 4 < 0] \Longleftrightarrow [x = 5 \text{ if } x \geq 2] \text{ or } [x = -1 \text{ if } x < 2] \Longleftrightarrow x = 5 \text{ or } x = -1.$ **23** $-2 \leq x \leq 3.$
25 $\sqrt{x+3} = 2 \Longleftrightarrow x + 3 = 4 \Longleftrightarrow x = 1.$ **26**   No real solutions.   **28** $\sqrt{4x+6} = 1 + \sqrt{2x+7} \Longleftrightarrow 4x + 6 = 1 + 2\sqrt{2x+7} + 2x + 7 \Longleftrightarrow x - 1 = \sqrt{2x+7} \Longleftrightarrow x^2 - 2x + 1 = 2x + 7$
and $x > 1 \Longleftrightarrow x^2 - 4x - 6 = 0 \text{ and } x > 1 \Longleftrightarrow x = 2 \pm \sqrt{10} \text{ and } x > 1 \Longleftrightarrow x = 2 + \sqrt{10}.$
29 $x = 9.$ **31** $(\sqrt{x-3})(\sqrt{x-4}) = \sqrt{6} \Longleftrightarrow (x-3)(x-4) = 6 \text{ and } x > 4 \Longleftrightarrow$
$x^2 - 7x - 18 = 0 \text{ and } x > 4 \Longleftrightarrow (x-9)(x+2) = 0 \text{ and } x > 4 \Longleftrightarrow x = -2, 9 \text{ and } x > 4 \Longleftrightarrow x = 9.$
32 $4 - 5x > 2x - 10 \Longleftrightarrow 14 > 7x \Longleftrightarrow 2 > x.$

34 $(4-x)/(x+1) > 0 \Longleftrightarrow [4 - x > 0 \text{ and } x + 1 > 0] \text{ or } [4 - x < 0 \text{ and } x + 1 < 0] \Longleftrightarrow$
$[4 > x \text{ and } x > -1] \text{ or } [4 < x \text{ and } x < -1] \Longleftrightarrow -1 < x < 4.$

35 $-4 \leq x \leq 1$.

37 $|-2x + 3| > 7 \Longleftrightarrow [-2x + 3 > 7 \text{ or } -2x + 3 < -7] \Longleftrightarrow [x < -2 \text{ or } x > 5]$.

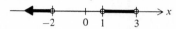

38 True for all x since $|a| \geq 0$ for every $a \in \mathbf{R}$. **40** $(x - 1)(x + 2)(x - 3) < 0 \Longleftrightarrow$ $[x - 1 > 0, \; x + 2 > 0, \text{ and } x - 3 < 0]$ or $[x - 1 > 0, \; x + 2 < 0, \text{ and } x - 3 > 0]$ or $[x - 1 < 0, x + 2 > 0, \text{ and } x - 3 > 0]$ or $[x - 1 < 0, \; x + 2 < 0, \text{ and } x - 3 < 0] \Longleftrightarrow 1 < x < 3 \text{ or } x < -2$.

Problem Set III ■ Applications

1 Let $d =$ increase in the dimension of the shorter end. Then $20 \cdot 30 = (30 - 4)(20 + d) \Longleftrightarrow$ $600 = 520 + 26d \Longleftrightarrow d = 40/13 = 3\frac{1}{13} = 3.08$ ft. **2** \$4500 at 4 percent and \$1500 at 3 percent. **4** Let x represent the time required to complete the job working together. Then the rate per hour is $1/7 + 1/9 = 1/x \Longleftrightarrow x = 63/16 = 3\frac{15}{16} = 3.94$ hours. **5** 3 hours. **7** Let $x =$ amount of antifreeze added. Then $(15)(0.30) + x = (15 + x)(0.45) \Longleftrightarrow x = 45/11 = 4\frac{1}{11}$ quarts. **8** $x = 20/3 = 6\frac{2}{3}$ quarts. **10** Let w represent the road width. Then $(100 - 2w)(100 - 2w) =$ 6400 and $2w < 100 \Longleftrightarrow w^2 - 100w + 900 = 0$ and $w < 50 \Longleftrightarrow [w = 90 \text{ or } w = 10]$ and $w < 50 \Longleftrightarrow$ $w = 10$ yards. **11** $t = 6$. **13** Let t represent the time for the sound to return. Because sound travels $1024t$ and the rock travels a distance $16(9 - t)^2$ and these distances are equal, we have $1024t = 16(9 - t)^2$ and $0 < t < 9 \Longleftrightarrow t^2 - 82t + 81 = 0$ and $0 < t < 9 \Longleftrightarrow [t = 81 \text{ or } t = 1]$ and $0 < t < 9 \Longleftrightarrow t = 1$ and $s = 1024$. **14** $4\frac{1}{2}$ percent.

Problem Set IV ■ Theoretical Developments

1 If $b' = b/2$, then $b = 2b'$. Substituting this into $x = \dfrac{-b \pm \sqrt{b^2 - 4ac}}{2a}$ gives the desired result.

4 T3.8.3. *Proof:* (1) $\sqrt{a} \geq 0$ and $\sqrt{b} \geq 0 \xrightarrow{\text{O-2}} \sqrt{a} \cdot \sqrt{b} \geq 0$. (2) $(\sqrt{a} \cdot \sqrt{b})^2 =$

$(\sqrt{a})^2(\sqrt{b})^2 \xrightarrow{\text{D3.8.1}} ab \xrightarrow{\text{TLE}} (\sqrt{a} \cdot \sqrt{b})^2 = ab \xleftrightarrow[\text{(1)}]{\text{D3.8.1}} \sqrt{a} \cdot \sqrt{b} = \sqrt{ab}$ QED.

CHAPTER 4, Section 4.7 (p. 142)

Problem Set I ■ Reading Comprehension

1 True. **2** False. A relation from A into B always has A as its domain. **3** False. The inverse of a relation defined by $p(x, y) = 0$ is obtained by interchanging the variables. **4** True. **5** False. The inverse of a function is not necessarily a function. **6** False. A function is 1-1 if, and only if, there is one and only one element of the domain associated with each range element.

7 True. If g is onto, then the range of g is B. **8** True. In fact the identity function maps each element of A into itself. **9** False. A constant function maps all elements of the domain into the same image element. **10** True. **11** False. No two distinct ordered pairs can have the same *first* coordinate. **12** False. No two points on the graph of a function can lie on the same vertical line. **13** False. Geometrically f^{-1} is the mirror image of f through the line defined by $y = x$.

Problem Set II ■ *Skills Development*

1

	function; 1-1, onto; constant, identity	domain	range	inverse	function; 1-1, onto; constant, identity
(a)	function	$\{1, 2, 3\}$	$\{a, c\}$	a → 1; c → 2, 3	not a function; onto
(b)	not a function; onto	$\{1, 2, 3\}$	$\{a, b, c\}$	a, b, c → 1, 2, 3	not a function; onto
(d)	not a function; onto	$\{1, 2\}$	$\{a, b, c\}$	a, b, c → 1, 2	function; onto
(e)	function; constant	$\{1, 2, 3\}$	$\{a\}$	a → 1, 2, 3	not a function; onto
(g)	not a function; onto	$\{-2, -1, 0, 3\}$	$\{-1, 2, 3\}$	$\{(2, -2),\ (-1, -1),\ (3, 0),\ (3, 3),\ (-1, 3)\}$	not a function; onto
(h)	function	R	$y \geq 0$	$x = (y + 1)^2$ or $y = -1 \pm \sqrt{x}$	not a function; onto
(j)	function	$\lvert x \rvert \geq 3$	$y \geq 0$	$x = \sqrt{y^2 - 9}$ or $y = \pm\sqrt{9 + x^2},\ x \geq 0$	not a function
(k)	not a function; onto	$x \geq 4$	R	$y = x^2 + 4$	function
(m)	not a function	$\lvert x \rvert \leq 2$	$\lvert y \rvert \leq 2$	$x^2 + y^2 = 4$	not a function

2(a) $\{(1,-1)\}$ and $\{(1,0)\}$. **(b)** Same as (a). **(d)** None.   **4** Fig. 4.7.13.   **5** $\{5,0,-3,-4\}$.

7

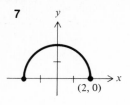

8

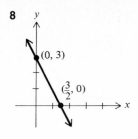

10

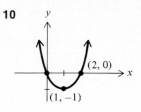

11

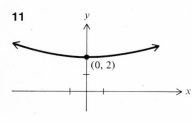

13

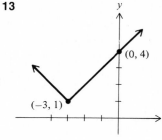

14

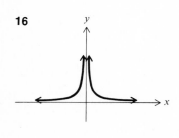

16

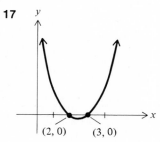

17

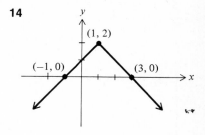

Problem Set III ■ *Theoretical Developments*

1 Yes. The letter we choose to represent a variable is irrelevant.   **4** Let f be a constant function of A into B. Then f is 1-1 if A contains only one element, and f is onto if B contains only one element.

Problem Set IV ■ *Applications*

1 $f = \{(r,c) \mid c = 2\pi r$ and $c, r \in \boldsymbol{R}_+\}$ or $\{(x,y) \mid y = 2\pi x$ and $x, y \in \boldsymbol{R}_+\}$. $\boldsymbol{R}(f) = \boldsymbol{D}(f) = \boldsymbol{R}_+$. $f^{-1} = \{(x,y) \mid x = 2\pi y$ and $x, y \in \boldsymbol{R}_+\}$ or $\{(x,y) \mid y = x/2\pi$ and $x, y \in \boldsymbol{R}_+\}$. Both f and f^{-1} are functions.

2 $g = \{(A,P) \mid P = 4\sqrt{A}$ and $P, A \in \boldsymbol{R}_+\}$ or $\{(x,y) \mid y = 4\sqrt{x}$ and $x, y \in \boldsymbol{R}_+\}$. $\boldsymbol{D}(g) = \boldsymbol{R}(g) = \boldsymbol{R}_+$. $g^{-1} = \{(x,y) \mid x = 4\sqrt{y}$ and $x, y \in \boldsymbol{R}_+\}$ or $\{(x,y) \mid y = x^2/16$ and $x, y \in \boldsymbol{R}_+\}$. Both g and g^{-1} are functions.

1

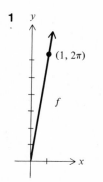

2

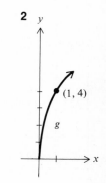

4 $h = \{(y, i) \mid i = (1/36)y \text{ and } y, i \in \mathbf{R}_+\}$ or $\{(x, y) \mid y = (1/36)x \text{ and } x, y \in \mathbf{R}_+\}$. $\mathbf{D}(h) = \mathbf{R}(h) = \mathbf{R}_+$. $h^{-1} = \{(x, y) \mid x = (1/36)y \text{ and } x, y \in \mathbf{R}_+\}$ or $\{(x, y) \mid y = 36x \text{ and } x, y \in \mathbf{R}_+\}$. Both h and h^{-1} are functions.

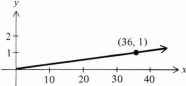

5 $d = 60t$.　　**7** $T = [\$0, \text{ if } 0 \leq I < \$600], [\$20, \text{ if } 600 \leq I < \$1200], [\$40, \text{ if } \$1200 \leq I < \$2400], \ldots, [\$(n \cdot 20), \text{ if } 2^{n-1} \cdot 600 \leq I < 2^n \cdot 600]$, where T is taxes and I is income.

8

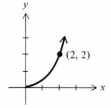

CHAPTER 4, Section 4.9 (p. 159)

Problem Set I ■ *Reading Comprehension*

1 False. $m = (y_2 - y_1)/(x_2 - x_1) = (y_1 - y_2)/(x_1 - x_2)$. Note the order of the subscripts.
2 True.　　**3** False. The slope of a horizontal line is zero.　　**4** False. Vertical lines do not represent functions.

Problem Set II ■ *Skills Development*

1(a) $y + 1 = -\dfrac{1}{3}(x - 2) \Longleftrightarrow y = -\dfrac{1}{3}x - \dfrac{1}{3}$

(b) $y = \dfrac{2}{5}x - \dfrac{4}{5}$

(d) $m = \dfrac{-2}{-3}, b = -2 \Longleftrightarrow y = \dfrac{2}{3}x - 2$

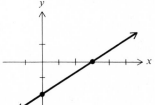

(e) $x = -2$

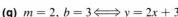

(g) $m = 2,\ b = 3 \Longleftrightarrow y = 2x + 3$

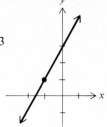

(h) $y = \dfrac{2}{3}x - \dfrac{5}{3}$

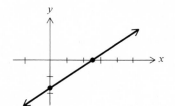

(j) $m = \dfrac{3}{2},\ b = -3 \Longleftrightarrow y = \dfrac{3}{2}x - 3$

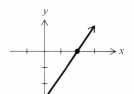

(k) $y = 3x + 3$

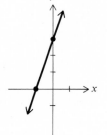

2(a) $(-4, 4)$ and $(-1, 2) \Longrightarrow m = -2/3$
$(-1, 2)$ and $(5, -2) \Longrightarrow m = -2/3$ $\left.\right\} \Longrightarrow$ The three points are collinear.

(b) The three points are collinear.

4

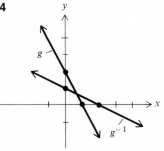

5

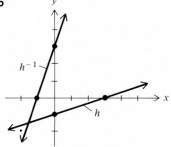

Problem Set III ■ Theoretical Developments

1 Since M is the midpoint of segment AB, the triangles are congruent. Thus, $\left[y_2 - y = y - y_1 \Longleftrightarrow y = \dfrac{y_1 + y_2}{2}\right]$ and $\left[x_2 - x = x - x_1 \Longleftrightarrow x = \dfrac{x_1 + x_2}{2}\right]$.

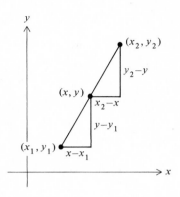

4 Place the triangle in a coordinate system and label as in the figure. The equations of the medians are $y = \dfrac{b}{a - 2c}(x - 2c)$, $y = \dfrac{b}{c - 2a}(x - 2a)$, $y = \dfrac{-2b}{a + c}x + 2b$. The point of intersection is $\left(\dfrac{2}{3}(c + a), \dfrac{2}{3}b\right)$.

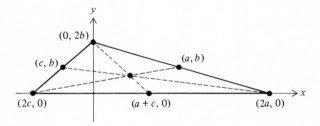

CHAPTER 4, Section 4.11 (p. 168)

Problem Set I ■ Reading Comprehension

1 True. **2** True. **3** False. Such a function always has either a maximum or a minimum value.

Problem Set II ■ Skills Development

1(a) $y = \left(x - \dfrac{3}{2}\right)^2 - \dfrac{1}{4}$

min: $\left(\dfrac{3}{2}, \dfrac{-1}{4}\right)$

intercepts: $(1, 0)$, $(2, 0)$

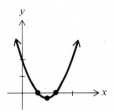

(b) $y = -(x-2)^2 + 2$
  max: $(2, 2)$
  intercepts: $x = 2 \pm \sqrt{2}$

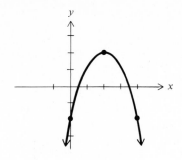

(d) $y = 2\left(x - \dfrac{5}{4}\right)^2 - \dfrac{25}{8}$
  min: $\left(\dfrac{5}{4}, \dfrac{-25}{8}\right)$
  intercepts: $(0, 0), \left(\dfrac{5}{2}, 0\right)$

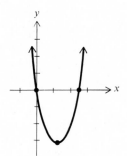

(e) $y = 2\left(x - \dfrac{7}{2}\right)^2 - \dfrac{1}{2}$
  min: $\left(\dfrac{7}{2}, \dfrac{-1}{2}\right)$
  intercepts: $(3, 0), (4, 0)$

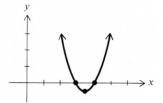

2

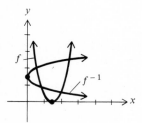

4	no x-intercept	one x-intercept	two x-intercepts
	$b^2 - 4ac < 0$	$b^2 - 4ac = 0$	$b^2 - 4ac > 0$
(a)	$k > 9/4$	$k = 9/4$	$k < 9/4$
(b)	$|k| < 2\sqrt{2}$	$|k| = 2\sqrt{2}$	$|k| > 2\sqrt{2}$

1 Increasing values for k shift the graph vertically upward along the y-axis.

Problem Set IV ■ *Applications*

1 $P = 40 = 2L + 2W$. $A = LW = (20 - W)(W) = -(W^2 - 20W + 100) + 100 = -(W - 10)^2 + 100$.
Thus the area reaches a maximum value of 100 when $L = W = 10$.   **2** $W = L = 15/2$.
4 Receipts = (price)(number of units sold). Let n equal the number of 5-cent price increases above
the basic 25-cent cost. $R = (25 + 5n)(500 - 50n) = 250(-n^2 + 5n + 50) = -250\left(n - \dfrac{5}{2}\right)^2 + \dfrac{28125}{2}$.

Thus R reaches a maximum of $140.63 when 375 units are marketed at $37\frac{1}{2}$ cents per unit.
5 The income reaches a maximum of $4500 when the selling price is $3.00, $n = 20$, and the
number of subscriptions is 1500.   **7** Let L represent the length and W the width. Then the
length of fencing is given by $L + 2W = 30$. Thus the area may be expressed as $A = LW =$
$(30 - 2W)(W) = -2W^2 + 30W = -2\left[W^2 - 15W + \left(\dfrac{-15}{2}\right)^2 - \left(\dfrac{-15}{2}\right)^2\right] = -2\left(W - \dfrac{15}{2}\right)^2 + \dfrac{225}{2}$.
Therefore, we obtain a maximum area of 225/2 square yards when $L = 15$ yards and $W = 15/2$ yards.
8 $y = \dfrac{7}{5}x - 100$. Break-even occurs at 71 units produced.

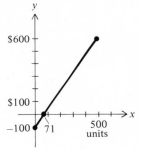

CHAPTER 4, Section 4.13 (p. 175)

Problem Set I ■ *Reading Comprehension*

1 False. An open sentence is an equation or inequality containing one or more variables.
2 True. The graph of the equation represents the boundary of the region, which is the solution.
3 False. The solution set of $2x - y < 4$ lies above the line $2x - y = 4$.

Problem Set II ■ *Skills Development*

1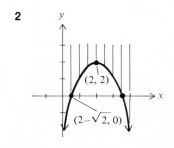

2

4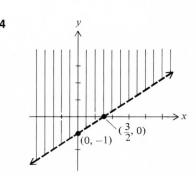

5

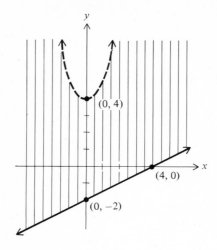

(0, 4)

(4, 0)

(0, −2)

Problem Set III ■ *Applications*

1 Let a = units of A, and b = units of B. Then $\mathbf{R} = \{(a, b) \mid 0 \le a \le 12 \text{ and } 0 \le b \le 8, a, b \in \mathbf{J}\}$.

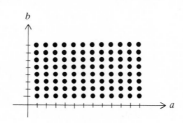

2 Let x = amount of A, and y = amount of B. Then $\mathbf{H} = \left\{(x, y) \ \middle| \ \dfrac{x}{4} + \dfrac{y}{16} \ge 1, x + y = 10, x \ge 0,\right.$

and $\left. y \ge 0\right\}$.

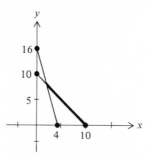

4 Let x = pounds of candy sold. Profit = sales − cost = $2.5x - (1500 + x) = 1.5x - 1500$. Thus, to make a profit, $1.5x - 1500 > 0 \Longleftrightarrow x > 1000$ pounds. If price is \$3.00 per pound, then $0 = 3x - (1500 - x) = 2x - 1500 \Longleftrightarrow x = 750$ pounds is the break-even point. Finally, if 2000 pounds can be produced, then $0 \le p(2000) - (1500 + 2000) = 2000p - 3500 \Longleftrightarrow p \ge 1.75$. Thus the price must be at least \$1.75 to guarantee no loss.   **5**   More than 12 bicycles to make a profit. It would pay to make the change since only 10 would be needed to make a profit.

CHAPTER 5, Section 5.3 (p. 187)

Problem Set I ■ *Reading Comprehension*

1 False. This definition applies only if x is a positive integer. **2** False. $a^{-n} = \dfrac{1}{a^n}$. **3** False. $a^0 = 1$ if $a \neq 0$; 0^0 is not defined. **4** False. $\sqrt[4]{16} = 2$. **5** False. The nth root of a negative number is not a member of the real numbers if n is even; however, if n is odd, then its nth root does exist; e.g., $\sqrt[3]{-8} = -2$. **6** True. $a^{m/n} = (a^m)^{1/n} = (a^{1/n})^m$. **7** All three are false. **8** True.

Problem Set II ■ *Skills Development*

1 $(-32)^{1/5} = -2$. **2** 3. **4** $(81)^{3/4} = ((81)^{1/4})^3 = 27$. **5** $-\dfrac{1}{32}$. **7** $x^{1/3}x^{2/5} = x^{5/15}x^{6/15} = x^{11/15}$. **8** $x^{5/4}$. **10** $\left(\dfrac{x^6}{y^2}\right)^{-1/2} = \dfrac{(y^2)^{1/2}}{(x^6)^{1/2}} = \dfrac{y}{x^3}$. **11** $\dfrac{5x^3y^{2/3}}{3}$. **13** $\dfrac{x^{3n} \cdot y^{2n-1}}{(x^n \cdot y^{2n})^{1/2}} = \dfrac{x^{3n} \cdot y^{2n-1}}{x^{n/2} \cdot y^n} = x^{3n} \cdot x^{-n/2} \cdot y^{2n-1} \cdot y^{-n} = x^{5n/2} \cdot y^{n-1}$. **14** $a^{n/6}b^{5/6}$. **16** $\dfrac{x^6y^{10}y^3}{x^3} = x^3y^{13}$. **17** $\dfrac{1}{a^2}$. **19** $a^{3/4}$. **20** $3a^2b^3$. **22** $\sqrt[3]{\dfrac{-27}{x^3y^6}} = \sqrt[3]{\dfrac{(-3)^3}{x^3(y^2)^3}} = \dfrac{-3}{xy^2}$. **23** $\dfrac{3x}{y^3z^2}$. **25** $\sqrt[3]{\dfrac{-125x^3z^6}{8y^6}} = \sqrt[3]{\left(\dfrac{-5xz^2}{2y^2}\right)^3} = \dfrac{-5xz^2}{2y^2}$. **26** $2^{7/6}$.

Problem Set III ■ *Theoretical Developments*

1(a) T5.2.1(II). *Proof:* $(a^m)^n \overset{\text{D5.2.1}}{=\!=\!=} \underbrace{a^m \cdot a^m \cdots a^m}_{n \text{ factors}} \overset{\text{T5.2.1(I)}}{=\!=\!=} a^{mn}$. **2(I)** $\sqrt[n]{a^n} \overset{\text{D5.2.5}}{=\!=\!=}$ $(a^n)^{1/n} \overset{\text{T5.2.2}}{\underset{\text{(II)}}{=\!=\!=}} a^{n/n} = a$. **3(a)** T5.2.2(I). *Proof:* Let $m = p/q$ and $n = r/s$. Then $a^m a^n \overset{\text{SubLE}}{=\!=\!=}$ $a^{p/q}a^{r/s} \overset{\text{T2.9.10}}{=\!=\!=} a^{ps/qs} \cdot a^{rq/qs} \overset{\text{D5.2.6}}{=\!=\!=} (a^{1/qs})^{ps}(a^{1/qs})^{rq} \overset{\text{T5.2.1(I)}}{=\!=\!=} (a^{1/qs})^{(ps+rq)} \overset{\text{D5.2.6}}{=\!=\!=}$ $a^{(ps/qs+rq/qs)} \overset{\text{T2.9.10}}{=\!=\!=} a^{(p/q+r/s)} \overset{\text{SubLE}}{=\!=\!=} a^{m+n}$. **(d)** T5.2.2(IV). *Proof:* $\left(\dfrac{a}{b}\right)^n \overset{\text{DDiv}}{=\!=\!=} (a \cdot b^{-1})^n$ $\overset{\text{T5.2.2(III)}}{=\!=\!=} a^n(b^{-1})^n \overset{\text{T5.2.2(II)}}{=\!=\!=} a^n(b^n)^{-1} \overset{\text{DDiv}}{=\!=\!=} \dfrac{a^n}{b^n}$. **(g)** T5.2.2(VII). *Proof:* $\left(\dfrac{a}{b}\right)^{-n} \overset{\text{T5.2.2(VI)}}{=\!=\!=}$ $\dfrac{1}{\left(\dfrac{a}{b}\right)^n} \overset{\text{T5.2.2(IV)}}{=\!=\!=} \dfrac{1}{\dfrac{a^n}{b^n}} \overset{\text{T2.9.13(VI)}}{=\!=\!=} \dfrac{b^n}{a^n}$.

CHAPTER 5, Section 5.5 (p. 192)

Problem Set I ■ *Reading Comprehension*

1 True. **2** False. $(-4)^{1/2}$ is not a real number. **3** True. **4** False. This base is not useful; $1^n = 1$ for each n. **5** False. $y = 3^{-x} = \left(\dfrac{1}{3}\right)^x$ is strictly decreasing.

Problem Set II ■ Skills Development

1 1, 1/4, 4. **2** −1/9, −9, −1, −1/27. (NOTE: $y = -3^x \neq (-3)^x$.) **4** 1/10, 1, 10, 100.
5 1, −1/2, 4, undefined.

7
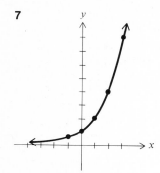

8
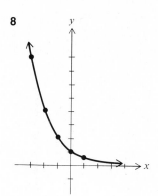

10
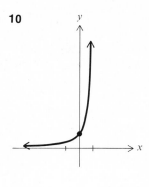

11 $y = 2^{-x} = \left(\frac{1}{2}\right)^x$. Hence graph is same as that for Problem 8.

13
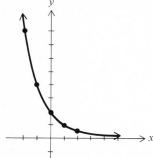

14
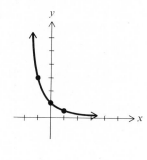

16 9.0250. **17** 0.013569. **19** 0.033373. **20** 148.41. **22** 7921. **23** 81.854.
25 $\sqrt{27 \cdot 10^{-6}} = (10^{-3})\sqrt{27} = 10^{-3}(5.1961) = 0.0051961.$ **26** 1369.

Problem Set III ■ Applications

1(a) $S = 200 - 100e^{-\frac{10}{20}} = 200 - 60.6531 = 139.3469.$ **(b)** 163.2121. **2** 0.6615.

CHAPTER 5, Section 5.9 (p. 208)

Problem Set I ■ Reading Comprehension

1 True. **2** False. The log function is also increasing. **3** True. The range of the exponential function is R_+ and thus the domain of the logarithmic function is also R_+. **4** False.
$f = \{(x, y) \mid x = 3^y\} = \{(x, y) \mid y = \log_3 x\}.$ **5** True. **6** False. $\log \frac{a}{b} = \log a - \log b \neq \frac{\log a}{\log b}.$

7 False. $\log(3.640 - 2) = \log 1.640$.   **8** False. $-3.6421 = -0.6421 - 3$.   **9** True.
10 False. $\log 4.76 = 0.6776$ is the mantissa.   **11** False. $\log 3.84 + 3 = \log 3840$.   **12** True.
13 True.   **14** True.   **15** False. $y = \text{antilog}_b x \Longleftrightarrow x = \log_b y \Longleftrightarrow b^x = y$.

Problem Set II ■ *Skills Development*

1 $\log_2 8 = 3$.   **2** $\log_3 9 = 2$.   **4** $\log_9 3 = 1/2$.   **5** $\log_8(\frac{1}{2}) = -\frac{1}{3}$.   **7** $\log_{10} 0.01 = -2$.
8 $\log_{10} 100{,}000 = 5$.   **10** $\log_2 32 = 5$.   **11** $\log_3(\frac{1}{9}) = -2$.   **13** $10^3 = 1000$.   **14** $2^3 = 8$.
16 $(1/2)^{-2} = 4$.   **17** $(\frac{1}{3})^{-3} = 27$.   **19** $10^1 = 10$.   **20** $10^0 = 1$.   **22** $\log_{1/2} x = -3 \Longleftrightarrow$
$(\frac{1}{2})^{-3} = x \Longleftrightarrow x = 8$.   **23** $1/9$.   **25** $\log_4(\frac{1}{2}) = x \Longleftrightarrow 4^x = \frac{1}{2} \Longleftrightarrow 2^{2x} = 2^{-1} \Longrightarrow 2x = -1 \Longrightarrow$
$x = -\frac{1}{2}$.   **26** -3.   **28** $\log_x(\frac{1}{8}) = -3 \Longleftrightarrow x^{-3} = \frac{1}{8} \Longleftrightarrow x^{-3} = (2)^{-3} \Longrightarrow x = 2$.   **29** $1/4$.
31 $\log_{10}(0.01) = x \Longleftrightarrow 10^x = 0.01 \Longleftrightarrow 10^x = (10)^{-2} \Longrightarrow x = -2$.   **32** -1.   **34** $\log_a a = x$
$\Longleftrightarrow a^x = a \Longrightarrow x = 1$.   **35** 0.

37

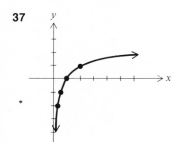

38

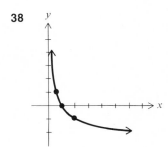

40

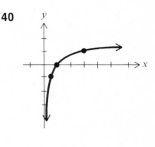

41

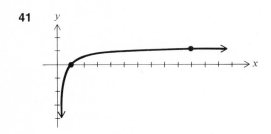

43 $\frac{1}{3}[2 \log x + \log z]$.   **44** $\frac{1}{4}[\log a + 3 \log b - \frac{1}{3} \log c]$.   **46** $\log 2 + \log \pi + \frac{1}{2}[\log L - \log g]$.
47 $\frac{1}{2}[\log 2 + \frac{1}{2} \log x - 2 \log y - \log z]$.   **49** $\log\left(\dfrac{xy^2}{z}\right)$.   **50** $\log\left(\dfrac{\sqrt{a} \ \sqrt[3]{c}}{(b+1)^3}\right)$.   **52** $\log \sqrt[3]{\dfrac{2\sqrt{5}}{3^2}}$.
53 $\log\left(\dfrac{20}{7^3}\right)$.   **55** $\log 467 = \log 4.67 + 2 = 2.6693$.   **56** 4.5729.   **58** $\log 0.00439 =$
$\log 4.39 - 3 = 0.6425 - 3$.   **59** $0.5672 - 1$.   **61** $\log 2763 = \log 2.763 + 3 = 3.4414$.   **62** 3.6845.
64 $\log x = 3.2923 = 0.2923 + 3 = \log 1.96 + 3 = \log 1960 \Longrightarrow x = 1960$.   **65** 0.003677.
67 $\log a = 4.4129 = 0.4129 + 4 = \log 2.588 + 4 = \log 25880 \Longrightarrow a = 25880$.   **68** $7.39 \cdot 10^{-11}$.
70 $\log \sqrt{0.0684} = \frac{1}{2}[\log 6.84 - 2] = \frac{1}{2}[0.8351 - 2] = 0.4176 - 1 = \log 2.616 - 1 = \log 0.2616 \Longrightarrow$
$\sqrt{0.0684} = 0.2616$.   **71** 0.6960.   **73** $\log(0.0245)^{-5} = -5[\log 2.45 - 2] = -5[0.3892 - 2] =$
$[-1.9460 + 10] = 8.0540 = 0.0540 + 8 = \log 1.132 + 8 = \log 1.132 \cdot 10^8 \Longrightarrow (0.0245)^{-5} = 1.132 \cdot 10^8$.
74 $2.775 \cdot 10^{-6}$.   **76** $\log \sqrt[3]{\dfrac{(64)^2}{(123)^3}} = \frac{1}{3}[2 \log 64 - 3 \log 123] = \frac{2}{3}[\log 6.4 + 1] - [\log 1.23 + 2] =$
$\frac{2}{3}[0.8062 + 1] - [2.0899] = 1.2042 - 2.0899 = 0.1143 - 1 = \log 1.301 - 1 = \log 0.1301 \Longrightarrow$

$\sqrt[3]{\dfrac{(64)^2}{(123)^3}} = 0.1301.$ **77** 934.8.   **79** $\log \dfrac{\sqrt{292}\ \sqrt[3]{486}}{\sqrt[4]{56.2}} = \frac{1}{2}\log 292 + \frac{1}{3}\log 486 - \frac{1}{4}\log 56.2 =$

$\frac{1}{2}(2.4654) + \frac{1}{3}(2.6866) - \frac{1}{4}(1.7497) = 1.6908 = 0.6908 + 1 = \log 4.907 + 1 = \log 49.07 \Longrightarrow$

$\dfrac{\sqrt{292}\ \sqrt[3]{486}}{\sqrt[4]{56.2}} = 49.07.$ **80** 0.04933.

Problem Set III ■ *Theoretical Developments*

1 $b^{\log_b x} = x \xLongleftrightarrow{\textbf{D5.6.1}} \log_b x = \log_b x.$ **4(a)** T5.6.1(IV). *Proof:* Let $x = \log_b a$ and $y = \log_b c.$

By D5.6.1, $b^x = a$ and $b^y = c.$ Then $\dfrac{a}{c} \xLongequal{\textbf{SubLE}} \dfrac{b^x}{b^y} \xLongequal{\textbf{T5.2.2(IV)}} b^{x-y}.$ Then using TLE and D5.6.1

we have $\log_b\!\left(\dfrac{a}{c}\right) = x - y \xLongequal{\textbf{SubLE}} \log_b a - \log_b c.$

Problem Set IV ■ *Applications*

1 $z = 276 \log \dfrac{a}{b} \Longleftrightarrow \dfrac{z}{276} + \log b = \log a \Longrightarrow \dfrac{250}{276} + \log(0.0404) = \log a \Longrightarrow 0.9058 + 0.6025 - 2 =$

$\log a \Longrightarrow 0.5083 - 1 = \log a \Longrightarrow \log 0.3223 = \log a \Longrightarrow a = 0.3223$ inches.   **2** 3.010.

4 $N = 10 \log \dfrac{250}{1} = 10 \log 250 = 10(2.3980) = 23.98$ decibels.

CHAPTER 5, Section 5.12 (p. 215)

Problem Set I ■ *Reading Comprehension*

1 False. We may use base 10 tables to compute tables or individual values for any base using

T5.10.1.   **2** True. Also called Napierian logarithms.   **3** False. $\dfrac{\log_b a}{\log_b d} \neq \log_b a - \log_b d.$

4 False. Natural or base e logs are more common in applications associated with natural
phenomena.   **5** False. $\log x$ is only defined for $x > 0.$

Problem Set II ■ *Skills Development*

1 $\log_3 84 = \dfrac{\log_{10} 84}{\log_{10} 3} = \dfrac{1.9243}{0.4771} = 4.0333.$ **2** $-6.2661.$ **4** $\log_7 0.086 = \dfrac{\log_{10} 0.086}{\log_{10} 7} =$

$\dfrac{0.9345 - 2}{0.8451} = \dfrac{-1.0655}{0.8451} = -1.2608.$ **5** 1.2211.   **7** $2^x = 6 \Longrightarrow x \log 2 = \log 6 \Longrightarrow x = \dfrac{\log 6}{\log 2} =$

$\dfrac{0.7782}{0.3010} = 2.5854.$ **8** 0.6310.   **10** $3x \log 10 = \log 3 - 2x \log 5 \Longleftrightarrow x(3 \log 10 + 2 \log 5) =$

$\log 3 \Longleftrightarrow x = \dfrac{\log 3}{3 \log 10 + 2 \log 5} = \dfrac{0.4771}{3 + 1.3979} = 0.1085.$ **11** $\pm 0.8788.$ **13** $\log x + 2 \log 3 =$

$3 \Longleftrightarrow \log 9x = 3 \Longleftrightarrow 9x = 10^3 \Longleftrightarrow x = \dfrac{1000}{9} = 111.1111.$ **14** $x = 498.$ **16** $\log x(x + 21) =$

$2 \Longleftrightarrow x^2 + 21x - 100 = 0$ and $x > 0 \Longleftrightarrow (x - 4)(x + 25) = 0$ and $x > 0 \Longleftrightarrow x = 4.$ **17** $x = 5.$

19 $\log_2 x = \log_3 5 \Longleftrightarrow x = 2^{\log_3 5}$.   **20** $x = 25$ or $x = \dfrac{1}{25}$.   **22** $|\log_x 9| = 2 \Longleftrightarrow [\log_x 9 = 2$ or $\log_x 9 = -2] \Longleftrightarrow [(x^2 = 9$ and $x > 0$ and $x \neq 1)$ or $(x^{-2} = 9$ and $x > 0$ and $x \neq 1)] \Longleftrightarrow [x = 3$ or $x = 1/3]$.

Problem Set III ■ Applications

1 $150 = 200 - 100e^{-t/20} \Longleftrightarrow e^{-t/20} = \dfrac{1}{2} \Longleftrightarrow e^{t/20} = 2 \Longleftrightarrow \dfrac{t}{20} = \log_e 2 = 0.6931 \Longleftrightarrow t = 13.862$ min.

2 $6.98 \cdot 10^4$.   **4** $\dfrac{1}{2} \cdot \dfrac{E}{R} = \dfrac{E}{R}(1 - e^{-Rt/L}) \Longleftrightarrow e^{-Rt/L} = \dfrac{1}{2} \Longleftrightarrow \dfrac{Rt}{L} = \log_e 2 \Longleftrightarrow t = \dfrac{L}{R}\log_e 2 = \dfrac{L}{R}(0.69315)$.

CHAPTER 6, Section 6.3 (p. 225)

Problem Set I ■ Reading Comprehension

1 True. $x^2 + 1 = 0$ has no real solution.   **2** False. The order of the pair is also significant — that is, $a + bi \neq b + ai$, unless $a = b$.   **3** True.   **4** True. But only because we *defined* $0i = 0$.
5 True.   **6** False. b, and not bi, is the imaginary part.   **7** False. $i^2 = -1$ but $(-1)^2 = +1$.

Problem Set II ■ Skills Development

1 $(-1 + 2i) + (3 + 4i) = (-1 + 3) + (2 + 4)i = 2 + 6i$.   **2** $-2 + 9i$.   **4** $-4 + (-3 - 5i) = (-4 - 3) + (-5)i = -7 - 5i$.   **5** $1 - i$.   **7** $(-1 + 2i)(3 + 4i) = [(-1)(3) - (2)(4)] + [(-1)(4) + (2)(3)]i = -11 + 2i$.   **8** -26.   **10** $(-4)(-3 - 5i) = 12 + 20i$.   **11** $6 + 2i$.

Problem Set III ■ Theoretical Developments

1 T6.2.1(I): $a + bi = a + bi$. *Proof:* By E-1 $[a = a$ and $b = b] \overset{\textbf{D6.2.2}}{\Longleftrightarrow} [a + bi = a + bi]$ QED.

CHAPTER 6, Section 6.5 (p. 231)

Problem Set I ■ Reading Comprehension

1 False. All of the field axioms are satisfied by complex numbers.   **2** False. The multiplicative identity for complex numbers is $1 + 0i$.   **3** False. The multiplicative inverse of (a, b) is $(a/(a^2 + b^2), -b/(a^2 + b^2))$ if $a, b \neq 0$.   **4** False. Division by $0 + 0i$ is the exception.
5 True.   **6** The main advantages are explained in Problem 5. Also, one system may be easier to work with than another because of its simple structure, yet their properties may be the same.

Problem Set II ■ Skills Development

1 $(-1 + 2i) - (3 + 4i) = (-1 + 2i) + (-3 - 4i) = -4 - 2i.$ **2** $6 - 3i.$ **4** $-4 - (-3 - 5i) =$
$-4 + (3 + 5i) = -1 + 5i.$ **5** $1 - 5i.$ **7** $(-1 + 2i)/(3 + 4i) = (-1 + 2i)(3 + 4i)^{-1} =$
$(-1 + 2i)(3/25 - (4/25)i) = (-3/25 + 8/25) + (6/25 + 4/25)i = 1/5 + (2/5)i.$ **8** $5/26 - (6/13)i.$
10 $-4/(-3 - 5i) = (-4)(-3 - 5i)^{-1} = (-4)(-3/34 + (5/34)i) = 6/17 - (10/17)i.$ **11** $-3/2 - (1/2)i.$
13 $(-2 + 3i) + (1 + 2i)(-3 - i) = (-2 + 3i) + (-1 - 7i) = -3 - 4i.$ **14** $8 + 6i.$
16 $[(3i) - (2 - i)]/[-2 + i] = [-2 + 4i]/[-2 + i] = (-2 + 4i)(-2 + i)^{-1} = (-2 + 4i)(-2/5 - (1/5)i) =$
$8/5 - (6/5)i.$ **17** $(-1 + 4i)/2.$ **19** $(-2 + 3i)/(1 + 2i) - (-3 - i)/(1 - 4i) =$
$[(-2 + 3i)(1 - 4i) - (1 + 2i)(-3 - i)]/(1 + 2i)(1 - 4i) = [(10 + 11i) - (-1 - 7i)]/(9 - 2i) =$
$(11 + 18i)/(9 - 2i) = (11 + 18i)(9 - 2i)^{-1} = (11 + 18i)(9/85 + (2/85)i) = 63/85 + (184/85)i.$
20 $(19 - 9i)/17.$ **22** $(-2 - 3i)\{(1 + 2i) + [(-1 + 4i) + (4 - i)]\} = (-2 - 3i)\{(1 + 2i) + (3 + 3i)\} =$
$(-2 - 3i)(4 + 5i) = 7 - 22i.$ **23** $-8 + 11i.$ **25** $-2x + yi = 4 - i \Longleftrightarrow -2x = 4$ and $y = -1$
$\Longleftrightarrow x = -2$ and $y = -1.$ **26** $x = -2, y = 0.$ **28** $(x + 3i)^2 = 2yi \Longleftrightarrow (x^2 - 9) + 6xi = 2yi$
$\Longleftrightarrow x^2 - 9 = 0$ and $6x = 2y \Longleftrightarrow x = \pm 3$ and $y = 3x \Longleftrightarrow (x = 3$ and $y = 9)$ or $(x = -3$ and
$y = -9).$

Problem Set III ■ Theoretical Developments

1 T6.4.1. *Proof:* $(a + bi) + (c + di) \overset{\textbf{D6.2.3}}{=\!=\!=\!=} (a + c) + (b + d)i$, which is complex because

$a + c$ and $b + d$ are real. $(a + bi)(c + di) \overset{\textbf{D6.2.4}}{=\!=\!=\!=} (ac - bd) + (ad + bc)i$, which is complex

because $ac - bd$ and $ad + bc$ are real. QED. **4** T6.4.4. *Proof:* $(a + bi) + (0 + 0i) \overset{\textbf{D6.2.3}}{=\!=\!=\!=}$

$(a + 0) + (b + 0)i \overset{\textbf{Id+}}{=\!=\!=} a + bi$ QED. **7** T6.4.8. *Proof:* $(a + bi)\left[\dfrac{a}{a^2 + b^2} + \left(\dfrac{-b}{a^2 + b^2}\right)i\right] \overset{\textbf{D6.2.4}}{=\!=\!=\!=}$

$\left(\dfrac{a^2}{a^2 + b^2} + \dfrac{b^2}{a^2 + b^2}\right) + \left(\dfrac{-ab}{a^2 + b^2} + \dfrac{ba}{a^2 + b^2}\right)i \overset{\textbf{In+}}{=\!=\!=} 1 + 0i$ QED.

9

	Real	Imag.	Conj.
(a)	-1	-3	$-1 + 3i$
(b)	0	2	$-2i$
(d)	$-1/3$	$-2/3$	$(-1 + 2i)/3$
(e)	-2	3	$-2 - 3i$

10(a) Let $z = a + bi$, then $[a + bi = a - bi] \overset{\textbf{D6.2.2}}{=\!=\!=\!\Longrightarrow} [-b = b] \Longrightarrow [b = 0] \Longrightarrow z = a + 0i = a$
$\Longrightarrow z$ is real. QED. **(d)** $(a + bi)(c + di) = (ac - bd) + (ad + bc)i = (ac - bd) - (ad + bc)i =$
$(a - bi)(c - di) = \overline{(a + bi)} \cdot \overline{(c + di)}$ QED.

CHAPTER 7, Section 7.4 (p. 238)

Problem Set I ■ Reading Comprehension

1 False. The factor x^{-1} renders it false. Only non-negative integral exponents are allowed.
2 False. $\sqrt{x} = x^{1/2}$, and $1/2$ is not a positive integer. **3** True.

Problem Set II ■ *Skills Development*

1 $x^3 + 3x^2 - 4x + 1.$ **2** $-6x^4 + x^3 - x^2 + 2x - 4.$ **4** $x^7 + x^6 - 2x^5 - x^4 - x^3 - 3x^2 + 2x + 3.$
5 $x^8 - 3x^7 + 4x^6 - 7x^5 + 7x^4 + 4x^3 - 6x^2 + 4x - 8.$ **7** $-x^4 - 3x^3 - 5x^2 + 7x + 4.$
8 $x^4 + 5x^3 + x^2 - 5x - 6.$

CHAPTER 7, Section 7.6 (p. 248)

Problem Set I ■ *Reading Comprehension*

1 False. $1/x$ is a rational algebraic expression. **2** True. But the degree of R may be zero.
3 False. Synthetic division is valid for all quotients of the form $P(x)/(x - a)$, where $P(x)$ is a
polynomial. **4** False. Synthetic division requires the divisor to be of degree one. **5** False.
Set up as $-2 \rfloor 1 \quad 0 \quad -3 \quad 2 \quad -1$. **6** False. Remainder after performing $P(x)/(x + c)$ is
$P(-c)$. **7** True. This is the factor theorem. **8** False. Direct substitution may be more
convenient.

Problem Set II ■ *Skills Development*

1
$$
\begin{array}{r|rrrr}
3 & 1 & -3 & 0 & 4 \\
  &   & 3 & 0 & 0 \\
\hline
  & 1 & 0 & 0 & 4
\end{array}
\Longrightarrow x^2 + \frac{4}{x - 3} \text{ and } P(3) = 3^3 - 3(3)^2 + 4 = 4.
$$

2 $x^4 + 2x^3 - 4x^2 + 7x - 26 + \dfrac{46}{x + 2}.$

4
$$
\begin{array}{r|rrrr}
2 & 1 & -1 & -5 & 6 \\
  &   & 2 & 2 & -6 \\
\hline
  & 1 & 1 & -3 & 0
\end{array}
\Longrightarrow P(2) = 0 \Longrightarrow \text{true.}
$$

5 False.

7
$$
\begin{array}{r|rrrrrr}
3 & 1 & -2 & 0 & -5 & -6 & 2 \\
  &   & 3 & 3 & 9 & 12 & 18 \\
\hline
  & 1 & 1 & 3 & 4 & 6 & 20
\end{array}
\Longrightarrow P(3) = 20 \Longrightarrow \text{false.}
$$

8 False.

10
$$
\begin{array}{r|rrrr}
-1 & 1 & -2 & 1 & -1 \\
   &   & -1 & 3 & -4 \\
\hline
   & 1 & -3 & 4 & -5 = P(-1),
\end{array}
\qquad
\begin{array}{r|rrrr}
4 & 1 & -2 & 1 & -1 \\
  &   & 4 & 8 & 36 \\
\hline
  & 1 & 2 & 9 & 35 = P(4),
\end{array}
\qquad
\begin{array}{r|rrrr}
6 & 1 & -2 & 1 & -1 \\
  &   & 6 & 24 & 150 \\
\hline
  & 1 & 4 & 25 & 149 = P(6).
\end{array}
$$

11 $-2610 = Q(1), -2652 = Q(-2),$ and $-5256 = Q(6).$

13
$$
\begin{array}{r|rrrr}
2 & 2 & -3 & 4 & -6 \\
  &   & 4 & 2 & 12 \\
\hline
  & 2 & 1 & 6 & 6 = f(2),
\end{array}
\qquad
\begin{array}{r|rrrr}
2i & 2 & -3 & 4 & -6 \\
   &   & 4i & -8-6i & 12-8i \\
\hline
   & 2 & -3+4i & -4-6i & 6-8i = f(2i),
\end{array}
\qquad
\begin{array}{r|rrrr}
i & 2 & -3 & 4 & -6 \\
  &   & 2i & -2-3i & 3+2i \\
\hline
  & 2 & -3+2i & 2-3i & -3+3i = f(i).
\end{array}
$$

14 $84 = g(3), 0 = g(\sqrt{2}),$ and $g(0) = 6.$

1

$$
1\,\overline{\big)\,1\ \ 0\ \ 0\ \ 0\ \ \cdots\ \ 0\ \ -1}
$$

$$
\ \ \ 1\ \ 1\ \ 1\ \ \cdots\ \ 1\ \ \ 1
$$

$$
\overline{\ 1\ \ 1\ \ 1\ \ 1\ \ \cdots\ \ 1\ \ \ 0}
$$

Thus $x - 1$ is a factor of $x^n - 1$, and the other factor is $x^{n-1} + x^{n-2} + \cdots + x + 1$.

4 $x + y = x - [-y]$. Let $x = -y$, and substitute into $x^n + y^n$ obtaining $(-y)^n + y^n$, which is zero if n is odd. **5** $k = 1$.

CHAPTER 7, Section 7.9 (p. 264)

Problem Set I ■ *Reading Comprehension*

1 True. **2** False. $n - 1$ or fewer turning points. **3** True. **4** False. Polynomials may have repeated factors. **5** False. If $P(x)$ has real coefficients and $P(c) = 0$, then $P(\bar{c}) = 0$. **6** False. There are, at *most,* three positive roots. **7** True. **8** True. (Refer to Theorem 9.8.6.)

Problem Set II ■ *Skills Development*

1 $y = x^3 - 3x^2 + x - 1$

$$
\begin{array}{r|rrrr}
 & 1 & -3 & 1 & -1 \\
\hline
-1 & 1 & -4 & 5 & -6 = P(-1) \\
1 & 1 & -2 & -1 & -2 = P(1) \\
2 & 1 & -1 & -1 & -3 = P(2) \\
3 & 1 & 0 & 1 & 2 = P(3)
\end{array}
$$

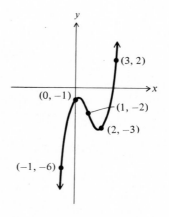

2

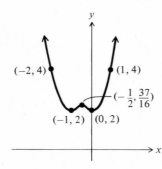

4 $P(x) = (x - 1)(x + 2)(x + 2) = x^3 + 3x^2 - 4.$ **5** $x^3 - 5x^2 + 8x - 6.$

7

$$
1 + i\,\overline{\big)\,1\ \ -2\ \ \ 3\ \ \ -2\ \ \ 2}
$$

$$
\ \ 1 + i\ \ -2\ \ +1 + i\ \ -2
$$

$$
1 - i\,\overline{\big)\,1\ \ -1 + i\ \ \ 1\ \ -1 + i\ \ \ 0}
$$

$$
\ \ 1 - i\ \ \ 0\ \ \ 1 - i
$$

$$
\overline{\ 1\ \ \ 0\ \ \ 1\ \ \ 0}
$$

$\Longrightarrow x^2 + 1 = 0 \Longrightarrow x = \pm i.$ Thus, $S = \{1 + i, 1 - i, i, -i\}.$

8 $S = \left\{ 3i, -3i, \dfrac{-3 + \sqrt{29}}{2}, \dfrac{-3 - \sqrt{29}}{2} \right\}.$

10 $P(x) = x^5 - 2x^4 + x^2 - x - 1 \Longrightarrow$ 3 changes, and $P(-x) = -x^5 - 2x^4 + x^2 + x - 1 \Longrightarrow$ 2 changes. Thus,

Pos.	Neg.	Complex
3	2	0
3	0	2
1	2	2
1	0	4

11

Pos.	Neg.	Complex
3	3	0
3	1	2
1	3	2
1	1	4

13

$$
\begin{array}{r|rrrrr}
3 & 1 & -2 & -1 & 1 & -2 \\
  &   & 3 & 3 & 6 & 21 \\
\hline
  & 1 & 1 & 2 & 7 & 19
\end{array}
\Longrightarrow \text{3 is an upper bound, and}
$$

$$
\begin{array}{r|rrrrr}
-2 & 1 & -2 & -1 & 1 & -2 \\
   &   & -2 & 8 & -14 & 26 \\
\hline
   & 1 & -4 & 7 & -13 & 24
\end{array}
\Longrightarrow \text{-2 is a lower bound.}
$$

14 2 is an upper bound and -6 is a lower bound.

16 Possible rational roots are $\{\pm 1, \pm 2\}$.

$$
\begin{array}{r|rrrr}
2 & 1 & -1 & -3 & 2 \\
  &   & 2 & 2 & -2 \\
\hline
  & 1 & 1 & -1 & 0
\end{array}
= P(2) \Longrightarrow x^2 + x - 1 = 0 \Longrightarrow x = \frac{-1 \pm \sqrt{5}}{2}. \text{ Thus, } S = \left\{2, \frac{-1 + \sqrt{5}}{2}, \frac{-1 - \sqrt{5}}{2}\right\}.
$$

17 $\{-1, -3/2, \sqrt{2}i, -\sqrt{2}i\}$.

19 $x^5 - x^4 - 7x^3 - 14x^2 - 24x = 0$ has 0 as a root, leaving $x^4 - x^3 - 7x^2 - 14x - 24 = 0$. Possible roots are

$\{\pm 1, \pm 2, \pm 3, \pm 4, \pm 6, \pm 8, \pm 12, \pm 24\}.$
$$
\begin{array}{r|rrrrr}
4 & 1 & -1 & -7 & -14 & -24 \\
  &   & 4 & 12 & 20 & 24 \\
\hline
  & 1 & 3 & 5 & 6 & 0
\end{array}
\Longrightarrow \text{4 is an upper bound and a root.}
$$

$$
\begin{array}{r|rrrr}
-3 & 1 & 3 & 5 & 6 \\
   &   & -3 & 0 & -15 \\
\hline
   & 1 & 0 & 5 & -9
\end{array}
\Longrightarrow \text{-3 is a lower bound.}
\qquad
\begin{array}{r|rrrr}
-2 & 1 & 3 & 5 & 6 \\
   &   & -2 & -2 & -6 \\
\hline
   & 1 & 1 & 3 & 0
\end{array}
\Longrightarrow \text{-2 is a root} \Longrightarrow
$$

$x^2 + x + 3 = 0 \Longrightarrow x = \dfrac{-1 \pm i\sqrt{11}}{2}.$ Thus, $S = \left\{0, 4, -2, \dfrac{-1 + i\sqrt{11}}{2}, \dfrac{-1 - i\sqrt{11}}{2}\right\}.$

20 $S = \left\{-2, \dfrac{7 \pm i\sqrt{95}}{6}\right\}.$

22
$$
\begin{array}{r|rrrr}
-1.2 & 1 & 1 & 2 & 3 \\
     &   & -1.2 & 0.24 & -2.688 \\
\hline
     & 1 & -0.2 & 2.24 & 0.312 > 0,
\end{array}
\qquad
\begin{array}{r|rrrr}
-1.3 & 1 & 1 & 2 & 3 \\
     &   & -1.3 & 0.39 & -3.107 \\
\hline
     & 1 & -0.3 & 2.39 & -0.107 < 0,
\end{array}
$$

$$
\begin{array}{r|rrrr}
-1.25 & 1 & 1 & 2 & 3 \\
      &   & -1.25 & 0.3125 & -2.890625 \\
\hline
      & 1 & -0.25 & 2.3125 & 0.109375 > 0.
\end{array}
$$
Hence, the root lies between -1.25 and -1.3 and to the

nearest tenth is -1.3.   **23** 1.7.

Problem Set III ■ Theoretical Developments

1 From Equation (10) of the proof of T9.86 given in the text, we have $-a_np^n = a_{n-1}p^{n-1}q + \cdots + a_1pq^{n-1} + a_0q^n$. Factoring q from the right side yields $-a_np^n = q(a_{n-1}p^{n-1} + \cdots + a_1pq^{n-2} + a_0q^{n-1})$. Since q is an integral factor of the right side, it must also be a factor of the left side. Because p and q are relatively prime, it follows that q must be a factor of a_n QED. **4** Conjugate pairs of complex roots can account for only an even number of roots, and therefore a real polynomial of odd degree has at least one real root. **7** $P(x) = x^3 - 6$; by the rational root theorem, the only candidates for rational roots are $\pm 1, \pm 2, \pm 3, \pm 6$. Since each of these fails to be a root, we conclude $\sqrt[3]{6}$ is not rational.

Problem Set IV ■ Applications

1 $w^4 - 5w^2 + 4 = 0 \Longleftrightarrow (w^2 - 4)(w^2 - 1) = 0 \Longleftrightarrow (w - 2)(w + 2)(w - 1)(w + 1) = 0 \Longrightarrow$
$S = \{1, -1, 2, -2\}$. **2** $S = \{\pm 1, \pm 2, \pm 1/2\}$. **3(a)** $s^3 + 6s^2 + 11s + 6 = 0 \Longrightarrow$

$$\begin{array}{r|rrrr} -1 & 1 & 6 & 11 & 6 \\ & & -1 & -5 & -6 \\ \hline & 1 & 5 & 6 & 0 \end{array} \Longrightarrow \text{factors are } (x + 1)(x^2 + 5x + 6) = (x + 1)(x + 2)(x + 3).$$

(b) $(x - 1)(x + 1)(x^2 + 4x + 5)$.

CHAPTER 7, Section 7.11 (p. 277)

Problem Set I ■ Reading Comprehension

1 False. A rational function is a function that can be written as a quotient of two polynomials.
2 False. Degree is defined for polynomial functions only. **3** False. Vertical asymptotes occur at the zeros of the denominator. **4** False. Horizontal asymptotes are found by examining the function-values after expressing $P(x)/D(x)$ in the form $Q(x) + R(x)/D(x)$. **5** False. Vertical asymptotes are never crossed by graphs of curves, while horizontal and oblique asymptotes may be crossed.

Problem Set II ■ Skills Development

1 $y = \dfrac{3}{x - 1} \Longrightarrow$ Vertical asymptote at $x = 1$,

horizontal asymptote at $y = 0$, y-intercept at -3, and no x-intercept. Utilizing this information, the graph may be sketched as shown at right.

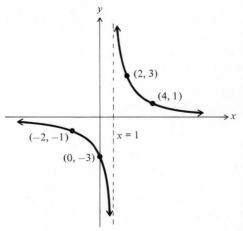

2

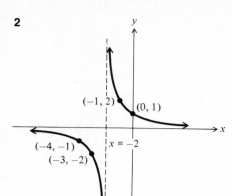

4 $y = \dfrac{x+1}{x+2} \Longrightarrow y = 1 + \dfrac{-1}{x+2}$. Thus, we have a vertical
asymptote at $x = -2$, a horizontal asymptote at $y = 1$, the
y-intercept at $(0, 1/2)$, and the x-intercept at $(-1, 0)$, since
$x + 1 = 0$ when $x = -1$. Utilizing this information, the graph
may be sketched as shown at right.

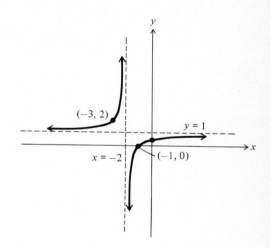

5

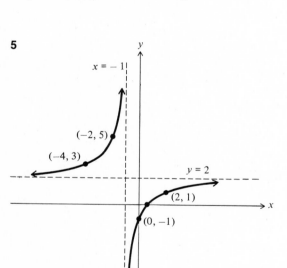

7 $y = \dfrac{x^2}{x^2 - 4} \Longrightarrow y = 1 + \dfrac{4}{x^2 - 4} = 1 + \dfrac{4}{(x-2)(x+2)} \Longrightarrow$
vertical asymptotes at $x = 2$ and $x = -2$, a horizontal
asymptote at $y = 1$, and the y-intercept at $(0, 0)$, which is
also the x-intercept. Utilizing this information, the graph is
sketched as shown at right.

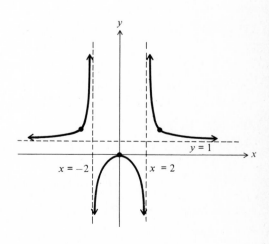

8

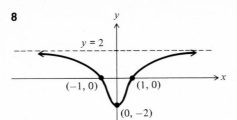

10 $y = \dfrac{x^3 - 1}{x^2} \implies y = x - \dfrac{1}{x^2} \implies$ a vertical asymptote

at $x = 0$, no horizontal asymptotes, an oblique asymptote at $y = x$, no y-intercepts, and an x-intercept at $x = 1$.

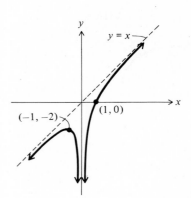

11

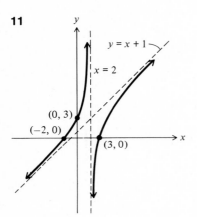

13 $y = \dfrac{1}{x^2 - 4} = \dfrac{1}{(x - 2)(x + 2)} \implies$ vertical

asymptotes at $x = 2$ and $x = -2$, a horizontal asymptote at $y = 0$, the y-intercept at $(0, -1/4)$, and no x-intercepts.

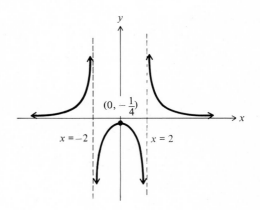

14

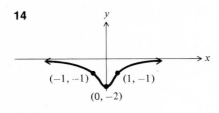

16 $y = \dfrac{3}{(x - 2)^2} \implies$ a vertical asymptote at $x = 2$,

a horizontal asymptote at $y = 0$, the y-intercept $(0, 3/4)$, and no x-intercepts.

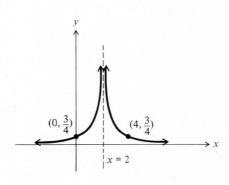

17

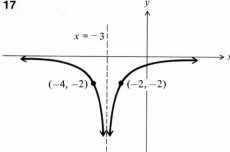

$x = -3$

$(-4, -2)$ $(-2, -2)$

CHAPTER 8, Section 8.3 (p. 287)

Problem Set I ■ *Skills Development*

1 $\begin{aligned} x + y + z &= 6 \\ x - 3y - 2z &= -9 \\ 2x - 2y + 3z &= 1 \end{aligned} \Bigg\} \Longrightarrow \begin{aligned} x + y + z &= 6 \\ -4y - 3z &= -15 \\ -4y + z &= -11 \end{aligned} \Bigg\} \Longrightarrow \begin{aligned} x + y + z &= 6 \\ -4y - 3z &= -15 \\ 4z &= 4 \end{aligned} \Bigg\} \Longrightarrow [z = 1, y = 3, x = 2]$
$\Longrightarrow S = \{(2, 3, 1)\}.$

2 $[x = 1/2, y = -3/2, z = 0] \Longrightarrow S = \{(1/2, -3/2, 0)\}.$

4 $\begin{aligned} x + y + z &= 0 \\ 2x - y - 4z &= 15 \\ x - 2y - z &= 7 \end{aligned} \Bigg\} \Longrightarrow \begin{aligned} x + y + z &= 0 \\ -3y - 6z &= 15 \\ -3y - 2z &= 7 \end{aligned} \Bigg\} \Longrightarrow \begin{aligned} x + y + z &= 0 \\ y + 2z &= -5 \\ 4z &= -8 \end{aligned} \Bigg\} \Longrightarrow [z = -2, y = -1, x = 3]$
$\Longrightarrow S = \{(3, -1, -2)\}.$

5 $[z = -14/3, y = -5/3, x = -4/3] \Longrightarrow S = \{(-4/3, -5/3, -14/3)\}.$

7 $\begin{aligned} x - y + z &= 1 \\ 2x + y + z &= 2 \\ -x - 2y - z &= -1 \end{aligned} \Bigg\} \Longrightarrow \begin{aligned} x - y + z &= 1 \\ 3y - z &= 0 \\ -3y &= 0 \end{aligned} \Bigg\} \Longrightarrow [y = 0, z = 0, x = 1] \Longrightarrow S = \{(1, 0, 0)\}.$

8 $[z = -1, y = 0, x = 1] \Longrightarrow S = \{(1, 0, -1)\}.$

10 $\begin{aligned} x + y + z &= 2 \\ x - y &= 5 \\ -x + 3y + z &= -10 \end{aligned} \Bigg\} \Longrightarrow \begin{aligned} x + y + z &= 2 \\ -2y - z &= 3 \\ 4y + 2z &= -8 \end{aligned} \Bigg\} \Longrightarrow \begin{aligned} x + y + z &= 2 \\ -2y - z &= 3 \\ 0 &= -2 \end{aligned} \Bigg\} \Longrightarrow \text{no solution exists.}$

11 No solutions exist.

Problem Set II ■ *Applications*

1 $\begin{aligned} x_1 &= 1500 + 0.10x_1 + 0.20x_2 \\ x_2 &= 2000 + 0.20x_1 + 0.30x_2 \end{aligned} \Bigg\} \Longrightarrow \begin{cases} x_2 = 3559.32 \\ x_1 = 2457.62 \end{cases}$

$\begin{aligned} x_3 &= 3000 + 0.40x_1 + 0.20x_2 \\ x_4 &= 4000 + 0.10x_1 + 0.20x_2 \\ x_5 &= 6000 + 0.20x_1 + 0.10x_2 \end{aligned} \Bigg\} \Longrightarrow \begin{cases} x_3 = 3000 + 0.40x_1 + 0.20x_2 = 4650.37 \\ x_4 = 4000 + 0.10x_1 + 0.20x_2 = 4957.62 \\ x_5 = 6000 + 0.20x_1 + 0.10x_2 = 6847.45 \end{cases}$

2 $x_1 = 23516.0, x_2 = 34716.91, x_3 = 1860.47, x_4 = 8883.76, x_5 = 7869.78, x_6 = 5260.47, x_7 = 4869.78.$

4 $\begin{aligned} 4I_1 - 3I_2 &= 1 \\ -3I_1 + 6I_2 - 2I_3 &= 0 \\ -2I_2 + 8I_3 &= 10 \end{aligned}\Bigg\} \Longrightarrow \begin{aligned} 12I_1 - 9I_2 &= 3 \\ 13I_2 &= 13 \\ -2I_2 + 8I_3 &= 10 \end{aligned}\Bigg\} \Longrightarrow \left[I_2 = 1, I_3 = \frac{3}{2}, I_1 = 1\right] \Longrightarrow S = \left\{\left(1, 1, \frac{3}{2}\right)\right\}.$

5 $S = \left\{\left(\dfrac{46}{83}, \dfrac{55}{83}, \dfrac{22}{83}\right)\right\}.$

CHAPTER 8, Section 8.5 (p. 301)

Problem Set I ■ Reading Comprehension

1 True. **2** False. a_{23} is in the second row, third column. **3** False. $A = B$ if and only if $a_{ij} = b_{ij}$ for every i, j. This also requires that $m = p$ and $n = q$. **4** False. The commutative law of multiplication fails. **5** True. **6** False. $cA = c(a_{ij}) = (ca_{ij})$. Every element of matrix A is multiplied by the scalar c. **7** False. $A \cdot B$ is defined only if the number of columns of A is equal to the number of rows of B. **8** False. If $C = AB$ then $c_{ij} = a_{i1}b_{1j} + a_{i2}b_{2j} + \cdots + a_{im}b_{mj}$.
9 False. The multiplicative identity matrix has a 1 for each element on the main diagonal and a 0 everywhere else.

Problem Set II ■ Skills Development

1 $\begin{bmatrix} 3 & -5 \\ -1 & 3 \end{bmatrix}$ **2** $\begin{bmatrix} -2 & -1 \\ 4 & 3 \end{bmatrix}$ **4** $\begin{bmatrix} 3 \\ 1 \\ -2 \end{bmatrix}$ **5** $\begin{bmatrix} 1 & 5 & 7 \\ -1 & 4 & 8 \end{bmatrix}$ **7** $\begin{bmatrix} -1 & 1 \\ 7 & 5 \end{bmatrix}$ **8** $\begin{bmatrix} 2 & -7 \\ 2 & 7 \end{bmatrix}$

10 $\begin{bmatrix} -1 \\ -5 \\ 8 \end{bmatrix}$ **11** $\begin{bmatrix} -3 & -2 & 8 \\ 7 & -3 & -3 \end{bmatrix}$ **13** $\begin{bmatrix} (2-4) & (-3+1) \\ (4+12) & (-6-3) \end{bmatrix} = \begin{bmatrix} -2 & -2 \\ 16 & -9 \end{bmatrix}$ **14** $\begin{bmatrix} -4 \\ 6 \end{bmatrix}$

16 $[(-2 - 8 + 15)] = [5].$ **17** $\begin{bmatrix} 2 & -1 \\ 3 & -4 \\ -1 & 3 \end{bmatrix}$ **19** $\begin{bmatrix} -3 & 2 & 1 \\ 4 & -1 & -2 \end{bmatrix}$ **20** $\begin{bmatrix} -3 & -2 & 0 \\ 1 & 2 & 0 \\ 0 & 1 & 1 \end{bmatrix}$

22 $\begin{bmatrix} -2 & 6 \\ 8 & -4 \end{bmatrix}$ **23** $\begin{bmatrix} -6 & 3 & -9 \\ 12 & 6 & -3 \end{bmatrix}$ **25** $\begin{bmatrix} 1 & -2 \\ -1 & 1 \end{bmatrix} \cdot \begin{bmatrix} 1 & -2 \\ -1 & 2 \end{bmatrix} = \begin{bmatrix} 3 & -6 \\ -2 & 4 \end{bmatrix}$ **26** $\begin{bmatrix} 7 & -17 \\ -3 & 21 \end{bmatrix}$

28 $X = \begin{bmatrix} -1 & 2 \\ 4 & -5 \end{bmatrix} - \begin{bmatrix} 3 & 4 \\ 2 & -6 \end{bmatrix} = \begin{bmatrix} -4 & -2 \\ 2 & 1 \end{bmatrix}$ **29** $\begin{bmatrix} -1 & 0 \\ 3 & -1 \end{bmatrix}$ **31** $x = 3$ and $y - 2 = -4 \Longrightarrow$
$x = 3$ and $y = -2.$

32 $y = -5, x = -8.$ **34** $x = -1, x^2 = 1, y = 1,$ and $y^2 = -1 \Longrightarrow x = -1$ and $y = 1.$
35 No solution exists.

Problem Set III ■ Theoretical Developments

1(I) $A = A \Longleftrightarrow a_{ij} = a_{ij}$ for each i, j. But this latter equality is true by the reflexive property of real numbers. $[A = B \Longrightarrow B = A] \Longleftrightarrow [a_{ij} = b_{ij} \Longrightarrow b_{ij} = a_{ij}].$ But this latter equality is true by

the symmetric property of real numbers. $[A = B \text{ and } B = C] \overset{\text{D8.4.1}}{\Longleftrightarrow} [(a_{ij}) = (b_{ij}) \text{ and}$

$(b_{ij}) = (c_{ij})]\overset{\textbf{TLE}}{\Longrightarrow}[(a_{ij}) = (c_{ij})]\Longleftrightarrow A = C$ QED.   **4(a)** T8.4.1(I). *Proof:* If A, B are of the same dimensions, then $A_{m \times n} + B_{m \times n} = (a_{ij} + b_{ij})_{m \times n}$.   **(d)** T8.4.1(V). *Proof:* $A + (-A) =$

$(a_{ij} + (-a_{ij})) = (0) = 0.$ **5(a)** T8.4.2(I). *Proof:* $a \cdot A \overset{\textbf{D8.4.4}}{=\!=\!=\!=\!=} (a \cdot a_{ij}).$ **(d)** T8.4.2(IV(a)).

Proof: $1 \cdot A \overset{\textbf{D8.4.4}}{=\!=\!=\!=\!=} (1 \cdot a_{ij}) \overset{\textbf{Id.}}{=\!=} (a_{ij}) = A.$ **(g)** T8.4.2(V(b)). *Proof:* $a \cdot 0 \overset{\textbf{D8.4.4}}{=\!=\!=\!=\!=}$

$(a \cdot 0_{ij}) = (0) = 0.$ **6(a)** T8.4.3(I). *Proof:* $A \cdot B \overset{\textbf{D8.4.5}}{=\!=\!=\!=\!=} (a_{i1}b_{1j} + a_{i2}b_{2j} + \cdots + a_{ip}b_{pj}) =$ $(c_{ij}).$ **7** T8.4.4. *Proof:* (a) Let $I_{m \times m} = (n_{ij})$, where $n_{ij} = 1$ if $i = j$, and $n_{ij} = 0$ if $i \neq j$.

(b) $I_{m \times m} A_{m \times n} \overset{\textbf{D8.4.5}}{=\!=\!=\!=\!=} (n_{i1}a_{1j} + n_{i2}a_{2j} + \cdots + n_{im}a_{mj}), \overset{\textbf{(1)}}{=\!=\!=} (n_{ii} \cdot a_{ij}) = (1 \cdot a_{ij}) = (a_{ij}) = A.$
The case with the identity on the right is similar.   **8(a)** $(A - B)(A - B) = A(A - B) - B(A - B) = A^2 - AB - BA + B^2$. However, $AB \neq BA$ in general; hence, $(A - B)^2 \neq A^2 - 2AB + B^2$, in general.

10(a) $(AB)C = \left(\begin{bmatrix} -2 & 3 \\ 4 & -1 \end{bmatrix} \cdot \begin{bmatrix} -1 & 4 \\ 0 & 3 \end{bmatrix}\right) \cdot \begin{bmatrix} 2 & 4 \\ -6 & 1 \end{bmatrix} = \begin{bmatrix} 2 & 1 \\ -4 & 13 \end{bmatrix} \cdot \begin{bmatrix} 2 & 4 \\ -6 & 1 \end{bmatrix} = \begin{bmatrix} -2 & 9 \\ -86 & -3 \end{bmatrix}$ and $A(BC) =$

$\begin{bmatrix} -2 & 3 \\ 4 & -1 \end{bmatrix} \cdot \left(\begin{bmatrix} -1 & 4 \\ 0 & 3 \end{bmatrix} \cdot \begin{bmatrix} 2 & 4 \\ -6 & 1 \end{bmatrix}\right) = \begin{bmatrix} -2 & 3 \\ 4 & -1 \end{bmatrix} \cdot \begin{bmatrix} -26 & 0 \\ -18 & 3 \end{bmatrix} = \begin{bmatrix} -2 & 9 \\ -86 & -3 \end{bmatrix} \Longrightarrow (AB)C = A(BC).$

11(a) $\begin{bmatrix} a & b \\ c & d \end{bmatrix} \cdot \begin{bmatrix} 2 & -3 \\ 0 & 1 \end{bmatrix} = \begin{bmatrix} 2a & -3a + b \\ 2c & -3c + d \end{bmatrix} = \begin{bmatrix} 2 & 5 \\ 8 & -7 \end{bmatrix} \Longrightarrow \begin{cases} 2a = 2, -3a + b = 5, \\ 2c = 8, -3c + d = -7. \end{cases} \Longrightarrow$

$\begin{cases} a = 1, b = 8, \\ c = 4, d = 5. \end{cases} \Longrightarrow A = \begin{bmatrix} 1 & 8 \\ 4 & 5 \end{bmatrix}.$ **(b)** No solution exists.   **12(a)** $\begin{bmatrix} x & 2 \end{bmatrix} \cdot \begin{bmatrix} 2 & -1 \\ 3 & 4 \end{bmatrix} \cdot \begin{bmatrix} x \\ -4 \end{bmatrix} =$

$\begin{bmatrix} 2x + 6 & -x + 8 \end{bmatrix} \cdot \begin{bmatrix} x \\ -4 \end{bmatrix} = 2x^2 + 6x + 4x - 32 \Longrightarrow 2x^2 + 10x - 13 = 0 \Longrightarrow x = \dfrac{-5 \pm \sqrt{51}}{2}.$

(b) $x = -1.$

Problem Set IV   ■   *Applications*

1 $\begin{bmatrix} 5 & 8 & 6 \end{bmatrix} \cdot \begin{bmatrix} 32 \\ 55 \\ 25 \end{bmatrix} + \begin{bmatrix} 3 & 4 & 2 \end{bmatrix} \cdot \begin{bmatrix} 32 \\ 55 \\ 25 \end{bmatrix} = \begin{bmatrix} 8 & 12 & 8 \end{bmatrix} \cdot \begin{bmatrix} 32 \\ 55 \\ 25 \end{bmatrix} = 1116$ cents or \$11.16.

2(a) $\begin{bmatrix} 2500 & 1800 & 1200 \end{bmatrix} \cdot \begin{matrix} \text{wood} \quad \text{cloth} \quad \text{steel} \quad \text{hours} \\ \begin{bmatrix} 7 & 8 & 2 & 20 \\ 9 & 9 & 2 & 25 \\ 12 & 11 & 3 & 32 \end{bmatrix} \end{matrix} = \begin{bmatrix} 48100 & 49400 & 12200 & 133400 \end{bmatrix}.$ **(b)** $\begin{bmatrix} 129.50 \\ 159.75 \\ 204.25 \end{bmatrix}.$

CHAPTER 8, Section 8.7 (p. 319)

Problem Set I   ■   *Reading Comprehension*

1   False. A matrix is a number composed of an array of real numbers whereas a determinant represents one real number only.   **2**   False.   **3**   True. $A_{ij} = (-1)^{i+j}M_{ij}.$ **4**   True.
5   False.   **6**   True.   **7**   True.

Problem Set II ■ Skills Development

1(a) $M_{11} = \begin{vmatrix} 2 & 0 \\ 0 & -2 \end{vmatrix} = -4 \Longrightarrow A_{11} = M_{11} = -4.$ **(b)** $M_{21} = 2,\ A_{21} = -2.$ **(d)** $M_{31} = \begin{vmatrix} -1 & 2 \\ 2 & 0 \end{vmatrix} =$

$-4 \Longrightarrow A_{31} = M_{31} = -4.$ **(e)** $M_{32} = -6,\ A_{32} = 6.$ **2(a)** $M_{21} = \begin{vmatrix} -1 & 0 & 3 \\ 3 & -1 & 2 \\ -1 & 1 & -1 \end{vmatrix} = \begin{vmatrix} -1 & 0 & 0 \\ 3 & -1 & 11 \\ -1 & 1 & -4 \end{vmatrix} =$

$-1 \begin{vmatrix} -1 & 11 \\ 1 & -4 \end{vmatrix} = 7 \Longrightarrow A_{21} = -M_{21} = -7.$ **(b)** $M_{22} = 1,\ A_{22} = 1.$ **(d)** $M_{43} = \begin{vmatrix} 2 & -1 & 3 \\ -2 & 1 & 0 \\ 0 & 3 & 2 \end{vmatrix} =$

$\begin{vmatrix} 2 & -1 & 3 \\ 0 & 0 & 3 \\ 0 & 3 & 2 \end{vmatrix} = 2 \begin{vmatrix} 0 & 3 \\ 3 & 2 \end{vmatrix} = -18 \Longrightarrow A_{43} = -M_{43} = 18.$ **(e)** $M_{23} = -13,\ A_{23} = 13.$ **4** Column 2 is all

zeros. **5** Row 1 = Row 3. **7** Column 3 $= -4 \cdot$ (Column 1). **8** Row 2 $= -3 \cdot$ (Row 1).
10 Factor a 3 from each element in column 2 of the left matrix. **11** 36. **13** $-22.$ **14** $-1.$

16 $\begin{vmatrix} 2 & 4 & -6 \\ 3 & 1 & -2 \\ -4 & 2 & 0 \end{vmatrix} = \begin{vmatrix} 10 & 4 & -6 \\ 5 & 1 & -2 \\ 0 & 2 & 0 \end{vmatrix} = -2 \begin{vmatrix} 10 & -6 \\ 5 & -2 \end{vmatrix} = -20.$ **17** 0. **19** $\begin{vmatrix} 3 & 2 & 1 & 0 \\ -2 & 3 & 0 & 1 \\ 0 & -3 & 1 & 2 \\ 1 & 0 & 3 & 2 \end{vmatrix} =$

$\begin{vmatrix} 0 & 2 & -8 & -6 \\ 0 & 3 & 6 & 5 \\ 0 & -3 & 1 & 2 \\ 1 & 0 & 3 & 2 \end{vmatrix} = -1 \cdot \begin{vmatrix} 2 & -8 & -6 \\ 3 & 6 & 5 \\ -3 & 1 & 2 \end{vmatrix} = - \begin{vmatrix} 2 & 0 & 0 \\ 3 & 18 & 14 \\ -3 & -11 & -7 \end{vmatrix} = -2 \cdot \begin{vmatrix} 18 & 14 \\ 11 & 7 \end{vmatrix} = -4 \cdot \begin{vmatrix} 9 & 7 \\ 11 & 7 \end{vmatrix} =$

$-28 \cdot \begin{vmatrix} 9 & 1 \\ 11 & 1 \end{vmatrix} = -28(9 - 11) = 56.$ **20** $x = 2.$ **22** $x \begin{vmatrix} 4 & 8 \\ -2 & 7 \end{vmatrix} = 0 \Longrightarrow x(44) = 0 \Longrightarrow x = 0.$

23 $x = -1.$

Problem Set III ■ Theoretical Developments

1 For a 3×3 determinant, evaluate three 2×2 determinants. For a 4×4, evaluate four 3×3 determinants; hence, twelve $(4 \cdot 3)$ 2×2 determinants. For a 5×5, evaluate five 4×4 determinants; hence, sixty $(5 \cdot 4 \cdot 3)$ 2×2 determinants. In general, for an $n \times n$ determinant, evaluate $\dfrac{n!}{2}$
2×2 determinants. **2(a)** Expanding the expressions of the left-hand and right-hand sides both result in $a_{11}a_{22}a_{33} + a_{12}a_{23}a_{31} + a_{13}a_{21}a_{32} - a_{13}a_{22}a_{31} - a_{11}a_{23}a_{32} - a_{12}a_{21}a_{33}.$ **4** $|A| \cdot |B| =$

$\begin{vmatrix} a_{11} & a_{12} \\ a_{21} & a_{22} \end{vmatrix} \cdot \begin{vmatrix} b_{11} & b_{12} \\ b_{21} & b_{22} \end{vmatrix} = (a_{11}a_{22} - a_{12}a_{21})(b_{11}b_{22} - b_{12}b_{21}).$ $|A \cdot B| = \begin{vmatrix} a_{11}b_{11} + a_{12}b_{21} & a_{11}b_{12} + a_{12}b_{22} \\ a_{21}b_{11} + a_{22}b_{21} & a_{21}b_{12} + a_{22}b_{22} \end{vmatrix} =$

$(a_{11}b_{11} + a_{12}b_{21})(a_{21}b_{12} + a_{22}b_{22}) - (a_{11}b_{12} + a_{12}b_{22})(a_{21}b_{11} + a_{22}b_{21}) = (a_{11}a_{22} - a_{12}a_{21})(b_{11}b_{22} - b_{12}b_{21}).$

Thus, $|A \cdot B| = |A| \cdot |B|.$ **7** $\begin{vmatrix} 1 & a & a^2 \\ 1 & b & b^2 \\ 1 & c & c^2 \end{vmatrix} = \begin{vmatrix} 1 & a & a^2 \\ 0 & b-a & b^2-a^2 \\ 0 & c-a & c^2-a^2 \end{vmatrix} = \begin{vmatrix} 1 & a & a^2 \\ 0 & 1 & b+a \\ 0 & 1 & c+a \end{vmatrix}(b-a)(c-a) =$

$$\begin{vmatrix} 1 & b+a \\ 1 & c+a \end{vmatrix} (b-a)(c-a) = [c+a-(b-a)](b-a)(c-a) = (c-b)(b-a)(c-a) =$$

$(a-b)(b-c)(c-a)$.   **10**   T10.6.5. *Proof:* Expanding by the ith row, we obtain

$$\begin{vmatrix} a_{11} & a_{12} & \cdots & a_{1n} \\ a_{21} & a_{22} & & a_{2n} \\ \vdots & & & \\ b_{i1}+c_{i1} & b_{i2}+c_{i2} & \cdots & b_{in}+c_{in} \\ \vdots & & & \\ a_{n1} & a_{n2} & \cdots & a_{nn} \end{vmatrix} = (b_{i1}+c_{i1})A_{i1} + (b_{i2}+c_{i2})A_{i2} + \cdots + (b_{in}+c_{in})A_{in} =$$

$$(b_{i1}A_{i1}+b_{i2}A_{i2}+\cdots+b_{in}A_{in}) + (c_{i1}A_{i1}+c_{i2}A_{i2}+\cdots+c_{in}A_{in}) = \begin{vmatrix} a_{11} & a_{12} & \cdots & a_{1n} \\ a_{21} & a_{22} & \cdots & a_{2n} \\ \vdots & \vdots & & \vdots \\ b_{i1} & b_{i2} & \cdots & b_{in} \\ \vdots & & & \\ a_{n1} & a_{n2} & \cdots & a_{nn} \end{vmatrix} + \begin{vmatrix} a_{11} & a_{12} & \cdots & a_{1n} \\ a_{21} & a_{22} & \cdots & a_{2n} \\ \vdots & & & \\ c_{i1} & c_{i2} & \cdots & c_{in} \\ \vdots & & & \\ a_{n1} & a_{n2} & \cdots & a_{nn} \end{vmatrix}$$

CHAPTER 8, Section 8.10 (p. 332)

Problem Set I ■ *Reading Comprehension*

1 True.   **2** True.   **3** False. To obtain the transpose of a matrix, interchange each row with the corresponding column.   **4** False. The transpose of a 3×4 matrix is a 4×3 matrix. **5** False. For a 1×1 determinant of -2 the notation $|-2|$ is not used since it would be confused with absolute value. The determinant of -2 is -2.   **6** False. $A^{-1} = \dfrac{1}{|A|}(A_{ij})^t$.   **7** True. **8** False. One and only one solution exists in this case.

Problem Set II ■ *Skills Development*

1 $A = \begin{bmatrix} -1 & 2 \\ 1 & 4 \end{bmatrix} \Longrightarrow A^{-1} = \dfrac{1}{-6}\begin{bmatrix} 4 & -2 \\ -1 & -1 \end{bmatrix} = \begin{bmatrix} -2/3 & 1/3 \\ 1/6 & 1/6 \end{bmatrix} \Longrightarrow (A^{-1})^{-1} = \dfrac{1}{-1/6}\begin{bmatrix} 1/6 & -1/3 \\ -1/6 & -2/3 \end{bmatrix} =$

$\begin{bmatrix} -1 & 2 \\ 1 & 4 \end{bmatrix} \Longrightarrow A = (A^{-1})^{-1}.$

Problem Set III ■ *Theoretical Developments*

1 The element in the ith column and jth row of $(AB)^t$ is $a_{i1}b_{1j} + a_{i2}b_{2j} + \cdots + a_{in}b_{nj}$. The jth row of B^t is $[b_{1j}b_{2j} \cdots b_{nj}]$, and the ith column of A^t is $\begin{bmatrix} a_{i1} \\ a_{i2} \\ \vdots \\ a_{in} \end{bmatrix}$. Therefore, the element in the ith column

and jth row of $B^t A^t$ is also $a_{i1}b_{1j} + a_{i2}b_{2j} + \cdots + a_{in}b_{nj}$.

4 $A \cdot A^{-1} = I \Longrightarrow [A \cdot A^{-1}] \cdot (A^{-1})^{-1} = I \cdot (A^{-1})^{-1} \Longrightarrow A = (A^{-1})^{-1}$. **7** $AB = I \Longrightarrow$
$A^{-1}(AB) = A^{-1}I \Longrightarrow B = A^{-1}$. $BA = I \Longrightarrow (BA)A^{-1} = IA^{-1} \Longrightarrow B = A^{-1}$. Thus, $AB = BA = I$
$\Longrightarrow B = A^{-1}$.

Problem Set IV ■ Applications

1 $(I - U)P = A \Longrightarrow P = (I - U)^{-1}A$. $I - U = \begin{bmatrix} 1 & 0 & -1/8 \\ -2 & 9/10 & -1/5 \\ -1 & -1/5 & 9/10 \end{bmatrix} \Longrightarrow (I - U)^{-1} = \begin{bmatrix} \frac{308}{243} & \frac{10}{243} & \frac{45}{243} \\ \frac{800}{243} & \frac{310}{243} & \frac{180}{243} \\ \frac{520}{243} & \frac{80}{243} & \frac{360}{243} \end{bmatrix}$.

Hence, $P = (I - U)^{-1}\begin{bmatrix} 1000 \\ 300 \\ 500 \end{bmatrix} = \begin{bmatrix} 1372.4 \\ 4045.3 \\ 2979.4 \end{bmatrix}$. **2(a)** ATTACK $\Longleftrightarrow$ 1 20 20 1 3 11 $\Longrightarrow \begin{bmatrix} 1 & 1 \\ 20 & 3 \\ 20 & 11 \end{bmatrix} = $

message matrix. $\begin{bmatrix} 1 & 2 & -1 \\ 3 & 0 & 1 \\ 2 & -1 & 1 \end{bmatrix} \cdot \begin{bmatrix} 1 & 1 \\ 20 & 3 \\ 20 & 11 \end{bmatrix} = \begin{bmatrix} 21 & -4 \\ 23 & 14 \\ 2 & 10 \end{bmatrix}$. **(b)** $\begin{bmatrix} 20 & 6 & -3 \\ 22 & 30 & 65 \\ 4 & 17 & 47 \end{bmatrix}$. **3(a)** $M = \begin{bmatrix} 1 & 2 & -1 \\ 3 & 0 & 1 \\ 2 & -1 & 1 \end{bmatrix}$

Mixer   Message   Coded   Coded
Message   Message

$\Longrightarrow M^{-1} = \frac{1}{2} \cdot \begin{bmatrix} 1 & -1 & 2 \\ -1 & 3 & -4 \\ -3 & 5 & -6 \end{bmatrix} \Longrightarrow \frac{1}{2} \cdot \begin{bmatrix} 1 & -1 & 2 \\ -1 & 3 & -4 \\ -3 & 5 & -6 \end{bmatrix} \cdot \begin{bmatrix} 14 & 23 \\ 44 & 35 \\ 22 & 11 \end{bmatrix} = \frac{1}{2} \cdot \begin{bmatrix} 14 & 10 \\ 30 & 38 \\ 46 & 40 \end{bmatrix} = \begin{bmatrix} 7 & 5 \\ 15 & 19 \\ 23 & 20 \end{bmatrix} \Longleftrightarrow$

$\begin{bmatrix} G & E \\ O & S \\ W & T \end{bmatrix} \Longrightarrow$ GO WEST. **(b)** WE WON.

CHAPTER 9, Section 9.3 (p. 348)

Problem Set I ■ Reading Comprehension

1 False. We must also show $P(1) = Q(1)$. **2** False. We cannot state the general
result with any degree of certainty based on any finite number of specific examples.
3 True. $[P(k + 1) \neq Q(k + 1) \Longrightarrow P(k) \neq Q(k)]$ is the contrapositive of the statement
$[P(k) = Q(k) \Longrightarrow P(k + 1) = Q(+1)]$ and is therefore logically equivalent to this statement.
4 False. $n! = n(n - 1)(n - 2) \cdots 3 \cdot 2 \cdot 1$.

Problem Set II ■ Skills Development

1 $1 + 3 + 5 + \cdots + (2n - 1) = n^2$. Proof: (a) $P(1) = Q(1) \Longleftrightarrow 1 = 1^2$. (b) $[P(k) = Q(k) \Longrightarrow$
$P(k + 1) = Q(k + 1)] \Longleftrightarrow [P(k) = 1 + 3 + 5 + \cdots + (2k - 1) = k^2 = Q(k) \Longrightarrow$ (Adding $(2k + 1)$
to both sides gives) $P(k + 1) = 1 + 3 + 5 + \cdots + (2k - 1) + (2k + 1) = k^2 + (2k + 1) = (k + 1)^2 = $

$Q(k + 1) \Longrightarrow P(k + 1) = Q(k + 1)]$ QED. **4** $1^2 + 2^2 + 3^2 + \cdots + n^2 = \dfrac{n(n + 1)(2n + 1)}{6}$.

Proof: (a) $P(1) = 1^2 = \dfrac{1(1+1)(2+1)}{6} = 1 = Q(1)$.   (b) $[P(k) = Q(k) \Longrightarrow P(k+1) =$

$Q(k+1)] \Longleftrightarrow \left[P(k) = 1^2 + 2^2 + 3^2 + \cdots + k^2 = \dfrac{k(k+1)(2k+1)}{6} = Q(k) \Longrightarrow \text{(Adding}\right.$

$(k+1)^2$ to both sides of this equation) $P(k+1) = 1^2 + 2^2 + 3^2 + \cdots + k^2 + (k+1)^2 =$

$\dfrac{k(k+1)(2k+1)}{6} + (k+1)^2 = \dfrac{k(k+1)(2k+1) + 6(k+1)^2}{6} = \dfrac{(k+1)(2k^2 + k + 6k + 6)}{6} =$

$\dfrac{(k+1)(k+2)(2k+3)}{6} = Q(k+1) \Longrightarrow P(k+1) = Q(k+1)\Big]$ QED.   **7** $1^3 + 2^3 + 3^3 + \cdots + n^3 =$

$\dfrac{n^2(n+1)^2}{4}$. *Proof:* (a) $P(1) = Q(1) \Longleftrightarrow 1^3 = \dfrac{1^2(1+1)^2}{4} = 1$.   (b) $[P(k) = Q(k) \Longrightarrow P(k+1) =$

$Q(k+1)] \Longleftrightarrow \left[P(k) = 1^3 + 2^3 + 3^3 + \cdots + k^3 = \dfrac{k^2(k+1)^2}{4} = Q(k) \Longrightarrow \text{(Adding } (k+1)^3 \text{ to}\right.$

both sides of this equation) $P(k+1) = 1^3 + 2^3 + 3^3 + \cdots + k^3 + (k+1)^3 = \dfrac{k^2(k+1)^2}{4} + (k+1)^3 =$

$(k+1)^2\left[\dfrac{k^2 + 4(k+1)}{4}\right] = \dfrac{(k+1)^2(k+2)^2}{4} = Q(k+1) \Longrightarrow P(k+1) = Q(k+1)\Big]$ QED.

10 $a + ar + ar^2 + \cdots + ar^{n-1} = \dfrac{a(1 - r^n)}{1 - r}; r \neq 1$. *Proof:* (a) $P(1) = Q(1) \Longleftrightarrow a = \dfrac{a(1 - r)}{1 - r} = a$.

(b) $[P(k) = Q(k) \Longrightarrow P(k+1) = Q(k+1)] \Longleftrightarrow \left[P(k) = a + ar + ar^2 + \cdots + ar^{k-1} = \dfrac{a(1 - r^k)}{1 - r} = \right.$

$Q(k) \Longrightarrow$ (Adding ar^k to both sides of this equation) $P(k+1) = a + ar + ar^2 + \cdots + ar^{k-1} + ar^k =$

$\dfrac{a(1 - r^k)}{1 - r} + ar^k = \dfrac{a(1 - r^k) + (1 - r)ar^k}{1 - r} = \dfrac{a - ar^k + ar^k - ar^{k+1}}{1 - r} = \dfrac{a(1 - r^{k+1})}{1 - r} = Q(k+1) \Longrightarrow$

$P(k+1) = Q(k+1)\Big]$ QED.   **13** $3^{n+7} < (n+7)!$. *Proof:* (a) $P(1) < Q(1) \Longleftrightarrow P(1) =$

$6561 = 3^8 < 8! = 40320 = Q(1)$.   (b) $[P(k) < Q(k) \Longrightarrow P(k+1) < Q(k+1)] \Longleftrightarrow$

$[P(k) = 3^{k+7} < (k+7)! = Q(k) \Longrightarrow$ (Multiplying both sides of the inequality by 3) $P(k+1) =$

$3^{k+7} \cdot 3 < 3 \cdot (k+7)! < (k+8)(k+7)! = (k+8)! = Q(k+1) \Longrightarrow P(k+1) < Q(k+1)]$ QED.

16 $(a - b)$ is a factor of $a^n - b^n$. *Proof:* (a) For $n = 1$, we have $(a - b)$ is a factor of $a - b$,

which is true.   (b) $[a - b$ is a factor $a^k - b^k \Longrightarrow a - b$ is a factor of $a^{k+1} - b^{k+1}] \Longleftrightarrow [(a - b)$ is

a factor of $a^k - b^k \Longrightarrow a^k - b^k = (a - b) P(a, b)$, where $P(a, b)$ is a polynomial in a and $b \Longrightarrow$

$a^{k+1} - b^{k+1} = (a^{k+1} - ab^k) + (ab^k - b^{k+1}) = a(a^k - b^k) + b^k(a - b) = a(a - b) P(a, b) + b^k(a - b) =$

$(a - b)(aP(a, b) + b^k) \Longrightarrow a - b$ is a factor of $a^{k+1} - b^{k+1}]$ QED.   **19** 2 is a factor of $n(n + 1)$

for each $n \in \mathbb{N}$. *Proof:* (a) 2 is a factor of $1(1 + 1)$.   (b) [2 a factor of $k(k + 1) \Longrightarrow k(k + 1) =$

$2 \cdot P(k)$, where $P(k)$ is a polynomial in $k \Longrightarrow (k + 1)(k + 2) = k(k + 1) + 2 \cdot (k + 1) =$

$2 \cdot P(k) + 2(k + 1) = 2[P(k) + (k + 1)] \Longrightarrow 2$ is a factor of $(k + 1)(k + 2)]$ QED.

CHAPTER 9, Section 9.5 (p. 357)

Problem Set I ■ **Reading Comprehension**

1 True.   **2** True.   **3** True.   **4** False. $\displaystyle\sum_{k=1}^{n} f_k = f_1 + f_2 + f_3 + \cdots + f_n$.   **5** False. A series is the sum of the terms of a sequence.   **6** False. A sequence of partial sums is a sequence of which each term is a finite series. That is, it may be written as $f_1, f_1 + f_2, f_1 + f_2 + f_3, \cdots,$ $f_1 + f_2 + f_3 + \cdots + f_n, \cdots.$ **7** True.   **8** True.

Problem Set II   ■   Skills Development

1(a) $f_1 = 2(-2) = -4, f_2 = 2(-1) = -2, f_3 = 2(0) = 0, f_4 = 2(1) = 2, f_5 = 2(2) = 4.$ **(b)** $-1/2, 1/3,$

$-1/4, 1/5, -1/6.$ **(d)** $f_1 = \dfrac{4}{1-1}$, which is undefined; $f_2 = \dfrac{4}{4-1} = \dfrac{4}{3}; f_3 = \dfrac{4}{9-1} = \dfrac{1}{2}; f_4 = \dfrac{4}{16-1} = \dfrac{4}{15};$

$f_5 = \dfrac{4}{25-1} = \dfrac{1}{6}.$ **(e)** $0, 1, 3, 6, 10.$ **2(a)** $-1, 1, 3, 5, \cdots, 2n - 3, \cdots.$ **(b)** $5n + 1.$

(d) $-1, 2, 8, 17, \cdots, [-1 + (1 + 2 + 3 + \cdots + n)3], \cdots; n = 0, 1, 2, \cdots.$ **(e)** $\dfrac{n-1}{n^2}.$

3	d	$L = a + (n-1)d$	$S_n = \dfrac{n}{2}[2a + (n-1)d] = \dfrac{n}{2}(a + L)$
(a)	3	$L = 2 + (13) \cdot 3 = 41$	$S = \dfrac{14}{2}(2 + 41) = 301$
(b)	-2	-19	-96
(d)	$\dfrac{4}{3}$	$L = \dfrac{1}{3} + 14\left(\dfrac{4}{3}\right) = \dfrac{57}{3} = 19$	$S = \dfrac{15}{2}\left(\dfrac{1}{3} + \dfrac{57}{3}\right) = 145$
(e)	$-\dfrac{3}{4}$	$-\dfrac{19}{4}$	$-\dfrac{55}{4}$

4	r	$L = ar^{n-1}$	$S_n = \dfrac{a - ar^n}{1 - r} = \dfrac{a - rL}{1 - r}$
(a)	2	$L = 3(2^9) = 1536$	$S = \dfrac{3 - 2(1536)}{-1} = 3069$
(b)	0.3	0.00002187	0.142848
(d)	0.01	$L = 20(10^{-2})^{19} = 2 \cdot 10^{-37}$	$S = \dfrac{20}{0.99} = 20.202020$
(e)	$\dfrac{1}{3}$	0.000001882	13.499999

5(a) $S = \dfrac{n[2a + (n-1)d]}{2} = \dfrac{40[6 + (39)2]}{2} = 1680.$ **(b)** 4100. **7** $L = a + (n-1)d \Longrightarrow 248 =$

$12 + (n-1)4 \Longrightarrow n = 60.$ **8** $n = 34.$ **9(a)** $S = \dfrac{n}{2}[2a + (n-1)d] \Longrightarrow S_{15} = \dfrac{15}{2}[6 + (14)3] =$

360. **(b)** 600. **10(a)** $L = a + (n-1)d = 1 + (29)(5) = 146.$ **(b)** 113. **11(a)** $L = ar^{n-1} =$

$1(3)^9 = 19683.$ **(b)** $2 \cdot 10^{-9}.$

12(a) $\displaystyle\sum_{k=1}^{20} (3k - 2) = 1 + 4 + 7 + \cdots + 58 \Longrightarrow S = \dfrac{n}{2}[2a + (n-1)d] = \dfrac{20}{2}[2 + 19(3)] = 590.$

(b) 266. **(d)** $\displaystyle\sum_{k=1}^{22} \left(\dfrac{1}{3}k - 2\right) = -\dfrac{5}{3} - \dfrac{4}{3} - \dfrac{3}{3} + \cdots + \dfrac{16}{3} \Longrightarrow S_n = \dfrac{n}{2}[2a + (n-1)d] =$

$\dfrac{22}{2}[-10/3 + (21)(1/3)] = \dfrac{121}{3}.$ **(e)** $-144.$

Problem Set III ■ Theoretical Developments

1 $\displaystyle\sum_{k=1}^{n} f_k + \sum_{k=1}^{n} g_k = (f_1 + f_2 + \cdots + f_n) + (g_1 + g_2 + \cdots + g_n) = (f_1 + g_1) + (f_2 + g_2) + \cdots + (f_n + g_n) =$

$\displaystyle\sum_{k=1}^{n} [f_k + g_k].$ **4** $\displaystyle\sum_{k=1}^{n} f_k = f_1 + f_2 + \cdots + f_n = \sum_{k=0}^{n-1} f_{k+1}.$

Problem Set IV ■ Applications

1 $a = \$500, d = \$200,$ and $n = 15 \Longrightarrow S = \dfrac{n}{2}[2a + (n-1)d] = \dfrac{15}{2}[1000 + (14)(200)] =$

$\$28,500.$ **2** 259.44 ft. **4** The first interest payment is $0.10(1000) = 100,$ the second
is $0.10(950) = 95,$ and so forth. After 20 payments the total interest is given by $S_{20} =$
$100 + [100 - 5] + [100 - (2)5] + \cdots + [100 - (19)5] \Longrightarrow a = 100, n = 20,$ and $d = -5 \Longrightarrow$
$S_{20} = \dfrac{20}{2}[2 \cdot 100 + (20 - 1)(-5)] = \$1050.00.$ **5** $(2^{23})a = 8388608a,$ where a is the amount
initially present.

CHAPTER 9, Section 9.7 (p. 363)

Problem Set I ■ Reading Comprehension

1 False. The binomial theorem has been stated here for positive integral exponents only.
2 True. However, this is not an efficient procedure for large values of n. **3** False. The
$(r + 1)$st term is $\dfrac{n!}{(n-r)!r!}a^{n-r}b^r.$ **4** False. The procedure suggested is only effective when the
number can be expressed as a sum of two numbers, one of which is easily raised to positive integral
powers and the other of which is less than 1.

1 $\dfrac{12!}{8!4!} = 495.$ **2** 560.   **4** $\dfrac{n!}{(n-2)!2!} = \dfrac{(n)(n-1)(n-2)!}{(n-2)!2!} = \dfrac{n(n-1)}{2}.$

5 $\dfrac{n(n-1) \cdots (n-r+1)}{r(r-1)(r-2) \cdots 3 \cdot 2 \cdot 1}.$ **7** $\dfrac{(2n+3)!}{(2n+1)!} = \dfrac{(2n+3)(2n+2)(2n+1)!}{(2n+1)!} = (2n+3)(2n+2).$

8 $\dfrac{n}{n+2}.$ **10** $(a-3b)^6 = a^6 - 6a^5(3b) + 15a^4(3b)^2 - 20a^3(3b)^3 + \cdots =$

$a^6 - 18a^5b + 135a^4b^2 - 540a^3b^3 + \cdots.$ **11** $x^8 + 16x^7y^2 + 112x^6y^4 + 448x^5y^6 + \cdots.$

13 $(a-3)^9 = a^9 + 9(a^8)(-3) + \dfrac{9!}{7!2!}a^7(-3)^2 + \dfrac{9!}{6!3!}a^6(-3)^3 + \cdots = a^9 - 27a^8 + 324a^7 - 2268a^6 + \cdots.$

14 $x^{12} - 6x^{11}y + \dfrac{33}{2}x^{10}y^2 - \dfrac{55}{2}x^9y^3 + \cdots.$ **16** 7th term is $\dfrac{11!}{6!5!}a^5\left(\dfrac{-b}{2}\right)^6 = \dfrac{231}{32}a^5b^6.$

17 $6435x^{15}y^{16}.$ **19** Each term is of the form $\dfrac{10!}{(10-r)!r!}\left(\dfrac{1}{x}\right)^{10-r}(x^2y)^r.$ Thus, the term

containing x^2 is determined by $\left(\dfrac{1}{x}\right)^{10-r}(x^2)^r = x^2 \Longleftrightarrow x^{2r+r-10} = x^2 \Longrightarrow 3r - 10 = 2 \Longrightarrow$

$r = 4.$ Therefore, the term is $\dfrac{10!}{6!4!}\left(\dfrac{1}{x}\right)^6(x^2y)^4 = 210x^2y^4$ and is the 5th term in the expansion.

20 The term is $924x^{18}$ and is the 7th term in the expansion.   **22** $(1 + 0.01)^{10} =$

$1 + 10(0.01) + \dfrac{10!}{8!2!}(0.01)^2 + \dfrac{10!}{7!3!}(0.01)^3 + \cdots = 1 + 0.1 + 0.0045 + 0.00012 + \cdots =$

1.105 (to the nearest thousandth).   **23** 0.851.   **25(a)** $(1 - 0.01)^{1/2} =$

$1^{1/2} + \dfrac{1/2}{1!}(1)^{-1/2}(-0.01)^1 + \dfrac{(1/2)(-1/2)}{2!}(1)^{-3/2}(-0.01)^2 + \cdots = 1 - 0.005 - 0.0000125,$ which,

to the nearest thousandth, is 0.995.   **(b)** 1.01.

Problem Set III ■ Applications

1(a) $P = 1000(1 + 0.05)^4 = 1000[1 + 4(0.05) + 6(0.05)^2 + 4(0.05)^3 + (0.05)^4] = \$1215.51.$

(b) \$8144.47.   **2(a)** $S = \dfrac{100(1+0.05)^{30} - 1}{0.05} = \dfrac{100\left[1 + 30(0.05) + \dfrac{30!}{28!2!}(0.05)^2 + \dfrac{30!}{27!3!}(0.05)^3 + \cdots\right] - 1}{0.05}$

$= \$8623.88.$ **(b)** \$29,677.27.

CHAPTER 10, Section 10.3 (p. 373)

Problem Set I ■ Reading Comprehension

1 True. The fixed point is the focus of the parabola and the fixed line is the directrix.   **2** False. The focal chord is a line segment parallel to the directrix that passes through the focus. Its endpoints lie on the parabola and its length is $4p$.   **3**   False. $(x - h)^2 = -4p(y - k)$ is the equation for a cup-down parabola.   **4**   True. It is p units from the directrix and p units from the focus.

Problem Set II   ■   Skills Development

1 $V = (3, 6)$ and $F = (3, 8) \Longrightarrow$ cup-up and $p = 2 \Longrightarrow (x - 3)^2 = 8(y - 6)$.   **2** $(x + 2)^2 = -16(y - 4)$.   **4** $V = (4, -1)$ and $F = (6, -1) \Longrightarrow$ cup-right and $P = 2 \Longrightarrow (y + 1)^2 = 8(x - 4)$.
5 $(y + 5)^2 = 16(x - 2)$.   **7** $V = (-4, 1)$ and directrix $y = -4 \Longrightarrow$ cup-up and $P = 5 \Longrightarrow (x + 4)^2 = 20(y - 1)$.   **8** $(y + 5)^2 = 24(x + 2)$.   **10** $F = (2, 4)$ and directrix $x = 3 \Longrightarrow$ $V = \left(\frac{5}{2}, 4\right)$, $P = \frac{1}{2}$, and cup-left $\Longrightarrow (y - 4)^2 = -2\left(x - \frac{5}{2}\right)$.   **11** $(y + 3)^2 = -4x$.   **13** $V = (3, 2)$ and distance from $(5, 6)$ to $(5, -2)$ is 8, hence $8 = 4p \Longrightarrow p = 2$, cup-right, and $F = (5, 2) \Longrightarrow$ $(y - 2)^2 = 8(x - 3)$.   **14** $(x + 1)^2 = -12(y - 4)$.

16 $y^2 - 6y + 8x - 23 = 0 \Longleftrightarrow y^2 - 6y + 9 = -8x + 23 + 9 \Longleftrightarrow (y - 3)^2 = -8(x - 4) \Longrightarrow$ $V = (4, 3)$, $p = 2$, $F = (2, 3)$, opens left, and directrix $x = 6$.

17

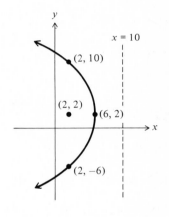

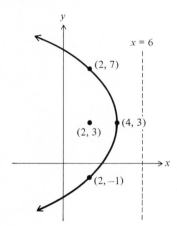

19 $y^2 + 4y + 32x - 92 = 0 \Longleftrightarrow y^2 + 4y + 4 = -32x + 92 + 4 \Longleftrightarrow$ $(y + 2)^2 = -32(x - 3) \Longrightarrow V = (3, -2)$, $p = 8$, opens left, $F = (-5, -2)$, and directrix $x = 11$.

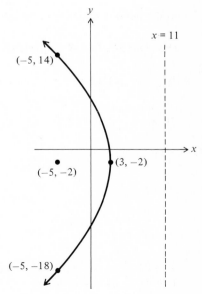

20

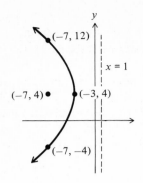

22 $x^2 + 4x + 4y - 12 = 0 \Longleftrightarrow x^2 + 4x + 4 = -4y + 12 + 4$
$\Longleftrightarrow (x + 2)^2 = -4(y - 4) \Longrightarrow V = (-2, 4)$, opens down,
$p = 1$, $F = (-2, 3)$, and directrix $y = 5$.

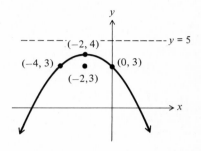

23

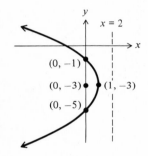

Problem Set III ■ *Theoretical Developments*

1(a) We have $d_1{}^2 = [x - (h + p)]^2$ and $d_2{}^2 = [x - (h - p)]^2 + (y - k)^2$. Setting $d_1{}^2 = d_2{}^2$ yields
$[x - (h + p)]^2 = [x - (h - p)]^2 + (y - k)^2 \Longleftrightarrow x^2 - 2(h + p)x + (h + p)^2 - [x^2 - 2(h - p)x + (h - p)^2] = (y - k)^2 \Longleftrightarrow -4p(x - h) = (y - k)^2$.

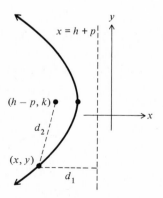

4 Parabolas that open up or open down are functions and may be written as $y = ax^2 + bx + c$.
They are two-to-one functions.

1 With the paraboloid on an axis system with a cross-section as described in the figure, the

equation is of the form $x^2 = 4py \Longrightarrow \dfrac{x^2}{4y} = p,\ y \neq 0$. Since $(8, 4)$ is a point on the curve, we have

$p = \dfrac{64}{4 \cdot 4} = 4$. Therefore, the focus is at $(0, 4)$ and hence is 4 inches above the vertex of the parabola.

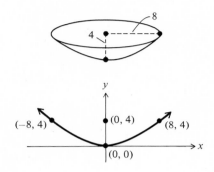

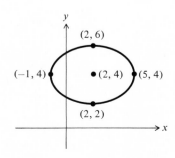

2 The focus is 9 feet above the vertex.

CHAPTER 10, Section 10.5 (p. 381)

Problem Set I ■ *Reading Comprehension*

1 False. An ellipse is the set of all points in a plane such that the sum of the distances from two fixed points called foci is a constant.   **2** True.   **3** False. $a^2 = b^2 + c^2$.   **4** True. This is the case where $a = b$.

Problem Set II ■ *Skills Development*

1 Center $(2, 4)$, $a = 3$, $b = 2$, and major axis horizontal $\Longrightarrow \dfrac{(x-2)^2}{9} + \dfrac{(y-4)^2}{4} = 1$; vertices

$(-1, 4)$, $(5, 4)$, $(2, 2)$, $(2, 6)$, and foci $(2 - \sqrt{5}, 4)$, $(2 + \sqrt{5}, 4)$.

2

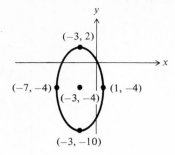

4 Center $(4, -2)$, $a = 10$, $b = 6$, and major axis horizontal $\Longrightarrow b^2 = a^2 - c^2 = 100 - 36 = 64 \Longrightarrow$

$b = 8 \Longrightarrow \dfrac{(x-4)^2}{100} + \dfrac{(y+2)^2}{64} = 1$.

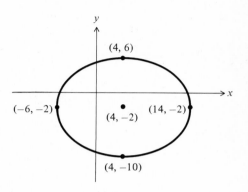

5

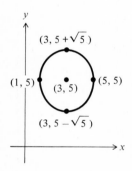

7 Vertices $(-4, -1)$, $(6, -1)$, and $F = (1 - \sqrt{21}, -1) \Longrightarrow$ major axis horizontal; $a = 5$, $F = (1 - \sqrt{21}, -1)$, and center at $(1, -1) \Longrightarrow c = \sqrt{21} \Longrightarrow b^2 = a^2 - c^2 =$

$25 - 21 = 4$. Thus, $\dfrac{(x-1)^2}{25} + \dfrac{(y+1)^2}{4} = 1$.

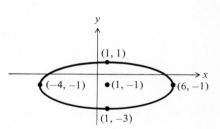

8

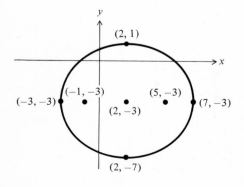

10 Vertices $(-4, 1)$, $(1, -3)$, $F = (-1, -3) \Longrightarrow$ major axis horizontal with $a - c = 2$ and

$b = 4 \Longrightarrow \left[a^2 - c^2 = b^2 \Longleftrightarrow (a-c)(a+c) = 16 \Longrightarrow a + c = \dfrac{16}{a-c} = \dfrac{16}{2} = 8 \right]$. Thus, $a + c = 8$ and

$a - c = 2 \Longrightarrow 2a = 10 \Longrightarrow a = 5 \Longrightarrow$ center $(-4, -3)$ and $\dfrac{(x+4)^2}{25} + \dfrac{(y+3)^2}{16} = 1$.

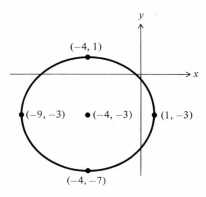

11 $\dfrac{(x-4)^2}{25} + \dfrac{(y-3)^2}{16} = 1$. **13** $9x^2 - 18x + 4y^2 + 16y - 11 = 0 \Longleftrightarrow 9x^2 - 18x + 4y^2 + 16y = 11$
$\Longleftrightarrow 9(x^2 - 2x + 1) + 4(y^2 + 4y + 4) = 11 + 9 + 16 \Longleftrightarrow 9(x-1)^2 + 4(y+2)^2 = 36 \Longleftrightarrow$
$\dfrac{(x-1)^2}{4} + \dfrac{(y+2)^2}{9} = 1$.

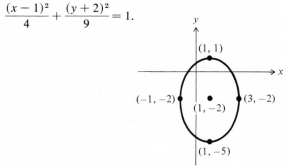

14 $\dfrac{(x+2)^2}{9} + \dfrac{(y+4)^2}{25} = 1$.

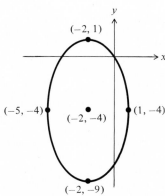

16 $4x^2 - 32x + 25y^2 - 300y + 864 = 0 \Longleftrightarrow$
$4(x^2 - 8x + 16) + 25(y^2 - 12y + 36) = -864 + 64 + 900$
$\Longleftrightarrow 4(x-4)^2 + 25(y-6)^2 = 100 \Longleftrightarrow$

$$\frac{(x-4)^2}{25} + \frac{(y-6)^2}{4} = 1.$$

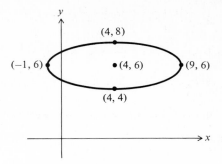

17 $\dfrac{x^2}{9} + \dfrac{(y-3)^2}{16} = 1.$

Problem Set III ■ *Applications*

1 $2a = 10{,}037 + 4{,}037 + 2(3{,}963) = 22{,}000 \Longrightarrow a = 11{,}000.$ $a - c = 4{,}037 + 3{,}963 \Longrightarrow$
$c = 11{,}000 - 8{,}000 = 3{,}000.$ $b^2 = a^2 - c^2 = (11{,}000)^2 - (3{,}000)^2 = 112{,}000{,}000.$ Therefore,

$$\frac{(x-3{,}000)^2}{(11{,}000)^2} + \frac{y^2}{112{,}000{,}000} = 1 \text{ is the path of the satellite with center of earth at center of}$$

coordinate system.   **2** $\dfrac{(x-15{,}624)^2}{(242{,}130)^2} + \dfrac{y^2}{583.83} = 1.$

CHAPTER 10, Section 10.7 (p. 387)

Problem Set I ■ *Reading Comprehension*

1 True.   **2** True.   **3** False. The minus sign on the y term indicates horizontal foci, and the minus sign on the x term indicates the foci lie on a vertical line.   **4** True. The length $2a$ is parallel to the line containing the foci and vertices.   **5** False. The cup up and down or cup left and right configuration is determined by the signs of the terms. (Refer to the "rule of thumb" stated in the answer to Problem 3 above.)

Problem Set II ■ *Skills Development*

1 Foci $(5, 4)$, $(-5, 4)$ and vertices $(2, 4)$, $(-2, 4) \Longrightarrow$ center $(0, 4)$; $c = 5$ and $a = 2$

$\Longrightarrow b^2 = c^2 - a^2 = 25 - 4 = 21 \Longrightarrow$ opens left and right and $\dfrac{x^2}{4} - \dfrac{(y-4)^2}{21} = 1.$

2 $\dfrac{-(x-5)^2}{9}+\dfrac{(y+1)^2}{16}=1.$ **4** $C=(1,3),\ F=(-4,3),$ and $V=(4,3)\Longrightarrow c=5$ and $a=3\Longrightarrow$

$b^2=c^2-a^2=25-9=16\Longrightarrow b=4.$ Thus, $\dfrac{(x-1)^2}{9}-\dfrac{(y-3)^2}{16}=1.$ **5** $\dfrac{(y+1)^2}{16}-\dfrac{(x+4)^2}{4}=1.$

7 $9x^2-36x-16y^2+128y-364=0\Longleftrightarrow 9(x^2-4x+4)-16(y^2-8y+16)=364+36-256\Longleftrightarrow$

$9(x-2)^2-16(y-4)^2=144\Longleftrightarrow\dfrac{(x-2)^2}{16}-\dfrac{(y-4)^2}{9}=1\Longrightarrow C=(2,4),\ a=4,\ b=3\Longrightarrow$

$c^2=a^2+b^2=25\Longrightarrow c=5\Longrightarrow$ foci $(7,4)$ and $(-3,4).$

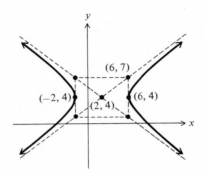

8 $\dfrac{(y+3)^2}{25}-\dfrac{(x-1)^2}{144}=1.$

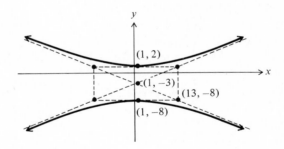

10 $4y^2-32y-x^2+6x=9\Longleftrightarrow 4(y^2-8y+16)-(x^2-6x+9)=9+64-9\Longleftrightarrow$

$\dfrac{(y-4)^2}{16}-\dfrac{(x-3)^2}{64}=1\Longrightarrow C=(3,4),\ a=4,\ b=8\Longrightarrow c^2=a^2+b^2=64+16=80\Longrightarrow$

$c=4\sqrt{5}.$

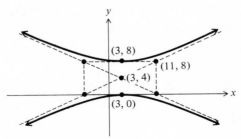

11 $\dfrac{(x-2)^2}{16} - \dfrac{(y+1)^2}{9} = 1.$

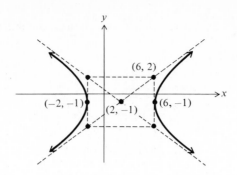

13 Foci $(-1,-3)$ and $(5,-3) \Longrightarrow$ center $(2,-3)$ and $c=3$. But $2a=4 \Longrightarrow a=2 \Longrightarrow$

$b^2 = c^2 - a^2 = 9 - 4 = 5$. Since the foci are horizontal, the equation is $\dfrac{(x-2)^2}{4} - \dfrac{(y+3)^2}{5} = 1.$

14 $\dfrac{-(x-5)^2}{7} + \dfrac{(y-4)^2}{9} = 1.$

Problem Set III ■ *Theoretical Developments*

1 $\dfrac{(x-h)^2}{a^2} - \dfrac{(y-k)^2}{b^2} = 1 \Longleftrightarrow y - k = \pm b\sqrt{\dfrac{(x-h)^2}{a^2} - 1} \Longleftrightarrow y - k = \pm \dfrac{b}{a}(x-h)\sqrt{1 - \dfrac{a^2}{(x-h)^2}}.$

But as x gets large, $\sqrt{1 - \dfrac{a^2}{(x-h)^2}}$ approaches 1. Thus, $y - k = \pm \dfrac{b}{a}(x-h)$ defines the asymptotes.

4 $\dfrac{x^2}{a^2} - \dfrac{y^2}{b^2} < 1.$ Graph $\dfrac{x^2}{a^2} - \dfrac{y^2}{b^2} = 1$ (the dotted curve in the figure). Then, by testing points, you find

the solution to the inequality to be the shaded area indicated in the figure.

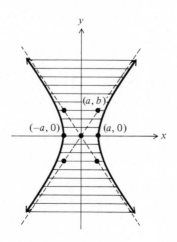

5(a) $\left|\dfrac{x^2}{2^2} - \dfrac{y^2}{3^2}\right| = 1 \Longleftrightarrow \left\{\dfrac{x^2}{4} - \dfrac{y^2}{9} = 1 \text{ or } \dfrac{-x^2}{4} + \dfrac{y^2}{9} = 1\right\}$. Thus, we obtain both vertical and horizontal

cups, as indicated in the figure.

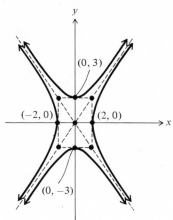

Problem Set IV ■ Applications

1 $F(5 - \sqrt{10}, 5)$ and $F(5 + \sqrt{10}, 5) \Longrightarrow$ center $(5, 5)$ and $c = \sqrt{10}$. To find a, we have $\dfrac{150}{25} =$

$6 \Longrightarrow 2a = 6 \Longrightarrow a = 3.$ $a = 3$ and $c = \sqrt{10} \Longrightarrow b^2 = 10 - 9 = 1 \Longrightarrow b = 1.$ Therefore the

equation of the hyperbola is $\dfrac{(x - 5)^2}{9} - (y - 5)^2 = 1.$ Using the second set of foci, we have

$(5 - \sqrt{5}, -10)$ and $(5 + \sqrt{5}, -10) \Longrightarrow$ center $(5, -10)$ and $c = \sqrt{5}$. To find a, we have $\dfrac{100}{25} =$

$4 \Longrightarrow 2a = 4 \Longrightarrow a = 2.$ $a = 2$ and $c = \sqrt{5} \Longrightarrow b^2 = 5 - 4 = 1 \Longrightarrow b = 1.$ Therefore the equation

is $\dfrac{(x - 5)^2}{4} - (y + 10)^2 = 1.$ Solving these two equations simultaneously, we obtain $y = -1$, or 35,

but we exclude the choice of 35 because it would not be physically feasible; hence, $y = -1$. Substituting this into one of the equations yields $x = 5 \pm \sqrt{333} \Longrightarrow x = 5 - \sqrt{333} \cong -13$ as the only possible choice for x. Thus, the coordinates of the ship would be $(-13, -1)$ or $(-325, -25)$ in actual miles.

CHAPTER 10, Section 10.9 (p. 390)

Problem Set I ■ Reading Comprehension

1 True.   **2** True. The equation becomes $x'^2 = 4(y')$.

1 $x^2 + 2x - 4y^2 + 8y - 7 = 0 \Longleftrightarrow x^2 + 2x + 1 - 4(y^2 - 2y + 1) = 7 + 1 - 4 \Longleftrightarrow$

$(x + 1)^2 - 4(y - 1)^2 = 4 \Longleftrightarrow \dfrac{(x + 1)^2}{4} - \dfrac{(y - 1)^2}{1} = 1.$ Let $x' = x + 1$, $y' = y - 1$. Then

$\dfrac{(x')^2}{4} - \dfrac{(y')^2}{1} = 1.$

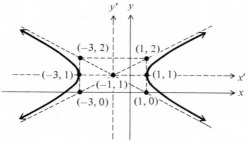

2 $(x')^2 + (y')^2 = 9$, where $x' = x - 1$ and $y' = y - 2$.

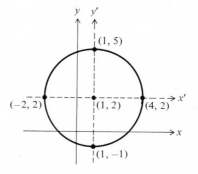

4 $-4x^2 + 9y^2 - 16x - 18y - 43 = 0 \Longleftrightarrow -4(x^2 + 4x + 4) + 9(y^2 - 2y + 1) = 43 - 16 + 9 \Longleftrightarrow$

$-4(x + 2)^2 + 9(y - 1)^2 = 36 \Longleftrightarrow -\dfrac{(x + 2)^2}{9} + \dfrac{(y - 1)^2}{4} = 1.$ Let $x' = x + 2$, $y' = y - 1$; then

$-\dfrac{(x')^2}{9} + \dfrac{(y')^2}{4} = 1.$

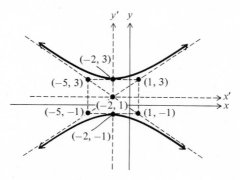

5 $\dfrac{(x')^2}{4} - \dfrac{(y')^2}{16} = 1$, where $x' = x + 3$ and $y' = y - 4$.

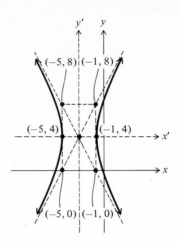

Index

A

a^0, 182
a^{-n}, 181
Abscissa, 126
Absolute value:
  in equations, 111–13
  in inequalities, 56–57
  of a real number, 55–56
Addition:
  associative law of:
    for complex numbers, 226
    for matrices, 292
    for real numbers, 65
  closure for, of real numbers, 65
  commutative law of:
    for complex numbers, 227
    for matrices, 292
    for real numbers, 66
  identity element for:
    of complex numbers, 227
    of matrices, 292
    of real numbers, 66
  law of:
    for equality, 70
    of complex numbers, 224
    of matrices, 291–93
    of quotients, 76
Additive inverse:
  law, 66
  of a complex number, 227
  of a matrix, 292
  of a real number, 66
  uniqueness of, 72
Algebraic expression, 95
Antecedent, 14
Antilogarithm, 204
Applied mathematics, 2
Arc, length of, 222
Arithmetic sequence (progression):
  common difference of, 354
  meaning of, 354
  nth term of, 354
  sum of n terms, 354–55
Associative law:
  of addition of complex numbers, 226
  of addition of matrices, 292
  of addition of real numbers, 65
  of multiplication of a matrix by a scalar, 295
  of multiplication of real numbers, 67
Asymptote(s):
  horizontal, 269–71
  oblique, 272–73
  of graph of a hyperbola, 272–74
  of graph of a rational function, 268ff

Asymptote(s): *cont.*
  vertical, 267–68
Axes:
  of a Cartesian coordinate system, 137
  rotation of, 390
  translation of, 388–390
Axioms, 62–63, 65ff
Axis of symmetry, 164, 371

B

Base of a logarithm, 198, 211–12
Base of a power, 180
Binomial:
  coefficient of $(r + 1)$st term in expansion of, 361
  expansion of, 361ff
  n factorial, 361
  Pascal's triangle, 360
  theorem, 360ff
Bounds, upper and lower for zeros, 259–260

C

Cartesian coordinate system, 137ff
  graph of an ordered pair on, 138
Characteristic of a logarithm, 201, 205
Circle, 379
Closure law:
  for addition of complex numbers, 226
  for addition of matrices, 292
  for addition of positive real numbers, 83
  for addition of real numbers, 65
  for multiplication of complex numbers, 226
  for multiplication of matrices, 298
  for multiplication of matrix by scalar, 295
  for multiplication of positive real numbers, 83
  for multiplication of real numbers, 66
  meaning of, 65
Coefficient, variation in sign of, of a polynomial, 258–59
Coefficient matrix of a system of linear equations, 328–29
Cofactor, of an element in a determinant, 307
Common logarithms, 198
Common ratio of a geometric sequence, 356
Commutative law:
  of addition of complex numbers, 227
  of addition of matrices, 292
  of addition of real numbers, 66
  of multiplication of complex numbers, 227
  of multiplication of real numbers, 67
Complement of a set, 38

Multiplication: *cont.*
  of matrices, 296
Multiplicative identity, 68, 228, 299
Multiplicative inverse, 68, 228, 324

N

Natural number, 46–48
Negation symbol, 17
Negative:
  of a complex number, 227
  of a matrix, 292
  of a real number, 66
Negative number(s):
  product of a, with a positive, 84
  product of two, 84
  real, 53
n factorial, 361
Non-Euclidean geometry, 3
Nonsingular matrix, 324
Null set, 33
Number(s):
  complex, 220
  matrix, 289
  real, 46–48
Number line, 52–54
Numerals:
  meaning of, 48
  modern Arabic, 48

O

Oblique asymptote, 272–73
One-to-one correspondence, 52, 82, 133
Open equations, 93–94
Open sentences, 92ff
Operations, on sets, 36ff
Orbits, elliptic, 380
Order:
  axioms for real numbers, 83
  properties of, 83–85
Ordered field, real numbers as, 81–83
Ordered pair(s):
  as complex numbers, 220–22
  components of, 128
  definition of, 128
  set of, as a field, 226
Ordered triple, 283
Ordinate, 126

P

Parabola, 161, 368ff
  definition of, 368
  directrix of, 368
  equations for, 369–70

Parabola: *cont.*
  focal chord, 370–71
  focus of, 368
  graph of, 165, 369–73
  latus rectum, 370
  maximum and minimum, 161ff
  zeros of, 161, 164
Parallel lines, 155
Parameter, 284
Parentheses, 39, 67
Partial sums, 353
Pascal's triangle, 360
Point-slope of a linear equation, 153
Polynomial(s):
  addition of, 237
  definition of, 236
  degree of, 236
  Descartes' rule of signs for, 258
  division of, 239ff
  exactly divisible, 239–40
  factor theorem for, 246–47
  function, 250
    graphs of, 250ff
  in a real variable, 105
  multiplication of, 238
  number of zeros, 255
  over R, 236
  rational algebraic expression, 239
  rational zeros, 261ff
  remainder theorem for, 245
  synthetic division of, 242ff
  upper and lower bounds theorem for zeros
    of, 259
  value of, 237
  zeros of, 254ff
Polynomial equation(s):
  approximating roots of, 263
  rational root of, 261ff
  real and complex zeros of, 254ff
Polynomial function(s):
  definition of, 250
  graphs of, 250ff
  isolating zeros of real, 263
  rational zeros of, 261
  real zeros of, 254ff
  turning points, 253
Positive real number(s):
  closure of set of, 83
  product of a, with a negative real number,
    84
Power(s):
  definition of, 180
  irrational, of a real number, 186
  rational, of a real number, 184
Principal *n*th root, 183
Principal square root, 117